Valerie Illingworth has a degree in physics from Bristol University and a master's degree in radiation physics from London University. She worked in reference book publishing before becoming freelance in 1976. She is the editor of several books including *New Penguin Dictionary of Electronics*, *Macmillan Dictionary of Astronomy* and Oxford's *Dictionary of Computing*.

THE PENGUIN DICTIONARY OF

PHYSICS

EDITOR: VALERIE ILLINGWORTH

MARKET HOUSE BOOKS LTD

SECOND EDITION

PENGUIN BOOKS

PENGUIN BOOKS

Published by the Penguin Group
Penguin Books Ltd, 27 Wrights Lane, London W8 5TZ, England
Viking Penguin, a division of Penguin Books USA Inc.
375 Hudson Street, New York, New York 10014, USA
Penguin Books Australia Ltd, Ringwood, Victoria, Australia
Penguin Books Canada Ltd, 2801 John Street, Markham, Ontario, Canada L3R 1B4
Penguin Books (NZ) Ltd, 182–190 Wairau Road, Auckland 10, New Zealand

Penguin Books Ltd, Registered Offices: Harmondsworth, Middlesex, England

First published 1977
Second edition, an abridgement of Longman's *Dictionary of Physics* (revised edition,
1991), edited by H. J. Gray and Alan Isaacs, with some entries from the first edition
(edited by H. J. Gray, 1958) and the second edition (Gray and Isaacs, 1975), 1990
1 3 5 7 9 10 8 6 4 2

Typeset by Market House Books Ltd
Printed in England by Clays Ltd, St Ives plc

CONTENTS

PREFACE

This dictionary is an abridged version of the recently revised *Dictionary of Physics* (Longman, 1991). Both books were prepared by Market House Books.

Although this dictionary is concerned primarily with the terminology of contemporary physics, words with a physical basis that are used in other scientific fields, such as physical chemistry, computing, astronomy, geophysics, medical physics, engineering subjects and music, have also been included. SI units are used throughout.

The dictionary should thus prove useful to students and teachers of physics and related subjects, to doctors and to scientists, technologists and technicians in research and industry. It contains many long entries in which a word of major importance is defined and discussed together with closely associated words. Shorter definitions supplement the longer entries.

The editor thanks Mr H. J. Gray and Dr Alan Isaacs, editors of the *Dictionary of Physics*, and also the contributors to the three editions of the Longman's dictionary, in particular Dr John Daintith, for entries that have passed into the *Penguin Dictionary of Physics*. The abridgements for this second edition of the *Penguin Dictionary of Physics* have been made by the editor without consulting the original contributors. Any changes of emphasis or style are therefore the responsibility of the editor.

VALERIE ILLINGWORTH, 1990

NOTES

An asterisk indicates a cross reference.

An entry having an initial capital letter is either a proper name or a trade name.

Syn. is an abbreviation for 'synonymous with'.

Symbols will be found in the Tables of SI units and the Table of Symbols for Physical Quantities (Appendix).

A

Abbe condenser A simple two-lens *condenser that has good light-gathering

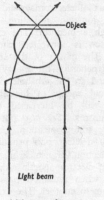

Abbe condenser

ability, the *numerical aperture being 1.25. It is therefore extensively used in general microscopy. Aberrations are not well corrected. A modified Abbe condenser called a *variable-focus condenser* is used to obtain a greater illuminated field area. The lower lens can be adjusted to bring light to a focus between the lenses. *See also* achromatic condenser.

Abbe criterion *See* resolving power.

Abbe number *Syn.* constringence; V-number. Symbol: *V*. The reciprocal of dispersive power. *See* dispersion.

aberration 1. A defect in the image formed by a lens or curved mirror, revealed as a blurring or distortion and possibly a false coloration. The four principal aberrations are *spherical aberration, *coma, and *astigmatism (which are due mainly to curvature of

the optical surface) and *chromatic aberration (which occurs only with lenses and is due to *dispersion). *Curvature of field and *distortion are other aberrations.

Aberrations occur when light rays do not pass close to the optic axis of the system but make a considerable angle to it (so that it is no longer accurate to replace the sine of the angle involved by the angle itself (in radians)). Of the six aberrations, only spherical and chromatic aberrations are found in the images of axial points. The other four occur only when off-axis points are involved. *See also* Seidel aberrations.

2. Of light. The apparent displacement in the positions of stars, attributable to the finite speed of light and to the motion of the observer; the latter results mainly from the earth's orbital motion around the sun.

3. A defect in the image produced by an *electron lens system.

ablation The removal of material from the surface of a moving body by decomposition or vaporization resulting from friction with the atoms or molecules of the atmosphere.

absolute expansion *See* coefficient of expansion.

absolute humidity *See* humidity.

absolute magnitude *See* magnitude.

absolute temperature Former name for *thermodynamic temperature.

absolute unit If a quantity y is uniquely defined in terms of quantities $x_1, x_2, \ldots$ by

$$y = f(x_1, x_2, \ldots),$$

the unit U_y of y can be obtained from the units U_{x1}, U_{x2} of x_1, x_2 from the equation

$$U_y \propto f(U_{x1}, U_{x2}, \ldots).$$

1

In any given system an absolute unit is one for which the constant of proportionality is unity. All units of the *SI system are absolute.

absolute zero The unattainable lower limit to temperature. It is the zero of *thermodynamic temperature: $0 \text{ K} = -273.15 \,°C = -459.67 \,°F$.

According to thermodynamics, if an ideal heat engine working in a *Carnot cycle has its lower-temperature isothermal process at absolute zero, no heat will be discharged and the efficiency of the engine will be one.

The ideal-gas absolute scale can be shown to be equivalent to the thermodynamic scale. Temperature on this scale is defined to be proportional to the product of pressure and volume in the limit as pressure tends to zero and volume to infinity. Absolute zero is that temperature at which this product is zero.

In quantum theory, absolute zero is interpreted as the temperature at which all particles are in the lowest-energy quantum states available. Generally the available states do not have zero energy, so there is still molecular energy at absolute zero (the *zero-point energy*). The kinetic energies of the molecules of an ideal gas would be zero at 0 K, but not those of a real substance.

absorbance *See* internal transmission density.

absorbed dose *See* dose.

absorptance *Syn.* absorption factor. Symbol: α. A measure of the ability of a body or substance to absorb radiation as expressed by the ratio of the absorbed *radiant or *luminous flux to the incident radiant or luminous flux. For radiant heat the absorptance of a body, measured against a vacuum, depends on the thermodynamics temperature T of

the body receiving the radiation and on the wavelength. The absorptance at a fixed frequency of radiation is called the *spectral absorptance*. *See also* internal absorptance.

absorption 1. Of radiation. Reduction in the flux of electromagnetic radiation, or other ionizing radiation, on passage through a medium. The energy of the radiation is converted into a different form. The nature of the absorbing process depends on the energy of the radiation and on the substance involved. The *internal energy of the absorbing medium will be increased by all these mechanisms. (*See also* photoelectric effect, photoionization, fluorescence, Compton effect, pair production.)

Reflection, transmission, and absorption of electromagnetic radiation can all occur when radiation passes from one medium to another. The extent to which they occur is given by the *reflectance, *transmittance, and *absorptance of the medium. The reduction in flux with distance traversed in a medium is given by the *linear absorption coefficient (for absorption alone) or by the *linear attenuation coefficient (when absorption and dispersion both occur).

2. Of sound. The reduction is *sound intensity when energy in the form of a sound wave passes from one medium to another. At the boundary of the media, part of the energy is reflected and part enters the second medium and may be regarded as being absorbed. The amount reflected is given by the *acoustic absorption coefficient.

The reduction in sound energy (and hence in sound intensity) is given by:
$$E = E_0 \exp(-\mu_\alpha x),$$
where E_0 is the incident energy, E the energy after a distance x, and μ_α is a constant called the *linear absorption coefficient*.

The absorption of sound energy is caused principally by viscous forces opposing the relative motion of the particles as the sound passes (involving transformation of acoustic energy into internal energy) and by heat being conducted from the compressed particles to the (lower-temperature) rarified ones. Heat radiated from compressions to rarefactions also causes some energy dissipation at low frequencies. In general, absorption of sound by gases is basically due to viscosity, the conduction and radiation effects becoming more important for waves of larger amplitude. Water vapour content (humidity) also affects the absorption.

3. The process in which a gas or liquid is taken up by another substance, usually a solid. The absorbed material thus permeates into the bulk of the material. *Compare* adsorption.

absorption bands (and lines) Dark bands or lines present in a *spectrum as a result of absorption by some intervening medium. *See also* absorption spectrum.

absorption coefficient 1. Of electromagnetic radiation. *See* linear absorption coefficient.

2. Of sound. *See* acoustic absorption coefficient.

absorption edge (discontinuity or limit) An abrupt discontinuity in the graph relating the *linear absorption coefficient of X-rays in a given substance with the wavelength of the radiation. At certain critical absorption wavelengths the absorption shows a sudden decrease in value. This occurs when the quantum energy of the radiation becomes smaller than the work required to eject an electron from one or other of the quantum states in the absorbing atom, and the radiation thus ceases to be absorbed by that state. Thus, radiation of wavelength greater than the K absorption edge cannot eject electrons from the K states of the absorbing substance.

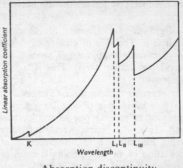

Absorption discontinuity

absorption factor *See* absorptance.

absorption hygrometer *See* chemical hygrometer.

absorption spectrum When light from a high-temperature source producing a continuous emission spectrum is passed through a medium into a spectroscope, the spectrum reveals dark regions where absorption has taken place (continuous, line, and band types). In general, the medium absorbs those wavelengths it would emit if its temperature were raised high enough. Solids and liquids show broad continuous absorption spectra; gases give more discontinuous types (line and band). *See* spectrum.

absorptivity 1. A measure of the ability of a substance to absorb radiation, as expressed by the *internal absorptance of a layer of substance when the path of the radiation is of unit length and the boundaries of the material have no influence.

2. Former name for *absorptance.

abundance Symbol: C. The number of atoms of a given isotope in a mixture of the isotopes of an element; usually expressed as a percentage of the total number of atoms of the element.

3

a.c. Abbreviation for alternating current.

acceleration **1.** Linear acceleration. Symbol: *a*. The rate of change of velocity, expressed in metres per second per second (or similar units). This is a vector quantity.
   2. Angular acceleration. Symbol: *α*. The rate of change of angular velocity, expressed in radians per second per second. This is a *pseudovector quantity having magnitude and the direction of orientation of an axis. In simple cases in which the axis of rotation is fixed, angular acceleration may be regarded as a scalar.

acceleration of free fall *See* free fall.

accelerator A machine for increasing the kinetic energy of charged particles or ions, such as protons or electrons, by accelerating them in an electric field. A magnetic field is used to maintain the particles in the desired direction. The particles can travel in a straight, spiral, or circular path. (*See* linear accelerator; cyclotron; synchrotron; proton synchrotron; synchrocyclotron; focusing.) At present, the highest energies are obtained in the proton synchrotron.

The Super Proton Synchroton at *CERN (Geneva) accelerates protons to 450 GeV. It can also cause proton–antiproton collisions with total kinetic energy (in centre-of-mass coordinates) of 620 GeV. In the USA the Fermi National Acceleration Laboratory proton synchrotron gives protons and antiprotons of 800 GeV, permitting collisions with total kinetic energy of 1600 GeV. The Large Electron Positron (LEP) system at CERN accelerates particles to 60 GeV.

All the aforementioned devices are designed to produce collisions between particles travelling in opposite directions. This gives effectively very much higher energies available for interaction than are possible when accelerated particles hit stationary targets. (*See also* intersecting storage ring.)

High-energy nuclear reactions occur when the particles (either moving or in a stationary target) collide. The particles created in these reactions are detected by sensitive equipment close to the collision site. New particles, including the tauon, W, and Z particles and requiring enormous energies for their creation, have been detected and their properties determined.

acceptor *See* semiconductor.

access time The mean time interval between demanding a particular piece of information from a computer *storage device and obtaining it.

accommodation The ability of the eye to alter its focal length and to produce clear images of objects at different distances. *See also* near point.

accumulator An electric cell that can be recharged after use. The common "lead" accumulator consists in principle of two plates coated with lead sulphate immersed in aqueous sulphuric acid. If connected to a suitable d.c. supply, current is sent through the cell and the anode is converted to lead(IV) oxide (lead peroxide) and the cathode reduced to metallic lead. If the two plates are then connected through an external circuit, the chemical action is reversed and current flows round the external circuit from the brown peroxide plate to the grey lead plate. The action may be summarized in the equation:

$$PbO_2 + Pb + 2H_2SO_4 \rightleftharpoons 2PbSO_4 + 2H_2O,$$

with the right-to-left reaction occurring on discharge and the left-to-right reaction occurring on charge.

An accumulator employing electrodes of nickel and iron (or cadmium) immersed in a 20% solution of potassium hydroxide is also employed for special purposes. (*See* Edison accumulator.)

achromat *See* achromatic lens.

achromatic colours Colours having no hue or saturation but only lightness. White, greys, and black are examples.

achromatic condenser A *condenser corrected for chromatic and spherical *aberrations, usually by having four elements, two of which are *achromatic lenses. It has a *numerical aperture of 1.4. It is used in *microscopes when high magnification is required. *See also* Abbe condenser.

achromatic lens *Syn.* achromat. A combination of two or more lenses using, if necessary, different kinds of glass, designed to remove the major part of *chromatic aberration. The elementary theory of simple achromatic doublets assumes two lenses of powers P_1 and P_2 placed in contact (with total power $P = P_1 + P_2$) made of glasses of *dispersive powers ω_1 and ω_2 so that the condition for achromatism ($\omega_1 P_1 + \omega_2 P_2 = 0$) is satisfied. To produce an achromatic converging lens, e.g. for telescope or photographic objectives, the dispersive power of the higher-power convergent lens must be less than that of the divergent lens of the combination. It is thus possible to bring two colours, say red and blue, to the same focus. There will still be some residual colour effects, known as a *secondary spectrum*. *See also* apochromatic lens.

achromatic prism A combination of two or more prisms that produces the same deviation of two or more colours so that objects viewed through them will not appear coloured (*see* chromatic aberra-

tion). As with thin lenses, "narrow angle" prisms are placed in contact in opposition, so that the *dispersive powers of the two glasses are inversely proportional to their angles of *deviation.

achromatism The removal of *chromatic aberration, or *chromatic differences of magnification, or both, arising from dispersion of light. Owing to nonlinearity of dispersion the correction is attempted for two colours in the first approximation, and for three colours in higher corrections. *See* achromatic lens; apochromatic lens.

acoustic absorption coefficient Symbol: α_a. The quantity $(1 - \rho)$, where ρ is the *reflection coefficient*, P_r / P_0; P_0 and P_r are the *sound power (or more generally the acoustic power) incident on and reflected from a body. *See also* absorption (2). The absorption and reflection coefficients both vary with frequency.

acoustic capacitance The imaginary component of acoustic *impedance due to the stiffness or elasticity (k) of the medium; it is equal to S^2/k, where S is the area in vibration.

acoustic delay line *See* delay line.

acoustic filters Systems using tubes and resonating boxes, in parallel and series, as acoustic *impedance elements to transmit high frequencies only (high-pass filter) or low frequency only (low-pass filter) or any given band of frequency (band-pass filter).

If any simple harmonic motion is impressed on equal impedances Z_1 connected in a conduit and separated by branches containing other equal impedances Z_2, it will not pass through unless the ratio Z_1 / Z_2 for the frequency of this SHM lies between certain values; i.e. all other frequencies which do not satisfy this condition will be rapidly attenuated and only those covering this range will get through.

acoustic grating A series of objects, such as rods of equal size, placed in a row a fixed distance apart. An acoustic grating has similar properties to an optical *diffraction grating. When a sound wave is incident upon an acoustic grating, secondary waves are set up that reinforce each other or cancel out according to whether or not they are in phase. The result for a sinusoidal sound wave is a series of maxima and minima spaced round the grating. When the incident sound is normal to the grating, the condition for a maximum diffracted sound at an angle θ to the normal is: $\sin \theta = m\lambda/e$, where λ is the wavelength of the sound, e is the width of a rod plus the space between it and the next, and m is an integer. e must be greater than λ for a diffraction pattern to be formed and this condition necessitates very large gratings for low-frequency sounds.

acoustic impedance *See* impedance.

acoustic inertance The imaginary component of acoustic *impedance, due solely to inertia. It corresponds to inductance in electric circuits. In the case of a mass (m) of gas in a conduit of cross section S, the inertance (L) is equal to m/S^2.

acoustic levitation *See* levitation.

acoustic mass *See* reactance (acoustic).

acoustic power *See* sound-energy flux.

acoustic pressure *See* sound pressure.

acoustic reactance *See* reactance (acoustic).

acoustics 1. The science concerned with the production, properties, and propagation of sound waves.

2. The characteristics of a room, auditorium, etc., that determine the fidelity with which sound can be heard within it. For a room, etc., to have good acoustics there should be no noticeable echoes, the loudness should be adequate and uniform, the *reverberation time should be near the optimum for the room, resonance should be avoided, and the room should be sufficiently soundproof.

acoustic stiffness *See* reactance (acoustic).

acoustic wave *Syn.* sound wave. A wave that is transmitted through a solid, liquid, or gas as a result of mechanical vibrations of the particles in the medium. The direction of motion of the particles is parallel to the direction of propagation of the wave (i.e. it is a *longitudinal wave), and the wave therefore consists of compressions and rarefactions of the medium. The term sound wave is sometimes limited to those with a frequency to which the human ear is sensitive, i.e. about 20–20 000 Hz. A travelling acoustic wave in a solid can be produced by applying mechanical stress to a crystal or as a result of *magnetostriction or the *piezoelectric effect. *See also* ultrasonics.

acoustoelectronics The study and use of devices in which electrical signals are converted into acoustic waves by *transducers and the acoustic signals are propagated through a solid medium.

actinic Of radiation, such as light or ultraviolet. Able to produce a chemical change in exposed materials.

actinium series *See* radioactive series.

action 1. The product of a component of momentum, p_i, and the change in the corresponding positional coordinate, q_i;

6

or more precisely the integral $\int p_i dq_i$. (*See* Hamiltonian function; quantum of action.)

2. Twice the time integral of the kinetic energy of a system, measured from an arbitrary zero time. (*See* least-action principle.)

activation analysis A sensitive analytical technique in which the sample is first activated by bombardment with slow neutrons, high-energy particles, or gamma rays and the subsequent decay of radioactive nuclei is then used to characterize the atoms present. For example, stable sodium nuclei can be activated by neutron capture:

$$^{23}Na + n \rightarrow\ ^{24}Na + \gamma.$$

The ^{24}Na nuclei decay to give γ-rays, electrons, and neutrinos:

$$^{24}Na \rightarrow\ ^{24}Mg + \gamma + e^- + \bar{\nu}.$$

The electrons have a characteristic energy spread and the γ-rays have energies of 2.75 and 1.37 MeV. Sodium can thus be detected by the presence of lines at these energies in the *gamma-ray spectrum of the irradiated material.

The usual method of irradiation is by neutron bombardment in a nuclear reactor. The technique is highly sensitive and can be used for a large number of elements. If the intensity of γ-ray emission is compared with that from a similarly treated standard, a quantitative analysis can be made.

activation cross section *See* cross section.

active aerial *See* directive aerial.

active component An electronic component such as a *transistor, that can be used to introduce *gain into a circuit. *Compare* passive component.

active current The component of an alternating current that is in *phase with the voltage, the current and voltage being regarded as vector quantities.

active voltage The component of an alternating voltage that is in *phase with the current, the voltage and current being regarded as vector quantities.

active volt-amperes The product of the current and the *active voltage or the product of the voltage and the *active current. It is equal to the power in watts.

activity 1. Symbol: A. The number of atoms of a radioactive substance that disintegrate per unit time $(-dN/dt)$. It is measured in *becquerel, or, formerly, in *curie.

2. A quantity used mainly in chemical thermodynamics to express the effective concentration of a substance in a solution or mixture.

3. *See* optical activity.

ADC Abbreviation for *analogue/digital converter.

Adcock direction-finder *Syn.* Adcock antenna. A radio direction-finder employing a number of spaced vertical aerials. It is designed so that any horizontally polarized components of the received waves have the minimum effect upon the observed bearings.

A/D converter *See* analogue/digital converter.

additive process A process by which almost any colour can be produced or reproduced by mixing together lights of three colours, called *additive primary colours*, usually red, green, and blue, the proportions of which determine the colour obtained; white light is obtained from approximately equal proportions of red, green, and blue light; yellow from a mixture of red and green light. *Colour television uses an additive process for final colour production. *See also* chromaticity. *Compare* subtractive process.

adhesion The interaction between the surfaces of two closely adjacent bodies that causes them to cling together. *Compare* cohesion.

adiabatic demagnetization A process used for the production of temperatures near *absolute zero. A paramagnetic salt is placed between the poles of an electromagnet and the field switched on, the resulting heat being removed by a helium bath. The substance is then isolated thermally and on switching off the field the substance is demagnetized adiabatically and cools.

adiabatic process A process in which no heat enters or leaves a system. For example, if gas is compressed or expanded in a cylinder by a piston it undergoes an adiabatic change if the cylinder does not allow transfer of heat between the gas and the surroundings. An adiabatic expansion results in cooling of a gas whereas an adiabatic compression has the opposite effect. If an *ideal gas undergoes a reversible adiabatic change in volume, the pressure (p) is related to volume (V) by the equation: $pV^\gamma = K$, where K is a constant and γ is the ratio of the *heat capacities C_p/C_V of the gas. Any reversible adiabatic change (*see* reversible change) is *isentropic*, i.e. during the change the entropy of the system remains constant. *Compare* isothermal process.

adiathermic Not transparent to heat.

admittance Symbol: Y. The reciprocal of *impedance; it is related to *conductance (G) and *susceptance (B) by:
$$Y^2 = G^2 + B^2.$$

adsorption The formation of a layer of foreign substance on an impermeable surface. *Compare* absorption.

advanced gas-cooled reactor (AGR) *See* gas-cooled reactor.

advection A process of transfer of atmospheric properties by horizontal motion in the atmosphere, such as the movement of cold air from polar regions. Advection is concerned with large-scale motions in the atmosphere; vertical, locally induced motions are *convection processes. In oceanography, advection is the flow of sea water as a current.

aerial *Syn.* antenna. That part of a radio system from which energy is radiated into (*transmitting aerial*), or received from (*receiving aerial*) space. An aerial with its *feeders and all its supports is known as an *aerial system*. The most important types of aerial are the *dipole aerial and the *directive aerial.

Although the word aerial is in general use, as in TV aerial, the word antenna (originally a US term) is now usually preferred in UK scientific and technical literature.

aerial array *Syn.* beam aerial. An arrangement of radiating or receiving elements so spaced and connected that directional effects are produced. With suitable design, very great directivity can be obtained and also, as a consequence, large *aerial gain. An array of elements along a horizontal line, which has marked directivity in the horizontal plane in a direction at right angles to the line of the array, is referred to as a *broadside array*. One that has directivity in the horizontal plane along the line of the array is called an *end-fire array* (or *staggered aerial*). Arrays are commonly designed for directivity in both horizontal and vertical planes. The horizontal directivity is influenced by the number of aerial elements that are arranged horizontally, whereas the vertical directivity depends upon the number of elements that are stacked in tiers (or stacks), one vertically above the other.

aerial gain The ratio of the signal power produced at the receiver input by the aerial to that which would be produced by a standard comparison aerial under similar conditions (i.e. similar receiving conditions and the same transmitted power). If the type of standard comparison aerial is not specified, a *half-wave dipole is implied.

aerial resistance The resistance that takes into account the energy consumed by an aerial system as a result of radiation and losses (e.g. I^2R losses in aerial wires, dielectric losses, earth losses, etc.). It is equal to the power supplied to the aerial divided by the square of the current at the aerial supply point.

aerial system *See* aerial.

aerodynamics The study of the motion of gases (particularly air) and the motion and control of solid bodies in air.

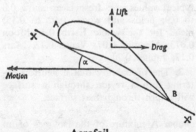

Aerofoil

aerofoil A body for which, when in relative motion with a fluid, the resistance to motion (drag) is many times less than the force perpendicular to motion (lift). The flight of aircraft depends on the use of aerofoils for wing and tail structure. The essential features of the aerofoil are the rounded leading edge A and the sharp trailing edge B (*see* diagram). The projection of the section on to the com-

mon tangent XX' is called the *chord*; the angle α is the *angle of incidence* or *attack*.

Due to viscosity, the layers of fluid passing over the upper and lower surfaces of the aerofoil arrive at the trailing edge with different velocities. This leads to the production of an eddy or vortex at the trailing edge accompanied by a counter-circulation around the aerofoil. This circulation is essential for the production of the lift force.

afterglow *See* persistence.

age equation or theory *See* Fermi age theory.

age of the earth A time estimated as 4.6 × 10^9 years. The earth is believed to share a common origin with the rest of the solar system.

age of the universe A time of between 10 × 10^9 to 20 × 10^9 years as determined by the *Hubble constant. This value is clearly uncertain and depends on current theories of *cosmology. *See also* expanding universe.

AGR Abbreviation for advanced *gas-cooled reactor.

Aharonov–Casher effect An effect in which an electrically neutral particle with a magnetic moment is influenced by an electric field. It can be demonstrated by diffraction of neutrons by a line of electric charge. There is a magnetic field associated with a moving neutron in the vicinity of an electric field, and this affects the neutron's phase. A similar effect on electrically charged particles moving close to a magnetic field is called the *Aharonov–Bohm effect*.

air Normal dry air has the following composition by volume:

Nitrogen	78.08%
Oxygen	20.94%
Argon	0.9325%
Carbon dioxide	0.03%
Neon	0.0018%
Helium	0.0005%
Krypton	0.0001%
Xenon	0.000 009%
Radon	6×10^{-18}%

For dry air:

Specific heat capacity at constant volume	$718 \, J \, kg^{-1} \, K^{-1}$
Specific heat capacity at constant pressure	$1006 \, J \, kg^{-1} \, K^{-1}$
Ratio of specific heat capacities	1.403
Boiling point (at atmospheric pressure)	$-193 \, °C$ to $-185 \, °C$ depending on age

Liquid air is produced by strong cooling under high pressure (*see* liquefaction of gases). It has a pale blue colour due to the presence of liquid oxygen.

air equivalent A measure of the efficiency of an absorber of nuclear radiation, expressed as the thickness of a layer of air at standard temperature and pressure that causes the same amount of absorption or the same energy loss.

air mass A part of the lower atmosphere in which the horizontal temperature gradient at all levels is very small. At the margins of an air mass the temperature gradients become steep; such a transition zone is called a *front*.

air pumps Air-exhausting pumps are used for withdrawing air or other gases from a closed chamber. Many of them are based upon the ordinary piston principle; others, e.g. the *filter pump, use a jet of mercury or water to trap and remove air from the vessel to be

exhausted. In the *Gaede molecular air pump*, a grooved cylinder rotates in a casing with very little clearance; a fixed comb projects into the grooves in the cylinder, the inlet for gas being on one side of the comb and the outlet on the other. On rotation of the cylinder, the gas is dragged from the inlet to the outlet; low pressures of the order of 0.001 mm of mercury can be achieved with a speed of revolution of the cylinder of 8000 to 12 000 revolutions per minute. *See also* pumps, vacuum.

Airy rings A diffraction pattern of alternate dark and light rings surrounding a bright circular patch (the *Airy disc*), obtained with a circular aperture. For example, it constitutes the "image" of a distant point object given by an astronomical telescope. The larger the aperture of the telescope, the small is the disc. *See* diffraction of light.

albedo 1. The fraction of incident light diffusely reflected from a surface. Some typical values are: fresh clean snow 0.8 to 0.9; fields and woods 0.02 to 0.15; mean for the whole Earth 0.4; Moon 0.073; Mercury 0.07; Venus 0.72, Mars 0.17; Jupiter 0.70.

2. The probability that a neutron entering into a region through a surface will return through that surface.

allobar A mixture of the isotopes of an element in proportions that differ from the natural isotopic composition.

allochromy The emission of electromagnetic radiation from atoms, molecules, etc. induced by incident radiation of a different wavelength, as in *fluorescence or the *Raman effect.

allotropy The existence of a solid, liquid, or gaseous substance in two or more forms (*allotropes*) that differ in physical rather than chemical properties.

allowed band *See* energy bands.

alloy A material, other than a pure element, that exhibits characteristic metallic properties. At least one major constituent must be a metal. A definite chemical compound of a metal with a nonmetallic element would not be counted as an alloy even if it showed some metallic properties. Most engineering alloys are still made by the procedure of melting a mixture of the constituents and then casting, but alloys are also produced by compacting metal powders, by simultaneous electrodeposition, by diffusion of constituents in the solid state, and by the condensation of vapours.

Steels are alloys of iron with the nonmetallic element carbon, often with other elements deliberately added; there is a wide range of alloy-steels used for special purposes, e.g. rustless chromium steels and high-permeability silicon steels.

Brass has copper and zinc as its major constituents. Most grades are about 5% denser than steel with a lower strength and melting point.

Bronze has copper and tin as its major constituents.

Amalgams are alloys involving the metal mercury.

Alnico A series of proprietary permanent-magnet alloys typically consisting of 18% Ni, 10% Al, 12% Co, 6% Cu, balance Fe.

alpha decay A radioactive disintegration process whereby a parent nucleus decays spontaneously into an *alpha particle and a daughter nucleus. The *mean life of the parent nucleus varies from 10^{-7} seconds to 10^{10} years. On classical theory the maximum height of the *potential barrier between the α-particle (atomic number Z_α) and the daughter nucleus (Z) is given by:
$$V_R = Z_\alpha Z e^2 / R,$$
where R is the radius of the nucleus. If a particle is inside a potential well with an energy $E_1 < V_R$, then the α-particle will be unable to escape. *Wave mechanics is necessary to explain the movement of α-particles through the barrier in terms of the *tunnel effect.

alpha particle (α-particle) The nucleus of a helium (^{4}He) atom, carrying a positive charge of $2e$. Its *proton number and *neutron number are both 2, a *magic number, so that it is a very stable particle. It has a *relative atomic mass of 4.001 506.

alpha rays (α-rays) A stream of *alpha particles ejected from many radioactive substances with speeds (of the order of 1.6×10^6 metres per second) characteristic of the emitting substance. Alpha rays have a penetrating power of a few centimetres in air but can be stopped by a thin piece of paper, the maximum range varying as the cube of the velocity. They produce intense *ionization along their track and can be detected by a *spark counter, *Geiger counter, or *bubble chamber, by their effect on a photographic plate, by the scintillations they produce on a fluorescent screen, etc.

altazimuth mounting *See* telescope.

alternating current (a.c.) An electric current that periodically reverses its direction in the circuit, with a *frequency, f, independent of the constants of the circuit. In its simplest form, the instantaneous current I varies with the time t in accordance with the relation:
$$I = I_0 \sin (2\pi f t),$$
where I_0 is the peak value of the current.

11

alternating-current generator A *generator for producing alternating e.m.f.s and currents. Examples are: *induction generator and *synchronous alternating-current generator.

alternating-gradient focusing *See* focusing.

alternator *See* synchronous alternating-current generator.

altimeter An aneroid *barometer measuring the decrease in atmospheric pressure with height above ground and calibrated to read the height directly.

altitude 1. The vertical distance above sea level.
2. One of a pair of coordinates, the other being *azimuth, giving the position of a star. (*See* celestial sphere.)

amalgam *See* alloy.

ambient 1. Surrounding, encompassing; e.g. ambient temperature is the temperature of an immediate locality.
2. Freely moving, circulating; e.g. ambient air.

Amici prism *See* direct-vision prism.

ammeter An instrument for measuring electric current. Common types are (*a*) the *moving coil*, (*b*) the *moving iron*, and (*c*) the *thermoammeter*. In (*a*), the current passes through a coil, pivoted with its plane parallel to the lines of a radial magnetic field produced by a permanent horseshoe magnet. The rotation is controlled by hair springs, and the deflection is directly proportional to the current. This is the most accurate type of ammeter, but only measures direct current. It can be adapted for alternating-current measurement by embodying a rectifier in the circuit. In (*b*), the current passing through a fixed coil magnetizes

two specially shaped pieces of soft iron, the mutual repulsion of which causes the rotation of a pointer over a scale. Since the effect depends on the square of the current, the instrument can be used both for a.c. and d.c. measurements. It is less accurate than (*a*) and the scale is nonuniform. The control may be either by gravity or by a spring. In (*c*), the current passes through a thin resistance wire (usually in a vacuum) that, in consequence, becomes heated. The rise in temperature is measured by a thermojunction soldered to the wire and connected to a sensitive moving-coil instrument. It is particularly useful for measuring high-frequency currents.

For many purposes, electronic instruments using digital display are replacing the earlier pointer instruments.

ammonia clock *See* clocks.

amorphous Devoid of crystalline form. True amorphous solids lack regular arrangement of their atoms. They are often supercooled liquids like glass and they may crystallize slowly in suitable conditions. Polycrystalline materials (e.g. metals) should not be confused with amorphous substances.

amount of substance Symbol: *n*. A dimensionally independent physical quantity that is not the same as *mass. It is proportional to the number of specified particles of a substance, the specified particle being an atom, molecule, ion, radical, electron, photon, etc., or any specified group of any of these particles. The constant of proportionality is the same for all substances and is known as the *Avogadro constant. Amount of substance is measured in *moles.

ampere Symbol: A. The *SI unit of electric current, defined as the constant current that, if maintained in two straight parallel conductors of infinite length, of negligible circular cross section, and placed one metre apart in a vacuum, would produce between these conductors a force equal to 2×10^{-7} newton per metre of length. This unit, at one time called the absolute ampere, replaced the international ampere (A_{int}) in 1948. The latter was defined as the constant current that, when flowing through a solution of silver nitrate in water, deposits silver at a rate of 0.001 118 grams per second. 1 A_{int} = 0.999 850 A.

ampere-hour The quantity of electricity conveyed across any cross section of a conductor when an unvarying current of one ampere flows in the conductor for 1 hour. The term is employed in stating the capacity of accumulators. One ampere-hour = 3600 coulombs.

Ampère–Laplace theorem A theorem for the magnetic field strength due to the current in a conductor. It is probably most useful in the form:

$$dB = \frac{\mu_0}{4\pi} \frac{I \sin \theta}{r^2} dl$$

where dB is the elemental magnetic flux density at a point distance r from an elemental length dl of conductor carrying a current I; θ is the angle between the direction of I and the radius vector r; μ_0 is the *magnetic constant.

Ampère's rule One of the numerous mnemonics for recalling the relation between the direction of an electric current and that of its associated magnetic field. If you imagine yourself to be swimming in the wire in the direction of the current and facing a magnetic needle placed below the wire, then the *north* pole of the needle is deflected towards your *left* hand.

Ampère's theorem If the *Ampère–Laplace theorem is applied to an infinitely long straight conductor carrying a current I, the flux density B at a point a distance r from the conductor is found to be given by $B = \mu_0 I/2\pi r$, where μ_0 is the *magnetic constant. From this Ampère showed that along any closed path around N conductors of elemental length dl, each of which carries a current I,

$$\oint B dl = \mu_0 NI,$$

This is Ampère's (circuital) theorem. He expressed it originally in terms of the work done in taking a unit *magnetic pole completely around a circuit.

ampere-turn The *magnetomotive force produced when a current of one ampere flows through one turn of a coil.

amplifier A device for reproducing an electrical input at an increased intensity. If an increased e.m.f. is produced operating into a high impedance, the device is a *voltage amplifier*, and if the output provides an appreciable current flow into a relatively low impedance, the device is a *power amplifier*. The most commonly used amplifiers operate by *transistors. Most practical amplifiers are a.c. amplifiers and consist of several small-gain *amplifier stages* coupled together to produce a substantial overall *gain. Negative *feedback is commonly used to provide stability and prevent the amplifier behaving as an oscillator. They are usually described by the range of frequencies amplified, e.g. wideband or tuned amplifiers, audiofrequency or radio frequency amplifiers. *See also* class A, class B, class C, class D amplifiers.

amplitude The *peak value of an alternating quantity in either the positive or negative direction. This term is applied particularly to the case of a sinusoidal vibration.

amplitude distortion *See* distortion.

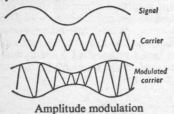

Signal

Carrier

Modulated
carrier

Amplitude modulation

amplitude modulation (a.m.) A type of *modulation in which the amplitude of the *carrier wave is varied above and below its unmodulated value by an amount that is proportional to the amplitude of the modulating signal and at a frequency equal to that of the modulating signal. An amplitude-modulated wave in which the modulating signal is sinusoidal may be represented by:

$$e = [A + B \sin pt] \sin \omega t,$$

where A = amplitude of the unmodulated carrier wave, B = peak amplitude variation of the composite, $\omega = 2\pi \times$ frequency of carrier wave, $p = 2\pi \times$ frequency of modulating signal.

a.m.u. Abbreviation for *atomic mass unit.

analogue computer A computer that performs computations (e.g. summation, multiplication, and integration) by manipulating continuously varying physical quantities, commonly voltage and time. The variables are analogues of the quantities involved in the computation.

analogue/digital converter (A/D converter; ADC) A device that samples an *analogue signal at fixed intervals and converts it into an equivalent digital signal. The digital signal comprises a series of binary values that are suitable for use by computer, for transmission over a data link, etc.

analogue delay line *See* delay line.

analogue signal A signal that varies continuously in amplitude and time, i.e. a smoothly varying value of voltage or current.

analyser A device (crystal, *Nicol prism, etc.) by which the direction of polarization of a beam of light can be detected; usually the light has been passed through a *polarizer before arriving at the analyser.

anamorphic lens An optical system of cylindrical lenses or of prisms in which image formation occurs on different scales in the horizontal and vertical directions, producing an image squeezed in one plane. It is used in wide-screen cinematography for compressing images on to the film, the image geometry being restored in a subsequent projection using another such system.

anaphoresis *See* electrophoresis.

anastigmat A photographic objective in which correction of spherical aberration, coma, radial astigmatism, curvature of image, and chromatic aberration, are attempted for large apertures and wide fields. The numerous designs, symmetrical and otherwise, involve the use of three- or four-lens cemented or uncemented objectives.

AND gate *See* logic circuit.

anechoic chamber *See* dead room.

anelasticity A property of a solid in which stress and strain are not uniquely related in the preplastic range.

anemograph A recording *anemometer.

anemometer A device for measuring the velocity of a fluid, specifically, of the air. Anemometers are of three types: (1) those that depend for their measurement on the difference of pressure between two points in the flow, such as the *venturi tube and the *Pitot tube; (2) those that use the cooling experienced by a heated body exposed to the fluid (*see* hot-wire anemometer); (3) those that use the momentum of the fluid to drive a small windmill, set of cups, or waterscrew facing the direction of flow.

The vane anemometer is a common type in the third group for recording the flow of a gas. It must be so mounted that the axis of rotation of the vanes points along the wind direction and may for this purpose be pivoted on a post with a tail fin, the aerodynamic force on which keeps the instrument facing upwind. The vanes are set at 45° to the axis. In the cup anomometer, three or four cup-shaped bodies are attached by radial supports to a vertical axle. The axle rotates at a speed related to the wind speed.

aneroid Not containing a liquid, as with the aneroid *barometer and aneroid manometer.

angle of friction The angle whose tangent is the coefficient of *friction. In measuring the coefficient of friction a body is placed on a plane and the latter tilted until the body will just slide down when gently tapped. The angle of the plane to the horizontal is then the angle of friction.

angstrom (or ångstrom) Symbol: Å. A length unit of 10^{-10} metre (0.1 nm), used in spectroscopy and to measure intermolecular distances. The use of this unit is discouraged.

Ångstrom pyrheliometer *See* pyrheliometer.

angular acceleration *See* acceleration.

angular dispersion *See* disperson.

angular displacement The angle through which a point, line, or body is rotated in a specified direction and about a specified axis.

angular frequency Symbol: ω. The *frequency of a periodic quantity expressed as the product of the frequency and the factor 2π.

angular impulse The time integral of the torque applied to a system, usually when applied for a short time. It is equal to the change in *angular momentum that it would cause on a free mass acting about a principal axis.

angular magnification *See* magnifying power.

angular momentum *Syn.* moment of momentum about an axis. *Symbol*: **L**. The product of *moment of inertia and *angular velocity ($I\omega$). Angular momentum is a *pseudovector quantity. It is conserved in an isolated system. *See* moment. *See also* atomic orbital.

angular velocity Symbol: ω. The rate at which a body rotates about an axis expressed in radians per second. It is a *pseudovector quantity equal to the linear velocity divided by the radius.

angular wavenumber *See* wavenumber.

angular wave vector *See* wavenumber.

anharmonic motion The motion of a body subjected to a restoring force that is not directly proportional to the displacement from a fixed point in the line of motion. *Compare* simple harmonic motion.

anhysteretic The magnetic state of a specimen when, in a constant magnetic field H, it has been subjected to an alternating field progressively reduced from a value greater than H to zero.

anion An ion that carries a negative charge, and in electrolysis moves towards the *anode. *Compare* cation.

anisotropic Not *isotropic; possessing or denoting a property that varies with direction.

annealing The process of heating a substance to a specific temperature lower than its melting point, maintaining that temperature for some time, and then cooling slowly. Slow crystallization thus takes place in the solid state under controlled temperature conditions. Annealing generally softens metals and stabilizes glass articles by allowing stresses produced during fabrication to disappear.

annihilation An interaction between a particle and its *antiparticle in which the two bodies disappear and photons or other elementary particles or antiparticles are created. Energy and momentum are conserved in the process.

At low energies, an *electron and a *positron annihilate to produce electromagnetic radiation. Usually the particles have little kinetic energy or momentum in the laboratory system before interaction; hence the total energy of the radiation is very nearly $2m_0c^2$, where m_0 is the rest mass of an electron. In nearly all cases two *photons are generated, each of 0.511 MeV, in almost exactly opposite directions to conserve momentum. Occasionally three photons are emitted, all in the same plane. Electron–positron annihilation at high ener-

gies has been extensively studied in particle *accelerators. Generally the annihilation results in the production of a *quark plus an antiquark (for example, $e^+e^- \rightarrow u\bar{u}$) or a charged *lepton plus an antilepton ($e^+e^- \rightarrow \mu^+\mu^-$). The quarks and antiquarks do not appear as free particles but convert into several *hadrons, which can be detected experimentally. As the energy available in the electron–positron interaction increases, quarks and leptons of progressively larger rest mass can be produced. In addition, striking resonances are present, which appear as large increases in the rate at which annihilations occur at particular energies. The *J/psi and similar resonances containing a *charm quark plus an anticharm antiquark are produced at an energy of about 3 GeV, for example, giving rise to abundant production of charmed hadrons. Bottom (b) quark production occurs at energies greater than about 10 GeV. A resonance at an energy of about 90 GeV, due to the production of the Z^0 *gauge boson involved in *weak interactions is currently under intensive study at the LEP and SLC e^+e^- colliders (*see* accelerator).

A *nucleon and an *antinucleon annihilating at low energy produce about half a dozen *pions, which may be neutral or charged. (An equal number of positive and negative pions is produced to conserve electric charge.) *See also* pair production.

annual parallax *See* parallax.

annular effect The phenomenon in fluid motion analogous to the *skin effect in alternating electric currents. With steady direct flow at low velocity in a tube, the velocity falls steadily from the centre towards the walls (*see* Poiseuille flow), but when the motion is alternating, as it is, for instance, when sound waves are being propagated in the tube, the mean alternating velocity rises from the centre towards the walls and finally falls within a thin laminar *boundary layer to zero at the wall itself. This is known as a *periodic boundary layer*. Its thickness increases as the square root of the frequency of the alternation. Similar conditions hold if the alternating flow overlays a direct flow in the tube.

anode The positive electrode of an electrolytic cell, electron tube, or solid-state rectifier. It is the electrode by which electrons leave a system. *Compare* cathode.

anode drop (or fall) The difference of potential, of the order of 20 volts, between the anode in a *gas discharge tube and a point in the gas close to the anode.

anode rays *See* gas-discharge tube.

anode saturation A condition arising in an electron tube when electrons are no longer attracted by the anode. This is due to a build up of electrons around the anode preventing further discharge. *See also* space charge.

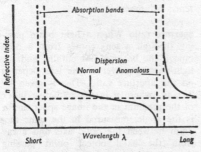

Anomalous dispersion

anomalous dispersion Rapid changes of refractive index with wavelength when the wavelength lies in the neighbourhood of *absorption bands of the ma-

terial. On the longer wavelength side of the absorption band, the refractive index is high, and on the shorter wavelength side, low. Normal *dispersion is such that shorter wavelengths are associated with higher refractive index.

anomalous viscosity A condition occurring in colloids, and in fact all fluids that consist of two or more phases present at the same time, such that the coefficient of *viscosity is not a constant but is a function of the rate at which the fluid is sheared as well as of the relative concentration of the phases; these fluids are called non-Newtonian. Usually the viscosity diminishes as the velocity gradient increases.

antenna The term now generally used in scientific and technical literature for *aerial. It was originally a US term.

antibonding orbital *See* molecular orbital.

anticoincidence circuit A circuit with two input terminals, designed to produce an output pulse if one input terminal receives a pulse, but not if both terminals receive pulses within a specified time interval.

anticyclone A rotary atmospheric disturbance with a spirally outward flow of air from a centre of high pressure. In the northern hemisphere the rotation is clockwise. *Compare* cyclone.

antiferromagnetism The property of certain materials that have a low positive magnetic *susceptibility (as in *paramagnetism) and exhibit a temperature dependence similar to that encountered in *ferromagnetism. The susceptibility increases with increasing temperature up to a certain point, called the *Néel temperature*, and then falls with increasing temperature in accordance with the *Curie–Weiss law. The material thus becomes paramagnetic above the Néel temperature, which is analogous to the *Curie temperature in the transition from ferromagnetism to paramagnetism.

Antiferromagnetism is a property of certain inorganic compounds such as MnO, FeO, FeF_2, and MnS. It results from interactions between neighbouring atoms leading to an antiparallel arrangement of adjacent magnetic *dipole moments.

antimatter Matter composed entirely of *antiparticles. An antihydrogen atom would consist of a nucleus containing an antiproton and an orbiting positron. The existence of antimatter in the universe has not been detected.

antinode A position in a *standing wave at which one of the types of disturbance in the wave has a maximum value. Generally there is another type of disturbance, the value of which will be a minimum at this point, and this position is said to be a *node of the other disturbance. Thus in standing electromagnetic waves there will be electric antinodes (i.e. places where the electric field has maximum values), which are nodes for the magnetic field.

antiparallel Parallel but pointing in opposite directions.

antiparticle *Syn.* conjugate particle. A particle that corresponds with another particle of identical mass and *spin but has such quantum numbers as charge (Q), *baryon number (B), *strangeness (S), *charm (C), and *isospin (I_3) of equal magnitude but opposite sign. Examples of a particle and its antiparticle include the electron and positron, proton and antiproton, the positively and negatively charged pions, and the up quark and up antiquark. The antiparticle corresponding to a particle with the symbol a is usually denoted ā. When a

particle and its antiparticle are identical, as with the photon and neutral pion, this is called a *self-conjugate particle. See also* annihilation.

antiresonance The condition in which a vibrating system responds with minimum amplitude to an alternating driving force, by virtue of the inertia and elastic constants of the system.

aperiodic 1. Nonperiodic (*see* period). Applied to a system (e.g. an electric circuit, instrument, etc.) that is adequately *damped.
2. Without frequency discrimination. Applied to a circuit designed for use at a frequency (or over a range of frequencies) sufficiently far removed from any of its natural or *resonant frequencies for its characteristics to be substantially independent of frequency within the required limits.

aperture The part of a lens through which light is allowed to pass, or the part of a mirror or other reflecting surface from which light or other radiation can be reflected. The aperture is also the diameter of such an area. *See also* apertures and stops in optical systems; f-number.

aperture ratio When a light beam passing through a lens comes from a near object, the beam is not parallel and the light-passing power of the lens depends on the aperture ratio. This is given by $2n \sin\alpha$, where n is the refractive index of the medium in the image space and α is the angle, measured in the same medium between the optic axis and a ray from the axial object point passing through the edge of the mechanical aperture. *Compare* f-number.

apertures and stops in optical systems The rays of light passing through an optical system will be limited by the

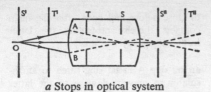

a Stops in optical system

*apertures of the various lenses or other components of the system. The mounts holding the components must therefore be considered as stops. In addition, there will probably be other stops introduced by the designer to cut out unwanted rays.

Consider an optical system containing the two stops (or lens apertures) S and T (Fig. *a*). Let S′ and T′ be the images of these stops formed in the "object space"–that is by rays of light passing through the system from right to left. If light is to pass through the aperture in T, it must pass through the corresponding image T′ before entering the system. Thus the light from the axial point O that will succeed in passing through the system will be limited to the cone of rays OAB. A possible subsequent course for these rays has been indicated by the dotted lines.

The amount of light from O which passes through the system is in this case limited by the aperture in the stop T. T is known as the *aperture stop* of the system. Its image in the object space, T′ is called the *entrance pupil*. T″, the image of the aperture stop in the image space (rays passing from left to right), is called the *exit pupil*.

Now consider light coming from the extra-axial point P (Fig. *b*). The ray PQ which passes through the centre of the entrance pupil is called the *principal ray*. The field of view in the plane of S′ will be limited by the diameter of the aperture in S′, and no light from points further from the axis than the point R will pass through the system. The actual stop S that limits the field of view in this way is known as the *field stop*. Its

image in the object space, S′, is the *entrance port* or *entrance window*. Its image in the image space, S″, is the *exit port* or *exit window*. The *field of view* is measured by the angle subtended by the entrance window at the centre of the entrance pupil.

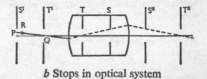

b Stops in optical system

aperture stop *See* apertures and stops in optical systems.

aperture synthesis A technique used in radio astronomy to obtain a high-resolution map of a radio source, which can cover a considerable area of sky. The *resolution corresponds to that obtained with a telescope of very large diameter, but is actually gained by using a number of small dish aerials that can be moved to cover the large aperture, or a few equally spaced strips of it. Pairs of dishes are set at different distances apart and connected as interferometers (*see* radio telescope). Their separate signals, recorded at all the necessary aerial spacings, are conveyed to a common receiver and the information processed to produce the map.

aphelion *See* perihelion.

aplanatic Free from *spherical aberration and *coma.

apochromatic lens A lens with a very high degree of correction of *chromatic aberration. The residual secondary spectrum of a lens achromatized for two colours is further reduced by using lens combinations with three or more different kinds of glass with appropriate partial dispersions. Such lenses are used as

19

microscope objectives and special photographic lenses.

apogee *See* perigee.

apostilb A unit of *luminance defined as the luminance of a uniformly diffusing surface that emits 1 *lumen per square meter.

apparent expansion *See* coefficient of expansion.

apparent magnitude *See* magnitude.

appearance potential 1. The potential difference through which an electron must be accelerated from rest to produce a given ion from its parent atom or molecule.
2. This potential difference multiplied by the electron charge, giving the least energy required to produce the ion. A simple ionizing process gives the *ionization potential of the substance; for example,

$$Ar + e \rightarrow Ar^+ + 2e.$$

Higher appearance potentials may be found for multiply charged ions:

$$Ar + e \rightarrow Ar^{++} + 3e.$$

Appleton layer *See* ionosphere.

arc A luminous electrical gas discharge characterized by high current density and low potential gradient. The intense ionization necessary to maintain the large current is provided mainly by thermionic emission from the cathode, which is raised to incandescence by the discharge. *See* conduction in gases; gas-discharge tube.

Archimedes' principle A body floating in a fluid displaces a weight of fluid equal to its own weight. *See* buoyancy.

arcing contacts Auxiliary contacts in any type of *circuit-breaker switch, designed

to close before and open after the main contacts thereby protecting the latter from damage by an *arc.

arcing ring A metal ring fitted to an insulator to prevent damage to the latter by a power arc.

arc lamp A type of lamp that utilizes the brilliant light accompanying an electric arc. The major portion of the light comes from the incandescent crater formed at the positive electrode.

arcover *See* flashover.

arc second Abbreviation: arc sec. Symbol: ″. A unit of angular measure equal to 1/3600 of a degree, used especially in astronomy. An *arc minute*, symbol: ′, is equal to 60 arc seconds.

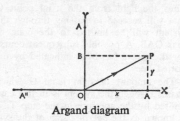

Argand diagram

Argand diagram A representation of complex numbers with reference to perpendicular axes: the horizontal axis of real quantities and a perpendicular axis of imaginary quantities. The complex quantity $x + iy$, where $i = \sqrt{(-1)}$, is represented by the line OP (or the point P) on the diagram. The length of OP is the *modulus* of the complex quantity $z = x + iy$ and is thus $r = \sqrt{(x^2 + y^2)}$, while the angle between OX and OP is the *amplitude* or *argument* θ of z and it can be shown that z may alternatively be written as:

$$re^{i\theta} = r \cos \theta + ir \sin \theta = x + iy.$$

armature 1. *Syn.* rotor. The rotating part in an electric *motor or *generator.

2. Any moving part in a piece of electrical equipment in which a voltage is induced by a magnetic field or which closes a magnetic circuit, e.g. the moving contact in an electromagnetic relay.

3. *See* keeper.

armature relay *See* relay.

arrow of time An effect that gives a direction to time, despite the fact that physical laws do not distinguish forward and backward directions. The direction of the psychological arrow is given by the observer's impression that time is something that passes, and the fact that one can remember the past but not the future. The direction of the thermodynamic arrow is given by the second law of *thermodynamics, i.e. the passage of time is always associated with an increase in entropy in a closed system. The direction of the cosmological arrow of time is given by the direction in which the universe is expanding. Some physicists believe that these three apparently separate arrows may be manifestations of the same phenomenon.

artificial radioisotopes *See* radioactivity.

asdic An acronym for *a*llied *s*ubmarine *d*etection *i*nvestigation *c*ommittee. Former name for sonar.

ASIC Abbreviation for application-specific integrated circuit. An *integrated circuit designed for a specific purpose rather than a generalized mass-produced one.

aspect ratio The ratio of the width of a *television picture to its height. An aspect ratio of 4:3 has been adopted in most countries including Britain and America.

aspherical lens or mirror A lens or mirror whose surface is part of a parabola, ellipse, hyperbola, etc., rather than part of a sphere, thus reducing optical *aberration, especially spherical aberration, to a minimum. The use of aspherical surfaces also leads to more simplified optical systems as the number of optical elements required is reduced.

Assmann psychrometer *See* psychrometer.

astable multivibrator *See* multivibrator.

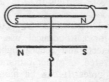

Astatic galvanometer

astatic system A system of magnets so arranged that there is no resultant directive force or couple on the system when placed in a uniform magnetic field. The simplest form consists of a pair of equal and parallel magnets mounted on the same axis, with their polarities in opposite directions. If a current-bearing coil encircles one of the magnets, its field affects mainly the magnet so encircled. This is the principle of the *astatic galvanometer.*

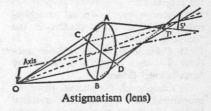

Astigmatism (lens)

astigmatism An optical *aberration in which, instead of rays converging to a single focus, they are caused to focus to two lines at right angles, separated by an interval known as the *conoid of Sturm*. (*See* diagram.) Astigmatism in the eye is caused either by curvature differences of the cornea or to a lesser extent because the crystalline lens is somewhat tilted. Correction involves the use of planocylindrical, spherocylindrical, or spherotoric lenses (*see* toric lenses). *Radial astigmatism* results from oblique incidence on a lens system.

Aston dark space *See* gas-discharge tube.

astrometry *See* astronomy.

astronomical telescope A *telescope designed for astronomical use. It can collect, detect, and record light, radio waves, infrared, or other radiation from celestial sources. It can be mounted in any of several ways (*see* telescope). *See also* reflecting telescope, refracting telescope, radio telescope.

astronomical unit (AU) A unit of length used in astronomy. It is defined as being equal to 149 597 870 km, which is very nearly equal to the mean distance between the centre of the earth and the centre of the sun (the original definition of the term).

astronomy The study of the universe and its contents. The main branches of the subject are:

(1) *Astrometry*, positional measurements of the stars and planets on the *celestial sphere;

(2) *Celestial mechanics*, relative motions of systems of bodies associated by *gravitational fields;

(3) *Astrophysics*, the internal structure, properties, and evolution of celestial bodies, and the production and expenditure of energy in such systems and in the universe as a whole. *Cosmology, *radio astronomy, *X-ray, *infrared, ultraviolet, and gamma-ray astronomy

are normally considered subsections of astrophysics.

astrophysics *See* astronomy.

asymptotic freedom A property of the *strong interaction by which, at high particle energies (i.e. shorter distances), the force is weakened and quarks and gluons behave almost as free particles. As the distance between particles tends to zero, the force tends to vanish. Asymptotic freedom is a consequence of certain *gauge theories with unbroken symmetries (i.e. non-Abelian gauge theories; *see* group).

asynchronous motor An a.c. motor whose actual speed bears no fixed relation to the supply frequency and varies with the load. An *induction motor is a typical example.

atmolysis A method of separating the constituents of a gas mixture by making use of their different rates of diffusion through a porous material.

atmometer An instrument for measuring the rate of evaporation of water into the atmosphere.

atmosphere 1. The *air. *See also* atmospheric layers.
2. Any gaseous medium.
3. *See* standard atmosphere.

atmospheric electricity The general electrical properties of the atmosphere, both under normal conditions and at the time of electric discharge (i.e. a *lightning flash).

The following data characterize the properties of the atmosphere at, or just above, sea level. They are mean fine-weather values:

direction of field	downward
potential gradient	130 volts per metre

total conductivity	3×10^{-4} siemens per metre
small ion mobility	1.4×10^{-4} $(m s^{-1})/(V\ m^{-1})$
air-earth current density	2×10^{-14} ampere per metre^2

An average lightning flash has a potential of about 4×10^9 volts. It provides a charge of 15 coulombs, and possesses about 2×10^{10} joules of energy. The average upward current is rather less than one ampere, but the peak covered in a flash may reach several tens of kiloamps.

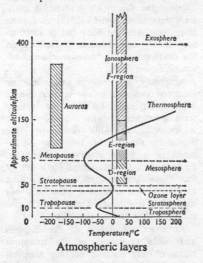

Atmospheric layers

atmospheric layers The gaseous layers into which the earth's atmosphere can be divided according to the change in physical properties, especially temperature. The altitude figures given in the diagram are approximate, as they vary over the earth's surface, and also show seasonal and diurnal changes. There are five main layers – the *troposphere*, *stratosphere*, *mesosphere*, *thermosphere*, and *exosphere*.

The troposphere contains some 75% of the atmosphere by mass. It is the layer of clouds and weather systems and is the most turbulent layer. It is heated from the ground by infrared radiation and convection, the temperature falling with increasing altitude to reach a minimum at the upper boundary of the layer, called the *tropopause*.

The temperature variation within the stratosphere leads to stability. The heating arises from the absorption of solar ultraviolet radiation by ozone molecules. These molecules are themselves formed by the action of UV on atmospheric oxygen. The greatest ozone concentration marks the upper limit of the *ozone layer*. The *ionosphere extends from the stratosphere into the exosphere.

The exosphere is the layer in which atmospheric constituents lose collisional contact with each other because of the very low density. It contains the *Van Allen belts and extends to the *magnetopause.

atmospheric pressure The pressure exerted by the atmosphere at the earth's surface, due to the weight of the air. The standard value at sea level is 1013 millibars or 101 325 newtons per square metre.

atmospherics Electromagnetic radiation produced by natural causes such as lightning. The term is also used to describe the disturbing effects that such radiation produces in a radio receiver.

atmospheric windows The gaps in atmospheric absorption that allow *electromagnetic radiations of certain wavelengths to penetrate the earth's atmosphere from space.

(1) *Optical window.* This allows through the whole visible spectrum (approx. 760–400 nm) and ultraviolet wavelengths down to about 320 nm. Radiation of wavelengths less than 320 nm

23

is absorbed by atoms and molecules in the atmosphere.

(2) *Infrared window*. This allows through wavelengths of 8–11 μm, corresponding to the region in which there is no absorption of infrared radiation by water vapour in the atmosphere. There are additional narrow-band infrared windows at several other micrometre wavelengths (1.25, 1.6, 2.2, 3.6, 5.0, and 21 μm). *See* infrared astronomy.

(3) *Radio window*. This allows through short-wave radio waves of wavelengths of approximately 1 mm to 30 m. *See* radio astronomy.

atom A concept originally introduced by the ancient Greeks (as a tiny indivisible component of matter), developed by Dalton (as the smallest part of an element that can take part in a chemical reaction), and made very much more precise by theory and experiment in the late-19th and early-20th century.

Following the discovery of the electron (1897), it was recognized that atoms had structure: since electrons are negatively charged, a neutral atom must have a positive component. The experiments of Geiger and Marsden on the scattering of alpha particles by thin metal foils led Rutherford to propose a model (1912) in which nearly all the mass of the atom is concentrated at its centre in a region of positive charge, the *nucleus, radius of the order 10^{-15} metre. The electrons occupy the surrounding space to a radius of 10^{-11} to 10^{-10} m. Rutherford also proposed that the nucleus has a charge of Ze and is surrounded by Z electrons (Z is the atomic number). According to classical physics such a system must emit electromagnetic radiation continuously and consequently no permanent atom would be possible. This problem was solved by the development of the *quantum theory.

The *Bohr theory of the atom (1913) introduced the concept that an electron in an atom is normally in a state of lowest energy (*ground state*) in which it remains indefinitely unless disturbed. By absorption of electromagnetic radiation or collision with another particle the atom may be excited – that is an electron is moved into a state of higher energy. Such excited states usually have short lifetimes (typically nanoseconds) and the electron returns to the ground state, commonly by emitting one or more *quanta of electromagnetic radiation. The original theory was only partially successful in predicting the energies and other properties of the electronic states. Attempts were made to improve the theory by postulating elliptic orbits (Sommerfeld 1915) and electron *spin (Pauli 1925) but a satisfactory theory only became possible upon the development of *wave mechanics after 1925.

According to modern theories, an electron does not follow a determinate orbit as envisaged by Bohr but is in a state described by the solution of a wave equation. This determines the *probability* that the electron may be located in a given element of volume. Each state is characterized by a set of four *quantum numbers, and, according to the *Pauli exclusion principle, not more than one electron can be in a given state. (*See* atomic orbital.)

An exact calculation of the energies and other properties of the quantum states is only possible for the simplest atoms but there are various approximate methods that give useful results. (*See* perturbation theory.) The properties of the innermost electron states of complex atoms are found experimentally by the study of X-ray spectra. The outer electrons are investigated using spectra in the infrared, visible, and ultraviolet. Certain details have been studied using microwaves (*see* Lamb shift). Other information may be obtained from magnetism, chemical properties, *appearance potentials, or *photoelectron spectroscopy.

atomic bomb *See* nuclear weapons.

atomic clock *See* clocks.

atomic force microscope A type of microscope, similar in operation to the *scanning tunnelling microscope but using mechanical forces rather than electrical effects. In the atomic force microscope the probe is a tiny chip of diamond held on a spring-loaded cantilever. The probe, which is in contact with the sample surface, is slowly moved and the tracking force between its tip and the sample is measured by deflections of the cantilever. The probe is raised and lowered in order to maintain the force constant, and a contour of the surface is thus produced. A raster scan of the whole sample allows a computer-generated contour map of the surface to be obtained. The atomic force microscope unlike the scanning tunnelling microscope, can produce images of non-conducting materials, such as biological specimens. The sample does, however, have to be fairly rigid.

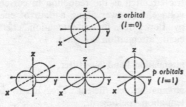

Atomic orbitals

atomic heat The former name for the *molar heat capacity of an element. *See* Dulong and Petit's law.

atomic mass constant Symbol: m_u. A fundamental constant equal to 1 unified *atomic mass unit.

atomic mass unit (unified) *Syn.* dalton. Abbreviation: a.m.u. Symbol: u. A unit of mass equal to 1/12 of the mass of an atom of carbon-12. It is equal to 1.6605 $\times$ 10^{-27} kg or approximately 931 MeV.

atomic number *Syn.* proton number. Symbol: Z. The number of protons in the nucleus of an atom or the number of electrons revolving around the nucleus. The atomic number determines the chemical properties of an element and the element's position in the *periodic table. All the isotopes of an element have the same atomic number although different isotopes have different *mass numbers.

atomic orbital An allowed *wave function of an electron in an atom obtained by a solution of *Schrödinger's wave equation. In a hydrogen atom, for example, the electron moves in the electrostatic field of the nucleus and its potential energy is $-e^2/r$, where e is the electron charge and r its distance from the nucleus. A precise orbit cannot be considered as in Bohr's theory of the *atom but the behaviour of the electron is described by its wave function, Ψ, which is a mathematical function of its position with respect to the nucleus. The significance of the wave function is that $|\Psi|^2 d\tau$ is the probability of locating the electron in the element of volume $d\tau$.

Solution of Schrödinger's equation for the hydrogen atom shows that the electron can only have certain allowed wave functions (eigenfunctions). Each of these corresponds to a probability distribution in space given by the manner in which $|\Psi|^2$ varies with position. They also have an associated value of the energy E. These allowed wave functions, or orbitals, are characterized by three quantum numbers similar to those characterizing the allowed orbits in the earlier quantum theory of the atom:

n, the *principal quantum number*, can have values of 1, 2, 3, etc. The orbital with *n* = 1 has the lowest energy. The states of the electron with *n* = 1, 2, 3, etc., are called *shells* and designated the K, L, M shells, etc.

l, the *azimuthal quantum number*, which for a given value of *n* can have values of 0, 1, 2, ... (*n*–1). Thus when *n* = 1, *l* can only have the value 0. An electron in the L shell of an atom with *n* = 2 can occupy two *subshells* of different energy corresponding to *l* = 0 and *l* = 1. Similarly the M shell (*n* = 3) has three subshells with *l* = 0, *l* = 1, and *l* = 2. Orbitals with *l* = 0, 1, 2 and 3 are called *s*, *p*, *d*, and *f* orbitals respectively. The significance of the *l* quantum number is that it gives the angular momentum of the electron. The orbital angular momentum of an electron is given by:
$$\sqrt{[l(l+1)(h/2\pi)]}.$$
m, the *magnetic quantum number*, which for a given value of *l* can have values $-l, -(l-1), \ldots, 0, \ldots (l-1), l$. Thus for a *p* orbital for which *l* = 1, there are in fact three different orbitals with *m* = –1, 0, and 1. These orbitals, with the same values of *n* and *l* but different *m* values, have the same energy. The significance of this quantum number is that it indicates the number of different levels that would be produced if the atom were subjected to an external magnetic field.

According to wave theory the electron may be at any distance from the nucleus but in fact there is only a reasonable chance of it being within a distance of $\sim 5 \times 10^{-11}$ metre. Indeed the maximum probability occurs when $r = a_0$ where a_0 is the radius of the first Bohr orbit. It is customary to represent an orbital by a surface enclosing a volume within which there is an arbitrarily decided probability (say 95%) of finding the electron. These are drawn in the diagram for some simple orbitals. Note, although *s* orbitals are spherical (*l* = 0),

orbitals with *l* > 0 have an angular dependence.

Finally, the electron in an atom can have a fourth quantum number, M_s, characterizing its *spin direction.

This can be $+\frac{1}{2}$ or $-\frac{1}{2}$ and according to the *Pauli exclusion principle, each orbital can hold only two electrons. The four quantum numbers lead to an explanation of the periodic table of the elements.

atomic stopping power *See* stopping power.

atomic volume The volume in the solid state of one mole of an element. Thus, atomic volume = atomic weight ÷ density of the solid.

atomic weight Former name for *relative atomic mass.

attenuation The reduction of a radiation quantity, such as *radiant intensity, particle *flux density, or energy flux density, upon the passage of the radiation through matter. It may result from any type of interaction with the matter, such as absorption, scattering, etc. In an electric circuit it is the reduction in current, voltage, or power along a path of energy flow. *See* linear attenuation coefficient; attenuation constant.

attenuation band *See* filter.

attenuation coefficient *See* linear attenuation coefficient.

attenuation constant Symbol: α. For a plane progressive wave at a given frequency, the attenuation constant is the rate of exponential decrease in amplitude of voltage, current, or field-component in the direction of propagation of the wave. For example,
$$I_2 = I_1 e^{-\alpha d},$$

where I_2 and I_1 are the currents at two points (I_1 being nearer the source of the wave) a distance d apart. α is usually expressed in *nepers or *decibels. See propagation coefficient.

attenuation distortion *See* distortion.

attenuation equalizer An electrical *network designed to provide compensation for attenuation *distortion throughout a specified band of frequencies.

attenuator An electrical *network or *transducer specifically designed to attenuate a wave without distortion. The amount of attenuation may be fixed or variable. A fixed attenuator is also called a *pad*. Attenuators are usually calibrated in *decibels.

atto- Symbol: a. The prefix 10^{-18}, e.g. 1 am = 10^{-18} metre.

audibility Ease of detection of a sound by ear. The sensitivity of the ear to a note depends on both its intensity and frequency. The intensity of a pure tone that is just audible is known as the *threshold of audibility*. Above this, the ear can detect greater intensities up to the threshold of feeling, beyond which the sensation changes to one of pain rather than hearing. These threshold intensities vary with frequency. At a frequency of 1000 hertz the maximum intensity that the ear can detect is about 10^{14} times the minimum. The average range of frequencies to which the ear is sensitive is from about 20 to 20 000 hertz at an rms sound pressure of 1 pascal. The frequency range becomes smaller at greater or lower sound intensities.

audiofrequency Any *frequency to which a normal ear can respond; it extends from about 20 to 20 000 hertz. In communication systems, satisfactory intelligibility of speech (i.e. of commercial quality) can be obtained if frequencies lying between about 300 and 3400 hertz are reproduced, and any frequency within this range is described as a *voice frequency*.

Auger effect *Syn.* Auger ionization. The spontaneous ejection of an electron by an excited positive ion to form a doubly charged ion: i.e.

$$A - e \rightarrow A^{+*} \rightarrow A^{2+} + e^-,$$

where A^{+*} represents an excited state of a singly charged ion and A^{2+} a doubly charged ion that may or may not be in its ground state. The first step may result from *internal conversion. Alternatively, it may be induced by an external stimulus such as bombardment by electrons or photons.

*Autoionization is very similar to the Auger effect and the terms are sometimes used synonymously.

Auger shower *Syn.* extensive air shower. A *shower of elementary particles produced by a primary *cosmic ray entering the atmosphere. Auger showers extend over large areas.

aurora An intermittent electrical discharge occurring in the rarefied *upper atmosphere. Charged particles in the *solar wind become trapped in the earth's magnetic field and move in helical paths along the lines of force, oscillating between the two magnetic poles. On entering the upper atmosphere the charged particles excite the air molecules. The resulting *emission of light takes many beautiful forms from slight luminosity to large streamers moving rapidly across the sky. The intensity of the aurora is greatest in polar regions (although it is sometimes seen in temperate zones) and during periods of high solar activity.

autodyne oscillator *See* beat reception.

autoemission *See* field emission; cold cathode.

autoionization A form of ionization involving two steps, first the excitation of an atom (or molecule) into a state with an energy above its *ionization potential and second, de-excitation from this state to give a positive ion with a lower energy and an ejected electron. The process is similar to the *Auger effect with the difference that the initial vacancy in an electron shell is caused by transfer of an electron from one orbital to another empty orbital. In the Auger effect the vacancy is formed by complete removal of the electron to give an ion. Autoionization results in a singly charged positive ion:

$$A \rightarrow A^* \rightarrow A^+ + e,$$

where A^* is an excited atom. The electron ejected has a characteristic energy equal to the difference between the energy of the excited atom and that of the ion.

automatic frequency control (a.f.c.) A device that automatically maintains the frequency of any source of alternating voltage within specified limits.

automatic gain control (a.g.c.) A method of automatically holding the output volume constant in a radio receiver, despite variations in the input signal.

autoradiograph A photograph of the distribution of a *radioisotope within a thin specimen of metal, biological tissue, etc. The specimen is *labelled with a *radioisotope and placed in contact with a photographic plate for a suitable exposure time, and an image is produced by the action of the radiation emitted. On developing the film, the autoradiograph can be seen.

autosynchronous motor *See* synchronous induction motor.

autotransformer A *transformer with a single winding, tapped at intervals, instead of two or more independent windings. The voltage drop across each tapped section is related to the total applied voltage in the same proportion as the number of turns of the selection is related to the total number of turns of the winding.

avalanche *Syn.* Townsend avalanche. A cumulative ionization process in which a single particle or photon ionizes several gas molecules within a device. Each electron liberated gains sufficient kinetic energy from an accelerating electric field to ionize other molecules, producing more free electrons and ions. Thus a large number of charged particles results from the initial event. The phenomenon is used, for example, in the *Geiger counter and *IMPATT diode.

avalanche breakdown A type of *breakdown in a *semiconductor diode caused by the cumulative multiplication of free charge carriers under the action of a strong electric field. Some free carriers gain enough energy to liberate new hole-electron pairs by collision, i.e. an *avalanche takes place. Avalanche breakdown is the breakdown that occurs in a reverse-biased p-n junction.

average *Syn.* mean. **1.** The sum of a set of observations divided by their number.

2. In the case of a continuous function $f(x)$, the mean value over the range x_1 to x_2 is

$$\frac{\int_{x_1}^{x_2} f(x)\, dx}{x_2 - x_1}.$$

Physically this corresponds to the mean ordinate of the graph of $f(x)$ (as ordinate) against x. Similar definitions exist for functions of more than one variable.

3. *See* weighted mean.

average life or lifetime *See* mean life.

Avogadro constant Symbol: L or N_A. The number of molecules contained in one mole of any substance. *Amount of substance is proportional to the number of specified entities of that substance, the Avogadro constant being the proportionality factor. It is the same for all substances and its value is 6.022 136 7 $\times$ 10^{23} mol^{-1}. *See also* Loschmidt constant.

Avogadro's hypothesis Equal volumes of all gases measured at the same temperature and pressure contain the same number of molecules, i.e. the volume occupied at a given temperature and pressure by a mole of a gas is the same for all gases (22.4 $\times$ 10^{-3} m^3 at STP). *See* ideal gas.

axial ratio The relative lengths of the three edges of the *unit cell of a crystal lattice, taking that of the b axis as unity.

axial vector *See* pseudovector.

axiom A self-evident proposition not requiring demonstration.

axion A hypothetical subatomic particle postulated to account for the fact that theories of *quantum chromodynamics predict the violation of *CP invariance in the strong interaction, but this has not been observed. Axions, if they exist, are light particles (less than 10^{-12} times the mass of the proton). Some physicists believe that large numbers of them may exist in the haloes of galaxies and that they may contribute to the *dark matter in the universe.

azimuth Position as measured by an angle round some fixed point or pole. *See* celestial sphere.

azimuthal quantum number *See* atomic orbital.

B

Babinet compensator An optical device that introduces a *phase difference of variable magnitude between the ordinary and extraordinary rays (*see* double refraction). It comprises two narrow-angle quartz wedges with parallel refracting edges and hypotenuse faces adjacent; the optic axes of the prisms are mutually perpendicular, and aligned parallel to the refracting edges. A light ray passing vertically downwards through some point on a refracting edge will traverse a distance d_1 in the upper wedge and d_2 in the lower wedge. The total relative phase difference is then proportional to $(d_1 - d_2)$. This can be varied by sliding one wedge over the other, maintaining parallel edges, and hence the desired relative phase difference can be obtained. *See also* half-wave plate; quarter-wave plate.

Babo's law The lowering of the *vapour pressure of a solvent by addition of a nonvolatile solute is proportional to the concentration of the solution.

back electromotive force An e.m.f. that opposes the normal flow of current in an electric circuit.

back focal length The distance from the last surface of an optical system to the second principal focus.

background noise *Syn.* random noise. *See* noise.

background radiation 1. The low-intensity radiation resulting from the bombardment of the earth by cosmic rays and from the presence of naturally occurring *radionuclides (such as ^{40}K, ^{14}C) in rocks, soil, air, building materials, etc. When measurements of radiation are being carried out a correction must be made for the background radiation.

2. *See* cosmic background radiation.

backing store Large-capacity computer storage devices, usually *disks, in which programs and data are kept when not required for processing by a *computer. A program and its associated data is copied from backing store into *main store only when the program is about to be executed.

back scatter The scattering process by which radiation emerges from the same surface of a material as that through which it enters. The term also applies to the radiation undergoing such a process.

backward-wave oscillator *See* travelling-wave tube.

baffle A partition used with a sound radiator to increase the path difference between sound originating from the front and back of the radiator. It is most commonly used to improve the frequency response of a loudspeaker.

balance An instrument whose primary function is to compare two masses. The balance with the widest usage is the equal-armed balance. If P and Q are two slightly different masses, placed one in each pan of the balance with arms of length a, and the balance comes to rest with the arms making a small angle θ with the horizontal then

$$\tan \theta = \frac{(P - Q)a}{(P + Q + 2w)h + Wk},$$

where w is the mass of a scale pan, and W is the mass of the balance beam, h is the height of the central knife-edge above the line joining the outer knife-edges, and k is the distance of the centre of gravity of the beam below the central knife-edge.

Other forms of balance are: (1) the *decimal balance*, which has arms in the ratio 10:1 and thus avoids the necessity for using heavy weights; (2) the *spring balance*, which consists essentially of a helical spring with its axis vertical. It may be used either by extending the spring or by compressing it as in the familiar use of household scales; (3) the *torsion balance*, which consists essentially of a vertical straight torsion wire, the upper point of which is fixed while the lower carries a horizontal beam; (4) the *modern balance* in which the deformation of a supporting system is measured using *strain gauges. *See also* microbalance.

balanced amplifier A push-pull amplifier. *See* push-pull operation.

ballast resistor A resistor constructed from a material having a high *temperature coefficient of resistance in such a way that, over a range of voltage, the current is substantially constant. It is connected in series with a circuit to stabilize the current in the latter by absorbing small changes in the applied voltage. The most common types are the *barretter and the *thermistor.

ballistic Of an instrument. Designed to measure an impact or brief flow of charge. *See* pendulum; ballistic galvanometer.

ballistic galvanometer A galvanometer adapted to measure the charge, Q, flowing through the instrument during the passage of a transient current, where $Q = \int_0^\infty I.\mathrm{d}t$. The period of the moving part of the galvanometer must be long compared with the duration of the current; the electromagnetic impulse due to the passage of the transient can then be deduced from the ballistic "throw", θ. For a suspended-magnet type galvanometer, $Q \propto \sin \frac{1}{2}\theta$; for a moving-coil instrument, $Q \propto \theta$.

Balmer series *See* hydrogen spectrum.

balun Acronym for balanced unbalanced. An electrical device used to couple a balanced impedance, such as an aerial, to an unbalanced *transmission line, such as a coaxial cable.

band 1. In communications, a range of frequencies within specified limits used for a definite purpose.
2. A closely spaced group of energy levels in atoms. (*See* energy band.)
3. *See* spectrum.

band-pass filter *See* filter.

band pressure level The *sound pressure level of a sound within a specified band of sound frequency.

band spectrum *See* spectrum.

band-stop filter *See* filter.

bandwidth 1. The difference between the upper and lower frequency limits of a *band, normally measured in hertz.
2. The range of frequencies over which a particular characteristic of an electronic device or system lies within specified limits.

bar Symbol: bar. A CGS unit of pressure equal to 10^5 pascal. It may be used with SI units, and SI prefixes may be attached to it. The millibar (Symbol: mbar or mb) is a commonly used unit of pressure in meteorology.

Barkhausen effect The magnetization of a ferromagnetic substance does not in-

crease or decrease steadily with steady increase or decrease of the magnetizing field but proceeds in a series of minute jumps. The effect gives support to the domain theory of *ferromagnetism.

Barlow lens A planoconcave lens placed between the *objective and *eyepiece in a *telescope to increase magnification.

barn Symbol: b. A unit of nuclear *cross section. 1 barn = 10^{-28} metre^2.

Barnett effect A long iron cylinder rotating at high speed about its longitudinal axis develops a slight magnetization proportional to the angular speed of rotation. This magnetization is due to the effect of the rotation on the electronic orbits in the atoms of the iron and on the electrons themselves, which have their own intrinsic *spin. *Compare* Einstein and de Haas effect.

barograph A recording *barometer. The common type consists of an aneroid barometer operating a pen that traces a line on a sheet of graph paper mounted on a slowly revolving drum.

barometer An instrument for measuring atmospheric pressure.
1. *Mercury barometers* consist of a glass tube about 80 cm long, closed at one end. The tube is filled with mercury and then placed, open end downward, in a reservoir of mercury. As the atmospheric pressure changes, the level of the mercury changes and, to a lesser degree, so does the level in the reservoir. The pressure is measured by the difference h between the levels (*see* pressure). The space at the top of the tube is known as a *Torricellian vacuum*.
In *Fortin's barometer*, the scale for measuring the height is fixed and the level of the mercury in the reservoir is adjusted (by moving the flexible bottom of the reservoir) to be at the zero of the

scale. The difference in height can then be read directly. There are also instruments in which the scale is moved so that its zero coincides with the lower mercury level.

2. The *aneroid barometer* consists basically of an evacuated flat cylindrical closed metal box with corrugated flexible faces that are kept from collapsing together by a spring. As the external pressure varies, the distance between the flat faces alters and a pointer is operated. Such an instrument is calibrated by comparison with a mercury barometer. It is also used as an *altimeter.

barostat A constant pressure device or pressure regulator, especially one that compensates for changes of atmospheric pressure, as in the fuel metering system of an aircraft engine.

barrel distortion *See* distortion (3).

barretter A device used for stabilizing voltage, consisting of a sensitive metallic resistor whose resistance increases with temperature. The resistor is usually enclosed in a glass bulb. When used in series with a circuit the voltage drop is kept constant over a range of variations in current. *See also* ballast resistor.

barycentric *See* centre of gravity.

baryon A collective name given to *hadrons (*elementary particles that can undergo strong interactions) with half integer spin. Baryons are thus *fermions. All baryons have a mass equal to or greater than the mass of the proton. An additive *quantum number, called the baryon number (B), may be defined such that baryons have baryon number $B = +1$, antibaryons $B = -1$, and all other particles $B = 0$. The total baryon number, which equals the number of baryons minus the number of antibaryons is conserved in all particle interactions. Baryons are composed of three *quarks while antibaryons contain three antiquarks. For example, the proton contains two up quarks (u) and a down quark (d) while the neutron contains an up quark and two down quarks. *See* Appendix, Table 7.

base The region in a bipolar junction *transistor that separates the emitter and collector and to which the *base electrode* is attached.

base units *See* SI units, also Appendix, Table 2; coherent units.

Bateman equations A set of equations that describe the decay of a chain of radioactive nuclides. If only the parent nuclide is initially present and there are N_1^0 atoms, then the number of atoms of the nth nuclide after a time t is given by the equation:

$$N_n(t) =$$

$$\sum_1^n \frac{\lambda_1 \lambda_2 \cdots \lambda_{n-1} N_1^0 e^{-\lambda_n t}}{(\lambda_1 - \lambda_n)(\lambda_2 - \lambda_n) \cdots (\lambda_{n-1} - \lambda_n)}$$

where λ_n is the *decay constant and N_n the number of atoms of the nth nuclide.

battery Two or more secondary cells, primary cells, or capacitors, electrically connected and used as a single unit.

BCS theory *See* superconductivity.

beam aerial *See* aerial array.

beam coupling The production in a circuit of an alternating current between two electrodes, when an intensity-modulated electron beam is passed.

beam current The current consisting of the beam of electrons arriving at the screen of a *cathode-ray tube.

beat-frequency oscillator An apparatus for generating electrical oscillations, the frequency of which can usually be varied over a range of audiofrequencies or *video frequencies. It incorporates two radio-frequency oscillators, one of which has a fixed frequency while the other has a frequency that can be varied at will. The output is obtained by the beating (*see* beat reception) of the two radio-frequency oscillations. The output voltage remains substantially constant at all output frequencies within the range covered.

beat oscillator *See* beat reception.

beat reception *Syn.* heterodyne reception. A method of radio reception in which beating is employed. The *beats (usually at an audiofrequency) are produced by combining the received radio-frequency oscillations with radio-frequency oscillations generated in the receiver by a separate oscillator (called a *beat oscillator*). The combined oscillations are then detected, amplified, and rendered audible. A beat oscillator that also functions as a detector and amplifier is called an *autodyne oscillator*. *Compare* superheterodyne receiver.

beats Fluctuations in sound intensity observed when two tones very nearly equal in frequency are sounded simultaneously. The phenomenon may be compared with that of *interference. At certain equal time intervals the wavetrains are in phase and reinforce each other; at intermediate periods they are in opposite phase and tend to neutralize each other. Combining the two tones of frequencies m and n, a tone of frequency $(m + n)/2$ is observed whose amplitude varies from $2A$ to 0 (A being the amplitude of each of the primary tones) at a *beat frequency* of $(m - n)$. If $(m - n)$ exceeds 20 beats per second, the beats merge into what is called a *difference tone*. The electrical counterpart of beats is used in *beat reception.

Beaufort scale A numerical scale used in meteorology in which successive values of wind velocities are assigned numbers ranging from 0 (calm) to 12 (hurricane) thus indicating wind forces. Numbers 13–17 are often added to indicate specific hurricane speeds.

Beckmann thermometer A *mercury in glass thermometer used for the accurate determination of small temperature changes. The lower bulb is much larger than that of an ordinary thermometer and the scale behind the capillary tube, which is about 30 cm long, is divided into hundredths of a degree and covers only about 5–6 °C. The temperature change to be measured can take place about any mean temperature in the range, say, 0 °C to 100 °C, by varying the amount of mercury present in the lower bulb. This is made possible by running in more mercury from the small reservoir bulb at the top of the capillary, or conversely by running some mercury from the lower bulb into the reservoir where it plays no further part in the production of the thermometer reading. The variable amount of mercury present in the bulb means that the scale graduations will only exactly represent true degrees Celsius at the setting for which the scale has been calibrated.

becquerel Symbol: Bq. The derived SI unit of *activity equal to one disintegration per second.

Becquerel effect The e.m.f. produced by illuminating the surface of one electrode in an electrolytic cell.

Beilby layer An amorphous layer about 5 nm thick produced by polishing a surface; the ordinary crystalline material is present below this layer. It has been found that this surface layer is produced by sliding friction dissipating enough energy to melt the surface. A substance can thus be polished by one of higher melting point. Running in of mechanical parts produces a deep Beilby layer.

bel *See* decibel.

bending moment The algebraic sum of the moments, about any cross section of a beam, of all the forces acting on the beam on one side of this section. It is immaterial which side of the section is considered.

Bernouilli's theorem In the steady frictionless motion of a fluid acted on by external forces which possess a gravitational *potential (V)* then

$$\int \frac{dp}{\rho} + \frac{1}{2} v^2 + V = C,$$

where p and ρ are the pressure and density of the fluid; v is the velocity of the fluid along a stream line; and C is a constant, depending on the particular stream line chosen, called *Bernouilli's constant*. The equation can be shown to agree with the principle of conservation of energy and may be expressed in a generalized form

$$\int \frac{dp}{\rho} - \frac{\partial \varphi}{\partial t} \pm \frac{1}{2} v^2 + V = A,$$

where ϕ is the velocity potential and A is a function of the time (t). For steady motion this reduces to the original equation.

Berthelot's equation of state The equation:

$$(p + a/TV^2)(V - b) = RT$$

This gives better agreement with experiment than *Van de Waal's equation at moderate pressures, but fails at the critical point. *See* equations of state.

34

Bessel functions A power series in x which are solutions of a linear differential equation of the form

$$x^2 \frac{d^2y}{dx^2} + x \frac{dy}{dx} + (x^2 - n^2)y = 0.$$

They have many applications in physics, for example in problems of heat conduction, etc. There are various kinds of Bessel functions, the most important being denoted by $J_n(x)$, the Bessel function of order n.

beta current gain factor Symbol: β. The short-circuit current-amplification factor in a bipolar *transistor with *common-emitter connection. It is given by:

$$\beta = (\partial I_C / \partial I_B),$$

the collector voltage, V_{CE}, being constant; I_C is the collector current and I_B the base current. β is always greater than unity and in practice takes value up to 500.

beta decay The spontaneous transformation of a nucleus into one of its neighbouring *isobars accompanied by the ejection of an electron or positron. The nucleus produced always has the same *mass number as the initial nucleus but differs in *atomic number by one. If an electron is emitted the number of nuclear protons increases by one, if a positron is emitted it decreases by one. Two examples of beta decay are

$$^{14}C \rightarrow {}^{14}N + e^- + \bar{\nu}$$
$$^{11}_6C \rightarrow {}^{11}_5B + e^+ + \nu$$

It is found that electrons and positrons ejected in beta decay always have a continuous distribution of energies and not a single energy equal to the energy released. The "missing energy" is carried away by the neutrino ν or antineutrino $\bar{\nu}$. The total energy carried by the electron and antineutrino or by the positron and neutrino is constant for a particular decay and thus the neutrino enables energy to be conserved. The neutrinos also enable *angular momentum and linear momentum to be conserved. Beta decay is now known to be a *weak interaction. *Parity is therefore not conserved. *See also* radioactivity.

beta particle (β-particle) An *electron or *positron emitted by the nucleus of a *radionuclide during *beta decay.

beta rays (β-rays) A type of *ionizing radiation consisting of a stream of *beta particles with a continuous distribution of kinetic energies up to a maximum characteristic of the source (*see* beta decay). Their absorption in matter depends mostly upon the mass/area of the absorber and only to a lesser degree upon the nature of the material. The rays from the most energetic emitters have maximum energies up to a few MeV, which penetrate matter up to a few tens of kilogram per metre squared. Some radionuclides give maximum energies of only a few tens of keV; these rays are absorbed by matter of the order of 10^{-2} kg m^{-2}, e.g. a few centimetres of air.

The spectrum of beta rays is sometimes accompanied by a line spectrum of electrons caused by *internal conversion.

betatron A cyclic *accelerator for producing high-energy electrons by means of magnetic induction. If an electron is describing a circular orbit of radius r in the magnetic field between the poles of an electromagnet, an increase in the magnetic flux through the orbit produces an acceleration of the electron. If the field at the circumference of the orbit is equal to half the average field inside the orbit, the radius r is unaltered, i.e. the particle continues in the same path. This is achieved by shaping the pole pieces.

In the betatron the magnet is excited by alternating current, and the electrons are injected into the field when the cur-

rent is beginning to rise from zero. They are deflected out of the field just before the current reaches its peak value, having completed several hundred thousand revolutions. The angular velocity ω of a particle moving in a fixed orbit in which the magnetic flux density is B is given by $\omega = eB/m$, where m is the mass. To maintain a constant angular frequency and orbit, B is increased by the same factor as m increases (due to the relativistic velocity of the particle). The functioning of the machine is therefore not affected by the relativistic mass increase. Energies up to 300 MeV have been produced; the electron beam is used to produce high quantum energy X-rays and also in particle research. *See also* synchrotron.

B/H loop *See* hysteresis loop.

bias *Syn.* bias voltage. A voltage applied to an electronic device to determine the portion of the *characteristic of the device at which it operates.

biaxial crystal A crystal in which there are two directions along which the polarized components of a ray of light will be transmitted with the same velocity. Such crystals belong to the orthorhombic, monoclinic, or triclinic systems.

biconcave lens A lens with both surfaces *concave. A *biconvex lens* has both surfaces *convex.

bifilar suspension A type of suspension used in electrical instruments in which the moving part is suspended on two parallel threads, wires, or strips.

bifilar winding A method of winding a wire to form a resistor or coil with negligible *inductance: the wire is doubled back on itself and wound double from the looped end (*see* diagram).

Bifilar winding

big-bang theory The theory in cosmology that all matter and radiation in the universe originated in a cataclysmic explosion that occurred $10-20 \times 10^9$ years ago. Since this initial state of extreme density and temperature, the universe has expanded and cooled.

Within a tiny fraction of a second from the big bang, a variety of elementary particles and antiparticles had been created and were undergoing interactions. Photons of radiation were produced by their *annihilation. Deuterium and helium nuclei were synthesized some 100 seconds after the big bang when the temperature was about 10^9 K. After about 10^4 years, when the temperature had fallen to about 10^4 K, the matter in the universe was composed mainly of an ionized gas of free electrons, protons, and helium nuclei.

Neutral hydrogen formed, by combination of free electrons and protons, when the temperature had dropped to about 3000 K. This process took place about 3×10^5 years after the big bang. Prior to this, the scattering of photons on electrons coupled matter and radiation so that they shared a common temperature. Since the combination of free electrons and protons, the radiation has cooled from 3000 K to 3 K, i.e. to the observed temperature of the microwave background (*see* cosmic background radiation). The matter, no longer coupled to the radiation, has interacted to form stars and galaxies.

The big-bang theory has been successful in explaining the expansion of the universe (detected from the observed *redshift of distant galaxies), the measured *cosmic abundance of helium, and the microwave background. *See also* inflationary universe.

Billet split lens A lens cut in two, so that the optical centres of the semi-lenses are slightly displaced laterally; in consequence, two real images of a slit are formed and in the overlapping region in front of these (coherent) images *interference takes place.

billion One thousand million (10^9). In American and French usage the term has always been used for 10^9. In Britain, a billion was formerly one million million (10^{12}). In the 1960s, the British Treasury started using the American sense (a thousand million) in its economic statistics, and this usage has now largely supplanted the original meaning.

bimetallic strip Two metals having different coefficients of expansion riveted together: an increase in temperature of the strip causes the strip to bend, the metal having the greater coefficient of expansion being on the outside of the curve. One end is rigidly fixed and movement of the other end can serve to open or close an electric circuit of a temperature control device, or to move the pointer of a pointer-type thermometer.

bimorph cell A device for converting electrical signals into mechanical motion using the *piezoelectric effect. It consists of two piezoelectric crystals (such as Rochelle salt) cut and joined together so that an applied voltage causes one to expand and the other to contract. The composite crystal thus bends as a result of the voltage across it. The converse effect of generation of an electric voltage by bending the cell is also used.

binary notation A method of expressing numbers by two digits, 0 and 1, rather than the ten digits used in decimal notation. The binary equivalents of some decimal numbers are as follows.

Decimal	Binary
0	0
1	1
2	10
3	11
4	100
5	101
6	110
7	111
8	1000
9	1001
10	1010
16	10000
32	100000
64	1000000
100	1100100

See also bit.

binary star A system of two stars that revolve around a common centre of gravity.

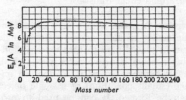

The binding energy for nucleon in MeV, as a function of the mass number A

binding energy 1. Symbol: E_B. The mass of a nucleus is slightly less than the mass of its constituent protons and neutrons. By Einstein's law of the *conservation of mass and energy ($E = mc^2$), this mass difference is equivalent to the energy released when the nucleons bind together. This energy is the binding energy. The graph of binding energy per nucleon, E_B/A, against *mass number, A, shows that as A increases E_B/A increases rapidly up to a mass number of 50–60 (iron, nickel, etc.) and then decreases slowly. There are therefore two ways in which energy can be released from a nucleus, both of which entail a rearrangement of nuclei occurring in the lower half of the curve to form nuclei in the upper, higher-energy part of the curve. Fission is the splitting of heavy atoms, such as uranium, into lighter atoms, accompanied by an enormous release of energy. Fusion of light nuclei, such as deuterium and tritium, releases an even greater quantity of energy.

2. The work that must be done to detach a single particle from a structure. *See* ionization potential.

binoculars *See* prismatic binoculars.

binomial distribution In a trial that can only have one of two results (say success or failure), the binomial distribution of probabilities, P_r, for obtaining r successes in n independent trials is given by:

$$P_r = \frac{n!}{r!(n-r)!}\, p^r q^{n-r},$$

where p is the probability of success in any one of the trials, and q is that of failure ($p = 1 - q$).

If c coins are thrown in one trial, the probability of a specific success (say heads only) is obtained from the binomial expansion of $(x + y)^c$, where x represents heads only and y tails only. The probability p of 1 head and 2 tails when 3 coins are tossed ($c = 3$) is the quotient of the coefficient of xy^2 and the sum of all the coefficients. Thus $p = \frac{3}{8}$ and $q = \frac{5}{8}$.

binomial expansion If $|x| < 1$, the expression $(1 + x)^n$, where n may be any positive or negative number and not necessarily an integer, is equal to:

$$1 + nx + \frac{n(n-1)x^2}{2} + \frac{n(n-1)(n-2)x^3}{3.2}$$

$$+ \frac{n(n-1)(n-2)(n-3)x^4}{4.3.2} + \cdots$$

37

biological half-life The time required for half the concentration of a particular substance to be removed from the body, or from a specific region in the body, by biological means when the rate of removal is approximately exponential.

biological shield A massive structure surrounding the *core of a nuclear reactor, provided to absorb most of the neutrons and gamma radiation in order to protect the operating personnel. Such shields are commonly of concrete and iron.

biomass energy *See* renewable energy sources.

biophysics Physics applied to biology. Biophysics is concerned with the physics of biological systems, with the use of physical methods in the study of biological problems, and with the biological effects of physical agents.

Biot and Savart's law The magnetic field due to current flowing in a long straight conductor is directly proportional to the current and inversely proportional to the distance of the point of observation from the conductor. The law is derivable from the *Ampère–Laplace theorem but was obtained experimentally by the authors.

Biot–Fourier equation An equation for heat conduction through a solid:
$$\partial T / \partial t = (\lambda / c\rho)\nabla^2 T,$$
where $\partial T / \partial t$ is the rate of change of temperature, ∇^2 is the *Laplace operator, λ the *thermal conductivity, c the *specific heat capacity and ρ the density. For heat flow in one dimension $\nabla^2 T$ becomes $\partial^2 T / \partial x^2$.

Biot's law The degree of rotation of the plane of *polarization of light propagated through an optically active medium, is inversely proportional (approximately) to the square of the wavelength of the light and proportional to the path length in the medium and to the concentration if the medium is a liquid.

bipolar electrode A metal plate in an electrolytic cell through which the current, or part of it, passes, but which is not connected either to the anode or the cathode of the cell. Since the current passes through the plate one face serves as a subsidiary cathode, the other as an anode.

bipolar integrated circuit *See* integrated circuit.

bipolar transistor A *transistor in which both electrons and holes play an essential part, e.g. a junction transistor. A bipolar junction transistor is commonly referred to simply as a transistor. *Compare* field-effect transistor.

biprism A prism with a very obtuse angle acting virtually as two narrow angle prisms placed base to base thereby splitting a beam into two parts with a small angle between the parts. A doubled image (small separation) of a single object can be formed. It can be used to produce *interference.

birefringence *See* double refraction.

bistable *Syn.* bistable multivibrator. A type of circuit having two stable states. *See* flip-flop.

bit A contraction of *binary digit*. Either of the digits 0 or 1, used in computing for the representation within a *computer of numbers (*see* binary notation), of letters, punctuation marks, and other characters (encoded in binary form), and also of machine instructions. The bit is thus the smallest unit of information in a computer. It is also the smallest unit of storage, since it can be represented as a physical state of a two-state system.

Examples include the two possible directions of magnetization of a spot on a magnetic disk or tape, and the high or low voltage that can be fed to a logic or memory circuit.

Bitter patterns Patterns demonstrating the presence of *domains in ferromagnetic crystals. They can be observed by coating the polished surface of the material with a colloidal suspension of ferromagnetic particles. The particles tend to gather at the domain boundaries where there is a strong magnetic field. The technique can also be used for detecting cracks and imperfections in ferromagnetic material.

black body *Syn.* full radiator. A body or receptacle that absorbs all the radiation incident upon it; i.e. a body that has both an *absorptance and *emissivity of 1 and that has no reflecting power. The radiation from a heated black body is called *black-body radiation. Stellar radiation can be described by assuming that stars are black bodies. While a black body is in fact only a theoretical ideal, it is in practice most nearly realized by the use of a small slit or hole in the wall of a *uniform temperature enclosure.

black-body radiation The thermal radiation from a *black body at a given temperature, having a spectral distribution of energy of the form shown in the diagram. The most striking feature of a set of distribution curves is that each has a definite maximum and that this shifts towards the region of shorter wavelengths as the temperature rises. The intensity of radiation of any given wavelength increases steadily as the temperature rises.

Many theoretical and empirical attempts were made to find a general formula to represent the black-body spectrum. Thermodynamic reasoning does

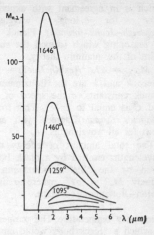

not give a complete answer, though it does predict two characteristic features of the radiation, firstly that for curves at different temperatures the value of $\lambda_{max}T$ is constant. This statement is known as the *Wien displacement law*. The second deduction is that the heights of corresponding ordinates vary directly as T^5. These two rules permit the construction of complete curves for all temperatures once any one is accurately known, but no further deductions can be made without making assumptions that are independent of any purely thermodynamic foundation. Wien deduced that

$$M_{e,\lambda} = c_1\lambda^{-5}\exp(-c_2/\lambda T),$$

where $M_{e,\lambda}$ is the *radiant exitance per unit wavelength range for wavelength λ and c_1 and c_2 are constants. This is known as the *Wien radiation law*. It is successful for short wavelengths, but it gives values somewhat too low for long wavelengths.

Rayleigh and Jeans applied the principle of equipartition of energy to a system of electromagnetic vibrations of different frequencies and gave

$$E_\lambda = CT\lambda^{-4},$$

which is in agreement with experiment only for long wavelengths (*Rayleigh-Jeans formula*). Planck gave, by reasoning which formed the starting point of the quantum theory,

$$M_{e,\lambda} = C\lambda^{-5}/[\exp(hc/\lambda kT) - 1],$$

where k and h are the Boltzmann and Planck constants, c the speed of light, and C is equal to $2\pi hc^2$. This formula (*Planck's formula*) agrees with experiment for all wavelengths.

The total amount of energy of all wavelengths emitted by a black body is given by the *Stefan-Boltzmann law, namely $M_e = \sigma T^4$, where σ is the Stefan-Boltzmann constant.

black-body temperature The temperature at which a *black body would emit the same radiation as is emitted by a given body. It is the temperature of a body as measured by a (radiation) *pyrometer. It is usually appreciably less than the true temperature of the body. For a temperature T_0 observed by a total radiation pyrometer, the true thermodynamic temperature T is given by $T_0^4 = \varepsilon T^4$, where ε is the emissivity of the source.

black hole An astronomical body with so high a gravitational field that the relativistic curving of space around it causes gravitational self-closure, i.e. a region is formed from which neither particles nor photons can escape, although they can be captured permanently from the outside.

The most promising candidates for black holes are massive stars that explode as *supernovae, leaving a core in excess of three solar masses. A core of this mass must undergo complete gravitational collapse because it is above the stable limit for *white dwarfs and *neutron stars. (*See also* Schwarzschild radius.) It has been suggested that black holes are the unseen components of certain binary systems. They could be detected through the gravitational fields.

Supermassive black holes, of 10^6 to 10^9 solar masses, probably lie at the centre of some galaxies. They may give rise to the *quasar phenomenon and the phenomena of other highly active galaxies. *See also* Hawking radiation.

black-out point *See* cut-off bias.

blanket A layer of *fertile material surrounding the *core of a nuclear reactor either for the purpose of breeding new fuel or to reflect some of the neutrons back into the core.

blazed grating A diffraction grating so ruled that the reflected light is concentrated into a few orders, or even into a single order of the spectrum. The flat grooves are each inclined at an angle, the *blaze angle*, to the grating surface so that the radiation is reflected in the direction of the diffraction *order intended to be bright.

blink microscope or comparator An instrument used to detect small differences in the *luminosity or position of stars between two photographs of the same part of the sky. The photographs are viewed alternately in rapid succession using a mechanical device.

Bloch functions The solutions of *Schrödinger's wave equation for an electron moving in a potential that varies periodically with distance. They have the form:

$$\psi = u_k(r) \exp(i k.r),$$

where u is a function depending on k, the wave vector, and k varies periodically with distance r. k has the same period as the potential and the lattice. Bloch functions are used in the mathematical formulation of the band theory of solids (*see* energy bands).

Bloch wall The transition layer between adjacent ferromagnetic *domains magnetized in different directions. It allows the spin directions to change slowly from one orientation to another, rather than abruptly.

blocking capacitor A *capacitor included in an electric circuit for the purpose of preventing the flow of direct current and low-frequency alternating current while permitting the flow of higher-frequency alternating current. Its *capacitance is usually chosen so that its *reactance is relatively small at the lowest frequency for which the circuit is intended to be used.

blocking oscillator A type of oscillator in which, after completion of (usually) one cycle of oscillation, blocking (i.e. cessation of oscillation) takes place for a predetermined period of time. The whole process is then repeated. It has applications as a pulse generator or as a time-base generator and is fundamentally a special type of *squegging oscillator.

blooming of lenses The process of depositing a transparent film (about one quarter wavelength) of lower refractive index on the surface of a lens of higher index, whereby, through destructive interference, surface reflection is eliminated. Calcium or magnesium fluoride is deposited by evaporation in a vacuum. Bloomed (or *coated*) surfaces have a pale purplish hue when examined by reflected light as the interference only occurs in the middle part of the spectrum.

body-centred The form of crystal structure in which the atoms occupy the centre of the lattice as well as the vertices. *Compare* face-centred.

Bohr magneton *See* magneton.

Bohr theory of the atom (1913) This was the first significant application of the quantum theory to atomic structure. Although the theory has been replaced (*see* quantum mechanics) it introduced several concepts that have remained as essential features of later theories.

The theory was applied in particular to the simplest atom, that of hydrogen, consisting of a nucleus and one electron. It was assumed that there could be a *ground state* in which an isolated atom would remain permanently, and short-lived states of higher energy to which the atom could be excited by collisons or absorption of radiation. It was supposed that radiation was emitted or absorbed in quanta of energy equal to integral multiples of hv, where h is the Planck constant and v is the frequency of the electromagnetic waves. (Later it was realized that a single quantum has the unique value hv.) The frequency of radiation emitted on capturing a free electron into the nth state (where $n=1$ for the ground state) was supposed to be $nh/2$ times the rotational frequency of the electron in a circular orbit. This idea led to, and was replaced by, the concept that the angular momentum of orbits is quantized in units of $h/2\pi$. The energy of the nth state was found to be given by:

$$E_n = -me^4/8h^2\varepsilon_0{}^2n^2,$$

where m is the *reduced mass of the electron. This formula gave excellent agreement with the then known series of lines in the visible and infrared regions of the spectrum of atomic hydrogen and predicted a series in the ultraviolet that was soon to be found by Lyman. (*See* hydrogen spectrum.)

The extension of the theory to more complicated atoms had some successes but raised innumerable difficulties, which were only resolved by the development of *wave mechanics. *See also* atom; atomic orbital.

boiling point The temperature of a liquid at which visible evaporation occurs throughout the bulk of the liquid, and at which the vapour pressure of the liquid equals the external atmospheric pressure. It is the temeprature at which liquid and vapour can exist together in equilibrium at a given pressure. The variation of boiling point with pressure may be obtained from the *Clausius–Clapeyron equation. The term is commonly restricted to that temperature at which liquid and vapour are in equilibrium at standard atmospheric pressure ($1.013\,25 \times 10^5$ Pa).

boiling-water reactor (BWR) A type of thermal *nuclear reactor in which water is used both as coolant and moderator, being allowed to boil by direct contact with the fuel elements. *See* pressurized-water reactor.

bolometer An instrument for the measurement of the total energy flux of electromagnetic radiation, especially that of microwave and infrared radiation. There are many types of bolometer, including semiconductor and specially cooled devices, but each is essentially a small resistive element capable of absorbing radiation. The resulting rise in temperature, which leads to a change in the resistance, is a measure of the power absorbed.

Boltzmann constant Symbol: k. A fundamental constant equal to R/L, where R is the *molar gas constant and L is the *Avogadro constant. It has the value $1.380\,658 \times 10^{-23}$ J K^{-1}.

Boltzmann distribution *See* Boltzmann's formula.

Boltzmann entropy theory *See* entropy.

Boltzmann's formula A formula showing the number of particles (n) having an energy (E) in a system of particles in thermal equilibrium. It has the form:
$$n = n_0 \exp(-E/kT),$$
where n_0 is the number of particles having the lowest energy, k the Boltzmann constant, and T the thermodynamic temperature.

If the particles can only have certain fixed energies, such as the energy levels of *atoms, then the formula gives the number of particles (n_i) in a state at an energy (E_i) above the *ground state energy. In certain cases several distinct states may have the same energy and the formula then becomes
$$n_i = g_i n_0 \exp(-E_i/kT),$$
where g_i is the *statistical weight of the level of energy E_i, i.e. the number of states having energy E_i. The distribution of energies obtained by the formula is called a *Boltzmann distribution*.

bomb calorimeter A device used for measuring the heat evolved by the combustion of a fuel.

bond energy The energy required to break a chemical bond between two atoms in a molecule. The bond energy depends on the type of atoms and on the nature of the molecule.

bonding orbital *See* molecular orbital.

bonding pad Metal pads usually arranged around the edge of a semiconductor *chip to which wires may be bonded to make electrical connection to the component(s) or the circuit(s) on the chip.

booster **1.** A generator or transformer inserted in an electric circuit to enable the voltage acting in the circuit to be increased (positive booster) or decreased (negative booster), or to change the phase of the voltage.
2. In broadcasting, a repeater station that receives the signal transmitted from

a main station, amplifies it, and then re-transmits it, sometimes with a change in frequency.

bootstrap (From the phrase, "pulling oneself up by the bootstraps".) A means or technique enabling a system to bring itself into some desired state. For example, it may be an electronic circuit in which positive *feedback of the output signals is used to control conditions in the input circuit.

bootstrap theory A theory that leads to or is concerned with the self-consistency of a more enveloping theory.

boron counter A radiation counter that uses a nuclear reaction with boron-10 for the detection of slow neutrons.

Bose condensation *Syn.* Bose–Einstein condensation. A phenomenon that occurs at low temperatures in systems consisting of large numbers of *bosons whose total number is conserved in collisions. Used in the explanation of *superfluidity, this phenomenon enables a significant fraction of the particles to occupy a single quantum state. No analogous phenomenon occurs for two or more *fermions, which are prohibited by the *Pauli exclusion principle from occupying the same quantum state.

Bose–Einstein statistics and distribution law *See* quantum statistics.

boson Any particle having integral *spin. Bosons obey Bose–Einstein statistics (*see* quantum statistics). Particles are either bosons or *fermions: *photons, *pions, and *kaons are all bosons.

Bouguer's law of absorption *See* linear absorption coefficient.

boundary layer If a fluid of low viscosity (air, water) has a relative motion with respect to solid boundaries, then at a large distance from the boundaries the frictional factors are negligible with regard to the inertia factors while near the boundaries the frictional factors are appreciable.

The fluid may be divided into two parts: first, a thin layer of fluid close to the solid boundaries in which the viscosity of the fluid is of major importance – this layer is called the *boundary layer*; secondly, the portion of the fluid that remains outside this boundary layer within which the fluid may be considered as nonviscid. It can be shown that the thickness of the boundary layer is directly proportional to $\sqrt{\nu}$ where ν is the *kinematic viscosity and that the normal pressure on the solid boundary is unaltered by the presence of the boundary layer.

Bourdon tube (and gauge) A curved tube of oval cross section with the longer diameter of the oval perpendicualr to the plane in which the tube is curved. If the volume of the tube is made to increase (by an excess pressure inside), the oval cross-section becomes more nearly circular and the tube tends to straighten out.

The Bourdon tube can be used as a recording thermometer. The tube is closed at both ends and completely filled with a liquid. The liquid expands more than the tube material with increase in temperature and causes the tube to straighten out. One end of the tube is fixed and the other, connected to a tracing point, draws a graph of temperature against time on a slowly moving surface. The Bourdon tube is also used as a *pressure gauge. The tube is closed at one end and the pressure applied at the other. One end of the tube is fixed and the other operates a pointer that indicates the pressure on a calibrated dial. A series of these Bourdon gauges can be used from vacuum to several tens of megapascals.

Boyle's law If a given mass of gas is compressed at constant temperature, the product of the pressure and volume remains constant. The law is found to be only approximately true for real gases, being exactly fulfilled only at very low pressure. An *ideal gas by definition obeys Boyle's law exactly.

Boyle temperature The *equation of state of one mole of a real gas can be written in the form:

$$pV = RT + Bp + Cp^2 + Dp^3 \ldots,$$

where RT, B, C, etc., are the virial coefficients (*see* virial expansion). R is the *molar gas constant and T the *thermodynamic temperature. For all gases the quantity B is negative at low temperatures and positive at high temperatures. Higher-order terms, which are generally positive, only become significant at very high pressures. At the Boyle temperature T_B the quantity B is zero so the gas obeys *Boyle's law almost exactly over a wide range of pressure.

Brackett series *See* hydrogen spectrum.

Bragg curve A curve on a graph produced by plotting the specific ionization caused in air by alpha particles (as ordinate) against the distance from the source (as abscissa). It has a characteristic shape in which the specific ionization increases with distance (as the energy falls) followed by a sudden drop at the point at which the energy of the alpha particles becomes so low that they capture electrons to become neutral helium atoms.

Bragg's law If a parallel beam of *X-rays, wavelength λ, strikes a set of crystal planes it is reflected from the different planes, *interference occurring between X-rays reflected from adjacent

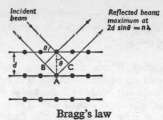

Bragg's law

planes. Bragg's law states that constructive interference takes place when the difference in pathlength, BAC, is equal to an integral number of wavelengths:

$$2d \sin \theta = n\lambda,$$

where n is an integer, d is the interplanar distance, and θ is the angle between the incident X-ray and the crystal plane. This angle is called the *Bragg angle* and a bright spot will be obtained on an interference pattern at this angle. A dark spot will be obtained if

$$2d \sin \theta = m\lambda,$$

where m is half-integral. The structure of a crystal can be determined from a set of interference patterns found at various angles from the different crystal faces.

branch and branch point Of an electrical network. *See* network.

branching The occurrence of competing decay processes (*branches*) in the *disintegration of a particular radionuclide. The *branching fraction* is given by the number of nuclei following a particular branch of a decay process to the total number of nuclei undergoing disintegration; it is usually expressed as a percentage. The *branching ratio* is the ratio of two specified branching fractions.

brass *See* alloy.

Bravais lattice *Syn.* space lattice. An indefinitely repetitive arrangement of points in space that fulfils the condition that the environment of each point is identically similar to that of every other point. There are fourteen such arrangements. *See* crystal systems.

breakdown 1. A sudden disruptive electrical discharge through an insulator, or between the electrodes of an *electron tube.

2. A sudden transition from high dynamic resistance in a semiconductor device to substantially lower dynamic resistance.

In both cases, the voltage at which breakdown occurs is called the *breakdown voltage*.

breakdown voltage (BDV) *See* breakdown.

breeder reactor A *nuclear reactor in which more *fissile material is produced than is consumed. In strict usage the term is restricted to reactors in which the nuclide produced is the same as that which is consumed. If they are different it is a *converter reactor. *See also* fast breeder reactor.

Breit–Wigner formula An equation giving the absorption *cross section, σ, of a particular nuclear reaction when the intermediate excited nucleus can decay in any of several ways. The cross section is a function of the energy, E, of the bombarding particle and when E is close to the energy E_c of the *compound nucleus then

$$\sigma = \frac{\sigma_0 \, \Gamma^2 (E_c/E)^{1/2}}{\Gamma^2 + 4(E - E_c)^2}$$

where σ_0 is the resonance *cross section and Γ is the width of the excited *energy level.

bremsstrahlung Electromagnetic radiation produced by the rapid deceleration of an electron during a close approach to an atomic nucleus. The radiative loss due to the "braking effect" increases rapidly with the energy of the electron, and for energies exceeding 150 MeV is responsible for most of the absorption of the electron's energy. The energy lost at each encounter is radiated as a single photon. Bremsstrahlung radiation forms an important constituent of *cosmic rays and is the continuous radiation that occurs in the production of *X-rays by electron bombardment. *See also* synchrotron radiation.

Brewster angle *See* polarizing angle.

Brewster's law *See* plane-polarized light; polarizing angle.

Brewster windows Reflecting surfaces, used in certain gas *lasers, to reduce reflection losses that would arise from using external mirrors. The surfaces are set at the Brewster angle to the incident light. *See* polarizing angle.

bridge A circuit made up of electrical elements (e.g. resistors, inductors, capacitors, rectifiers, etc.) arranged in the form of a quadrilateral, or its electrical equivalent. Two opposite corners of the quadrilateral are made the input and the other pair the output of the circuit. (*See* bridge rectifier.) Bridges are most commonly used in a variety of measuring instruments in which the output is connected to a current detector and the circuit adjusted until the bridge is balanced and no current is detected. In this way an unknown resistance, capacitance, or inductance can be compared with known standards.

bridge rectifier A *full-wave rectifier consisting of a bridge with a rectifier in each arm, as shown in the diagram.

Bridgman effect The absorption or the liberation of heat arising from a nonuniform current distribution which occurs when an electric current passes through an *anisotropic crystal.

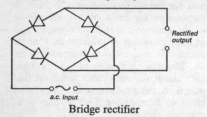

Bridge rectifier

Brillouin zone If *Schrödinger's equations for electronic energies are solved with a periodic function $u(k)$ to give the energies of an electron in a solid, the solutions fall into permitted bands (*see* energy bands). If the solutions are plotted in the *reciprocal lattice of the crystal being considered, the zones enclosing the solutions for $k = 1,2,\ldots n$, are called Brillouin zones.

Brinell hardness *See* hardness.

British thermal unit (Btu). The amount of heat required to raise the temperature of 1 lb of water by 1 °F. The International Tables Btu, defined in terms of SI units, is equal to 1055.06 joules.

brittleness The property of solids whereby they separate into pieces without first passing through a plastic stage. The breakdown point is usually sharply defined, although in the case of some metals the rate of growth of the stress is important; thus, a material subjected to sudden deforming forces may be brittle but it may yield plastically when the stress is built up slowly.

bronze *See* alloy.

brown dwarf A type of astronomical object with a mass between that of a large planet and a small star. Brown dwarfs are faint objects with a mass too small

to sustain fusion but great enough to generate energy by gravitational pressure. Their mass can be several times the mass of Jupiter (up to a limit of about 80 times the mass of Jupiter). It has been suggested that brown dwarfs (sometimes called 'jupiters') are significant contributors to the *dark matter in the universe.

Brownian movement or motion The increasing and irregular motion in all directions of small particles, about 1 μm in diameter, when held in suspension in a liquid. It is a visible demonstration of molecular bombardment by the molecules of the liquid. (*See* kinetic theory.) The smaller the suspended particles, the less likely are the molecular impacts on opposite sides to balance out simultaneously and the more noticeable the motion. It can also be observed in particles of smoke.

Brownian motion sets a theoretical limit to the sensitivity of a chemical balance at 10^{-9} g and sets a limit to galvanometer measurements at 10^{-11} A.

brush A conductor that serves to provide electrical contact with a conducting surface moving relatively to the brush, usually between the stationary and moving parts of an electrical machine. Brushes are made of specially prepared carbon with or without copper.

brush discharge A luminous discharge from a conductor that occurs when the electric field near the surface exceeds a certain miniumum value but is not sufficiently high to cause a true spark. It appears as a large number of intermittent luminous branching threads, penetrating some distance into the gas surrounding the conductor; the distance is greater for an anode than for a cathode at the same potential. A nonuniform field is essential for the effect.

bubble chamber An instrument in which the tracks of an ionizing particle (*see* ionizing radiation) are made visible as a row of tiny bubbles in the liquid inside a large chamber. The liquid, usually hydrogen, helium, or deuterium, is maintained under pressure so that it can be heated without boiling to a temperature slightly above its normal boiling point. Immediately before the passage of a particle, the pressure is reduced, and the liquid would normally boil after about 50 milliseconds. However, the release of energy resulting from ionization of atoms along the track of the moving particle causes rapid localized boiling along this path. After about 1 ms, the bubbles are big enough to photograph and a record is obtained of the particle's track and that of any decay or reaction products. The pressure in the chamber is then increased again to prevent the bulk of the liquid from boiling.

buffer An isolating circuit used to minimize reaction between two circuits. Usually it has a high input *impedance and low output impedance. An *emitter-follower is an example.

build-up time The time taken for a current in an electronic circuit or device to rise to its maximum value.

bulk lifetime The average time interval between the generation and recombination of *minority carriers in the bulk material of a *semiconductor.

bulk modulus *See* modulus of elasticity.

bulk strain *See* strain.

bumping In the absence of nuclei, bubbles do not form until the temperature of a liquid is above the boiling point so that when formed the vapour pressure inside the bubbles greatly exceeds the applied pressure. The consequent rapid expansion of the bubbles causes violent motion (bumping) of the containing vessel. Small pieces of porous pot placed in the liquid, by providing nuclei, prevent this occurring.

bunching *See* velocity modulation.

Bunsen burner A gas burner in which a regulated amount of air mixes with the gas stream at the bottom of the tube of the burner, the flame being at the top. The air is drawn in by the suction effect of the fine gas jet, a consequence of *Bernouilli's theorem.

Bunsen cell A primary cell, much used before the introduction of accumulators.

Bunsen ice calorimeter A calorimeter for measuring the specific heat capacity of a solid or liquid of which only a small quantity is available.

buoyancy The tendency of a fluid to exert a lifting effect on a body wholly or partly immersed in it. *Archimedes principle states that such a body experiences an upward force equal to the weight of the fluid that would fill the space occupied by the immersed part of the body. This force acts through the centre of gravity of that fluid which would replace the immersed part of the body and this point is the *centre of buoyancy* of the body. The plane in which the liquid surface intersects the stationary floating body is the *plane of flotation*. For the body to be in equilibrium (*a*) the upthrust must be equal to the weight of the body, (*b*) the centre of gravity of the body and the centre of buoyancy must be in the same vertical line.

burial *Syn*. graveyard. A place where highly radioactive products from *nuclear reactors may be safely deposited and stored, usually in noncorrosive containers.

burn-up 1. The significant reduction in the quantity of one or more *nuclides arising from neutron absorption in a *nuclear reactor. The term can be applied to fuel or other materials.

2. *Syn.* fuel irradiation level. The total energy released per unit mass of nuclear fuel.

bus Contraction of busbar. A set of conducting wires connecting several components of a computer and allowing these components to send signals to each other. It is thus a pathway for signals.

busbar 1. *Syn.* bus. Generally, any conductor of low *impedance or high current-carrying capability relative to other connections in a system. It is usually used to connect many like points in a system, as with an *earth bus*. Busbars frequently feed power to various points.

2. *See* bus.

BWR Abbreviation for boiling-water reactor.

bypass capacitor A shunt capacitor connected in a circuit in order to provide a path of comparatively low *impedance for alternating current. The frequency of the alternating current passed depends on the magnitude of the capacitance. Such a capacitor is commonly used to prevent a.c. signals from reaching a particular point in a circuit or to separate out a desired a.c. component.

byte A fixed number of *bits (now almost always 8 bits) that can be handled and stored as a single unit in a computer.

C

cable *See* coaxial cable; paired cable.

cadmium cell *See* Weston standard cell.

cadmium sulphide cell A compact photoconductive cell (*see* photoconductivity) consisting of a layer of cadmium sulphide sandwiched between two electrodes. The high electrical resistance of the CdS drops when light falls on the cell. A current flowing through the cell will vary according to the amount of incident light. A battery is required to provide the current in the cell. It is used in *exposure meters and can be built into cameras. It has a much higher sensitivity than the *selenium cell.

cadmium ratio Symbol: R_{Cd}. The ratio of the neutron-induced radioactivity in a sample to the radioactivity induced under identical conditions when the sample is covered with cadmium, which has a high capture cross section for thermal neutrons. For large values of the ratio, it is therefore a measure of the ratio of *thermal to *fast neutrons.

caesium clock *See* clocks.

calcite *Syn.* Iceland spar. A crystalline form of calcium carbonate easily breaking along cleavage planes into rhombohedrons, each face being a parallelogram with angles 78° 5′ and 101° 55′. The crystal exhibits *double refraction. The ordinary and extraordinary rays show the light polarized at right angles. Optically it is classed as a negative *uniaxial crystal.

calculus of variations A mathematical method for solving those physical problems that can be stated in the form that a certain definite integral shall have a stationary value for small changes of the functions in the integrand and of the limits of integration.

calendar year *See* time.

calibration Determination of the absolute values of the arbitrary indications of an instrument.

calomel electrode A *half-cell consisting of a mercury electrode in contact with a solution of potassium chloride saturated with calomel (mercury(1) chloride, Hg_2Cl_2). It is used as a reference electrode in physical chemistry.

caloric theory The theory of the nature of heat widely held up to 1850, according to which heat was an imponderable, self-repellant fluid, caloric. It was unable to account for the production of an unlimited supply of heat by friction as occurred in the experiments of Rumford and was abandoned when Joule showed that heat was a method of transfer of energy and determined the value of the *mechanical equivalent of heat.

calorie A unit of heat and internal energy no longer employed in scientific calculations. Formerly defined as the quantity of heat required to raise the temperature of one gram of water from 14.5 °C to 15.5 °C at standard pressure, the calorie (symbol: cal_{IT}) is now formally defined as 4.1868 joules.

calorific value The amount of heat liberated by the complete combustion of unit mass of a fuel, the water formed being assumed to condense to the liquid state. The determination is carried out in a *bomb calorimeter and the value is usually expressed in $J\,kg^{-1}$ or similar units.

calorimeter Any vessel or apparatus in which quantitative thermal measurements may be made. The simplest form, used for the *method of mixtures, consists of a copper can containing water and resting on insulating feet inside a water jacket at a definite temperature, which enables the radiation correction to be calculated. Through an insulating lid, used to prevent evaporation, passes a thermometer, to record the temperature changes of the water in the copper can, together with a stirrer.

calorimetry The measurement of quantities of heat, including the *specific heat capacity and *specific latent heat of objects, the *calorific value of fuels, and the heat of formation and of solution of chemical compounds.

camera 1. A light-tight chamber containing a lens system, a shutter, and a film or sometimes a plate at the back. The lens system forms a sharp real image of distant or near objects on the film. Usually the shutter has a variable diaphragm so that the *f-number can be altered. The exposure time is also a variable quantity. Film is sensitive to light, ultraviolet, and X-radiation, and can be made sensitive to infrared radiation. (*See also* photography.) The lens-system material must be transparent to the radiation incident on the film. Glass or plastic lenses are used in cameras for normal use. Additional lens systems are the *zoom lens and *telephoto lens. Various types of camera are used in scientific work for recording and measuring purposes. The *Polaroid (Land) camera is a means of obtaining an immediate scientific record of an oscilloscope trace, microscope image, etc.
2. *See* television camera.

camera lucida A microscope accessory for attachment at the eyepiece end, whereby a virtual image is formed in the plane of drawing paper, permitting the simultaneous view of a hand-manipulated pencil to draw the outline.

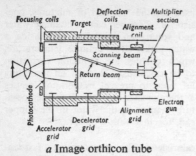

a Image orthicon tube

camera tube The *transducer device in a *television camera that converts the optical image of the scene to be transmitted into electrical video signals. Most camera tubes are *electron tubes: the two basic types are the *image orthicon* and the *vidicon*, from which many other tubes have been developed. There is also a solid-state camera tube in which the transducer is an array of *CCDs (charged-coupled devices), and is thus much smaller and lighter than devices containing electron tubes.

In the image orthicon (Fig. *a*) light from a scene is focused on the *photocathode, which consists of a light-sensitive material deposited on a thin sheet of glass. Electrons are emitted from the photocathode in proportion to the intensity of the light and focused onto a target consisting of a thin glass disc with a fine mesh on the photocathode side of the disc. The impact of electrons from the photocathode causes secondary emission of electrons from the target, greater than, but proportional to, the original electron density from the photocathode. These secondary electrons are collected by the mesh screen and drained off to a power supply. The target is left with a positive static-charge pattern corresponding to the original light image. The reverse side of the disc is scanned with a low-velocity electron beam. The electrons are turned back at the target

glass into the multiplier section. Areas of positive charge on the target are neutralized by electrons from the beam, so the beam returning to the multiplier section varies in density in proportion to the charge on the plate: the return beam is thus modulated with video information.

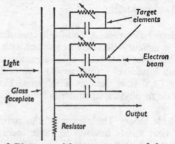

b Photosensitive target area of the vidicon

The vidicon type of camera tube (Fig. *b*) is widely used in closed-circuit television and as an outside broadcast camera as it is smaller, simpler, and cheaper than the image-orthicon type. The photosensitive target area of the vidicon consists of a transparent conducting film placed on the inner surface of the thin glass faceplate, and a thin photoconductive layer deposited on the film. The photoconductive layer may be considered as an array of discrete elements consisting of a light-dependent resistor with a parallel capacitor (Fig. *b*). A positive voltage is applied to the conducting layer and this has the effect of charging the capacitive elements. The amount of charge in each element will depend on the value of the parallel resistor – the lower the resistance the more charge will be stored. Normally the resistors have a high value, but when light strikes the target area, the resistance drops. If the target is scanned with a low-velocity electron beam, the capacitors will be discharged and a current will flow in the conducting layer

(Fig. *c*). The magnitude of the current developed is a function of the charge on the target area and hence of the illumination from the optical lens system.

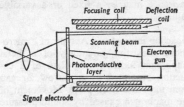

Focusing coil Deflection coil

Scanning beam

Electron gun

Photoconductive layer

Signal electrode

c Vidicon tube

In the *plumbicon*, a modern development of the vidicon tube, the photoconductive material is replaced by a layer of semiconductor material. The mode of operation is similar to the vidicon tube, but the target elements may be considered as semiconductor current sources controlled by light energy. Such tubes have very low dark current and good sensitivity and light-transfer characteristics.

The performance of camera tubes depends very greatly on the scanning system. Beam alignment is achieved by using small coils to ensure that the electron beam emerging from the electron gun is central. Deflection of the beam is provided by deflection coils controlling the horizontal and vertical directions; these coils are supplied with a *sawtooth voltage causing a linear scan with very rapid return to the start of the scanning position. Focusing coils are also provided to ensure a small cross section when the beam reaches the target. *See also* colour television.

Campbell's bridge A *bridge for measuring a mutual inductance by comparison with a standard capacitance C. The usual arrangement is illustrated in the diagram. I is an indicating instrument, such as a microphone or *oscilloscope, L the self-inductance of the coil between A and B, and M the mutual inductance of

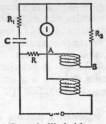

Campbell's bridge

the pair of coils. The resistances are varied until the bridge is balanced, when:

$$\frac{L}{M} = \frac{R + R_1}{R} \quad \text{and} \quad \frac{M}{C} = RR_2.$$

candela Symbol: cd. The *SI unit of *luminous intensity; the luminous intensity in a given direction of a source that emits monochromatic radiation of frequency 540×10^{12} Hz and that has a radiant intensity in that direction of $(1/683)$ watt per steradian.

This definition replaces the former definition: the luminous intensity, in the perpendicular direction, of a surface of $1/600\,000$ m^2 of a *black body at the temperature of freezing platinum at standard atmospheric pressure.

candle power Former name for *luminous intensity.

canonical Furnishing, or according to a general rule or formula.

canonical distribution A term used in statistical mechanics, expressed by:
$$f = A \exp(-\text{energy}/\Theta)$$
$$dp_1 \ldots dp_n dq_1 \ldots dq_n,$$
in which f means the fraction of the systems in an assemblage (molecules in a gas for example) whose momenta lie between $p_1 \ldots p_n$ and $p_1 + dp_1 \ldots p_n + dp_n$ and whose associated coordinates lie between $q_1 \ldots q_n$ and $q_1 +$

$dq_1 \ldots q_n + dq_n$. A is a constant and Θ is the modulus of the distribution; Θ can be identified with kT, the product of the *Boltzmann constant and the thermodynamic temperature. Maxwell's law of distribution of velocities among the molecules in a gas is a limiting case of a canonical distribution.

canonical equations Equations of classical mechanics as expressed in the form of *Hamilton's equations, namely
$$dp_i/dt = -\partial H/\partial q_i;$$
$$dq_i/dt = \partial H/\partial p_i,$$
p_i and q_i being respectively the momenta and associated coordinates, while H is the energy of the system expressed as a function of p_i, q_i, and time t.

capacitance Symbol: C. The property of an isolated *conductor, or set of conductors and *insulators, to store electric charge. If a charge Q is placed on an isolated conductor, the voltage is increased by an amount V. The capacitance of the conductor is defined as Q/V; for a given conductor it is a constant and depends on the size and shape of the conductor. Two conductors, or a conductor and a semiconductor, together form a *capacitor, and the capacitance C is defined as the ratio of charge on either conductor to the potential difference between them. The unit of capacitance is the *farad. *See also* mutual capacitance.

capacitive coupling *See* coupling.

capacitive reactance *See* reactance.

capacitive tuning *See* tuned circuit.

capacitor An electronic component that has an appreciable *capacitance. It consists of at least one pair of conductors, or of a conductor plus semiconductor, each pair separated by a *dielectric (an insulator). For most types of capacitor,

the value of the capacitance depends on the geometry of the device and the electrical properties of the dielectric, which may be solid, liquid, or gaseous. The capacitance may be a fixed or a variable value.

capacitor microphone A type of *microphone consisting essentially of a metal diaphragm forming one plate of a capacitor, separated by a narrow air gap from the other fixed plate. Movement of the diaphragm caused by sound-pressure variations alters the capacitance of the device. These variations in capacitance modulate the potential difference applied across the capacitor.

capacity The amount of information that can be held in a computer storage device.

capillarity An obsolete term to describe the effects of surface tension. The term is derived from the most prominent of these effects – the rise or fall of liquids in vertical capillary tubes.

capillary Having a minute or hair-like bore.

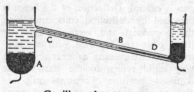

Capillary electrometer

capillary electrometer An electrolytic cell, one electrode of which is a pool of mercury A, while the other is the meniscus B of a thread of mercury in a capillary tube, CD. If a potential difference is applied to the electrodes, a small charging current flows through the cell, producing *polarization at the meniscus B. (The polarization at A is negligible on account of its much greater area.)

The electric field due to this polarization alters the *surface tension of the mercury and the meniscus, therefore, moves to a new position of equilibrium. The apparatus is very sensitive, but cannot be used to measure potential differences greater than 0.9 volt. The movement of the mercury is not strictly proportional to the applied p.d., and the electrometer is most conveniently used as a null instrument.

capture Any process by which an atom, ion, molecule, or nucleus acquires an additional particle. *Radiative capture* involves the emission by a nucleus of *capture gamma rays* immediately following a capture process. The radiative capture of a neutron produces an isotope (usually radioactive) of the original element with the *mass number increased by unity. The *cross sections for the process vary very greatly between nuclides, and for a given nuclide usually decrease as the neutron energy increases, except for peaks of characteristic energies. *Electron capture* involves capture, by a nucleus, of one of its orbital electrons. For example a $^{7}_{4}Be$ nucleus can acquire one of the electrons in the K-shell of its atom and turn into a $^{7}_{3}Li$ nucleus. The process is $p + e \rightarrow n + \nu$, the neutrino ($\nu$) being emitted. Usually the K-shell electron is captured in this process (*K-capture*) although other shells are sometimes involved.

The capture of an electron into an orbit of an atom, molecule, or ion is more usually called *electron attachment.

Carathéodory's principle A theorem in thermodynamics that can be used to derive the second law, without making reference to thermodynamic cycles. It states that it is impossible to reach every state in the neighbourhood of any arbitrary initial state by means of adiabatic processes only.

carbon cycle A cycle of *nuclear reactions resulting in the formation of one helium nucleus from four hydrogen nuclei, with the emission of gamma rays (γ), positrons ($\bar{e}$), and neutrinos (ν):

$$^{12}C + {}^{1}H \rightarrow {}^{13}N + \gamma$$
$$^{13}N \rightarrow {}^{13}C + \bar{e} + \nu$$
$$^{1}H + {}^{13}C \rightarrow {}^{14}N + \gamma$$
$$^{1}H + {}^{14}N \rightarrow {}^{15}O + \gamma$$
$$^{15}O \rightarrow {}^{15}N + \bar{e} + \nu$$
$$^{1}H + {}^{15}N \rightarrow {}^{12}C + {}^{4}He$$

The carbon-12 is reformed at the end of the cycle and therefore acts as a catalyst. This cycle is believed to be the major source of energy in hot massive stars. *See also* proton–proton chain; thermonuclear reaction.

carbon-14 dating *See* radiocarbon dating.

carbon microphone A *microphone that makes use of the decrease of *contact resistance of carbon granules as the applied pressure is increased. Variations in sound pressure are transmitted to the carbon granules by means of a diaphragm, and the corresponding changes in resistance are detected as fluctuations in a current passing through the granules.

carbon resistor *See* resistor.

cardinal points *See* centred optical system.

Carey–Foster bridge A modification of the *Wheatstone bridge designed to measure the difference in resistance between two nearly equal resistances in terms of the resistance per unit length of the bridge wire. Two resistances are placed in the ratio arms of a Wheatstone bridge and the balance point found on the resistance wire. The resistances are then switched and a new balance point found; the distance between the points on the resistance wire is proportional to the difference between the resistances. If the resistance wire has been precalibrated, the difference in length is equal to the difference between the resistances.

Carnot–Clausius equation For a reversible closed cycle (*see* reversible change), an equation giving the total change in the *entropy of the system as $dq/T = 0$ where dq is the quantity of heat taken in by the system during an infinitesimal reversible change of state and T is the thermodynamic temperature of the system during this change.

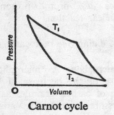

Carnot cycle

Carnot cycle A reversible cycle in which the working substance is compressed adiabatically from T_2 to T_1, expands at T_1 isothermally, then expands adiabatically from T_1 to T_2 and finally is compressed isothermally at T_2. This brings the pressure, volume, and temperature back to their initial values. It represents the cycle of the ideal heat engine. *See* thermodynamics.

Carnot's theorem No engine can be more efficient than a reversible engine working between the same temperatures. Hence all reversible engines working between the same temperatures are equally efficient, the efficiency being independent of the nature of the working substance, depending only on these temperatures. *See* thermodynamics.

carrier 1. An electron or *hole that can move through a metal or *semiconductor. Carriers enable charge to be trans-

ported through a solid and are responsible for conductivity. (*See also* majority carrier; minority carrier.)

2. A substance used to provide a bulk quantity of material containing traces of *radioisotopes for use in physical and chemical operations. It is used in *radioactive tracer studies and in the preparation of chemical compounds containing radioisotopes.

3. The wave or signal whose characteristics are modified in the process of *modulation. Modulation produces spectral components falling into frequency bands on either side of the carrier frequency. These are called the upper or lower *sidebands according to whether the frequency range is above or below the carrier frequency.

carrier concentration The number of *carriers – electrons or holes – per unit volume of *semiconductor.

carrier storage *See* storage time.

carrier wave *See* carrier (3).

Cartesian coordinates *See* coordinate.

cascade A chain of electronic circuits or elements connected in series, so that the output of one is the input of the next.

cascade liquefaction A chain process by which a gas can be liquefied. A starter gas with a high *critical temperature (such as chloromethane, CH_3Cl) is liquefied by increase of pressure, and evaporation of this liquid cools a second gas (such as ethene, C_2H_4) below its critical temperature so that it, too, may be liquefied by increase of pressure. The process continues until the final gas (such as oxygen) is cooled and liquefied under pressure.

cascade shower *See* cosmic rays.

Cassegrain telescope *See* reflecting telescope.

catadioptric system An optical system, such as a telescope, that uses both reflecting and refracting components to form the final image. Examples include the *Schmidt and *Maksutov telescopes.

cataphoresis *See* electrophoresis.

catastrophe theory A theory of dynamic systems based on analogy with topographical form. If a system depends on *n* variables, a state of the system can be represented by a region in this space. Catastrophe theory considers the topological classifications of such regions, and, in particular, the conditions under which a discontinuous 'catastrophic' change can occur. Originally developed for biology, the theory has applications in physics (e.g. in optics and mechanics) as well as uses in social sciences.

catching diode *Syn.* clamping diode. A *diode used to limit the voltage at some point in a circuit. A diode will start to conduct at the *diode forward voltage, V_d, typically 0.7 V, and will therefore prevent the voltage applied in the forward direction from rising above this value.

cathetometer A device for measuring vertical heights consisting of a vertical scale along which a horizontally mounted telescope or microscope may be moved.

cathode The negative electrode of an electrolytic cell or electron tube. It is the electrode by which electrons enter a system. *Compare* anode.

cathode follower *See* emitter follower.

cathode-ray oscilloscope Usually shortened to oscilloscope. An instrument

that enables a variety of electrical signals to be examined visually. Any variable that can be converted into an electrical signal can be studied, making the oscilloscope an extremely valuable tool. The signal of interest is fed, after amplification, to one set of deflection plates of a *cathode-ray tube, usually the vertical deflection plates. The beam is moved horizontally across the screen by the voltage from a sweep generator (usually called a *time-base generator) incorporated in the oscilloscope. The resultant trace seen on the screen is a composite of the two voltages, and suitable choice of sweep speed in the horizontal direction allows easy visualization of the input signal. The simplest type of time base is a constantly variable sweep generator producing a *sawtooth waveform, so that the trace moves slowly and uniformly across the screen, then returns almost instantaneously to the starting point. A more sophisticated type of sweep-trigger circuit may be employed when the sweep is initiated by an external trigger pulse (often the presented signal), so that each sweep is started in synchronism with the trigger pulse.

Extra facilities usually found on a modern oscilloscope include a delayed trigger, access to the X-deflection plates allowing an external time base or other modulating signal to be used, and often facilities for beam-intensity modulation.

cathode rays *See* gas-discharge tube.

cathode-ray tube (CRT) A funnel-shaped *electron tube that permits the visual observation of electrical signals. A CRT always includes an *electron gun for producing a beam of electrons, a grid to control the intensity of the electron beam and thus the brightness of the display, and a luminescent screen to con-

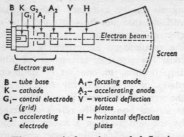

B – tube base
K – cathode
G_1 – control electrode (grid)
G_2 – accelerating electrode
A_1 – focusing anode
A_2 – accelerating anode
V – vertical deflection plates
H – horizontal deflection plates

a Electrostatic focusing and deflection

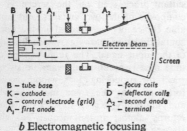

B – tube base
K – cathode
G – control electrode (grid)
A_1 – first anode
F – focus coils
D – deflector coils
A_2 – second anode
T – terminal

b Electromagnetic focusing and deflection

vert the electron beam into visible light. *Focusing of the beam of electrons and the deflection of the beam according to the electrical signal of interest, may be done either electrostatically or electromagnetically (*see* Figs. *a* and *b*), or by a combination of both methods. In general, electrostatic deflection is employed when high-frequency waves are to be displayed, as in most *cathode-ray oscilloscopes, and electromagnetic deflection is employed when high-velocity electron beams are required to give a bright display, as in *television or *radar receivers.

cation An ion that having lost one or more electrons has a net positive charge and thus moves towards the cathode of an electrolytic cell.

catoptric power Of a mirror. *See* power.

catoptric system An optical system in which the principal optical components are reflecting surfaces. *See also* catadioptric system.

Cauchy dispersion formula A formula for the dispersion of light of the form:

$$n = A + (B/\lambda^2) + (C/\lambda^4),$$

where n is the refractive index, λ the wavelength, and A, B, and C are constants. It gives a reasonable agreement with experiment for many substances over limited regions of the spectrum. Sometimes only the first two terms are necessary.

causality The principle that every effect is a consequence of an antecedent cause or causes. For causality to be true it is not necessary for an effect to be predictable as the antecedent causes may be too numerous, too complicated, or too interrelated for analysis.

According to the *uncertainty principle, however, events on the subatomic scale are neither predictable nor can they be shown to obey causal laws. If both the position and momentum of an electron, say, cannot be established precisely, consecutive observations of what may be thought of as the same electron may in fact be observation of two different electrons. Therefore individual particles cannot be identified. In *quantum theory the classical certainty of causality is replaced by *probabilities that specific particles exist in specific positions and take part in specific events.

caustic curve (and surface) Rays in a meridian plane from an object after reflection or refraction at spherical surfaces in general do not focus at one point. Consecutive rays as one moves away from the axis, intersect at points lying on a curved line (the caustic), possessing an apex or *cusp* lying at a parax-

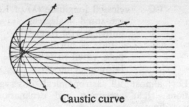

Caustic curve

ial focus. Reflected and refracted rays are tangential to the caustic.

cavity absorbent A device that may take the form either of a narrow tube or cavity through which sound is passed, or a hollow resonator placed in the sound field. In the case of a resonant cavity, large vibrations are set up in it at its natural frequency and sound energy is absorbed from the field in the neighbourhood of the resonator. Most absorbent materials are porous so that they actually consist of large numbers of small cavities.

cavity resonator *Syn.* resonant cavity. When suitably excited by external means, the space contained within a closed or substantially closed conducting surface will maintain an oscillating electromagnetic field, and the complete device, the cavity resonator, displays marked electrical *resonance effects. It has several resonant frequencies that are determined by its dimensions. Cavity resonators are used in place of tuned resonant circuits for high-frequency applications.

CCD Abbreviation for charge-coupled device. A semiconductor device through which packets of charge can be moved in a controlled manner. It is essentially an analogue *shift register. The CCD can perform a wide variety of functions. Its ability to provide a precise predetermined time delay to an analogue signal allows it to be used for sig-

nal processing. For example, *CCD multiplexers* are used for *time division multiplexing and *CCD filters* can perform many simple and complex filtering functions. One or more shift registers formed from CCDs can be used to store digital information; this *CCD memory* is a form of sequential-access memory. An array of CCDs forms the light-sensitive target area in a solid-state TV camera known as a *CCD camera*. Such arrays are also now used in highly sensitive electronic equipment associated with telescopes to detect light and ultraviolet radiation from space.

CD Abbreviation for compact disc.

celestial mechanics *See* astronomy.

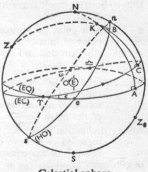

Celestial sphere

celestial sphere A sphere of infinite radius, with its centre at the centre of the earth E, that rotates once in 24 hours of sidereal *time. It is used for positional astronomy.

N	north celestial pole: point of projection of earth's north pole
S	south celestial pole
EQ	celestial equator: circle of projection of earth's equator

EC ecliptic: circle of projection of apparent path of the sun around the earth; the sun moves anticlockwise as viewed from N

♈ vernal equinox: point of intersection of equator and ecliptic, where sun crosses from south to north of equator

♎ autumnal equinox: point of intersection of equator and ecliptic, where sun crosses from north to south of equator

ε obliquity of ecliptic: 23.4°

O observer on earth

Z zenith: point of projection of O

Z_0 nadir

HO horizon: great circle having Z, Z_0 as poles

n north point (of horizon): point of intersection of ZN extended and horizon

K celestial object

BK altitude (*a*) of K

nB azimuth (*k*) of K: measured in degrees east of the north point

AK declination (δ) of K: regarded as positive if K is north of the equator

♈A right ascension (α) of K: measured in hours and minutes (24 hours = 360°) anticlockwise from ♈

ZNn meridian for observer at O: when K lies on meridian it is said to transit and to have an *hour angle H* of zero; H increases after transit, and is equal to the difference between local sidereal time and the right ascension of the body

CK celestial latitude (β) of K: regarded as positive if K is north of the ecliptic

♈C celestial longitude (λ) of K: measured in degrees anticlockwise from ♈

nN altitude of north celestial pole, equal to terrestrial latitude of observer

For almost all astronomical observations, right ascension and declination coordinates are employed.

cell 1. A pair of plates in an electrolyte from which electricity is derived by chemical action; a unit of a battery. A *primary cell* (or *voltaic cell*) is one in which the current is produced directly from chemical action by the solution of one of the plates. Current can be drawn at once from a primary cell as soon as it is made. A *secondary cell* has to be "charged" by passing a current through it in the reverse direction to its discharge, the chemical actions in the cell being reversible (*see* accumulator). The potential difference between the poles of a cell in a closed circuit depends on its internal resistance, and on the external resistance through which it is maintaining a current. The p.d. (*U*) between the terminals of a cell is given by:

$$U = ER/(r + R),$$

where *E* is the e.m.f. on open circuit, *r* is the internal resistance, and *R* the external resistance. (*See* polarization.)

2. *See* unit cell.

cellular telephone A radiotelephone network for mobile subscribers, in which the country is divided into adjacent cells, each containing a receive/transmit station. As the user moves from one cell to the other, the signals are automatically switched.

Celsius scale The official name of the centigrade temperature scale with the ice point as 0° and the boiling point as 100°. The degree Celsius (symbol: °C) is equal in magnitude to the *kelvin. A Celsius temperature *t*, in °C, is converted to a *thermodynamic temperature *T*, in kelvin, by the relationship

$$t = T - T_0 = T - 273.15,$$

where T_0 is a thermodynamic temperature fixed as 0.01 K below the triple point of water (273.16 K). On the *International Practical Scale of Temperature (1968), temperatures are expressed in both kelvin and degrees Celsius.

centi- Symbol: c. A prefix denoting 10^{-2}, e.g. 1 centimetre (cm) = 10^{-2} metre.

centigrade scale *See* Celsius scale.

central force A force on a moving body that is always directed towards a fixed point or towards a point moving according to known laws.

central processing unit (CPU) *Syn.* central processor. The principal operating part of a *computer, in which program instructions are interpreted and executed. In larger and more complex computers, processing tasks and functions are now distributed among various units, each acting independently. These units are simply referred to as *processors.*

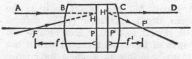

a Centred optical system

centred optical system A system consisting of a number of spherical refracting or reflecting surfaces having their centres on a common axis. Maxwell considered a perfect system in which there was complete point, line, and plane correspondence between object and image. Such correspondence is found in practice provided the rays forming the image are restricted to those passing near the axis of the system.

A ray AB (Fig. *a*) entering the (converging) system parallel to the axis will in general cross the axis after emergence at some point F′ (the *second focal point*). Similarly the ray CD, at the same height above the axis as AB, will have passed into the system after crossing the axis at some point such as F (the *first focal point*). The two incident rays shown fix an object point H for which the corresponding image point must be H′. The plane HP drawn through H perpendicular to the axis is the *first principal plane*. The plane H′P′ similarly drawn through H′, is the *second principal plane*. The principal planes thus have the properties of being *conjugate (object and image planes), and of yielding unit magnification (since HP = H′P′). P and P′ are the first and second *principal points*.

The *first focal length* of the system, f, is defined as the distance from the first principal point to the first focal point. The *second focal length*, f', is the distance from the second principal point to the second focal point. If there is the same medium on both sides of the system the two focal lengths are equal. If the media on the object and image side have different refractive indices n and n', then the relation between the numerical values of the focal lengths is that $n/f = n'/f'$.

Provided object distances are measured from the first principal plane and image distances from the second, the results of simple thin-lens theory can be applied to any centred optical system. In a thin lens the principal planes coincide in the plane of the lens.

In Fig. *b*, the rays entering the system directed towards the axial point N leave as though from the axial point N′ and make the same angle with the axis. The points N and N′ that have this property are called the first and second *nodal points* of the system. They are conjugate points of unit angular magnification.

The distance from the first focal point to the first nodal point is equal to the second focal length, while the distance

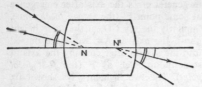

b Nodal points of centred optical system

from the second focal point to the second nodal point is equal to the first focal length. With the same medium on both sides of the system the two focal lengths become equal. The principal points must then coincide with the nodal points.

The three pairs of points – focal, principal, and nodal – are called the *cardinal points* of the system. If the positions of the cardinal points of a system are known, the position, nature, and size of the image of any object can be calculated without reference to details of the system.

centre of buoyancy *See* buoyancy.

centre of curvature *See* curvature.

centre of gravity 1. Of a body in a uniform gravitational field (e.g. a body small compared with the earth in the earth's gravitational field). The force on the body (its weight) is the resultant of the forces on the individual particles. As these forces are all parallel, their resultant passes through a particular point fixed with respect to, but not necessarily on, the body; however it may be turned relative to the field. This point is the centre of gravity and it coincides with the *centre of mass of the body.
2. Of a body in a nonuniform gravitational field. The forces on the particles of the body (no longer a system of parallel forces) are reducible, in general, to a single force and a couple. This single force does not, in general, pass through a single point fixed with respect to the

body, as the body is turned in the field. If the matter in the body is distributed with spherical symmetry, the couple reduces to zero and the force always passes through the centre of mass; only such a body has a centre of gravity in a nonuniform field, and is said to be *centrobaric* or *barycentric*.

centre of inertia *See* centre of mass.

centre of mass *Syn.* centroid; centre of inertia. A point such that if any plane passes through it, the sum of the products of the masses of the constituent particles by their perpendicular distances from the plane (the sum of the *mass moments*) is zero. In common usage, centre of mass and *centre of gravity are synonymous since when the latter exists it coincides with the former.

centre of pressure The point on a plane surface, immersed in a fluid, at which the resultant pressure on the surface may be taken to act. If the surface is horizontal in the liquid, the centre of pressure coincides with the *centre of gravity; otherwise it is below the centre of gravity but gets nearer to it as the liquid depth increases. (It is defined with respect to a plane area because the system of forces on a curved area is not always reducible to a single force.) *See* buoyancy.

centrifugal force *See* force.

centrifuge A rapidly rotating bar or flywheel on a vertical axle from the rim of which a series of tubes are suspended so that their lower closed ends are free to tilt upwards and outwards. At high speed the centrifugal force outwards is far greater than gravity, and suspensions put in the tubes settle out much more quickly than in the ordinary way. It is also used for measuring sizes, shapes, and weights of particles. *See also* ultra-centrifuge.

centripetal force *See* force.

centrobaric *See* centre of gravity.

centroid *See* centre of mass.

centrosymmetry Symmetry with respect to a point. Crystals that are centrosymmetrical have their faces arranged in parallel pairs that are alike or enantiomorphous in surface characteristics. Centrosymmetry is equivalent to twofold rotation about an axis plus reflection across a plane perpendicular to the axis.

Cerenkov detector A sensitive device in which the *Cerenkov radiation, produced by a fast particle usually moving through water, is detected by one or more *photomultipliers, so allowing the path of the particle to be reconstructed.

Cerenkov radiation The bluish light emitted by a beam of high-energy charged particles passing through a transparent medium at a speed, v, that is greater than the speed of light c' in that medium. The light is emitted at all angles, θ, to the direction of motion of the particles, thus forming a conical wave front of angle 2θ, where $\cos \theta = c' / v$; $c' = c/n$, where n is the refractive index of the medium and c is the speed of light in a vacuum. The angle θ can be used to measure the speed of the particles, and hence the kinetic energy if the nature of the particles is known. Cerenkov radiation is analogous to the shock wave produced by a *sonic boom.

CERN The European Laboratory for Particle Physics, previously known as the European Organization for Nuclear Research. The research centre for high-energy physics situated at Geneva, Switzerland. In 1990 there were 14 member countries, including the United Kingdom. *See* accelerator.

CerVit A proprietary glass-ceramic material that alters very little in size or shape when subjected to normal temperature changes. It has thus been used in telescope optics.

CGS system of units A system of units based on the centimetre as unit of length, the gram as unit of mass, and the second as unit of time. Although strictly applicable to mechanical measurements only, the system was extended to cover thermal measurements by the addition of the inconsistently defined *calorie. In extending the system further to enable electrical measurements to be carried out, it was recognized that a further fundamental quantity needed definition. This idea gave rise to two alternatives:

(a) the CGS-electromagnetic units, based on the *permeability of free space having unit size;

(b) the CGS-electrostatic units, based on the *permittivity of free space having unit size.

Because, as Maxwell proved, the product of the permeability and permittivity of free space is c^{-2}, where c is the speed of light, systems (a) and (b) are mutually exclusive.

The *Gaussian* (or symmetric) *system of units* uses units from system (a) to measure magnetic quantities and those from system (b) to measure electric quantities. In consequence, some equations of electromagnetic relationships contain c explicitly. All versions of the CGS system have now been superseded by *SI units for general scientific purposes; however, Gaussian units are still used in particle physics and in relativity. *See also* Heaviside–Lorentz units. *See* Conversion Factors (Appendix, Table 1).

chain reaction A series of nuclear transformations initiated by a single nuclear fission. For example, the fission of a ^{235}U nucleus is accompanied by the emission of one, two, or three neutrons, each of which is capable of causing further fission of ^{235}U nuclei.

When each transformation causes an average of one further transformation the reaction is said to be *critical*. If the average number of further transformations is less than one, the reaction is *subcritical*; if it exceeds one, it is *supercritical*. *See also* critical mass.

Chandrasekhar limit *See* white dwarf.

channel 1. A route along which information can be sent in a computer or communications system. It may, for example, be a telephone link between two computers or a band of frequencies assigned for transmission and reception of radio or TV broadcasts.

2. In a *field-effect transistor, the region between *source and *drain, whose conductivity is modulated by the voltage applied to the *gate.

chaos theory The theory of the unpredictable behaviour that can arise in systems obeying deterministic laws as a result of their sensitivity to variations in the initial conditions or to an excessive number of variables. Although deterministic laws enable the condition of a system to be predicted at any time in the future, to do so often depends on an ability to specify with great precision a set of parameters at an exactly specified moment. An example of chaos theory occurs in long-term weather forecasting. The meteorological laws may be well understood, but obtaining exact parameters to use with them may not be possible. In the *butterfly effect*, for example, it is postulated that the flap of a butterfly's wings can so upset the sensitive meteorological dynamics that an unfore-

cast tornado may be set off by it. In fact, simulation of weather systems has played an important role in the development of chaos theory. Other fields in which this form of unpredictability occurs include turbulent fluid flow, oscillations in electric circuits, reaction kinetics in chemistry, and many situations in biology, astronomy, and economics.

characteristic A relation between two magnitudes that characterizes the behaviour of any device or apparatus. The relations are most frequently used for *transistors, and are usually plotted in the form of a family of graphs, called *characteristic curves*. These relate the currents obtained to the voltages applied for a range of operating conditions.

characteristic curve 1. *See* characteristic.

2. A graph relating the internal transmission density (formerly optical density) to the common logarithm of the light exposure for a photographic film.

characteristic equations *See* equations of state.

characteristic function One of a set of functions satisfying a particular equation with specified boundary conditions. For example, the functions $A_n \sin n\pi x/l$ (where n is any integer) all satisfy the differential equation of transverse wave motion on a uniform flexible string of length l with both ends fixed.

In *wave mechanics, characteristic functions are *well-behaved* (i.e. physically possible) solutions of *Schrödinger's wave equation for an atomic particle, and the corresponding values of the energy of the particle are known as *characteristic values*. If there is more than one solution of the differential equation corresponding to a particular characteristic value, the system is said to be *degenerate*.

Characteristic functions and values occur in *matrix mechanics also. In quantum mechanics particularly, characteristic values and functions are often called *eigenvalues* and *eigenfunctions* (from the German).

characteristic impedance *See* transmission line.

characteristic temperature *See* Debye theory of specific heat capacities.

characteristic value *See* characteristic function.

characteristic X-radiation *See* X-rays; X-ray spectrum.

charge Symbol: Q. A property of some *elementary particles that causes them to exert forces on one another. The natural unit of negative charge is that possessed by the electron and the proton has an equal amount of positive charge. The use of the terms negative and positive are purely conventional and are used to differentiate the types of forces that charged particles exert on each other. Like charges repel and unlike charges attract each other. The charge of a body or region arises as a result of an excess or deficit of electrons with respect to protons. Charge is the integral of electric current with respect to time and is measured in coulombs. The electron has a charge of $1.602\,177\,33 \times 10^{-19}$ coulomb. *See also* electromagnetic interaction.

charge conjugation parity *Syn.* C-parity. Symbol: C. A quantum number associated with those elementary particles (such as π^0 and η) that have zero charge, baryon number, and strangeness. It is conserved in *strong and *electromagnetic interactions. In simple terms, it shows whether the *wave function describing the particle is unchanged ($C = +1$) or changes sign ($C = -1$) when the particle is replaced by its *antiparticle. *See also* CP invariance; CPT theorem.

charge-coupled device *See* CCD.

charge density 1. *Volume charge density.* Symbol: ρ. The electric charge per unit volume of a medium or body. It is measured in coulombs per metre cubed.
2. *Surface charge density.* Symbol: σ. The electric charge per unit area of a surface. It is measured in coulombs per metre squared.

charge-transfer device A *semiconductor device in which discrete packets of charge are transferred from one location to the next. Such devices can be used for the short-term storage of charge in a particular location. Several different types exist, one major classification being *CCDs (charge-coupled devices).

Charles's law (Also known as Gay-Lussac's law.) All gases and unsaturated vapours have the same mean thermal expansivity at constant pressure over the range 0 °C to 100 °C. The law in fact applies approximately over a much wider range of temperatures but is not exact except at very low pressures.

The law is often stated incorrectly in the form of a linear relationship between volume and temperature, and this is sometimes extended to a direct proportionality between volume and thermodynamic temperature. Such statements cause difficulty because of failure to specify the temperature scale used and to distinguish between real and *ideal gases.

charm A *quantum number used in the theory of *quarks and *hadrons. The charm quark (c) has charm $+1$ and its antiquark c̄ has charm -1. All other quark flavours have charm 0. The charm of a particle is the sum of the number of charmed quarks minus the number of anticharmed quarks.

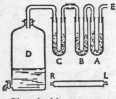

Chemical hygrometer

chemical hygrometer *Syn.* absorption hygrometer. A *hygrometer in which water flowing from the aspirator D (*see* diagram) causes a known volume of air to be drawn through E, any moisture contained in the air being removed in the tubes A, B, and C, which contain a drying agent. The process is repeated with the tube RL, containing water, attached at E. Air passing through RL becomes saturated with water vapour. The ratio of the increase in weight of A and B before RL is attached and after gives the relative *humidity of the atmosphere.

chemical shift A change in the position of a spectrum peak resulting from a small change in energy of a given state caused by a chemical effect. Thus, in the *Mossbauer effect there is a difference between the energy of a given state for nuclei in the pure element and the energy when the element is combined with other elements in a compound. This appears as a shift in the spectrum.

chemiluminescence *See* luminescence.

chief ray The central or representative ray of a pencil of rays from an object point on or off the axis, to the centre of the entrance pupil.

chip A small piece of a single crystal of *semiconductor material containing either a single electronic component or device or an *integrated circuit. Chips are commonly sliced from a large single crystal of semiconductor – a *wafer* – that is used as the substrate on which the integrated circuits, etc., are fabricated. The chips are then packaged in various ways, e.g. as a *dual in-line package (DIP).

chirality For any spin 1/2 (Dirac) particle u, two *chiral states*, u_L and u_R (known as 'left-handed' and 'right-handed' states), can be defined. For a massless particle, such as the *neutrino, the left-handed and right-handed chiral states correspond to the particle having helicity -1 and helicity $+1$ respectively, i.e. to the particle spin being oriented opposite to or along the particle's direction of motion. For a particle with non-zero mass, such as the *electron, this is true only in the limit of very high energy. The chiral states are useful in theories of the *weak interaction and in *electroweak theories. Weak interactions involving the charged $W^\pm$ bosons can be expressed purely in terms of left-handed particles and right-handed antiparticles, while photon and Z^0 interactions involve both chiral states. Only left-handed neutrinos and right-handed antineutrinos are found to exist in nature, i.e. the neutrino always has helicity -1 and the antineutrino always has helicity $+1$.

Chladni plates Flat plates used to investigate vibrations in solids. Clamped at one point (a *node), a plate is vibrated, as by bowing, at another point. This causes fine sand, sprinkled on the surface, to collect along the nodal lines. A great variety of patterns (*Chladni figures*) can be obtained by clamping and exciting the plate at different positions.

choke 1. An *inductor that presents a relatively high impedance to alternating current. It is often used in audiofrequency and radio frequency circuits, to impede the audiofrequency or radio frequency signals, or to smooth the output of a rectifying circuit.

2. A groove cut into the metal surface of a *waveguide, approximately one-quarter of a wavelength deep, to prevent the escape of microwave energy.

chroma The attribute of a visual sensation by which the amount of pure colour can be judged, irrespective of the amount of white or grey present.

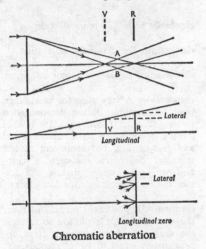

Chromatic aberration

chromatic aberration *Syn.* chromatism. An *aberration of lenses. Since the refractive index of a refracting medium depends on the wavelength (*see* dispersion), the focal length of a lens varies according to the colour of the incident light. The image of a point source of white light is therefore blurred and appears coloured; tinged with a surround of blue or violet at the focus for red, and with red at the blue focus. At an intermediate position a white circle AB

occurs – the *circle of least confusion*. For standards of comparison, the colours corresponding with the C (red) and F (blue-green) lines of hydrogen are chosen. The distance between the foci for these colours is the *longitudinal chromatic aberration*. The reciprocals of the principal focal lengths are the powers; the difference of these powers is commonly referred to as the *chromatic aberration*. For a thin lens, the last-mentioned chromatic aberration is ωP, where ω is the dispersive power of the glass and P is the power of the lens for yellow (sodium D) light.

The sizes of the images for different colours will be different; the difference in size is called the *lateral chromatic aberration* for the object considered. When chromatic aberrations have been corrected for two colours (*see* achromatic lens), owing to nonlinearity of dispersion there is a residual chromatic aberration referred to as the *secondary spectrum*. *See also* apochromatic lens.

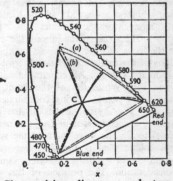

Chromaticity diagram and two colour triangles.

chromaticity An objective description of the *colour quality* of a visual stimulus, such as a coloured light or a surface, irrespective of its *luminance (*see* colour system). Chromaticity and luminance completely specify a colour stimulus.

65

The colour quality is defined in terms of its *chromaticity coordinates*. These three coordinates, *x*, *y*, *z* are equal to the ratio of each of the *tristimulus values* of a light to their sum. The tristimulus values, *X*, *Y*, *Z*, are the amounts of the three reference or matching stimuli required to match exactly the light under consideration in a given trichromatic system. Hence

$$x = \frac{X}{X + Y + Z} \text{ (redness)}$$

$$y = \frac{Y}{X + Y + Z} \text{ (greenness)}$$

$$z = \frac{Z}{X + Y + Z} \text{ (blueness)}$$

Thus all colour can be reduced to a common function since $x + y + z = 1$. When *x*, *y*, and *z* all approximately equal $1/3$, the colour is almost white.

A *chromaticity diagram* is obtained when *x* is plotted against *y*, the graph being horseshoe-shaped and the locus of all monochromatic colours (*see* diagram). The straight line joining the ends is the locus of pure purple, i.e. combinations of the extreme red and blue monochromatic colours. All colours lie within these loci. White lies at the point, C, (the white point), having coordinates *x* = *y* = $1/3$. Any colour lies on the line joining a spectral colour (on the horseshoe) to C. The wavelength of the spectral colour used is the *dominant wavelength* for the colour under consideration. The position of this colour on the line depends on the proportions of the spectral colour and white required to obtain the colour. The *excitation purity* of the colour is the ratio of the distances of the colour and the spectral colour from the white spot. The dominant wavelength is roughly equivalent to hue, and excitation purity to saturation (*see* colour).

One or more *colour triangles* can be drawn on the chromaticity diagram. They represent the entire range of chromaticities that can be obtained from a combination, by a *subtractive process, of three dyes of the *primary colours cyan (blue-green), magenta (red-blue), and yellow. Triangle (a) shows the colours obtained by a combination of the three dyes above; triangle (b) is obtained by mixing the three dyes and a white dye.

chromaticity diagram *See* chromaticity.

chromatic resolving power *See* resolving power.

chromatic scale *See* musical scale.

chromatism *See* chromatic aberration.

chrominance signal *See* colour television.

chronometer A very accurate timekeeper, used for example in the determination of longitude on board ship. Its rate of motion is controlled by a balance wheel and hair spring, but there are several technical differences between a chronometer and a watch, making the former the more accurate. *See also* clocks.

chronon A hypothetical particle of time; the time taken for a photon to traverse the diameter of an electron. It is approximately equal to 10^{-24} seconds.

chronoscope An electronic instrument for measuring very short time intervals.

circle of least confusion *See* chromatic aberration; spherical aberration.

circuit A number of electrical components connected together to form a conducting path and fulfilling a desired function such as amplification or oscillation. A circuit may consist of discrete components or may be an *integrated circuit.

If the components form a continuous closed path through which a current can circulate, the circuit is said to be *closed*; when the circuit is broken, as by a switch, it is said to be *open*. *See also* magnetic circuit.

circuital *See* curl.

circuit-breaker A device for making and breaking an electric circuit under normal or under fault conditions. *See* contactor; switch; tripping device.

circular polarized radiation *See* polarization.

civil year *See* time.

cladding 1. The process of bonding one metal to another to prevent corrosion of one of the metals. It is used in *nuclear reactors to prevent corrosion of a *fuel element by the *coolant and the escape of fission products. **2.** *See* fibre-optics system.

clamping diode *See* catching diode.

Clark cell A voltaic cell, formerly adopted as a standard of e.m.f. It consists of a mercury electrode surrounded by a paste of mercury sulphate, the negative electrode being a rod of pure zinc in a saturated solution of zinc sulphate. Its e.m.f. was defined to be 1.4345 volts at 15 °C. It has now been superseded by the *Weston standard cell.

class A amplifier A *linear amplifier operated under such conditions that the output current flows over the whole of the input cycle. The output wave shape is essentially a replica of the input wave shape. Class A amplifiers have low distortion and low efficiency.

class AB amplifier A *linear amplifier operated so that, in general, the output current flows for more than half but less than the whole of the input cycle. Class AB amplifiers tend to operate as *class A for low-input signal levels and *class B at high-input signal levels.

class B amplifier A *linear amplifier operated to produce a half-wave rectified output, i.e. the output current is cut off at zero input signal. In order to duplicate the input waveform successfully, two transistors are required each conducting for one half of the input cycle. Class B amplifiers have high efficiency but suffer from crossover *distortion.

class C amplifier An *amplifier in which output current flows for less than half of the input cycle. The output waveform is not a replica of the input waveform for all amplitudes and class C amplifiers are therefore nonlinear. Class C amplifiers are more efficient than other types, but introduce more distortion.

class D amplifier An *amplifier operating by means of pulse-width modulation. (*See* pulse modulation.) The input signal is used to modulate a square wave with respect to its *mark-space ratio. The modulated square wave then operates *push-pull switches so that one switch operates when the input is high, and the other when the input is low. The resultant current in the output load is proportional to the mark-space ratio and hence the input signal. Class D amplifiers are theoretically highly efficient, but to avoid distortion the switches must be operated faster than is generally practicable.

classical physics The long-established part of physics, excluding *relativity and *quantum theories.

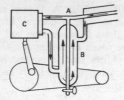

Claude process

Claude process A process for the liquefaction of air that depends on the cooling produced when a gas undergoes
adiabatic expansion and performs external work. The gas under pressure divides at A (*see* diagram) into two parts,
one part going to the expansion chamber C where it performs external work
(used in the compressor) and cools during the adiabatic expansion. This cooled
gas cools the other part from A in the
interchanger B and as this gas is under
pressure it eventually liquefies. Lubrication of moving parts at these low
temperatures is achieved by the use of
the liquid air itself as lubricant.

Clausius–Clapeyron equation The equation:

$$\frac{dp}{dT} = \frac{L}{T(v_2 - v_1)},$$

where v_1 and v_2 are the specific volumes
of the substance in two different physical states, L is the specific latent heat
for the change from one of these states
to the other. It gives the variation of
boiling point or freezing point with applied pressure.

Clausius's equation The equation:

$$c_2 - c_1 = T \frac{d}{dT}\left(\frac{L}{T}\right),$$

where c_1 and c_2 are the specific heat capacities of the liquid and vapour respectively and L is the specific latent heat
of vaporization at the thermodynamic
temperature T. The specific heat capacities are defined under conditions such
that the two phases remain in equilibrium. The value for the liquid is almost
the same as that measured at constant
pressure but that for the vapour differs
greatly and may even be negative for
some substances, including steam.

Clausius's virial law *See* virial law.

cleavage The easy separation of a crystal
into two parts owing to a relative weak-
ness of bonds along a particular direction. Normal to this direction the two
parts of the crystal show clean good
surfaces parallel to the *cleavage plane*.

clinical thermometer A mercury in glass
thermometer used for the accurate determination of the temperature of the human body and graduated from
35–46 °C (or 95–110 °F). The mercury
in the thin-walled bulb B expands past
the constriction S into the capillary
tube. On removing the thermometer
from the patient, the mercury beyond S
cannot recede into the bulb because of
the constriction. The thermometer is reset by shaking.

clock frequency *Syn.* clock rate. The
master frequency delivered by an electronic device, called a *clock*, at fixed intervals to synchronize operations in a
*computer. The clock, generally a stable
oscillator, generates an extremely regular
series of fixed-width pulses – *clock
pulses*. The reciprocal of the pulse repetition rate is the clock frequency, normally given in megahertz. Because of its
constant rate, the clock signal can be
used to synchronize the operations of
related pieces of computer equipment so
that events take place in sequence at

fixed times. For example, the clock signal is used to initiate actions within a *logic circuit and to synchronize the activities of a number of such circuits.

clock pulses *See* clock frequency.

clock rate *See* clock frequency.

clocks The earliest clocks were based on processes that take place at a constant rate, such as the apparent movement of the sun, the rate at which a candle burns, or the fall of sand in an hourglass. More advanced devices use periodic processes of constant frequency.

(1) *Pendulum clock.* The pendulum clock employs Galileo's discovery that the period of a *pendulum is a function only of its length and not its mass or initial displacement. Each swing should take place under the same conditions as its predecessor; good clocks have *compensated pendulums and are kept in airtight cases under constant temperature conditions. The highest precision pendulum clocks are accurate to about 0.01 seconds per day.

(2) *Crystal clock.* For precise scientific measurements a higher degree of accuracy is obtained from the crystal clock in which a quartz crystal is made to oscillate at about 100 000 hertz by *electrostriction. Such clocks are accurate to about 0.001 seconds per day.

(3) *Atomic clock.* Even greater accuracy can be obtained from atomic clocks in which the periodic process is a molecular or atomic event associated with a particular spectral line. For example, when energy is supplied to an ammonia molecule, the molecule can exist in a vibrationally excited state in which the nitrogen atom passes through the plane of the three hydrogen atoms to an equivalent position on the opposite side. This oscillation has a frequency of 23 870 hertz and ammonia therefore strongly absorbs microwave radiation of this frequency. This is the basis of the *ammonia clock*. A quartz oscillator can be used to supply energy to ammonia gas at this frequency. When the oscillator supply varies from this value, the energy is no longer absorbed and is used in a feedback circuit to correct the crystal oscillator. An ammonia *maser can also be used as a frequency standard.

(4) *Caesium clock.* A similar device in which the frequency is defined by the energy difference between two different states of the caesium nucleus in a magnetic field (*see* nuclear magnetic resonance). A beam of caesium atoms is split by a nonuniform magnetic field into distinct components. Atoms in the lower energy state are directed into a cavity and fed with radio frequency radiation at a frequency of 9 192 631 770 hertz, which corresponds to the energy difference between the two states. Some caesium atoms are raised to the higher energy state by absorption of this radiation and the mixture of caesium atoms is analysed by a further magnetic field. A signal from the atom detector is fed back to the r.f. oscillator supply to prevent it from drifting from the resonant frequency. In this way the supply is locked to the spectral line frequency and the accuracy is better than one part in 10^{13}. The caesium clock is used in the international (SI unit) definition of the *second.

See also synchronous clock; clock frequency.

close-packed structure A crystalline arrangement in which similar atoms, supposed spherical, are packed as economically of space as is possible. The two common arrangements are the *face-centred cubic and hexagonal close-packed structures (*see* crystal systems), but combinations of these also occur. The essential condition is that each atom shall be symmetrically surrounded by twelve others.

cloud chamber *Syn.* Wilson cloud chamber. An apparatus for making visible the tracks of ionizing particles. It consists of a gas-filled chamber containing a saturated vapour, which can be made supersaturated by the sudden cooling produced in an adiabatic expansion. The excess moisture is deposited in drops on the trail of ions left behind by the passage of a particle. The drops, suitably illuminated, can then be photographed. *See also* diffusion cloud chamber.

cloud-ion chamber An instrument combining the functions of an ionization chamber (utilizing free-electron collection) and the Wilson *cloud chamber, in the same gas volume.

Clusius column A device used for separating two isotopes by means of thermal diffusion. It is a long vertical column with a radial temperature gradient produced by an electrically heated wire along its axis. The lighter isotope tends to concentrate around the wire and the heavier one concentrates near the cool walls of the column. Convection currents carry the lighter isotope to the top of the tube.

CMOS *See* complementary transistors.

coated lenses *See* blooming of lenses.

coaxial cable *Syn.* coax. A cable that consists of two or more coaxial cylindrical conductors, insulated from each other. The outermost conductor is often earthed. Coaxial cables do not produce external fields and are not affected by them. They are thus frequently used for transmission of high-frequency signals, as in television, radio, etc.

Cockcroft–Walton generator or accelerator A high-voltage direct-current *accelerator especially for the acceleration of protons. The d.c. voltage is produced from cascaded rectifier circuits and capacitances to which a low a.c. voltage is applied.

Coddington lens A powerful magnifying glass; in effect, a complete sphere with a central stop.

coefficient of absorption *See* absorption coefficient.

coefficient of contraction The ratio of the area of the *vena contracta of a jet of fluid to the area of the orifice through which it is discharging; values lie between 0.5 and 1.

coefficient of coupling *See* coupling (1).

coefficient of expansion 1. For a solid, the coefficient of expansion is given by the expression $\Delta X/X \times 1/t$, where t is the rise in temperature producing an increase ΔX in the magnitude of the quantity X. The unit is simply $(°C)^{-1}$. If X is the length of the solid, the coefficient of *linear expansion* is obtained, symbol: α_l; if X is the volume of the specimen, the coefficient of *cubic expansion* is obtained, symbol: α_V. In general,
$$X_t = X_0(1 + \alpha t),$$
where X_0 is the original value of the quantity. Since the coefficients are small, the coefficient of cubic expansion may be taken as three times the coefficient of linear expansion.

(2) For a liquid, there are two coefficients of cubic expansion. The coefficient of *apparent expansion* is the coefficient calculated from $\Delta V/V \times 1/t$ without account being taken of the expansion of the containing vessel. The coefficient of *real* or *absolute expansion* is the coefficient obtained when allowance is made for the expansion of the

containing vessel, and is equal to the sum of the coefficient of apparent expansion and the coefficient of cubic expansion of the material of the containing vessel.

(3) For a gas, the expansion is considerable. The coefficient of increase of volume of a gas at constant pressure is the ratio of the change in volume per degree Celsius change in temperature to the volume at 0 °C, the pressure remaining constant. The coefficient of increase of pressure of a gas at constant volume is the ratio of the change in pressure per degree Celsius change in temperature to the pressure at 0 °C, the volume remaining constant. For an ideal gas both these coefficients (α) are equal to 0.003 6608 per °C so that

$$V = V_0(1 + \alpha t).$$

This means that the volume V would become zero at a temperature of $-1/\alpha$ or -273.15 °C. This temperature is also the absolute zero of *thermodynamic temperature.

coefficient of friction *See* friction.

coefficient of restitution If two spheres collide directly, the relative velocity after impact is in a constant ratio to the relative velocity before impact, and in the opposite direction. If the bodies collide obliquely, the same result holds for the relative velocity components along the line of centres at the instant of collision. This constant ratio, which depends on the materials of the spheres, and is unity for perfectly elastic bodies and zero for completely inelastic ones, is called the coefficient of restitution. The coefficient becomes slightly dependent on the relative velocity if this is very high.

coefficient of viscosity *See* viscosity.

coercive force The reversed magnetic field required to reduce the *magnetic flux density in a substance from its remanent value to zero. *See* hysteresis loop; ferromagnetism.

coercivity The value of the coercive force for a substance that has been initially magnetized to saturation.

coherence The degree to which electromagnetic radiation or any other oscillating quantity maintains a near-constant phase relationship, both temporally and spatially. The time over which the phase relationship remains nearly constant is called the *coherence time*. It is approximately equal to $1/\Delta\nu$, where $\Delta\nu$ is the bandwidth of the source. There is said to be *correlation* or *temporal coherence* between the oscillations during this period. The pathlength corresponding to the coherence time is called the *coherence length*. *Spatial coherence* is related to the spatial extent of the source, i.e. to the angle made by the beam emitted from the source. *Lasers are capable of producing radiation with considerable temporal and spatial coherence.

coherence length *See* coherence.

coherence time *See* coherence.

coherent radiation Radiation exhibiting a high degree of *coherence, *laser radiation being a practical example.

coherent units A system of units, such as *SI units, in which the quotient or product of any two units gives the unit of the resultant physical quantity. For example, in SI units the unit of length is the metre and the unit of time is the second, the coherent unit of velocity is therefore the metre per second. The *base units* of a coherent system (such as the metre and second in SI units) are an arbitrarily defined set of physical quantities: all the other units in the system are derived from the base units by

71

defining relationships and are called *derived units*.

cohesion The property of a substance that enables it to cling together in opposition to forces tending to separate it into parts; the tendency of the different parts of a body to maintain their relative positions unchanged. *Compare* adhesion.

coil A conductor or conductors wound in a series of turns. Coils are used to form *inductors or the windings of *transformers, *motors, and *generators.

coincidence circuit A circuit with two input terminals that is designed to produce an output pulse only when both input terminals receive a pulse within a specified time interval.

cold cathode The cathode of an electron tube that is caused to emit electrons by having a sufficiently high voltage gradient, instead of the cathode being operated at high temperature. Electrons may be ejected from the cathode by *secondary emission caused by positive ions produced from residual gas in the tube, or in the case of very highly evacuated tubes, autoemission may occur.

cold emission *See* field emission.

cold fusion Nuclear fusion occurring at normal temperatures, rather than at the high temperatures necessary to overcome electrostatic repulsive forces between nuclei. There have been two main approaches to producing fusion at low temperatures. One is an electrolytic method; it has been suggested that, under certain conditions, electrolysis of deuterium oxide using a palladium cathode can produce low-temperature nuclear fusion. Deuterium ions liberated at the cathode are absorbed in the crystal lattice of the electrode, where they are forced together, thus overcoming the repulsive electrostatic force. However, claims that high-energy outputs using this method have been obtained have not been reproduced; the necessary output of neutrons for a genuine fusion reaction has not been detected.

The other approach to cold fusion has been to shield one of the deuterium atoms by binding it with a negative muon. In this technique a muon replaces an electron in a deuterium atom. Because the muon is 207 times heavier than the electron, the resulting muonic atom of deuterium is much smaller and is able to approach another deuterium atom more closely, allowing nuclear fusion to occur. The muon is then released to form another muonic atom, and so on; i.e. the muon acts as a catalyst for the fusion reaction. One problem with this approach is the short lifetime of the muon, which restricts the number of fusion reactions it can catalyse.

cold trap A tube, cooled with liquid air, or dry-ice (frozen carbon dioxide) in acetone, that will condense vapour passing into it.

collective excitations Quantized modes in a many-body system that arise when cooperative motion of the system as a whole is considered. This type of *excitation arises as a result of the interactions between particles. *Plasmons and *phonons in solids are examples of collective excitations. Collective excitations obey Bose–Einstein statistics (*see* quantum statistics).

collector The region in a bipolar junction *transistor into which *carriers flow from the *base. The electrode attached to this region is called the *collector electrode*.

collector-current multiplication factor In a junction *transistor, minority *carriers entering the *collector from the *base region can carry sufficient energy to create electron-hole pairs in the collector thus causing an increase in minority carrier current in the collector. The collector-current multiplication factor is the ratio of enhanced current flow in the collector as a result of the flow of minority carriers from the base into the collector, to the current carried by the minority carriers at collector voltage. Under normal operating conditions this factor is unity, but under high-field conditions it rapidly increases to infinity as the *avalanche breakdown voltage is reached.

collector ring *See* slip ring.

colligative property A property of a system that depends on the number or concentration of molecules present rather than their nature. The *osmotic pressure of a solution and the *pressure of a gas at low pressures are colligative properties.

collimator 1. An optical system that produces a beam of parallel light. It is used in *spectrometers, *telescopes, etc.

2. *Syn.* finder. A small fixed telescope attached to a larger one in order to set the line of sight of the large instrument.

3. An apparatus, usually in the form of a cylindrical tube of a heavy material, that is used to limit the size of a beam of charged particles or X- or gamma-radiation to the required dimensions.

collision In *kinetic theory, the mutual action of molecules, atoms, etc., when they encounter one another. A collision is thought of as being one of three kinds.

(1) *Elastic collision.* One in which the total kinetic energy of translation is un-

changed after the collision, none being translated into other forms. In nuclear physics, an elastic collision is one in which the incoming particle is scattered without exciting or breaking up the struck nucleus.

(2) *Inelastic collision* (or *inelastic collision of the first kind*). One in which the total kinetic energy of translation is decreased by the collision while some other form of energy is increased. For example, a neutron may undergo an inelastic collision with a nucleus, which is thereby raised to an excited state that decays by gamma emission. The most extreme case of an inelastic collision is one in whch the colliding particles do not separate after impact. Some writers (especially engineers) reserve the term "inelastic" for this special case.

(3) *Superelastic collision* (or *inelastic collision of the second kind*). One in which the total kinetic energy of translation is increased by the collision while some other form of energy is decreased. For example, molecules of a gas in contact with a solid at higher temperature on the average recoil with higher kinetic energy, at the expense of the vibrational energies of the molecules of the solid.

In all kinds of collision total energy, mass, momentum, and angular momentum are conserved.

collision density The total number of a specified type of collision occurring per unit time per unit volume of material.

colloid A substance consisting of particles of ultramicroscopic size (1–100 nm), intermediate between those of a true solute and those of a suspension. They exhibit Brownian movements, and owing to their electrical charge are subject to *cataphoresis. Though frequently used, the term is a loose one since most substances can be brought to the colloid state by a suitable technique.

colorimetry The science that aims at specifying and reproducing colours as a result of measurement. Colorimeters of three types: (*a*) colour album or filter samples for comparison – essentially empirical; (*b*) monochromatic colorimeters that match colours with a mixture of monochromatic and white lights; (*c*) trichromatic colorimeters in which a match is effected by a mixture of three colours.

colour 1. The sensation normally experienced when light of sufficient intensity and with a spectral distribution significantly different from normal daylight (or other forms of *white light) enters the eye. The colour sensation arises most strongly from radiation covering only a small part of the visible spectrum. Besides possessing *luminosity, colours have *hue* and *saturation*. Saturation is the degree to which a colour departs from white and approaches a pure spectral colour. Hue is determined by wavelength – a pure continuous spectrum shows a continuous variation of saturated hues. When a hue is diluted with white light (desaturated, impure), the colour is classed as a *tint*. The *shade* of a colour refers to its luminosity.

Observers with normal *colour vision group the hues of the spectrum in the six colours (wavelengths given in nm): red (740–620), orange (620–585), yellow (585–575), green (575–500), blue (500–435), violet (435–390). Such subjects observe typically 100 steps of hue difference across the spectrum. As a hue is diluted with white, about 20 tints are possible.

By mixing colours, other colours emerge and a sharp distinction must be drawn between combining coloured lights, the mixing of pigments, and the transmission of light by colour filters. Mixing coloured lights is an *additive process. The other two are *subtractive processes. When lights of different colours are arranged to illuminate a white screen viewed by eye, or if they are arranged to produce overlapping illumination on the retina of the eye, the ensuing sensation has another colour whose hue and saturation depend on the relative proportions of the mixing colours (Newton) and its luminosity is the sum of the separate luminosities (Abney). Pairs of pure spectral hues produce colours of different hues with differing degrees of saturation: certain pairs, called *complementary colours, combine to form white light.

By using lights of three *primary colours in various proportions it is possible in general to produce any other colour, i.e. a colour match can be made with another colour. (*See* chromaticity.) Slightly fewer colours can be obtained by mixing dyes and inks as the three colours used do not correspond exactly to the three light primaries. *See also* surface colour; colour system.

2. A quantum number of *quarks. *See* quantum chromodynamics.

colour blindness *See* colour vision.

colour equation An algebraic equation that expresses the results of an additive mixture (*see* additive process) of three *primary colours in terms of another colour or white. *See also* colour system; chromaticity.

colour filter *See* filter.

colour mixture *See* additive process; subtractive process.

colour photography *See* photography.

colour picture tube A type of *cathode-ray tube designed to produce the coloured image in *colour television. Varying the intensity of excitation of three different *phosphors, which produce the three primary colours red, green, and

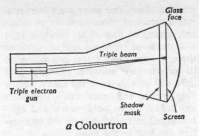

a Colourtron

blue, reproduces the original colours of the image by an *additive process.

The conventional colour picture tube consists of a configuration of three *electron guns tilted slightly so that the electron beams intersect just in front of the screen. Each electron beam has a focusing system and is directed towards a different colour phosphor. There are several different types of colour picture tube, the main differences being in the configuration of electron guns and arrangement of the phosphors on the screen.

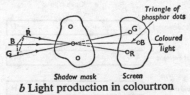

b Light production in colourtron

One main type of tube (e.g. the *colourtron*) has a triangular arrangement of electron guns, and the phosphors arranged as triangular sets of coloured dots. A metal shadow mask is placed directly behind the screen, in the plane of intersection of the electron beams, to ensure that each beam hits the correct phosphor, the beams being blanked out by the mask while moving from one position to the next (Figs. *a*, *b*).

The other main type has the three electron guns arranged in line horizontally, an aperture grille of vertical wires, and the phosphors arranged as vertical stripes on the screen. The latter type

has advantages in focusing the beams but has a smaller field of view than the former.

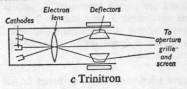

c Trinitron

The *Trinitron* is a type of colour picture tube that has a single electron gun with three cathodes aligned horizontally, an aperture grille, and vertically striped phosphors. The cathodes are tilted towards the centre so that the electron beams intersect twice, once within the electron-lens focusing system and once at the aperture grille (Fig. *c*). This allows a single electron-lens system to be used for all three beams, needing fewer components and thus making the system much lighter and cheaper. It also gives a greater effective diameter of the lens system than is found in conventional tubes, hence producing sharper focusing. Horizontal alignment of the cathodes means that misconvergence of the beams only occurs in the horizontal direction.

colour quality *See* chromaticity.

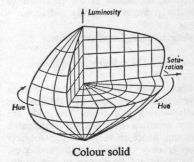

Colour solid

colour system The representation of a colour in terms of a specific set of coordinates. For objective colour systems, the coordinates dominant wavelength, excitation purity, and luminance (*see* chromaticity) are frequently used. For subjective colour systems, the coordinates are usually luminosity *L*, saturation or chroma *S*, and hue *H*. If these form a system of cylindrical coordinates, the colour is found inside a roughly elliptical solid.

colour television Television, based on an *additive colour process, in which a composite signal is produced from the three video outputs from *camera tubes, each of which is sensitive to red, blue, or green light. A colour decoder in the receiver extracts the video information for each colour and a coloured image is produced (*see* colour picture tube).

The composite signal is broadcast in two parts to achieve compatibility with black-and-white (monochrome) receivers. The *luminance signal* contains brightness information and produces a black-and-white image. The *chrominance signal* contains the colour information and is transmitted as a subcarrier of suitable frequency.

colour temperature Of a nonblack body. The temperature of a *black body that has approximately the same energy distribution as occurs in the spectrum of the body. *See also* selective radiation.

colour triangle *See* chromaticity.

colourtron *See* colour picture tube.

colour vision The *retina of the human eye contains two sorts of light-sensitive cells – *rods* and *cones*. The rods are very sensitive but do not distinguish colours, hence in very dim light objects appear only in black, white, and grey. The cones are insensitive to low intensities but discriminate fine detail and distinguish colours, provided there is sufficient illumination. Colour vision can be explained by the *trichromatic theory*,

which assumes that there are three separate systems of cones sensitive to either red, green, or blue light. Incident light will therefore stimulate one or more of these systems to an extent depending on its colour (*see* additive process). Red light will stimulate the red cones, yellow light the red and green cones. The cones are linked to the optic nerve and electric impulses resulting from the stimulation are sent to the brain.

Defective colour vision – popularly and misleadingly known as *colour blindness* – is hereditary. It can be explained by the number of cones of one or more kinds being much less than normal, possibly zero in some cases.

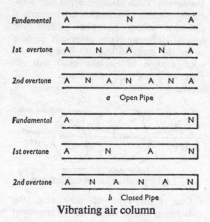

Vibrating air column

column of air The vibrations of air columns are the sources of sound in the organ as well as in the different types of wind instruments.

The simplest case of a vibrating mass of air is that in a hollow cylindrical pipe, the ends of which may be open or closed. To put forward a simple theory, the following assumptions must be made: (1) The motion in the tube is uniform, i.e. the viscosity of the medium inside the tube is neglected so that only plane waves need be considered. The diameter of the tube should therefore be sufficiently great yet small with respect to the length of the pipe and with the wavelength of the sound. The walls of the pipe are assumed to be rigid. (2) Vortices or rotatory motions are not set up inside the tube. (3) Oscillations are so rapid that the changes may be considered adiabatic.

Under these conditions, when a cylindrical air column is set into resonant vibration by some means, *standing waves are set up due to *progressive and retrogressive waves. In the open pipe (both ends open, Fig. *a*), there must be a displacement antinode A at each end and therefore, in the fundamental mode of vibration, a *node N in the middle. The frequency of the fundamental is then $c/2l$, where c is the speed of sound in the medium inside the pipe. and l the pipe length. In the next possible mode, there must be two nodes in the pipe and an antinode at the centre, and so on. The frequencies of the different modes are in the ratio 1:2:3:..., representing a full *harmonic series of wavelengths.

For a closed pipe (one end closed and the other open, Fig. *b*), there must always be a displacement node at the closed end and an antinode at the open end. In the fundamental mode, the frequency is $c/4l$. In the next mode of vibration, the frequency is $3c/4l$ and so on. In general, the frequencies bear the ratios 1:3:5:..., representing the odd harmonic series of wavelengths. *See also* end correction.

coma An *aberration of a mirror or lens in which the image of a point lying off the axis presents a comet-shaped appearance. While the rays from the central zone Z_A (*see* diagram) focus to a point A, the zones Z_B and Z_M form annular rings or *comatic circles*, B and M of progressively varying diameter and centration. The overlap of these circles

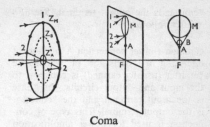

Coma

produces the comatic patch AM, which is called the *tangential coma* and the radius of the comatic circle M the *sagittal coma*: the latter is practically one-third of the former. For freedom from coma, the lateral magnification for all zones should be constant, which demands the fulfilment of the *sine condition. For a single lens, the lens with least *spherical aberration has the least coma.

combination tones *Syn.* resultant tones. If two tones of frequencies f_1, f_2, are sounded together very loudly, at least two other rather faint tones, of frequencies $(f_1 + f_2)$ and $|f_1 - f_2|$ (the absolute value), may be heard. The first is an example of a *summation tone* and the second an example of a *difference tone*; collectively they are known as combination tones. Other audible combination tones are: $2f_1$, $3f_1$, $2f_2$, $3f_2$, $|f_1 - 2f_2|$ $|f_2 - 2f_1|$. *Compare* beats.

common base connection A method of operating a *transistor in which the *base (usually earthed) is common to the input and output circuits; the *emitter is the input terminal and the *collector is the output terminal. This type of connection is frequently used as a voltage amplifier stage.

common collector connection A method of operating a *transistor in which the *collector (usually earthed) is common to both input and output circuits; the *base is the input terminal and the *emitter is the output terminal. *See also* emitter follower.

common emitter connection A method of operating a *transistor in which the *emitter (usually earthed) is common to the input and output circuits; the *base is the input terminal and the *collector is the output terminal. This type of connection is used for power amplification with a nonsaturated transistor and for switching with a transistor in saturation.

common impedance coupling *See* coupling.

communications channel A *channel used for the transfer of data. *See also* communications system.

communications line or link Any physical medium, such as a telephone line, cable, radio beam, or optical fibre, used to carry information between different locations.

communications network *See* communications system; network.

communications satellite *See* satellite.

communications system Any system whereby information can be conveyed efficiently and reliably from source to destination. With more than one source and/or destination, the system is called a *network. Information in a communications system is sent by means of a *communications channel. It is encoded in a digital form before transmission and decoded at its destination(s). This reduces to a minimum the errors arising in the information as a result of *noise in the channel.

commutator 1. A device for reversing the direction of the current in an electric circuit or in some part of a circuit.

2. A device employed in electrical machines to connect in turn each of the sections of an armature winding with an external electric circuit. It may be used as a simple current-reverser, or to convert alternating current into direct current (or vice versa). In a d.c. machine connection is made with the external circuit by means of carbon *brushes, which are kept in contact with the outer surface of the commutator.

compandor *See* volume compressors (and expanders).

comparator 1. A device for measuring the difference in length between two line standards or for measuring horizontal distances by comparison with a standard scale.

2. A circuit, such as a differential amplifier, that compares two signals and produces an output that is a function of the result of the comparison.

compass A magnet freely pivoted horizontally so that it can set itself along the lines of force of the earth's magnetic field. It usually carries a scale divided into degrees and marked with the cardinal points, or these may be printed on a circular card to which the magnet, or system of magnets, is fixed. *See also* gyrocompass.

compensated pendulum A *pendulum so constructed that the distance between the support and the centre of gravity of the bob is independent of temperature so that the time period does not vary with temperature.

complementarity The principle that a system, such as an electron, can be described either in terms of particles or in terms of wave motion (*see* de Broglie equation). According to Bohr these views are complementary. An experiment that demonstrates the particle-like nature of electrons will not also show their wave-like nature, and vice versa.

Colour	$\lambda/10^{-9}$ m	Complementary	$\lambda/10^{-9}$ m
Red	656	Green-Blue	492
Orange	607	Blue	489
Golden Yellow	585	Blue	485
Yellow	567	Indigo Blue	464
Green Yellow	563	Violet	433

complementary colours Two pure spectral *colours that when mixed produce white light. The negative of a colour photograph is in the complementary colours of the original scene. The table gives the complementary colours of lights (not pigments). There is no complementary spectral colour to green.

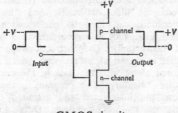

CMOS circuit

complementary transistors A pair of transistors of opposite type, i.e. n-p-n and p-n-p bipolar junction *transistors. *Class B push-pull amplifiers often employ complementary bipolar junction transistors.

Complementary MOS *field-effect transistors (abbreviation: *CMOS*) are used for *logic circuits with low heat dissipation and power consumption. The basic *inverter is shown in the illustration. When the input is low, the p-channel device conducts and the output is high. If the input is high, the n-channel device conducts and the output is low. The output of such a device will, in general, be driving like stages with essentially capacitive input impedance and

the d.c. current flowing will therefore be zero.

compound microscope *See* microscope.

compound nucleus The highly excited nucleus, of short lifetime, formed immediately after a nuclear collision.

compound pendulum *See* pendulum.

compound-wound machine A d.c. machine in which the *field magnets are provided with both series and shunt excitation windings. The series windings carry the load current of the machine and if the series field assists the shunt field, the machine is said to be *cumulatively compound-wound*; if the series field opposes the shunt field, the machine is said to be *differentially compound-wound*.

compressibility Symbol: κ. The reciprocal of the bulk modulus (*see* modulus of elasticity); it is volume strain divided by pressure change:
$$\kappa = -(1/V)(\partial V/\partial p).$$
The value of the partial derivation is that for constant temperature or constant entropy according to whether conditions are isothermal or adiabatic. For solids, values are typically of the order 10^{-11} Pa^{-1}, and for liquids 10^{-9} Pa^{-1}. For a gas the isothermal compressibility is very nearly equal to the pressure, so at one atmosphere it is approximately 10^{-5} Pa^{-1}.

compression A process occurring in *longitudinal waves of sound traversing an elastic medium and resulting in a greater density than for normal pressure. The particles of the medium start vibrating in the direction of wave propagation, about their normal position. As a result, local changes of density occur due to the relative juxtaposition of neighbouring particles. Compressions occur at points of maximum density and *rarefactions* at points of minimum density. A continuous variation of pressure in the medium results as the wave passes.

compressor *See* volume compressors (and expanders).

Compton effect *Syn.* Compton scattering. An interaction between a photon of electromagnetic radiation and a free electron (or other charged particle) in which some of the energy of the photon is transferred to the particle. As a result, the wavelength of the photon is increased by an amount $\Delta\lambda$, where

$$\Delta\lambda = (2h/m_0 c) \sin^2 \tfrac{1}{2}\phi.$$

This is the *Compton equation*; h is the Planck constant, m_0 the rest mass of the particle, c the speed of light, and ϕ the angle between the directions of the incident and scattered photon. The quantity $h/m_0 c$ is known as the *Compton wavelength*, symbol: λ_C, which for an electron is equal to 0.002 43 nm.

The outer electrons in all elements and the inner ones in those of low atomic number have *binding energies negligible compared with the quantum energies of all except very soft X- and gamma rays. Thus most electrons in matter are effectively free and at rest and so cause Compton scattering. In the range of quantum energies 10^5 to 10^7 electronvolt this effect is commonly the most important process of attenuation of radiation. The scattering electron is ejected from the atom with large kinetic energy and the ionization that it causes plays an important part in the operation of detectors of radiation.

In the *inverse Compton effect* there is a gain in energy by low-energy photons as a result of being scattered by free electrons of much higher energy. As a consequence, the electrons lose energy.

Compton equation *See* Compton effect.

Compton scattering *See* Compton effect.

Compton wavelength *See* Compton effect.

computer Any device by which data, received in the appropriate form, can be manipulated so as to produce a solution to some problem. In general the word now refers to a *digital computer*, the other basic form, the *analogue computer, being far less versatile.

The digital computer accepts and performs operations on discrete data, i.e. data represented in the form of combinations of characters. Before being fed into the computer, the characters consist of letters, numbers, punctuation marks, etc. Once inside the computer, the characters are represented as combinations of *bits. Arithmetic and *logic operations can then be performed on these groups of bits according to a set of instructions. The instructions form what is known as a *program, which is stored along with the data in the computer *memory. The program instructions are interpreted and executed by one or more *processors. Data and programs are generally input to a computer by means of a *keyboard. The information is usually output on a *VDU, *printer, or *plotter. *See also* microcomputer; mainframe.

computer graphics *See* graphics.

concave Curving inwards. *Concave mirrors* are converging in action. *Concave lenses* are those that are thinner at the centre and diverging in action – *biconcave, planoconcave,* and *concave* or *diverging meniscus.* (*See* diagrams under *lens.*) The *concave grating* is a *diffraction grating ruled on a spherical metal reflecting surface, eliminating *chromatic aberration and ultraviolet absorption by the media (glass, etc.) otherwise required to produce dispersion, to focus the light, etc.

concentration cell A cell in which two electrodes of the same metal are immersed in solutions of different concentrations of one salt of the same metal. The solutions may be separated by a porous partition. Metal dissolves in the weaker and is deposited from the stronger solution. The e.m.f. depends on the substances and the concentrations, but is usually a few hundredths of a volt.

concentric lens A convexoconcave lens whose surfaces have the same centre of curvature; its central thickness is equal to the difference in the radii of curvature.

condensation 1. The process in which a vapour or gas is transformed to a liquid, accompanied by a release of latent heat.
2. (sound) The ratio of the instantaneous excess of density to the normal density at a point in a medium transmitting longitudinal sound waves.

condensation pump *See* pumps, vacuum.

condenser 1. (heat) A device for the continuous removal of heat, e.g. a stream of cold water that removes the latent heat evolved when a vapour condenses to a liquid as in distillation. In a heat engine it is the system or reservoir to which the working substance rejects that part of the heat not converted into work.
2. (light) A mirror or lens combination used in optical instruments (e.g. projectors, compound microscopes) to concentrate light from a source into a defined beam so that the light source can be focused on to an object (which may be an opaque object or a transparent slide). It is usually a planoconvex lens or a pair of planoconvex lenses

conduction in gases

with plane sides facing out. The design differs widely according to the purpose. A *Fresnel lens is often used as a condenser in a projector. The *microscope condenser* is an elaborate substage lens or mirror combination, for use with higher-power objectives, so that rays converge without *aberration and fill the aperture of the objective uniformly. Its *numerical aperture should not be less than that of the objective. The term is also applied in microscopy to special designs for dark-ground illumination and for simple condensing lens devices. *See also* Abbe condenser.
3. (electric) Obsolete term for *capacitor.

conductance Symbol: G. The reciprocal of resistance (for a direct current) or the real part of the *admittance Y, where $Y = G - iB$ and B is the *susceptance (for an alternating current). It is measured in siemens.

conduction The transmission of thermal, electric, or acoustic energy through a medium without any transfer of mass. The conduction of electricity results from the action of an applied electric field on charge carriers in a medium. (*See* energy bands; electrolysis; conduction in gases.) *See also* conduction of heat.

conduction band *See* energy bands.

conduction current *See* current.

conduction electrons *See* energy bands.

conduction in gases The passage of an electric current through a gas. The conduction process is dependent on the presence of ions (*see* gaseous ions). Natural ionization caused by cosmic rays, the presence of small quantities of radioactive substances, ultraviolet light, etc. gives rise to a continuous minute

81

conduction of heat

current when a small electric field is maintained between two electrodes in the gas. This is due to the collection of ions of opposite sign at each electrode.

With increasing field strength, there is an initial rise in the current, and it then remains constant, until the applied potential difference is sufficiently high to allow the electrons to create further ions by collision. With further increased field strength, the current strength increases regularly and rapidly until a critical p.d. known as the *breakdown voltage* is reached. At this point additional phenomena suddenly occur depending on the pressure, and the distance between the electrodes. The process is known as an electrical *discharge*.

At pressures of the order of atmospheric pressure a spark passes (*spark discharge*), while at lower pressures a regular series of glowing masses of gas appear in the tube, the colour of the glow depending on the gas (*see* glow discharge). With small distances between the electrodes a brilliant light is emitted (*arc discharge*). Immediately after the onset of discharge the p.d. across the discharge tube is reduced to a definite value known as the *maintenance potential*; the current strength remains constant. The curve relating current to applied p.d. for a given separation of electrodes in a gas at a particular pressure is known as the *discharge characteristic*. In a discharge in gases at low pressures (below one pascal), cathode rays (*see* gas-discharge tube) are produced. At lower pressures *X-rays are produced by the impact of the cathode rays on the anode of the discharge tube.

The brilliant colourful glow of the glow discharge is utilized in display lighting tubes using neon, mercury vapour, etc. as the gas. The spark and arc discharges provide a rich source of ultraviolet and this is utilized in making ultraviolet lamps and sources.

82

conduction of heat The transfer of heat through a body not involving radiation or the flow of the material. In gases transfer is caused by collisions between the higher-energy molecules in regions at higher temperatures with lower-energy molecules in adjacent cooler regions. In dielectric solids the vibrations of molecules are transmitted through the body as waves as a result of the elastic bonding between the particles. These waves are of the same nature as sound but have very much higher frequencies (of order 10^{12} Hz), and their energies are quantized as *phonons. On balance more energy is transferred by phonons from the hotter to the colder parts than is transferred the opposite way. Theories of heat conduction treat the phonons as if they were gas molecules moving within the space occupied by the solid, being scattered by irregularities in the material. Conduction in nonmetallic liquids is intermediate in character between that in solids and that in gases.

In solid and liquid metals conduction is normally almost entirely by means of the valence electrons. These are treated theoretically as gas molecules. *See also* thermal conductivity; Wiedemann–Franz–Lorenz law.

conductivity 1. Symbol: γ or σ. The reciprocal of *resistivity. It is also defined as the current density divided by the electric field strength: this definition is often more useful when considering solutions; it is then known as the *electrolytic conductivity*, symbol: κ. Conductivity is measured in siemens per metre.

2. *See* thermal conductivity.

conductor (electrical) A substance, or body, that offers a relatively small resistance to the passage of an electric current. *See also* energy bands.

cone *See* colour vision.

cone of friction The resultant force of one flat surface on another is the resultant of the normal force and the frictional force. This resultant must lie in the cone whose axis is the normal to the surfaces and whose semi-apical angle has its tangent equal to the coefficient of limiting friction.

confinement 1. *See* containment. **2.** *See* quark confinement.

conic sections *Syn.* conics. The family of curves formed as sections when a right circular double cone is cut by a plane. The possible sections are: the *ellipse* (with the circle as a special case), the *parabola*, and the *hyperbola* (with a pair of lines as a limiting case). Alternatively, a conic is the locus of a point moving so that its distance from a fixed point (the *focus*) bears to its distance from a fixed line (the *directrix*) a constant ratio (the *eccentricity*, symbol e). If $e < 1$, the conic is an ellipse ($e = 0$ for a circle); if $e = 1$, it is a parabola; if $e > 1$, it is a hyperbola. The parabola has only one finite focus and corresponding directrix; the other conics have symmetrical pairs of foci and corresponding directrices and a centre of symmetry and hence are *central conics*.

conjugate 1. (general) Joined in a reciprocal relation, as two points, lines, quantities, things that are interchangeable with respect to the properties of each.
2. (light) Of foci, planes, and points. Relating to interchangeable properties of object and image. Thus, if I is the image of O, then if I is made the object, its image would be at O. The principal points, nodal points, symmetric points of a system are in turn conjugate pairs; the principal focal points are conjugate with infinity. The points at which object and image are coincident (centre of curvature of a mirror; pole of a mirror or

thin lens) are *self-conjugate*. *See* Newton's formula.

conjugate impedances Impedances that have equal resistance components, and also equal reactance components, the latter having opposite signs. For example, two impedances,
$$Z_1 = R + iX \text{ and } Z_2 = R - iX$$
are conjugate impedances.

conjugate particle *See* antiparticle.

conjunction *See* opposition.

conservation of charge The principle that the total net charge of any system is constant.

conservation of energy *See* conservation of mass and energy.

conservation of mass and energy The *principle of conservation of energy* states that the total energy in any system is constant. The *principle of conservation of mass* states that the total mass in any system is constant. In classical physics these two laws were independent. Although conservation of mass was verified in many experiments the evidence for this was limited. In contrast the great success of theories assuming the conservation of energy established this principle with the highest degree of certainty, and Einstein assumed it as an axiom in his theory of *relativity. According to this theory the transfer of energy E by any process entails the transfer of mass $m = E/c^2$, hence the conservation of energy ensures the conservation of mass.

In Einstein's theory inertial and gravitational mass are assumed to be identical and energy is the total energy of a system. Some confusion often arises because of idiosyncratic terminologies in which the words mass and energy are given different meanings. For example, some particle physicists use "mass" to mean the rest-energy of a particle and "energy" to mean "energy other than rest-energy". This leads to alternative statements of the principles, in which terminology is not generally consistent. *See* Einstein's law.

conservation of matter *Syn.* conservation of mass. Matter can be neither created nor destroyed. One of the basic principles of 19th-century chemistry. *See* conservation of mass and energy.

conservation of momentum 1. In any system of mutually interacting or impinging particles, the linear momentum in any fixed direction remains unaltered unless there is an external force acting in that direction.
2. Similarly, the angular momentum is constant in the case of a system rotating about a fixed axis provided that no external torque is applied.

conservative field A field of force in which the work done in taking a test particle from one point to another is independent of the path taken between them (e.g. a scalar potential field such as an electrostatic or gravitational field).

constantan An alloy of about 50% copper and 50% nickel with a comparatively high resistivity and low temperature coefficient of resistance. It is used extensively in winding electrical resistors, and with copper, iron, silver, etc., in forming thermocouples with a comparatively large e.m.f.

constant pressure gas thermometer A thermometer in which the volume occupied by a given mass of gas at a constant pressure is used for the measurement of the temperature of the bath in which the bulb containing the gas is immersed. A temperature of t_p on this scale is defined as:

$$t_p = \frac{V_t - V_0}{V_{100} - V_0} \times 100°C$$

where V_t, V_{100}, and V_0 are the volumes occupied by the gas at the temperature t_p, the steam point, and the ice point respectively.

The scale given by employing an actual gas in the thermometer will differ from the *ideal gas scale, due to the fact that the gas does not obey Boyle's law exactly except at infinitely low pressure. A correction must therefore be applied to the readings.

constant volume gas thermometer A thermometer in which the pressure exerted by a constant volume of gas is used for the measurement of the temperature of the bath in which the bulb containing the gas is immersed. A temperature of t on this scale is defined as:

$$t = \frac{p_t - p_0}{p_{100} - p_0} \times 100°C$$

where p_0, p_{100}, and p_t are the pressures exerted by the gas when at the ice point, the steam point, and at the temperature t respectively. Using hydrogen or nitrogen gas in a platinum-iridium or platinum-rhodium bulb, temperatures from $-260 °C$ to $1600 °C$ may be standardized.

There are two chief errors causing the temperature on the gas scale to differ from the *thermodynamic temperature: (1) that the gas is not ideal so that the product pV is equal to $(A + Bp)$ and only becomes independent of the pressure if the pressure is very small. (2) The volume of the gas is not constant and not all at the same temperature.

constrain To limit to a predetermined position or path. If the motion of a system is subject to frictionless constraining forces (*constraints* or reactions) then these forces exist in equal and opposite pairs (Newton's 3rd law of motion) and do no net work. Such constraints include: (*a*) The reaction on a movable body in smooth contact with a fixed body. (*b*) The pair of reactions at a smooth contact. (*c*) The reaction of a body rolling on a fixed body. (*d*) The pair of reactions at a rolling contact. (*e*) The pair of reactions between two particles of a rigid body.

Constraints reduce the number of *degrees of freedom of the system. *See* virtual work principle.

constringence *See* Abbe number.

constructive interference *See* interference.

contactor A type of switch for making and breaking an electric circuit, designed for frequent use.

contact potential The difference of potential that arises when two conductors of different material are placed in contact. Thus a metal will be at a different potential from a conducting liquid in which it is immersed; an electric field will exist in the space between plates of different metals, when the plates are electrically connected. The contact potential is usually of the order of a few tenths of a volt. It results from the difference between the *work functions of the two metals.

contact resistance The resistance at the surface of contact of two conductors.

containment 1. *Syn.* confinement. The process of preventing the *plasma from coming into contact with the walls of the reaction vessel in a controlled *thermonuclear reaction. The time for which ions are trapped in the plasma is called the *containment time*.

2. The prevention of the release of unacceptable quantities of radioactive material beyond a controlled zone in a nuclear reactor.

3. The containment system of a nuclear reactor.

continuity principle For continuous motion, the increase of mass of fluid in any time interval δt within a closed surface drawn in the fluid is equal to the difference between the mass flow in and the mass flow out through the surface. This statement is expressed mathematically in the *equation of continuity*:

$$\frac{\partial \rho}{\partial t} + \frac{\partial(\rho u)}{\partial x} + \frac{\partial(\rho v)}{\partial y} + \frac{\partial(\rho w)}{\partial z} = 0$$

where ρ is the density of the fluid at time t and (u, v, w) are the Cartesian components of the vector velocity V at the space point (x, y, z).

continuous flow calorimeter A type of calorimeter in which heat is supplied at a constant rate to fluid flowing at a constant rate. A steady state is eventually reached when all temperatures remain constant with time so that small temperature differences may be accurately determined without any error due to lag of the thermometers. The heat capacity of the calorimeter does not enter into the thermal equation and since all temperatures are steady the external loss of heat by radiation or other means is more regular and certain.

continuous spectrum *See* spectrum.

continuum A continuous series of components or elements that together form a reference system. The three dimensions of space and the dimension of time together form a *four-dimensional continuum.

contrast 1. The ratio of the difference in intensity between the lightest and darkest areas in a subject or its reproduction to the sum of these intensities.

2. The rate at which the photographic density of a film changes with exposure, measured as the slope, γ, of the *characteristic curve of the film. A rapid change corresponds to a high-contrast film.

control electrode An electrode to which the input signal voltage is applied to produce changes in the currents of one or more of the other electrodes. In a bipolar *transistor with *common-emitter connection the control electrode is the base electrode, in a *field-effect transistor it is the gate electrode, in a *cathode-ray tube it is the *modular electrode, in a *thermionic valve it is the control grid.

control rod One of a number of rods that can be moved up or down along its axis into the *core of a *nuclear reactor to control the rate of the *chain reaction. The rods usually contain a neutron absorber such as cadmium or boron.

convection (of heat) The process of transfer of heat in a fluid by the movement of the fluid itself. There are two distinct types of convection:

(1) *Natural* (or *free*) *convection*, when the motion of the fluid is due solely to the presence of the hot body in it giving rise to temperature and hence density gradients, the fluid thus moving under the influence of gravity.

(2) *Forced convection*, in which a relative motion between the hot body and the fluid is maintained by some external agency (e.g. a draught), the relative velocity being such as to make the contribution of the gravity currents negligible.

A theoretical treatment of convection is best achieved by means of dimensional analysis using mass, length, time, and temperature as primary dimensions.

For dynamically similar bodies it is shown that for natural convection

$$\left(\frac{hl}{\lambda\theta}\right) = f_1\left(\frac{l^3 g\alpha\rho^2\theta}{\eta^2}\right)f_2\left(\frac{C\eta}{\lambda\rho}\right).$$

The expression contains three dimensionless groups, namely: $(hl/\lambda\theta)$, the *Nusselt number; $(l^3 g\alpha\rho^2\theta/\eta^2)$, the *Grashof or free convection number; $(C\eta/\lambda\rho)$, the *Prandtl number. The form of the functions f_1 and f_2 must be assumed to be dependent on the shapes of the bodies, etc. involved.

In the case of forced convection the expression takes the form

$$\left(\frac{hl}{\lambda\theta}\right) = F_1\left(\frac{lv\rho}{\eta}\right)F_2\left(\frac{C\eta}{\lambda\rho}\right),$$

introducing in addition to the Nusselt and Prandtl numbers, the *Reynolds number $(lv\rho/\eta)$.

In the case of a solid surface losing heat by natural convection in a fluid in contact with it, there is always a layer of fluid at rest relative to the solid, heat being transferred through this layer by thermal conduction into the bulk of the moving fluid. In free convection this film is stationary while for forced convection it is continuously being removed and renewed.

convection current 1. (heat) A stream of fluid, warmer or colder than the surrounding fluid and in motion because of the buoyancy forces arising from the consequent differences in density.

2. (electricity) A moving electrified body constitutes an electrical convection current, and this type of current can flow without potential difference or energy change, and produces no heat. It can nevertheless produce a magnetic effect.

convectron An instrument that gives an electrical indication of deviation from coordinate

the vertical. It is based on the fact that the convection cooling of a straight, fine wire is much greater when the wire is horizontal than when vertical.

conventional current The concept of a current that flows from positive to negative, i.e. in the opposite direction to the electron flow. This convention originated before the electronic nature of a current was understood.

converging lens A *lens that can bring a parallel light beam passing through it to a point. *Compare* diverging lens.

conversion *See* converter reactor.

conversion electron *See* internal conversion.

conversion factor *Syn.* conversion ratio. In nuclear physics, the ratio of the number of *fissile atoms produced from the fertile material in a *converter reactor to the number of fissile atoms of fuel destroyed in the process.

converter reactor A *nuclear reactor in which *fertile material is transformed by a nuclear reaction into *fissile material. This process is known as *conversion*. A converter reactor can also be used to produce electric power. *See also* conversion factor; fast breeder reactor.

convex Curving outwards. A *convex mirror* is diverging in action. A *convex lens* is thicker at the centre. Thin lenses are classed at *biconvex*, *planoconvex*, and *convex meniscus*. (*See* diagram under lens.) While thin convex lenses are converging in action, thicker lenses may be telescopic, diverging, or convergent according to thickness.

coolant A fluid used to reduce the temperature of a system by conducting away heat produced by the system.

In a *nuclear reactor the coolant transfers heat from the *core to the steam-raising plant or to an intermediate heat exchanger. In *gas-cooled reactors the coolant is usually carbon dioxide. In *boiling-water and *pressurized water reactors, water acts as both coolant and *moderator: in *heavy-water reactors, *heavy water fulfils this dual role. In *fast breeder reactors the need to transfer a large quantity of heat through a small surface area necessitates a liquid-metal coolant (e.g. sodium).

Coolidge tube An early type of X-ray tube.

cooling curve A temperature/time curve used for the determination of melting points (constant-temperature portion) or for applying a radiation or cooling correction.

cooling method The determination of the specific heat capacity of a liquid by comparing the time taken for the liquid and an equal volume of water to cool in identical vessels through the same range of temperature.

Cooper pair *See* superconductivity.

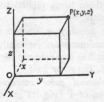

a Cartesian coordinates. (*x* is negative in this figure and *y* and *z* positive)

coordinate 1. One of the quantities used to define the position of a point relative to a *frame of reference. There are three main coordinate systems:

87

(1) *Cartesian coordinates.* Three mutually perpendicular lines OX, OY, OZ (Fig. *a*) are drawn through a point O known as the *origin*. These lines and O are fixed in the frame of reference, and in abstract work may themselves be the frame of reference. The position of a point P relative to these axes is given by the perpendicular distances of P from the three coordinate planes ZOY, XOZ, YOX; these distances x, y, z, are the coordinates of P. The axes are "right-handed" or "left-handed" according to whether turning OX towards OY (through the smaller angle) would drive a right-handed screw along the z-axis in a positive or negative direction respectively. Right-handed axes are most often used.

b Cylindrical polar coordinates

(2) *Cylindrical polar coordinates.* The position of a point P is specified by three coordinates: radial distance r, azimuthal angle θ, and axial distance z (Fig. *b*), these being related to the Cartesian system by

$$x = r\cos\theta, \ y = r\sin\theta, \ z \equiv z$$

These coordinates are especially useful if the system has some degree of symmetry about OZ, the *polar axis*.

(3) *Spherical polar coordinates.* The coordinates of P (Fig. *c*) are the radius r, the angle of colatitude (or azimuthal angle) θ, and the angle of longitude ϕ. These are related to the Cartesian system by

$$x = r\sin\theta\cos\phi, \ y = r\sin\theta \ \sin\phi, \ z = r\cos\theta$$

This system is useful if the system has some symmetry about the point O.

In two-dimensional problems only two coordinates need to be specified; these are (x,y) in Cartesian, or (r,θ) in plane polar coordinates. r is often called the *radius vector* or *radius* in spherical and plane polar coordinates.

Curvilinear coordinates are any sets of parameters that define the position of a point as the intersection of curves or of curved surfaces. Thus, (2) and (3) are special cases. Latitude and longitude are curvilinear coordinates for the surface of the earth.

2. Generalized coordinates. *See* degrees of freedom.

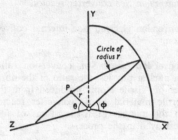

c Spherical polar coordinates

coordination lattice A crystal lattice in which each ion bears the same relation to the neighbouring ions in all directions, so that the identity of the molecules becomes ambiguous.

copper loss *Syn.* I^2R loss. The power loss in watts due to the flow of electric current in the windings of an electrical machine or transformer. It is equal to the product of the square of the current and the resistance of the winding.

Corbino effect If a current is passed from the centre to the circumference of a metal disc, the surface of which is normal to a magnetic field, a current will flow round the circumference.

core 1. The ferromagnetic portion of the magnetic circuit of an electromagnetic device. A simple *ferrite core* is a solid piece of ferromagnetic material in the shape of a ring, cylinder, etc. A *laminated core* is composed of *laminations of ferromagnetic material. A *wound core* is constructed from strips of ferromagnetic material wound spirally in layers.

2. The central part of a *nuclear reactor in which the *chain reaction takes place. In a thermal reactor it includes the fuel assembly (*see* fuel element) and the *moderator, but not the *reflector.

3. The central iron-rich portion of the earth, radius about 3500 km. The temperature and pressure are extremely high, possibly reaching 5000 kelvin and 400 gigapascals respectively. The inner core is solid, the outer core liquid. The source of the earth's magnetic field is though to lie in a complex dynamo action in the outer core. *See* geomagnetism.

4. A small ferrite ring formerly used in a computer *memory to store one *bit of information. The direction of magnetization of the core was sensed and read as 0 or 1.

core loss *Syn.* iron loss. The total power loss in the iron core of a magnetic circuit when subjected to cyclic changes of magnetization such as, for example, that which occurs in the core of a transformer. The loss is due to magnetic *hysteresis and *eddy currents. It is usually expressed in watts at a given frequency and value of the maximum flux density.

core-type transformer A *transformer in which the windings enclose the greater part of the laminated *core. The windings are formed around the *yoke*, which is built up from a stack of laminations (*see* diagram). Additional laminations form the *limbs* around each winding and complete the core. *Compare* shell-type transformer.

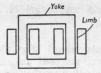

Core-type transformer

Coriolis force *See* force; Coriolis theorem.

Coriolis theorem The acceleration of a particle with respect to a Newtonian frame of reference, N, is the vector sum of (*a*) its acceleration with respect to some other frame of reference S, which is in motion relative to N, (*b*) the acceleration of N relative to S, and (*c*) the *Coliolis acceleration*, which equals twice the vector product of the angular velocity of S with respect to N and the linear velocity of the particle with respect to S.

The quantity (mass × Coriolis acceleration) has the dimensions of force and the Coriolis force has this magnitude but is oppositely directed from the Coriolis acceleration. It is an inertial force. *See* force, for a simple example.

corkscrew rule A rule for determining the direction of lines of magnetic force around a wire carrying a current. If a corkscrew is imagined to be turned in the manner necessary to drive it along in the direction of the current, then the lines of force are in the same sense as the rotation of the head of the corkscrew.

corona 1. An electric discharge appearing round a conductor when the potential gradient at the surface is raised above a critical value, so that a partial breakdown of the surrounding gas takes place.

2. The outermost region of the sun's atmosphere: it is visible as a faint halo during a solar *eclipse. It has an approximate temperature of one million degrees.

corpuscular theory The theory, which has been proposed in various forms from time to time, that light consists of particles. A luminous body was supposed to emit small elastic particles with the speed of light. They travelled in straight lines in isotropic media, were repelled on reflection, suffered change of direction by attraction on refraction. Since the corpuscular theory required a faster rate of travel in optically denser media, it was supplanted by the *wave theory*. This offers a readier explanation of interference, diffraction, and polarization but fails to explain the interaction of light with matter, the emission and absorption of light, photoelectricity, dispersion, etc.; these can only be explained by a quasi-corpuscular theory involving packets of energy – light quanta or photons. It thus appears that two models are required to explain the phenomenon of light, according to Bohr's principle of *complementarity. *See* quantum theory.

correcting plate *Syn.* corrector. A thin lens or lens system used to correct *spherical aberration in spherical mirrors and *coma in parabolic mirrors. *See* Schmidt corrector.

correlation 1. A reciprocal relationship between two variables x and y. If measurements indicate in a vague way that x and y are related, statistical methods may then be applied to the results to determine whether or not this apparent connection is significant. A quantity called the *correlation coefficient* r is evaluated; $r = 0$ corresponds to no connection whatever, and $r = 1$ to perfect correlation.

2. *See* coherence.

correspondence principle The principle, due to Bohr, that since the classical laws of physics are capable of describing the properties of macroscopic systems, the principles of *quantum mechanics, which are applicable to microscopic systems, must give the same results when applied to large systems. For example, the electrons in Bohr's theory of the *atom can only occupy certain orbits. It is found that for larger orbits the behaviour of the atom becomes more like the behaviour expected from classical mechanics.

cosmic abundance The relative proportion of each element found in the universe, measured in terms of mass or numbers of atoms. In terms of mass, there is approximately 73% hydrogen, 25% helium, 0.8% oxygen, 0.3% carbon, 0.1% neon and nitrogen, 0.07% silicon, 0.05% magnesium.

cosmic background radiation Diffuse radiation from space, detected at many wavebands throughout the electromagnetic spectrum. At radio, infrared, X-ray, and possibly γ-ray wavelengths, it is thought to be the cumulative contribution of many unresolved and individually weak sources. The most important form, however, is the *microwave background radiation*. This background peaks at a wavelength of about 1 mm and is *black-body radiation characteristic of a temperature of 2.9 K. It is very nearly equal in intensity from all directions of the sky, i.e. it is very nearly isotropic. It is now assumed to be the remnant of the radiation content of the very early hot phase of the universe. *See* big-bang theory.

cosmic rays Highly energetic particles that move rapidly through space and continuously bombard the earth's atmosphere from all directions. Cosmic rays consist mainly of nuclei of the most abundant elements in the universe, primarily protons. Also present are a small number of electrons, positrons, antiprotons, neutrinos, and gamma-ray photons. These particles are known collectively as *primary cosmic rays*, and have a wide range of energies, from about 10^8 to over 10^{20} electronvolts.

The primary particles collide with oxygen and nitrogen nuclei in the atmosphere and these events, together with subsequent decays and interactions of the resulting particles, lead to large numbers of *elementary particles and photons constituting *secondary cosmic rays*. A large number of particles can be formed from one primary particle and this is known as an *air shower*. The initial products are principally charged and neutral *pions. Muons and neutrinos are formed from the subsequent decay of charged pions:

$$\pi^+ \rightarrow \mu^+ + \nu, \; \pi^- \rightarrow \mu^- + \bar{\nu}$$

Muons do not interact strongly with matter and a large proportion are detected at the earth's surface and even in deep mines. A neutral pion (π^0) decays into two very energetic gamma-ray photons with energies of about 70 MeV. These each can produce one electron and one positron in passing near the nucleus of an atom (*see* pair production) . Each particle then loses energy and emits *bremsstrahlung radiation – more photons – which produce yet more electrons and positrons, and so on. Thus the original neutral pion leads to a large number of electrons and positrons, called a *cascade* or *cascade shower*.

Occasionally a single particle with an energy greater than 10^{15} eV may enter the atmosphere and lead to a large number of secondary cosmic rays, extending over a large area. This is called an *extensive air shower* or *Auger shower*.

The flux of secondary particles at sea level is very low, being greatest at lower energies – several thousand particles per square metre per second. The flux varies with latitude because charged particles are affected by the earth's magnetic field, and is a minimum at the equator. Particles must have a minimum energy before they can overcome the earth's field and enter the atmosphere. There is also an *east-west effect* in that more particles approach from the west than the east. The effect shows that cosmic rays have a positive charge.

The origin of cosmic rays is still uncertain. It is thought that almost all cosmic rays with energies less than 10^{18} eV are generated by sources within our Galaxy: *supernova explosions may produce medium- and low-energy particles, solar flares produce only very low-energy particles. It is also thought that these particles are confined within the Galaxy for millions of years by the weak magnetic field of the Galaxy, their directions of travel becoming almost uniformly scattered.

cosmic string A one-dimensional flaw in space-time postulated in certain *grand unified theories of particle physics. Cosmic strings may also have important consequences in cosmology.

cosmology The branch of *astronomy concerned with the evolution, general structure, and nature of the universe as a whole. *See also* big-bang theory.

Cotton–Mouton effect Some isotropic transparent solids and liquids show slight double refraction towards light when in a strong magnetic field. *See* Kerr effects; magnetic effects.

coudé system *See* telescope.

coulomb Symbol: C. The *SI unit of electric *charge, defined as the charge transported in one second by an electric current of one ampere.

Coulomb field The *electric field around a point charge.

Coulomb force A force of attraction or repulsion resulting from the interaction of the *electric fields surrounding two charged particles. The magnitude of the force is inversely proportional to the square of the distance between the particles.

coulombmeter An instrument in which the electrolytic action of a current is used for measurement of the quantity of electricity passing through a circuit.

Coulomb scattering The *scattering of charged particles, such as alpha particles, by nuclei as a result of the electrostatic forces between them. If the incident beam contains one alpha particle per unit area, then the number of particles, w, per unit solid angle that suffer a deflection ϕ is given by:

$$w(\varphi) = \left(\frac{Z_1 Z_2 e^2 m}{4\pi\varepsilon_0 p^2} \right)^2 \frac{1}{\sin^4(\varphi/2)}$$

where $Z_1 e$ and $Z_2 e$ are the charges of the scattered and scattering particles and m and p are the mass and momentum of the scattered particle.

Coulomb's law The mutual force F exerted by one electrostatic point charge Q_1 on another Q_2 is proportional to the product of the charges divided by the square of their separation d:
$$F = Q_1 Q_2 / 4\pi\varepsilon d^2,$$
where ε is the absolute *permittivity of the medium.

Coulomb's theorem The intensity E of an electric field near a surface possessing a surface density of charge σ is given by:
$$E = \sigma/\varepsilon,$$
where ε is the absolute *permittivity of the medium.

counter 1. Any device for detecting and counting individual particles and photons. The term is used for the detector and for the instrument itself. Most detectors work by multiplication of the number of ions or electrons formed by a single particle or photon; each ionizing event leads to a pulse of current or voltage and these are electronically counted. (*See* Geiger counter, crystal counter, proportional counter, semiconductor counter, and scintillation counter.)
2. Any electronic circuit that records and counts pulses of current or voltage. *See also* counter/frequency meter.

counter/frequency meter An instrument containing a frequency standard, usually a piezoelectric oscillator, that can be used as a counter or frequency meter by counting the number of events, or cycles, in a specified time. It may also be used to measure the time between events by counting the number of standard pulses occurring during a given number of events or cycles.

couple A system composed of, or equivalent to, two equal and antiparallel forces. The *moment is equal to the product of either force by the perpendicular distance between them and is the same about any axis perpendicular to the plane of the forces. It is an axial *vector.

coupled systems Two or more mechanical vibrating systems connected so that they react on one another. In such systems there is a transfer of energy from one system to another involving a change in the natural frequency of the individual systems. There are conditions, particularly at *resonance, when the energy drain from one system to another is sufficiently great for one system to be unable to maintain maximum amplitude.

Most musical instruments may be considered to be coupled systems, some elements being sharply tuned, e.g. the strings of a piano, and some having a very broad resonance curve, e.g. the soundboard of the piano.

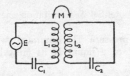

a Mutual-inductance coupling

coupling 1. Of two oscillating electric circuits. The means by which the circuits interact so that energy is transferred from one to the other. In Fig. *a*, the circuits are coupled by mutual inductance between their individual inductances. This is *mutual-inductance coupling*. The coupling is sometimes made by means of an impedance that is common to both circuits, examples of which are shown in Figs. *b* and *c*. This is *common-impedance coupling*. The *coupling coefficient*, *K*, may be defined by:

$$K = X_m / \sqrt{X_1 X_2},$$

where X_m is the reactance common to both circuits, X_1 and X_2 are respectively the total reactances of the two circuits, both of the same kind as X_m. Thus, in Fig. *a*:

$$K = \omega M / \sqrt{(\omega L_1 \times \omega L_2)}$$
$$= M / \sqrt{L_1 L_2}$$

In Fig. *b*:

$$K = \omega L_m / \sqrt{[\omega(L_1 + L_m) \times \omega(L_2 + L_m)]}$$
$$= L_m / \sqrt{[(L_1 + L_m)(L_2 + L_m)]}$$

In Fig. *c*:

$$K = \sqrt{\{C_1 C_2 / [(C_1 + C_m)(C_2 + C_m)]\}}$$

Sometimes the coupling is mixed, e.g. the types shown in Figs. *a* and *c* are applied simultaneously.

2. An interaction between different properties of a system or an interaction between two or more systems. There are two extreme types of coupling for atomic or nuclear particles:

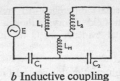

b Inductive coupling

In *Russell–Saunders coupling* (or *L–S coupling*), the resultant, *L*, of the *orbital angular momentum of all particles interacts with the resultant, *S*, of the *spin of all particles. In *j-j coupling*, the total angular momenta (orbital + spin) of individual particles interact with each other.

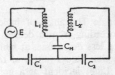

c Capacitive coupling

coupling coefficient 1. *See* coupling.
2. The numerical assessment between 0 and 1 characterizing the degree of coupling between two mechanically vibrating systems or electrical circuits. Maximum coupling is 1, no coupling 0. *See* coupled systems.

C-parity *See* charge conjugation parity.

CP invariance The simultaneous conservation of *charge conjugation (C) and *parity (P). *CP violation* occurs in the weak interaction of kaon decay to an extent of about one part per thousand. *See also* CPT theorem.

CPT theorem The theorem that the simultaneous operation of *charge conjugation (C), *parity (P), and time reversal (T) is a fundamental symmetry of relativistic *quantum field theory. If *C*, *P*, or *T* are violated singly or in pairs the principles of relativistic quantum field theory are not affected; however, violation of *CPT invariance* would make a

93

fundamental difference to relativistic quantum field theory, although no experimental evidence exists for its violation.

CPU Abbreviation for central processing unit.

cradle guard *See* guard wires.

creep The slow permanent deformation of a crystal or other specimen under sustained stresses.

critical *See* chain reaction.

critical angle The angle of incidence of light, proceeding from a denser medium towards a less dense one, at which grazing refraction occurs (angle of refraction = 90°). Light incident at a greater angle suffers total internal reflection. The critical angle, C, is given by $\sin C = n'/n$ in which $n > n'$, where n is the refractive index of one medium and n' of the other.

critical angle refractometers *Refractometers in which grazing incidence (*see* critical angle) is arranged between a medium whose refractive index is required and another of known index; the position of the boundary of light transmitted enables the refractive index to be calculated or to be observed directly on a scale.

critical constants *See* critical pressure; critical temperature; critical volume.

critical damping *See* damped.

critical isothermal The isothermal curve relating the pressure and volume of a gas at its critical temperature.

critical mass The minimum mass of a *fissile material that will sustain a

*chain reaction. *See also* nuclear weapons.

critical point *See* critical state.

critical potential *See* excitation energy.

critical pressure The saturated *vapour pressure of a liquid at its critical temperature.

critical reaction *See* chain reaction.

critical state The state of a substance when it is at its *critical temperature, pressure, and volume. Under these conditions the density of the liquid is the same as that of the vapour. The point corresponding to this state on an isotherm, is the *critical point*.

critical temperature The temperature above which a gas cannot be liquefied by increase of pressure. *See* equations of state.

critical velocity The velocity of fluid flow at which the motion changes from *laminar to turbulent flow.

critical volume The volume of a certain mass of substance measured at the critical pressure and temperature.

Crookes dark space *See* gas-discharge tube.

Crookes radiometer An instrument used to detect the presence of heat radiation. Four vertical vanes V, each blackened on one side, are free to rotate about a vertical axis in an evacuated glass vessel. When radiation falls on the blackened face of a vane it is absorbed and the temperature of the black face rises above that of the clear face. Molecules striking the blackened face carry away on an average, more momentum than those rebounding from the clear face, and thus the vane tends to rotate in such a direction that the black face continually recedes from the source of radiation. The gas pressure must be low enough to prevent a large number of intermolecular collisions, which would quickly equalize the velocities.

cross coupling *See* decoupling.

crossed cylinder A thin lens with cylindrical surfaces whose axes are crossed obliquely or at right angles. More particularly, a weak lens that has the effect of equal concave and convex cylinders crossed at right angles.

crossed lens A form of spherical lens that shows minimum spherical *aberration in parallel light. For glass of refractive index 1.5, this occurs in the biconvex form in which the second surface has about six times the radius of curvature of the first.

crossed polarizers *See* Malus's law.

crossover network A type of filter circuit designed to pass frequencies above a specified value through one path, and frequencies below that value through another. The value of specified frequency is the *crossover frequency*, and the circuit is so designed that at that frequency the output of the two channels is equal. Such a network is widely used with a loudspeaker to separate the bass and treble components.

cross section Symbol: σ. A measure of the probability of a particular collision process, stated as the effective area particles present to incident particles for that process. For example, if a beam of neutrons is passed through matter, one possible reaction is neutron *capture in which a nucleus retains the neutron on collision. It is supposed that capture occurs when the neutron is some minimum distance, d, from the centre of the nucleus. For capture, the nuclei appear to present an effective cross-sectional area (called the *capture cross section*) σd^2 to the incident neutrons. σ is not the physical cross-sectional area of the nucleus but the effective cross section for neutron capture. It is sometimes called the *activation cross section*. Its value depends on the energy of the incident neutrons as well as the nuclei considered. In particular, when the kinetic energy in centre of mass coordinates equals the energy difference between the *ground state of the nucleus and some higher state, the cross section is high and is called the *resonance cross section*.

The use of cross sections is also applied to other nuclear reactions, as well as to interactions between atoms, electrons, ions, etc. Cross section has units of m^2 and is often measured in *barns.

crosstalk Interference in the form of an unwanted signal in part of a circuit due to the presence of a signal in an adjacent circuit. It is a very common type of interference occurring in telephone, radio, and many data systems.

CRT Abbreviation for *cathode-ray tube.

cryogenics The study of the production and effects of very low temperatures. A *cryogen* is a refrigerant used for obtaining very low temperatures.

cryometer A thermometer designed for the measurement of very low temperatures.

cryostat A vessel that can be maintained at a specified low temperature; a low-temperature thermostat.

cryotron A type of switch that operates at very low temperatures and depends on *superconductivity. One form consists of a wire surrounded by a coil in a liquid-helium bath. Both the wire and the coil are superconducting and a low voltage can produce a current in the wire. If a current is also passed through the coil, its magnetic field alters the superconducting properties of the wire and switches off the current, thus the presence or absence of a current in the coil determines the ability of the wire to conduct. Cryotrons can be made very small and have low current requirements.

crystal 1. A three-dimensionally periodic arrangement of atoms in solids. Partial crystallinity can exist in two or one dimensions. *See* crystal structure; crystal systems; crystal texture.
2. In electronics, an element specially cut from a *piezoelectric crystal, such as quartz or barium titanate.

crystal analysis *See* X-ray crystallography.

crystal base The entire content of the *unit cell, whether considered as a symmetrical arrangement of atoms, or as a symmetrical distribution of electron density.

crystal class Crystals that are brought to self-coincidence by the operations of a *point group.

crystal clock *See* clocks.

crystal-controlled oscillator A type of *piezoelectric oscillator that can be designed to have a very high degree of frequency stability. *See* clocks.

crystal counter A device for detecting and counting subatomic particles that depends on their ability to increase the conductivity of a crystal. If a potential difference is applied across a crystal that is struck by a particle or photon,

the electron-ion pairs produced by the impact cause a transient increase in its conductivity. The pulses of current resulting from successive impacts are electronically counted. The operation of a crystal counter is analogous to that of a *Geiger counter in which the radiation induces conductivity in a gas.

crystal cut Crystal sections (usually thin plates or bars) may be cut in particular crystallographic directions, e.g. for use as *piezoelectric oscillators; these directions are specified as *cuts* and designated by letters such as X-cuts, AT cut, and so on.

crystal detector A *detector that depends for its action upon the rectifying properties of certain crystals when placed in contact with one another, or of a crystal in contact with a metal. It was used extensively in the earliest types of radio receiver, and more recently as a detector and mixer of microwaves.

crystal diffraction The constructive and destructive *interference of waves scattered by the periodic arrangement of electrons, nuclei, or field of force in a crystal, to give a pattern of discrete spectra.

crystal dynamics The study of the movements of atoms or of the variations of electron density in crystals.

crystal filter A filter that uses one or more piezoelectric crystals to provide its resonant or antiresonant circuits.

crystal grating The symmetrical arrangement of the atoms in a crystal in a series of parallel planes that enables the crystal to act as a three-dimensional *diffraction grating for X-rays. This is the basis of the method of *X-ray crystallography.

crystallography The science of the forms, properties, and structure of crystals, that is, of solids in which physical properties may vary regularly with direction, being the same along all parallel directions. *See also* X-ray crystallography.

crystal microphone *Syn.* piezoelectric microphone. A device making use of the *piezoelectric effect to convert the mechanical strain produced by sound pressure into electrical signals. *Piezoelectric crystals cut in a certain direction show electrical charges on their faces when pressure is applied. Thus, the varying pressures caused by a sound wave striking the crystal face will produce alternating e.m.f.s across the crystal.

crystal oscillator *See* piezoelectric oscillator.

crystal structure The specification both of the geometric framework (*see* unit cell; space group) to which the crystal may be referred, and of the arrangement of atoms or electron-density distribution relative to that framework.

crystal systems A classification of crystals based on their *unit cells. The 14 *Bravais lattices, and 32 *point groups, can be referred to seven crystal systems of three axes: (1) *triclinic*, in which the axes need be neither equal nor mutually perpendicular; (2) *monoclinic*, in which the axes need not be equal, but one is perpendicular to the other two; (3) *orthorhombic*, in which the axes need not be equal, but they are all mutually perpendicular; (4) *tetragonal*, in which two axes must be equal, and all are mutually perpendicular; (5) *rhombohedral*, in which all axes are equal, and equally inclined to each other at an angle of less than 120°; (6) *hexagonal*, in which two axes are equal and inclined to each other at 120°, being both perpendicular to the third unique axis; (7) *cubic*, in which all axes are equal and are mutually perpendicular. Since a crystal built on a rhombohedral lattice can in fact be referred also to hexagonal axes, it is sometimes the practice to amalgamate (5) and (6).

crystal texture A crystal may be, in theory, an ideally periodic arrangement of atoms throughout the whole of its volume. In practice, however, such perfect crystals almost never occur. Even apparently single crystals consist of a conglomerate or mosaic of smaller crystallites that may be parallel (but with discontinuities at their mutual boundaries) or slightly disorientated. A massive crystalline specimen may consist of crystallites of larger or smaller grain size, partially or completely disorientated in one, two, or three directions. All these questions of crystal perfection, crystallite size, and orientation are covered by the word texture, and profoundly affect many crystalline properties.

cubic expansion coefficient *See* coefficient of expansion.

cubic system *See* crystal systems.

cumulatively compound wound *See* compound-wound machine.

curie Symbol: Ci. A former unit of *activity of a radioactive nuclide corresponding to 3.7×10^{10} disintegrations per second (approximately equal to the activity of 1 g of radium). It has been replaced by the *becquerel, where 1 Ci $= 3.7 \times 10^{10}$ Bq.

Curie constant The product of the magnetic *susceptibility per unit mass and the thermodynamic temperature; this quantity is approximately constant for many paramagnetic substances. *See* Curie's law.

Curie point *See* Curie temperature.

Curie's law The susceptibility (χ) of a paramagnetic substance is universely proportional to the *thermodynamic temperature* (T): $\chi = C/T$. The constant C is called the *Curie constant* and is characteristic of the material. This law is explained by assuming that each molecule has an independent magnetic *dipole moment and that the tendency of the applied field to align these molecules is opposed by the random motion due to the temperature.

Curie temperature *Syn.* Curie point. Symbol: θ_C or T_C. *See* Curie–Weiss law.

Curie–Weiss law A modification of *Curie's law, followed by many paramagnetic substances (*see* paramagnetism). It has the form:

$$\chi = C/(T - \theta).$$

The law shows that the susceptibility is proportional to the excess of temperature over a fixed temperature θ: θ is known as the *Weiss constant* and is a temperature characteristic of the material.

For ferromagnetic solids (*see* ferromagnetism) there is a change from ferromagnetic to paramagnetic behaviour above a particular temperature and the paramagnetic material then obeys the Curie–Weiss law above this temperature; this is the *Curie temperature* for the material. Below this temperature the law is not obeyed. Some paramagnetic substances, such as gadolinium, obey the Curie–Weiss law above the temperature θ_C and do not obey it below, but are not ferromagnetic below this temperature. The value θ in the Curie–Weiss law can be thought of as a correction to Curie's law reflecting the extent to which the magnetic dipoles interact with each other. In materials exhibiting *anti-

ferromagnetism the temperature θ corresponds to the *Néel temperature*.

curl *Syn.* rotation. A vector quantity associated with a vector field, F. It is the vector product (*see* vector) $\nabla \times F$, where ∇ is the differential operator *del. Thus:

$$\text{curl } F = \nabla \times F =$$
$$i \times \partial F/\partial x + j \times \partial F/\partial y +$$
$$k \times \partial F/\partial z,$$

where i, j, k are *unit vectors along the x-, y-, and z-axes respectively.

The magnitude of curl F at a point is the maximum value of the line integral of the vector, per unit area, taken round the bounding edge of an infinitesimally small area at the point when the small area is oriented to produce the maximum line integral. The direction of the vector is that of the normal to the area and its sense is such that when looking along the vector one sees the line integral taken clockwise round the area boundary. (E.g. in the motion of liquids, if the curl of the velocity is not zero the particles are in rotation.)

A field that has a curl is generally described as *rotational*, *vortical*, or *circuital*; if the curl is zero everywhere, the field is irrotational or nonvortical. *Compare* gradient; divergence.

current Symbol: I. A flow of electric *charge in a substance – solid, liquid, or gas. The charge carriers may be electrons, *holes, or ions. The magnitude of a current is given by the amount of charge flowing in unit time; it is measured in *amperes. The direction is by convention from a point of higher potential to one of lower potential.

A *conduction current* is a current flowing in a conductor, the electricity being conveyed by the motion of electrons or ions through the material of the conductor. A conduction current of 1 ampere is equivalent to the flow of about 10^{18} electrons per second. A *displacement current is due to a change in the electric flux density in a dielectric; e.g. the current through a capacitor when connected in series with an alternating p.d. A current flowing always in the same direction in a circuit is called *unidirectional*; it may or may not be pulsating. A *direct current* flows always in the same direction, without sensible pulsations. *Compare* alternating current.

current balance An instrument for accurately determining a given current or, more fundamentally, the size of the *ampere by measuring the force between current-carrying conductors.

current density Symbol: *j* or *J*. The ratio of the current to the cross-sectional area of the current-carrying medium. The medium may be a conductor, or a beam of charged particles. The ratio may be specified either as a *mean current density* or as density at a point.

current transformer *Syn.* series transformer. An instrument transformer in which the primary winding is connected in series with the main circuit and the secondary winding is closed through an instrument (e.g. ammeter) or other device. The ratio of primary and secondary currents is approximately the inverse of the primary to secondary turns-ratio of the transformer. Current transformers are extensively used to extend the range of a.c. instruments, to isolate instruments from high-voltage circuits and to operate protective relays in a.c. power systems.

curvature For a spherical lens or mirror or a wavefront, the radius of the sphere on which such surfaces lie is the *radius of curvature*, *r*. The centre of the sphere is called the *centre of curvature*. The reciprocal of the radius of curvature is the curvature of the surface, *R*. If *r* is in metres then *R* is in *dioptres. A plane

wavefront (zero curvature) incident on a lens or mirror will be changed into a spherical wavefront having a curvature of $1/f$ impressed on it, where f is the focal length of the lens or mirror. This ratio gives the *power of the lens or mirror.

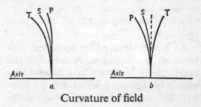

Curvature of field

curvature of field *Syn.* curvature of image. In general, a plane object at right angles to the axis of an optical system will not give rise to a plane image. Instead, in the absence of *astigmatism, the image would lie on a paraboloid surface known as the *Petzval surface*. This aberration is called curvature of field.

The effects of astigmatism will be superimposed on those of curvature of field with the result that the tangential and sagittal focal planes T and S are displaced from the Petzval surface P. Two possible cases are indicated in the diagram.

In the case shown in Fig. *b*, the effects of astigmatism have been used to offset the curvature of the Petzval surface, and a flat screen placed in the position indicated by the dotted line would show a reasonably focused image. Variations of this sort are achieved in practice using optical systems with more than one lens, altering the spacing of the various component lenses, and adding suitably positioned stops.

curvilinear coordinates *See* coordinate.

cut-off frequency Of a passive electrical or acoustical network as a nondissipative system – both characterized by possessing no internal source of energy. The frequency, reached by varying the frequency of the applied voltage or driving force, at which the attenuation quickly changes from a small value to a much higher value. For example, a typical loudspeaker with a cut-off frequency of 6000 hertz will fail to respond appreciably to frequencies of this value and above.

An active electrical or acoustical network as a dissipative system has the same cut-off frequency as a passive network with the same inductance (or inertia) and capacitance (or elastic) components.

The term is also applied to the limiting frequencies of acoustic and electric filters. *See* filters.

cybernetics The theory of control systems, concerned with the common characteristics of diverse systems including computers, automated factory processes, and the nervous systems of living organisms (such as man). It enables a comparison to be made of problems of control, communication, and feedback of information in biological and engineering systems. The theory has been applied to the design of automatic control mechanisms.

cybotaxis The special arrangement of molecules in a *liquid crystal.

cycle 1. An orderly set of changes regularly repeated.

2. One complete set of changes in the value of a periodic function; one vibration, one oscillation, etc.

cyclone A large area of atmospheric disturbance with an inward spiral rotation of air about a point of low pressure. In the northern hemisphere its rotation is anticlockwise. *Compare* anticyclone.

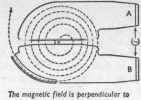

The magnetic field is perpendicular to the plane of the paper

Cyclotron

cyclotron An *accelerator in which charged particles describe a spiral path of many turns at right angles to a constant magnetic field, and are given an acceleration, always in the same sense, from an alternating electric field each time they cross the gap between the two conductors, A and B (*see* diagram). The cyclotron depends on the fact that the time t taken by a particle of mass m and charge e to describe a semicircle in a plane at right angles to a uniform magnetic flux density B is $\pi m / Be$ and is thus independent of the velocity. The radius r of the semicircle described by the particle increases as the velocity v increases: $v = Ber/m$. Thus large pole pieces and very powerful magnets are required. As the beam approaches the circumference of the conductors, an auxiliary electric field deflects the particles from the circular path and they leave through a thin window. The energy that can be obtained from such a device is limited by the relativistic increase in mass of the particle as the velocity increases, the maximum being about 25 MeV.

cylindrical lens A lens with one face a portion of the curved surface of a cylinder. Thin lenses for correcting astigmatism of the eye require one surface to be cylindrical and the other spherical – a *spherocylindrical lens*. (*See also* toric lens.) Reflecting surfaces may also be a portion of the curved surface of a cylinder.

cylindrical polar coordinates See coordinate.

cylindrical winding A type of winding used in *transformers. The coil is helically wound and may be single-layer or multilayer. Its axial length is usually several times its diameter.

D

d'Alembert's (or Alembert's) principle See force.

DALR Abbreviation for dry adiabatic *lapse rate.

dalton See atomic mass unit.

Dalton's law of partial pressures The total pressure of a mixture of gases is equal to the sum of the *partial pressures that would be exerted by the gases if they were present separately in the container. See ideal gas.

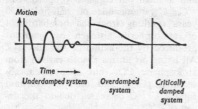

Underdamped system   Overdamped system   Critically damped system

damped 1. Of a free oscillation. Progressively dying away due to an expenditure of energy by friction, viscosity, or other means. The word *damping* is used for both the cause of the energy loss and the progressive decrease in the ampli-

tude of oscillation. If the damping is such that the system just fails to oscillate, the system is *critically damped* and for greater or lesser degrees of damping than this, it is *overdamped* or *underdamped* respectively. In electrical indicating instruments, three systems of damping are in common use: (1) air friction, (2) fluid friction (oil), and (3) eddy-current, and for such instruments the damping is usually designed to be slightly less than critical. (1) and (2) rely on viscous forces, and in all three systems, resistance is nearly proportional to velocity of motion. See damper; deadbeat.

2. Of a vibration. Decreasing in amplitude with time due to the resistance of the medium to the vibration. For small amplitudes of vibration the resistive force is approximately proportional to the velocity.

The force equation for a damped simple harmonic motion is:
$$m\ddot{x} = -kx - \mu\dot{x},$$
where x is the displacement, and m, k, and μ are inertia, elastic, and resistive terms respectively. The solution of this equation is:
$$x = a\,e^{-\alpha t}\sin(\omega t - \delta),$$
where the *decay* or *damping factor* $\alpha = \mu/2m$, and the frequency $n = \omega/2\pi$, where

$$\omega = \sqrt{\left(\frac{k}{m} - \frac{\mu^2}{4m^2}\right)}.$$

damper A device incorporated in an electrical or other indicating instrument to provide the necessary damping (*see* damped).

damping factor 1. *Syn.* decrement. The ratio of the amplitude of any one of a series of damped oscillations to that of the following one.

2. *Syn.* decay factor. *See* damped.

101

Daniell cell A primary *cell, now used only for demonstration purposes. The positive pole is of copper immersed in a saturated solution of copper(II) sulphate and the negative an amalgamated zinc rod in a solution of dilute sulphuric acid. The two solutions are separated by a porous partition. The cell has a fairly constant e.m.f. of about 1.08 volts and an internal resistance of a few ohms. This form of cell must be dismantled when not in use.

dark-field illumination *Syn.* dark-ground illumination. *See* microscope.

dark matter Matter that could comprise a large percentage of the mass of the universe but is undetectable except by its gravitational effects. The concept was first proposed in connection with large clusters of galaxies, when it was found that the mass required to keep the galaxies gravitationally bound was on average ten times greater than the mass actually observed in the cluster. The nature of the dark matter is unknown, but it is thought that it cannot all consist of normal baryonic matter, i.e. protons and neutrons. *See* axion.

dark space The comparatively nonluminous portion of an electrical discharge through a gas. *See* gas-discharge tube.

d'Arsonval galvanometer An instrument in which the current to be measured is passed through a narrow rectangular coil, suspended so as to be free to turn about a vertical axis in the magnetic field between the poles of a permanent horseshoe magnet. A cylinder of soft iron is usually supported inside the coil. The instrument combines reasonably high sensitivity (currents of the order nanoamp may be detected) with a comparatively low resistance and high degree of damping. The name is often applied to any form of moving-coil galvanometer.

dash-pot A mechanical device that prevents any sudden or oscillatory motion of a moving part of any piece of apparatus. It depends for its action upon the viscous resistance of air or of a liquid (e.g. oil), and in a simple form consists of a piston loosely fitted in a cylinder filled with oil.

dating Any of several methods for determining the age of archaeological and fossil remains, rocks, etc., by measuring some property of the organic or inorganic material that changes with time. This property may be dependent on some aspect of nuclear decay, such as the decay of radiocarbon or the uranium series, *thermoluminescence, or *electron spin resonance. These aspects are studied by *radiometric dating* techniques. The property may alternatively be dependent on a chemical change with a time-dependent rate constant, such as amino acid racemization.

(1) *Radiocarbon dating* (or *carbon-14 dating*) is a method for determining the age of objects up to 35 000 years old containing matter that was once living, such as wood. Atmospheric carbon consists mainly of the stable isotope ^{12}C and a small but constant proportion of ^{14}C, a radionuclide of half-life 5730 years resulting from the bombardment of atmospheric nitrogen by neutrons produced by the action of *cosmic rays. All living organisms absorb carbon from atmospheric CO_2, but after death, absorption ceases and the once-constant ratio $^{14}C/^{12}C$ decreases due to the decay of ^{14}C:

$$^{14}C \rightarrow {}^{14}N + e + \bar{\nu}.$$

The ^{14}C concentration in a sample, found by using a sensitive *counter of particles, gives an estimate of the time elapsed since death of the living organism.

The age of geological specimens, which can be many millions of years old, is determined from the proportion of a natural radionuclide (with a very long half-life) and its daughter nuclide contained in a sample of rock or mineral.

(2) *Potassium-argon dating.* Potassium, in combination with other elements, occurs widely in nature especially in rocks and soil. Natural potassium contains 0.001 18% of the radioisotope ^{40}K, which decays, with a half-life of 1.28 × 10^9 years, partly to the stable isotope of argon, ^{40}Ar. Determination of the ratio of $^{40}K/^{40}Ar$ gives an estimation of ages up to about 10^7 years.

(3) *Rubidium-strontium dating.* Natural rubidium, a much rarer element than potassium, contains 27.85% of the radioisotope ^{87}Rb, which decays with a half-life of 4.8 × 10^{10} years into the stable isotope of strontium, ^{87}Sr. Determination of the ratio $^{87}Rb/^{87}Sr$ gives an estimate of age of up to four thousand million years.

(4) *Uranium-lead* and *thorium-lead dating.* The long-lived nuclides thorium-232, uranium-238, and uranium-235 decay through radioactive series to lead-208, lead-206, and lead-207 respectively. Determination of the abundances of uranium and/or thorium and the isotopic composition of lead in rocks permits dating up to about 4 × 10^9 years.

daughter product Any nuclide that originates from a given *nuclide, the *parent*, by radioactive *decay.

Davisson–Germer experiment (1927), The first experiment to demonstrate *electron diffraction, and hence the wavelike nature of particles. A narrow pencil of electrons from a hot filament cathode was projected *in vacuo* onto a nickel crystal. The experiment showed the existence of a definite diffracted beam at one particular angle, which depended on the velocity of the electrons. Assuming this to be the Bragg angle (*see* Bragg's law) the wavelength of the electrons was calculated and found to be in agreement with the *de Broglie equation.

Dawes' rule *See* resolving power.

day A unit of time equal to 24 hours, i.e. 86 400 seconds. *See* time *for* sidereal day.

d.c. Abbreviation for direct current.

deadbeat Indicating an instrument that is *damped so that any oscillating motion of its moving parts dies away very rapidly.

dead room *Syn.* anechoic chamber. A room that absorbs practically all the incident sound. For a room to be completely dead it must be made soundproof by insulating its floor, walls, and ceiling from the rest of the building and by using heavy soundproof doors. All the surfaces, including the floor, are then covered with several centimetres of highly absorbent material such as rock wool. The possibility of the formation of standing waves is further reduced by using an asymmetrical room or by placing absorbent deflectors in suitable places in the room. Most modern dead rooms employ large numbers of inward-pointing pyramids covered with an absorbing material to minimize possible reflections of sound.

dead time In any electrical device, the time interval immediately following a stimulus during which it is insensitive to another stimulus.

de Broglie equation A particle of mass m moving with a velocity v will under suitable experimental conditions exhibit the characteristics of a wave of wavelength λ given by the equation $\lambda = h/mv$, where h is the Planck constant. The equation is the basis of *wave mechanics. *See* de Broglie waves.

de Broglie waves *Syn.* matter waves; phase waves. A set of waves that represent the behaviour, under appropriate conditions, of a particle (e.g. its diffraction by a crystal lattice). The wavelength is given by the *de Broglie equation. They are sometimes regarded as waves of probability, since the square of their amplitude at a given point represents the probability of finding the particle in unit volume at that point. These waves were predicted by de Broglie in 1924 and observed in 1927 in the *Davisson–Germer experiment.

debye A former unit of electric dipole moment equal to $3.335\ 64 \times 10^{-30}$ coulomb metre.

Debye length The maximum distance at which the *Coulomb field of charged particles in a *plasma can interact.

Debye–Scherrer ring The circular diffraction ring, concentric with the undeflected beam, formed when a narrow pencil of monochromatic X-radiation is passed through a mass of finely powdered crystal. Since the orientation of the powdered fragments is entirely random, the incident radiation must fall on some of them at the Bragg angle (*see* Bragg's law) for some one given set of the crystal planes, and hence a diffracted beam will be formed. Since everything is symmetrical about the axis of the incident pencil, the diffracted rays will lie on the surface of a cone. Each set of crystal planes will form its own diffracted cone, and on intercepting the emergent radiation by a photographic plate, the resulting negative will show a series of dark rings surrounding the central spot due to the undeflected

beam. The vertical angle of the cone is 4θ, where θ is the Bragg angle for the wavelength of the radiation employed, and the given set of planes. This pattern is the basis of the *Debye–Scherrer method* used extensively in *X-ray crystallography.

Debye–Sears effect An effect used for studying and measuring the speed of sound waves in a transparent liquid. If a *piezoelectric crystal is placed in the liquid and vibrated at a fixed frequency it sets up acoustic waves – alternate regions of compression and rarefaction in the liquid with nodes every half wavelength. The liquid is held in a parallel-sided glass cell through which a beam of light of known wavelength is passed. The regions of compression and rarefaction in the liquid act as a *diffraction grating with a grating interval equal to the wavelength of the acoustic waves. This wavelength can be measured from the position of the diffracted light beam and the speed of sound in the liquid can thus be obtained from the product of its frequency and wavelength.

Debye theory of specific heat capacities (1912) Debye applied the *quantum theory to the independent vibrations of a solid considered as a continuous elastic body with an atomic structure such that the frequencies stop abruptly at a maximum frequency (v_m). The molar heat capacity at constant volume is then given by the *Debye function*:

$$C_v = 9R\left(\frac{4}{x^3}\int_0^x \frac{\xi^3}{e^\zeta - 1}\,\mathrm{d}\xi - \frac{x}{e^x - 1}\right)$$

where $x = hv_m/kT$; $\zeta = hv/kT$, k is the Boltzmann constant and h is the Planck constant. The *Debye characteristic temperature* Θ_D, is defined as hv_m/k, so that C_v is a function of (Θ_D/T), which is in fair agreement with experimental results.

Debye T³ law At low temperatures the specific heat capacity is proportional to the cube of the thermodynamic temperature. At such temperatures the Debye function becomes

$$C_v = \tfrac{12}{5} \pi^4 R T^3 / \Theta_D{}^3$$

so that, since R and Θ_D are constants for the substance, $C_v \propto T^3$. *See* Debye theory of specific heat capacities.

deca- Symbol: da. A prefix meaning 10. For example one decameter (1 dam) = 10 metres.

decay 1. The transformation of a radioactive *nuclide, the parent, into its *daughter product by disintegration, resulting in the gradual decrease in the *activity of the parent. (*See also* alpha decay; beta decay; radioactivity.)
2. The gradual decline of brightness of an excited *phosphor.
3. *See* damped.

decay constant *Syn.* disintegration constant. Symbol: λ. The probability per unit time of the radioactive decay of an unstable nucleus. It is given by the formula $\lambda = -dN/dt.1/N$, where $-dN/dt$ is the *activity, A, of the nuclide and N is the number of undecayed nuclei present at time t. The exponential decrease with time of the activity of a radionuclide in which there are N_0 nuclei at time $t =_0 0$ is found from the formula $N = N_0 e^{-\lambda t}$. The time required for half the original number of nuclei to decay ($N = \tfrac{1}{2}N_0$) is the *half-life, $T_{1/2}$, given by $T_{1/2} = 0.693\,15/\lambda$. The reciprocal of the decay constant is the *mean life.

deci- Symbol: d. A prefix meaning 0.1. For example one decimeter (1 dm) = 0.1 metre.

decibel Symbol: dB. A unit used especially in communications and acoustics to measure *power level differences and *sound pressure levels. It is a more

practical unit than the *bel*, symbol: B, which is equal to 10 decibels.

The power level difference, L_p, is equal to n decibels when, for two powers, P_1 and P_2,

$$10 \log_{10}(P_2/P_1) = n.$$

Thus 1 dB represents an increase of P_2 over P_1 of 26%, 10 dB a 10 times increase, 50 dB a 10^5 times increase, and so on. In the case of sound power, P_1 is a reference power. One decibel is about the smallest change that the ear can detect. The decibel is not a unit of loudness. In the case of electric power, if P_1 is the input power of a network and P_2 the corresponding output power, then if n is positive, i.e. $P_2 > P_1$, there is a gain in power; if n is negative, i.e. $P_2 < P_1$, there is a power loss.

The sound pressure level, L_p, is equal to n decibels when for two pressure, p and p_0 (a reference pressure),

$$20 \log_{10}(p/p_0) = n.$$

The decibel is related to the *neper (symbol: Np) by the equation:

$$1 \text{ dB} = (\log_e 10)/20 \text{ Np} = 0.1151 \text{ Np}.$$

decimal balance *See* balance.

declination 1. (magnetic) The angle between the magnetic meridian and the geographical meridian at a particular point. Its value depends on the position of the point on the earth's surface, and at a given point changes slowly with time.

2. (astronomical) *See* celestial sphere.

declinometer An instrument for determining the magnetic *declination. It consists essentially of an arrangement by which the angle between the magnetic axis of a compass needle and the direction of some heavenly body can be read on a horizontal circular scale. An approximate value can be obtained with a prismatic compass: for accurate measurements the *Kew magnetometer is used.

decomposition voltage The maximum potential difference that can be applied to the electrodes of an electrolytic cell without giving rise to a permanent current through the cell.

decoupling The removal of any unwanted a.c. components from a circuit or circuit element. These unwanted components are sometimes caused by *coupling between circuits, particularly those with a common power supply (in which case it is known as *cross coupling*). Decoupling is usually achieved by using a series inductance or shunt capacitor.

decrement *See* damping factor.

decrement gauge *See* molecular gauge.

de-emphasis *See* pre-emphasis.

defect All crystalline solids consist of regular periodic arrangements of atoms or molecules. Departures from regularity are known as *defects* and they can be classified in two types.

Frenkel defect   *Schottky defect*

a Point defects in crystal lattice

(1) *Point defects* are defects involving one single atom or molecule. A *Frenkel defect* is a vacant lattice site with an associated interstitial atom – i.e. an atom that is not in a normal lattice position. Such atoms are called *interstitials* and the vacant lattice site is called a *vacancy* (Fig. *a*). The Frenkel defect is both the vacancy and the interstitial and it can be formed by movement of an atom from its normal lattice position. A *Schottky defect* is simply a vacant lattice

point. It requires energy to create a point defect by moving an atom from its position but, since the defects introduce disorder into a crystal, this is offset by an increase in configurational *entropy (i.e. entropy caused by disorder). Consequently, all crystals above absolute zero have a certain number of such defects. The number of Schottky defects in a crystal at equilibrium is:

$$n = N \, e^{-E/kT},$$

where N is the total number of lattice sites and E the energy required to produce the defect. The number of Frenkel defects is given by:

$$n = \sqrt{(NN' \, e^{-E/2kT})},$$

N' being the number of interstitial sites. The equilibrium number of point defects rises exponentially with temperature; typically, for a metal at about 700 °C, 1 in 10^5 sites are vacant. Point defects are responsible for *diffusion in solids. They can be produced in high concentrations by heating the solid to a high temperature and cooling it, by straining it, or by treatment with *ionizing radiation.

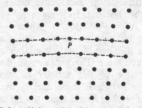

b Edge dislocation in crystal lattice

(2) *Line defects* are extended departures from regularity in crystals and they are often called *dislocations*. There are two basic types, *edge dislocations* and *screw dislocations*. An edge dislocation is shown in Fig. *b*. It corresponds to an extra plane of atoms introduced in one part of the crystal. The *dislocation line* extends into the crystal perpendicular to the plane of the paper at P. Screw dislocations are more difficult to visualize and illustrate. If a cylinder is

taken, as in Fig. *c*, and cut along a plane ABCD and displaced as shown, then a screw dislocation results. The atoms thus have a helical arrangement around the axis of the cylinder, which is the dislocation line. Many dislocations in crystals have features of both edge and screw dislocations. The dislocations in solids are responsible for plastic deformation above the *elastic limit. They are also formed by deformation.

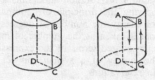

c Screw dislocation

defect conduction In a *semiconductor, conduction due to the presence of *holes in the *valence band. The presence of the holes is due to point *defects in the crystal lattice.

deflector coils, plates *See* cathode-ray tube.

deformation potential The electric potential caused by mechanical deformation of the crystal lattice of *semiconductors and conductors. *See* piezoelectric effect.

degaussing 1. Neutralization of the magnetization of a ship by surrounding it with a system of current-carrying cables that set up an exactly equal and opposite field.
2. In colour television, the use of a system of coils to neutralize the earth's magnetic field thus preventing the formation of colour fringes on the image.

degeneracy 1. A condition that arises when an atomic or molecular system with a number of possible quantized states has two or more distinct states of the same energy. *See* statistical weight.

2. The state of matter when it is at such a high density that all the electrons have been stripped from the atoms. The material consists of atomic nuclei and electrons in a closely packed and highly dense form. *See* degenerate gas.

degenerate *See* characteristic function; statistical weight.

degenerate gas A gas in which the concentration of particles is sufficiently high for the Maxwell–Boltzman distribution (*see* distribution of velocities) not to hold; the behaviour of the gas is then controlled by *quantum statistics.

The pressure in a degenerate gas consisting of *fermions is called the *degeneracy pressure*; this exceeds the thermal pressure because according to the *Pauli exclusion principle particles very close together must possess different momenta and according to the *uncertainty principle the difference in momentum is inversely proportional to the distance between them. Thus, in a high-density gas the relative momentum of the particles is high and unlike thermal pressure does not tend to zero as the temperature tends to absolute zero. *White dwarfs and *neutron stars are thought to be supported against collapsing gravitationally by the degeneracy pressure of their electrons and neutrons, respectively.

degenerate level An *energy level of a quantum mechanical system that corresponds to more than one quantum state.

degenerate semiconductor A *semiconductor with the *Fermi level located inside either the valence or conduction band (*see* energy bands). The material is essentially metallic in behaviour over a wide temperature range.

degradation 1. The decrease in the availability of energy for doing work, as a result of the increase of *entropy within a closed system. (*See* thermodynamics.)

2. The loss of energy of a beam of particles or an isolated particle passing through matter as a result of the interaction of the particles with the matter.

degree 1. A unit of temperature difference. The Celsius and Fahrenheit degrees were formerly defined as 1/100th and 1/180th respectively of the temperature difference between the ice and steam points, so that 1 °C = $\frac{9}{5}$ °F. The unit of *thermodynamic temperature, no longer called a degree, is the *kelvin.

2. (math.) The rank of an equation or expression as determined by the highest power of the unknown or variable quantity. The degree of a curve or surface is that of the equation expressing it.

3. *See* electric degree.

degree Celsius Symbol: °C. A unit used in expressing temperatures on the *Celsius scale. It is now an *SI unit, defined in terms of thermodynamic temperature, one degree Celsius being equal to one *kelvin. (*See also* degree.) It was formerly called the *degree centigrade*.

degrees of freedom 1. Of a mechanical system. The number of independent variables needed to describe the system's configuration; e.g. a system consisting of two particles connected by a rigid bar has 5 degrees of freedom since 5 coordinates (3 of the mass centre or of either particle, together with 2 angles) are needed to specify its state. The smallest number of coordinates needed to specify the state of the system is called its *generalized coordinates* and since these specify the state of the complete system, they also specify the state of any individual particle of the system. The generalized coordinates may be chosen in more than one way – as in the example quoted. The number of degrees of freedom depends only on the possibilities of motion of the various parts of the system and not on the actual motions. For a monatomic gas the number is 3. For a diatomic gas with rigid molecules it is 6, made up of 3 degrees of freedom of the centre of gravity to move in space, 2 degrees of freedom of the line joining the two atoms to change its direction in space and 1 for rotation about this axis. In applying the principle of *equipartition of energy the number of degrees of freedom is taken as the number of independent squared terms in the expression for the energy of a system, subject to certain reservations.

2. *Syn.* degrees of variance. In the *phase rule, the variable factors, such as temperature, pressure, and concentration, needed to define the condition of a system in equilibrium.

degrees of variance *See* degrees of freedom.

dekatron A type of cold-cathode *scaler with usually ten sets of electrodes that function in turn. When an impulse is received, a glow discharge is transferred from one set of electrodes to the next. The tubes may be used for switching or for visual display of counts in the decimal system.

del *Syn.* nabla. Symbol: ∇. The differential operator

$$i(\partial/\partial x) + j(\partial/\partial y) + k(\partial/\partial z),$$

where *i*, *j*, and *k* are *unit vectors along the x-, y-, and z-axes respectively. The *Laplace operator is ∇².

delayed neutrons Neutrons arising from nuclear *fission that are not directly formed in the fission process but are produced from excited fission products by *beta decay.

delay line A transmission line or any other device that introduces a known delay in the transmission of a signal. An *acoustic delay line* is a device in which the signal is converted to acoustic waves, usually by means of the *piezo-electric effect, and these waves are delayed by circulating them in a liquid or solid medium. They are then reconverted to electrical signals. Fully electronic *analogue delay lines* are based on *CCDs while *digital delay lines* use CCDs or *shift registers.

delta connection A particular example of the *mesh connection employed in *three-phase a.c. circuits in which three conductors, windings, or phases are connected in series to form a closed circuit, which may be represented by a triangle (Δ), the main terminals of the circuit being the junctions between the three separate circuits. *Compare* star connection.

delta function (δ-function) *See* Dirac function.

delta radiation (δ-radiation) Secondary electrons emitted by the impact of *ionizing radiation on matter. Their energies are of the order of only 10^3 eV; they can cause further-ionizations.

Demagnetizing field

demagnetizing field The magnetic field due to the free poles developed on a specimen of ferromagnetic material during the process of magnetization. The demagnetizing field (*see* diagram), in the medium between the free poles, opposes the applied magnetic field, *H*, the effective strength of which is thus reduced by an amount proportional to the existing intensity of magnetization of the specimen.

demodulation The reverse of *modulation, i.e. the extraction or separation of the modulating signal from a modulated carrier wave. Circuits or devices used for this purpose are called demodulators or detectors.

demodulator *See* detector.

demultiplexer *See* multiplex operation.

denaturant An isotope added to a *fissile material to make it unsuitable for use in nuclear weapons.

densitometer An instrument for measuring the optical transmission or reflection of a material. It is often used for converting the images on a photographic plate into quantitative form.

density 1. Symbol: ρ. The mass per unit volume of a substance. In SI units it is measured in $kg\,m^{-3}$. The *relative density* (symbol: *d*) is the density of a substance divided by the density of water; this quantity was formerly called *specific gravity*. At the *maximum density of water, $\rho = 1000$ kg m^{-3}; therefore the relative density of any substance is one-thousandth of its density.

2. Vapour density. The density of a gas or vapour divided by the density of hydrogen, both being at *STP.

3. Density, in general, expresses the closeness of any linear, superficial, or space distribution, e.g. electron density = number of electrons per unit volume; *see also* charge density.

4. *See* reflection density; transmission density.

depletion layer A *space-charge region in a *semiconductor in which there is a net charge due to insufficient mobile charge carriers. Depletion layers are formed, for example, at the interface between a p-type and n-type semiconductor in the absence of an applied field. They are also formed at the interface of a metal and a semiconductor.

depletion mode *See* field-effect transistor.

depolarizer *Syn.* depolarizing agent. A substance used for removing the effects of *polarization in a primary cell, by reacting either chemically or electrolytically with the hydrogen ions liberated at the positive pole.

depth of field If the lens in a camera or other optical instrument is focused on a particular object, the image of the object will be in focus but the images of objects on either side will be slightly out of focus. The zone in which the blurring of the image cannot be noticed is the depth of field of the lens. It depends not only on what standard of sharpness is acceptable, but also on the aperture and focal length of the lens and the distance from object to lens. A small aperture and short focal length produce a large depth of field. *Compare* depth of focus.

depth of focus The range of distance lying along the axis of a lens over which an image remains in acceptable focus. It is directly proportional to *f-number and object distance and inversely proportional to the focal length of the lens. *Compare* depth of field.

derived units *See* SI units, also Appendix, Table 4; coherent units.

desorption The removal of adsorbed gas from a solid surface during which process heat is taken from the surface. The method is used for the liquefaction of helium.

Destriau effect *See* electroluminescence.

destructive interference *See* interference.

detector 1. *Syn.* demodulator. In communications, a circuit or apparatus used to separate the original information from the modulated *carrier wave.

2. Any device or apparatus used to detect or locate the presence of radiation, particles, etc.

deuteron A nucleus of an atom of deuterium, having a unit positive charge.

deviation 1. *Syn.* variation. The difference between an observation and its true value. The latter has often to be replaced by its nearest known value, i.e. the mean or *average of all the observations, in which case the difference is often called the *residual*. The *mean deviation* is the average of the deviations when all are given a positive sign. The *standard deviation* is the square root of the average of the squares of the deviations of all the observations. (*See* frequency distribution.) **2.** In *frequency modulation, the amount by which the carrier frequency is changed by modulation.

deviation (angle of) The angle between the incident ray and the reflected (or refracted) or emergent ray. A ray after a single reflection is deviated ($\pi - 2i$), where i is the angle of incidence; after successive reflection at two plane mirrors the deviation is ($2\pi - 2A$), where A is the angle between the mirrors. The angle of deviation by a prism depends on the angle of incidence and the angle of the prism. *Minimum deviation* by a prism occurs when the refraction is symmetrical. The minimum deviation D appears in the equation:

$$n = \sin \tfrac{1}{2}(A + D)/\sin \tfrac{1}{2}A,$$

where A is the principal refracting angle of the prism and n the refractive index. For narrow-angle prisms,

$$D = (n - 1)A$$

(approximately), and over a moderate range of angle of incidence there is an approximately constant deviation.

Dewar vessel *Syn.* vacuum flask; *UK tradename*: Thermos flask. A glass vessel consisting of a double-walled flask with the interspace completely evacuated to prevent gain or loss of heat by the contents of the flask through gaseous conduction and convection. Transfer of heat by radiation is reduced by silvering the inside walls so that a body placed in such a vessel is thermally isolated from the atmosphere outside and its temperature will remain practically unchanged for long periods.

dew point The highest temperature a surface may have in order that dew may condense on the surface from a humid atmosphere. *See* humidity.

dextrorotatory Capable of rotating the plane of polarization of polarized light in a clockwise direction, as viewed against the direction of motion of the light. *See* optical activity.

diamagnetism A property of substances that have a negative magnetic *susceptibility; thus the relative *permeability is less than that of a vacuum and lies between 0 and 1. Diamagnetism is caused by the motion of electrons in atoms around the nuclei. An orbiting electron produces a magnetic field in the same way as an electric current flowing in a coil of wire. If an external magnetic field is applied, the electrons change their orbits and velocities so as to produce a magnetic field that opposes the applied field, in accordance with *Lenz's law.

Lines representing the flux in a uniform magnetic field become more separated when passing through the material; similarly, if a diamagnetic substance is placed in a nonuniform field, there is a force acting from the stronger to the weaker part of the field. If a bar of diamagnetic material is placed in a uniform magnetic field, it tends to orientate itself so that the longer axis is at right angles to the flux.

Diamagnetism is a very weak effect: the relative permeability is only slightly less than 1. All substances possess diamagnetism but in some cases it is totally masked by stronger *paramagnetism or *ferromagnetism. Examples of purely diamagnetic substances are copper, bismuth, and hydrogen. The diamagnetic properties of materials are not affected by their temperature.

diaphragm An opaque screen containing a circular aperture centred and normal to the axis of an optical system; it controls the amount of light passing through the system.

diathermic Transparent to heat.

diatonic scale *See* musical scale.

dichroism The property possessed by some crystals, such as tourmaline or Polaroid, of selectively absorbing light vibrations in one plane, while allowing the vibrations at right angles to pass through.

dielectric A substance that can sustain an electric field and act as an insulator.

dielectric constant *See* permittivity.

dielectric heating The heating effect that occurs when a high-frequency alternating electric field is applied across a nonconducting material, i.e. a dielectric. It is a result of *dielectric hysteresis. Nor-

111

mally the power is provided by some form of oscillator. Dielectric heating is used extensively for the preheating of plastic materials.

dielectric hysteresis A phenomenon, akin to magnetic *hysteresis, as a result of which the *electric displacement in a dielectric depends not only on the applied electric field strength, but also on the previous electrical history of the specimen. It entails a dissipation of energy from the field when the specimen is subjected to an alternating electric flux.

dielectric loss The total dissipation of energy that occurs in a dielectric when it is subject to an alternating electric field.

dielectric polarization *Syn.* electric polarization. Symbol: *P*. Stress set up in a dielectric owing to the existence of an electric field, as a result of which each element of the dielectric functions as an electric dipole. It measures the increased flux present in the dielectric due to the presence of the latter, and is defined as the function $(D - \varepsilon_0 E)$, where E is the applied field strength, D is the electric displacement, and ε_0 is the electric constant. *See also* displacement current.

dielectric strength *Syn.* disruptive strength. The maximum electric field that an insulator can withstand without breakdown, under given conditions. It is usually measured in $V\,mm^{-1}$.

Diesel cycle A *heat engine cycle in which air is the working substance and the fuel a heavy oil. From A to B the air is compressed adiabatically to a very high temperature. From B to C the burning fuel causes expansion at constant pressure, while CD is the remainder of the working stroke, being an adiabatic expansion. At D a valve opens and the pressure falls to atmospheric.

AE and EA represent the exhaust and charging strokes. A separate fuel pump is necessary to inject the oil into the cylinder at high pressure.

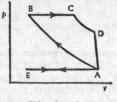

Diesel cycle

Dieterici equation of state *See* equations of state.

difference tone *See* combination tones.

differential air thermometer A simple instrument for the detection of radiant heat. Two equal closed bulbs A and B, one clear, the other blackened, contain air at atmospheric pressure. Radiation falling on the apparatus is more readily absorbed by the blackened bulb, and so the pressure inside B rises so that the liquid stands at a higher level in the left-hand connecting tube than in the right-hand tube.

Differential air thermometer

differential amplifier A type of *amplifier with two inputs, whose output is a function of the difference between the inputs.

differentially compound-wound *See* compound-wound machine.

differential resistance The ratio of a small change in the voltage drop across a resistance to the change in current producing the drop. It is the resistance of a device or component part that is measured under small-signal conditions.

differentiator A circuit designed so that the output is the differential with respect to time of the input.

diffraction analysis The study of crystal structure by means of the diffraction of a beam of electrons, neutrons, or X-rays. *See* electron diffraction; X-ray analysis; neutron diffraction.

diffraction grating A device for producing spectra by diffraction and for the measurement of wavelength. Commonly it consists of a large number of equidistant parallel lines (of the order 7500 per cm) ruled with a diamond point on glass, speculum metal, or an evaporated layer of aluminium (ruled gratings) or of a plastic cast taken from a ruled surface (replica grating). Diffracted light after "reflection" or "transmission" produces maxima of illumination (spectral lines) according to the equation:

$$d(\sin i + \sin \theta) = m\lambda,$$

where d is the grating interval, i.e. the distance between corresponding points of adjacent lines ($= 1/N$, in which N is the number of lines per unit distance), i the angle of incidence, θ the direction of the diffracted maximum with the normal corresponding with the "order" m of the spectrum ($m = 0$, for the central image).

The concave grating of Rowland removes the necessity of using achromatic collimation and telescopic objectives. Transmission gratings can be used in the visible, near ultraviolet and near infrared. Reflection gratings are needed for the far ultraviolet. Reflective gratings used at grazing incidence permit absolute measurements of X-ray wavelengths. *See also* blazed grating; echelon grating.

diffraction of light If the shadow of an object cast on a screen by a small source of light is examined, it is found that the boundary of the shadow is not sharp. The light is not propagated strictly in straight lines, and peculiar patterns are produced near the edges of the shadow, which depend on the shape and size of the object. This breaking up of the light, which occurs as it passes the object, is known as *diffraction* and the patterns observed are called *diffraction patterns*. The phenomenon arises as a consequence of the wave nature of light. Apertures in objects produce a similar effect.

It is usual to distinguish between two classes of diffraction phenomena. In *Fresnel diffraction*, only the simple arrangement of source, diffracting object, and screen are involved. In *Fraunhofer diffraction*, a parallel beam of light passes the diffracting object and the effects are observed in the focal plane of a lens placed behind it.

(1) *Fresnel diffraction* Certain of the effects observed in Fresnel diffraction can be explained if the wavefront falling on the obstacle is assumed to be divided into a number of concentric annular zones, the distances of the peripheries of the zones from the observation point on the screen increasing by one half wavelength from zone to zone. The zones are called *Fresnel zones*. Each point on the wavefront can be considered as the source of a secondary wave (*see* Huygens' principle) and each of these secondary waves will make its own contribution to the light reaching the observation point. It can be shown that all the Fresnel zones have roughly the same area, so that each zone can be considered to contain the same number of secondary sources; hence each zone will contribute the same amount of light to the observation point. However, because each zone is a further half wavelength away than the next, the contributions

from adjacent zones will be out of phase. Thus the total contribution can be represented as the sum of a series of terms alternately positive and negative, each term representing the contribution from one Fresnel zone. Since the outer zones are directed more obliquely, the magnitude of the terms representing the contributions of the zones falls off steadily from term to term. The sum of the series can be shown to be equal to half the first term, so that the amplitude of the light reaching the observation point from the whole unrestricted wavefront is half the amplitude that would result if all but the first Fresnel zone were blocked out. The intensity of the light is proportional to the square of the amplitude and will thus be one quarter of that due to the first zone alone.

If the obstacle has a circular aperture, the amount of light reaching the central point of the diffraction pattern will depend on the number of Fresnel zones that fill the aperture. If all but the first zone is blocked out by the obstacle, the intensity at the central point is four times greater than that observed if no obstacle were in place. If the first two zones are effective, the resulting intensity at the central point will be very small, since the contributions from the two zones are nearly equal but are out of phase. In this case, the diffraction pattern consists of a bright circle of light with a central dark spot. In general, if an odd number of zones is effective, the centre of the pattern is bright, while an even number results in a dark central point. The general pattern consists of concentric light and dark rings.

If a circular object is used, the central Fresnel zones will be blocked out. The sum of the series of terms that represent the effective zones will still be equal to half the first term, so that this resulting amplitude at the centre of the pattern is half the amplitude due to the first effec-

tive zone. It follows that there will always be some light reaching the central point of the diffraction pattern so there is always a relatively bright spot at the centre of the shadow of a circular obstacle.

If the obstacle consists of alternately opaque and transparent annular zones, it is possible to arrange that every other Fresnel zone is effective for a particular observation point. The result will then be a high intensity of illumination at that point, since the light from the effective zones will all arrive in phase. Such an obstacle (called a *zone plate*) will produce at the observation point a bright "image" of the point source and will in this sense act as a lens. It is capable of producing, by diffraction, an image of any small bright object.

In practical cases the radii of the Fresnel zones are very small. Thus diffraction effects are seen only with small obstacles or apertures and at the edges of the shadows of larger obstacles.

In considering the diffraction patterns produced by obstacles consisting of a straight edge, a slit, or a wire, it is convenient to divide the wavefront into strips parallel to the edges of the obstacle rather than into circular zones. The distances of the edges of the strips from the observation point are again made to increase by one half wavelength from strip to strip. The diffraction patterns can then be predicted by considering the total effects at the observation point of all the half-period strips not blocked out by the obstacle.

If the obstacle is a straight edge, then at the edge of the geometrical shadow the obstacle will block out all the strips over one half of the wavefront. Passing into the shadow, more and more strips of the other half of the wavefront will be blocked out and there is a gradual diminution of light received until almost complete darkness prevails. Outside the shadow, the full effect of one half of the wavefront is seen and in addition more and more strips of the other half are uncovered as one moves outwards. Since adjacent strips are out of phase with each other, the total effect will be a minimum if an even number of additional strips is uncovered and a maximum if the number is odd. Thus outside the geometrical shadow there are alternate bright and dark bands of decreasing contrast.

In the case of a slit there will be a bright central line to the diffraction pattern if an odd number of strips of each half of the wavefront is uncovered, and a dark line if the number is even. In general the pattern consists of an unsharp shadow of the slit crossed by dark lines. The case of the wire is rather similar to that of a circular obstacle: there is always a relatively bright line in the centre of the shadow.

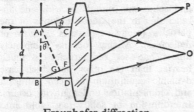

Fraunhofer diffraction

(2) *Fraunhofer diffraction.* In this class of diffraction, a parallel beam of light falls on the diffracting object and the effects are observed in the focal plane of a lens placed behind it. Thus in the diagram, AB represents a slit whose length is perpendicular to the plane of the paper, and on which falls a parallel beam of light. According to *Huygens' principle, each point in the slit must be considered as a source of secondary wavelets that spread out in all directions. Now the wavelets travelling straight forward along AC, BD, and so on, will arrive at the lens in phase and will produce a strong illumination at O. Secondary wavelets spreading out in a

direction such as AE, BF, and so on will arrive at the lens with a phase difference between successive wavelets, and the effect at P will depend on whether this phase difference causes destructive *interference or not.

For example, if the distance BG in the diagram is equal to one wavelength of the light used, then there is a path difference of one whole wavelength between light reaching the lens from opposite ends of the slit. There is accordingly a path difference of half a wavelength between light from any point in the upper half of the slit and the corresponding point in the lower half, and light from these two points will interfere destructively. The result will be that the light from the whole of the upper half of the slit interferes destructively with that from the whole of the lower half, and no light reaches P. A similar effect occurs if BG is two wavelengths, three wavelengths, and so on. Between these positions of zero illumination, there will be bright regions, and the resulting pattern seen in the plane OP is of alternate dark and light bands running parallel to the length of the slit (i.e. perpendicular to the plane of the paper). It will be noticed that (unlike the corresponding case of Fresnel diffraction) there is always a bright line at the centre of the diffraction pattern. The separation of the diffraction bands increases as the width of the slit is reduced; with a wide slit the bands are so close together that they are not readily noticeable. The separation also depends on the wavelength of the light, being greater for longer wavelengths. If white light is used, a few coloured bands are therefore produced.

In the case of a circular aperture, the diffraction pattern consists of a central bright patch (called the *Airy disc*) surrounded by alternate dark and light circular bands. Again the bands are close together if a large aperture is used but

115

their separation increases as the diameter of the aperture used is reduced.

In the case of the slit shown in the diagram, the first dark line at P is in a direction θ such that BG is one wavelength, λ. If d is the width of the slit, then $\theta = \lambda/d$ (since θ is small). In the case of a circular aperture it can be shown that the direction of the first dark circle is given by a similar expression, $\theta = 1.22\lambda/d$.

diffraction of sound Sound waves only cast sharp shadows when intercepted by a solid or when made to pass through a slit, if the solid or slit is of large dimension compared with the wavelength. To a certain degree, however, the rays are bent or diffracted giving an interference pattern of regions of varying intensity just inside the shadow or shadows. This diffraction is more pronounced as the dimension of the solid or slit is made smaller, and in the extreme case where the dimension is less than the wavelength the sound is reradiated from the obstacle with uniform intensity in all directions; the slit radiates only on the side away from the incident rays and thus produces hemispherical waves, but the solid radiates in all directions and therefore produces spherical waves.

The analysis of the general case is facilitated by Huygens' construction, i.e. by regarding the wavefront as consisting of an infinite number of point sources radiating spherical waves. These secondary waves interfere with each other to produce the interference patterns obtained. (*See* diffraction of light.)

diffraction pattern *See* diffraction of light.

diffractometer An instrument used in *diffraction analysis to measure the intensities of diffracted beams of X-rays or neutrons at different angles. An *ionization chamber or *counter is usu-

ally used and the beam to be diffracted is usually monochromatic.

diffused junction A junction between two regions of different conductivity within a *semiconductor, formed by diffusion of the appropriate impurity atoms into the material. The material is heated to a predetermined temperature in an atmosphere containing the desired impurities in gaseous form. Atoms that condense on the surface diffuse into the semiconductor material in both the vertical and horizontal directions. The numbers of impurity atoms and the distance travelled at any given temperature is well defined by *Fick's law.

Modern diffusion techniques use the *planar process whereby selective diffusions are made into well-defined areas.

diffusion 1. The process by which fluids and solids mix intimately with one another due to the kinetic motions of the particles (atoms, molecules, groups of molecules). Mixing occurs completely unless one set of particles is much heavier than the other, in which case dynamic equilibrium between diffusion and sedimentation under gravity occurs. Interdiffusion of solids (e.g. gold into lead) also occurs. *See* Fick's law; Graham's law.

2. The scattering of a beam of light on reflection or transmission. A light beam reflected from a rough surface does not obey the laws of reflection but is scattered in many directions (diffuse reflection). Similarly, light transmitted by certain materials does not obey the laws of refraction and is scattered in the medium (diffuse transmission).

3. The degree to which the directions of propagation of sound waves vary over the volume of a reverberant sound field.

diffusion cloud chamber A type of *cloud chamber in which supersaturation is achieved by diffusion of a vapour from a hot to a cold surface through an inert gas. As the vapour supply is continually replenished by diffusion, the chamber can be made almost continuously sensitive to ion tracks. There are no moving parts.

diffusion coefficient *See* Fick's law.

diffusion current *See* limiting current.

diffusion length In a *semicondcutor, the average distance travelled by *minority carriers between generation and recombination.

diffusion pump *See* pumps, vacuum.

diffusivity Symbol: α. A measure of the rate at which heat diffuses through a substance. It is equal to the thermal conductivity divided by the specific heat capacity at constant pressure and the density, i.e. $\alpha = \lambda/\rho c_p$. It is measured in $m^2 s^{-1}$.

digital audio tape (DAT) Magnetic tape used for digital recording of sound and also for storing computer information. In the case of digital recording, the audiofrequency signal is normally sampled 48 000 times per second and the characteristics of the sampled signal are converted into 16-bit words − as occurs in compact disc systems. The recording method for DAT is derived from video recording, the tape being wrapped helically around a rotating drum. This enables one or more tape heads in the drum to record the digital signal on slanted tracks on the slow-moving tape at a very high density. The sound reproduced on replay of DAT is of very high quality.

digital circuit Any circuit designed to respond to discrete values of input voltage and produce discrete output voltage levels. Usually only two values of voltage are recognized, as in binary *logic circuits. *Compare* linear circuit.

digital computer *See* computer.

digital inverter *See* inverter.

digital recording A means of recording sound whereby the audiofrequency signals are converted to a digital form − a series of discrete numbers − that is then stored on a system such as a compact disc or digital audio tape.

digital voltmeter (DVM) A voltmeter that displays the measured values as numbers composed of digits. The voltage to be measured is usually supplied as an analogue signal, and the voltmeter samples the signal repetitively and displays the voltage sampled.

digitron *Syn.* Nixie tube. A type of cold cathode scaling tube in which the cathodes are shaped into the form of characters, usually the digits 0 to 9. A switching connection to one side of the power supply selects the cathode required. These tubes have been used for display purposes in *counters, calculators, etc.

dilation *Syn.* dilatation. A change of volume.

dilatometer An apparatus for studying thermal expansion.

dimensional analysis A technique whose main uses are: (*a*) to test the probable correctness of an equation between physical quantities; (*b*) to provide a safe method of changing the units in a physical quantity; (*c*) to assist in recapitulating important formulae; (*d*) to solve partially a physical problem whose direct solution cannot be achieved by normal methods; (*e*) to predict the behaviour of a full-scale system from the behaviour of a model; (*f*) to suggest relations between fundamental constants.

The basis of the technique is that the various terms in a physical equation must have identical *dimensional formulae if the equation is to be true for all consistent systems of units. For example, in the equation $s = ut + \frac{1}{2}at^2$ (which applies to uniformly accelerated motion in a straight line), all the terms have the dimension of length. The method does not check pure numbers (e.g. the $\frac{1}{2}$ in the last term).

Mechanical quantities can be represented in terms of three independent dimensional fundamentals: mass M, length L, and time T. For example, area $= L^2$, velocity $= LT^{-1}$, force $= MLT^{-2}$, energy $= ML^2T^{-2}$. Electric and magnetic quantities require an additional independent dimension, namely current I or charge Q. Since power is the product of current and voltage, the dimensions of voltage are $ML^2T^{-3}I^{-1}$ or $ML^2T^{-2}Q^{-1}$. *See also* dynamic similarity.

dimorphism The existence of a substance in either of two possible crystalline forms. Carbon, for example, can crystallize as graphite or diamond. *See also* allotropy.

dineutron An unstable system consisting of a pair of *neutrons; it is assumed to have a transient existence in certain *nuclear reactions.

diode Any electronic device with only two electrodes. There are several different types of diode and their applications depend on their voltage characteristics. Diodes are usually used as rectifiers. Although *thermionic valves were originally used, diodes are now semiconductor devices. Semiconductor diodes consist of a single p-n junction. Current flows when forward voltage is applied to the

diode (*see* diode forward voltage), and increases exponentially with voltage, becoming substantially constant after a few tenths of a volt (*see* illustration). If voltage is applied in the reverse direction, only a very small leakage current flows until the *breakdown voltage is reached. *See also* Gunn diode; IMPATT diode; light-emitting diode; photodiode; tunnel diode; varacter diode; Zener diode.

diode forward voltage *Syn.* diode drop; diode voltage. The voltage across the terminals of a semiconductor *diode when current flows in the forward direction. Because of the exponential nature of the semiconductor-diode current characteristic, the diode voltage is approximately constant over the range of currents commonly used in practical circuits. A typical value is about 0.7 V at 10 mA.

diode laser *See* semiconductor laser.

diode transistor logic (DTL) An early family of integrated *logic circuits, with the inputs through *diodes and the output taken from the collector of an inverting transistor. The basic circuit is a NAND gate. The output transistor is designed to operate under saturated conditions, and the speed of DTL logic circuits is therefore less than *emitter-coupled logic circuits, because of the longer delay time. DTL circuits have been largely replaced by *transistor-transistor logic circuits.

diode voltage *See* diode forward voltage.

dioptre A unit used to express the power of a spectacle lens, equal to the reciprocal of the focal length in metres. It is also applicable to *vergence and curvature, convergence being regarded as positive.

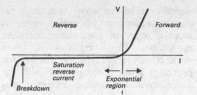

V/I curve for semiconductor diode

dioptric system An optical system in which the principal optical components are refracting elements, such as lenses. *Compare* catoptric system.

dip *Syn.* inclination. The angle made with the horizontal by the direction of the earth's local magnetic field. It varies from 0° at the magnetic equator to 90° at the magnetic poles. A magnet freely suspended at its centre of gravity would set with its magnetic axis in the magnetic meridian and inclined to the horizontal at the angle of dip.

dip circle An instrument for determining the angle of *dip. It consists essentially of a thin magnet supported so as to be free to rotate about a horizontal axis through its centre of gravity, its inclination to the horizontal being read on a vertical circle, divided into degrees. For accuracy numerous adjustments are required, including the reversal of the needle on its bearings, and the reversal of the magnetism.

dipole A system of two equal and opposite charges placed at a very short distance apart. The product of either of the charges and the distance between them is known as the *electric dipole moment* (symbol: p). A small loop carrying a current I behaves as a magnetic dipole. It has a *magnetic dipole moment* (symbol: p or μ) equal to IA, where A is the area of the loop.

dipole aerial An *aerial very commonly

used for frequencies below 30 MHz. It consists of a centre-fed horizontally mounted conductor, the length of which is usually one-half wavelength of the transmitted or received radio wave (a *half-wave dipole*). The ends may be folded back and joined together at the centre (a *folded dipole*).

dipole moment *See* dipole.

diproton An unstable system consisting of a pair of *protons; it is assumed to have a transient existence in certain *nuclear reactions.

Dirac constant *Syn.* rationalized Planck constant. Symbol: $\hbar$ (called *h*-bar or crossed-*h*). The *Planck constant divided by 2π, $1.054\,5888 \times 10^{-34}$ J s.

Dirac equation An equation for the wave functions of fermions used in relativistic *quantum mechanics. It can be regarded as a version of the *Schrödinger wave equation that takes relativity into account. There are a number of ways of expressing the equation; one form is:

$$i\alpha.\nabla\psi + (mc/\hbar)\beta\psi = (i/c)\partial\psi/\partial t,$$

where m is the mass of a free particle, c the speed of light, t the time, and $\hbar$ the rationalized Planck constant. The wave function is ψ, i is $\sqrt{-1}$, and α and β are square matrices satisfying certain symmetry rules. The Dirac equation can be used to show that fermions must have spin 1/2 and it also predicts the existence of antiparticles.

Dirac function *Syn.* δ-function. A function of x defined as being zero for all values of x other than $x = x_0$ and having the definite integral from $x = -\infty$ to $+\infty$ equal to unity. It is much used in quantum mechanics and can be used, for example, to represent an impulse in dynamics. The graph of the function is that of a single, infinitely high and infinitesimally wide peak of total area unity placed at $x = x_0$.

direct broadcast by satellite (DBS) A method of broadcasting in which a communications *satellite in geostationary orbit is used as the main transmitter. The signal to be broadcast is transmitted from its point of origin on the earth to the satellite, where it is amplified and retransmitted to cover a wide area of the earth's surface. It is detected directly by individual receivers using a suitable *dish aerial tuned to the DBS signals.

direct-coupled amplifier *Syn.* d.c. amplifier. An *amplifier in which the output of one stage is coupled directly to the input of the next stage, or through a chain of resistors. It is capable of amplifying direct current.

direct current (d.c.) An electric current that flows in one direction only and is substantially constant in magnitude.

direct-current restorer *Syn.* d.c. restorer. A device to restore or impose a given d.c. or low-frequency component to a signal after passing through a circuit that has low impedance to fast variations in current, but high impedance for d.c. or low-frequency current.

direct-gap semiconductor A semiconductor, such as a gallium arsenide, in which an electron with an energy E_g can make a direct transition across the forbidden band between the valence and conduction bands (*see* energy bands), the energy gap being E_g; it does so by absorbing a photon of energy E_g or by emitting a photon of energy E_g. In an *indirect-gap semiconductor*, such as silicon, an electron with energy E_g cannot be excited directly across the forbidden band but requires a change in momentum.

directive aerial An aerial that, as a radiator or receiver of radio waves, is more effective in some directions than in others. The directivity is often obtained by employing a *passive* (or parasitic) aerial in conjunction with an *active* aerial. The latter is an aerial connected directly to the transmitter or receiver. The former is an aerial that influences the directivity but, in transmission, is excited by the e.m.f. induced in it by the nearby active aerial, or, in reception, reacts with the active aerial by virtue of the mutual impedance between them. A passive aerial is called a *reflector* or a *director* according to whether it is placed respectively behind or in front of the active aerial.

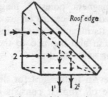

Direct-vision prism

direct-vision prism *Syn.* Amici prism; roof prism. A prism combination that produces dispersion without deviation of the central part of the spectrum (yellow *D* line). It is used in the construction of the *direct-vision spectroscope.*

disappearing filament pyrometer An *optical pyrometer in which an image of the hot source is focused by a telescope objective lens onto the filament of an electric lamp, which is viewed through a red filter by an eyepiece. The observer varies the current through the lamp filament by means of a rheostat until the filament becomes indistinguishable against the background of the image of the source. The red filter enables the matching to be made for a small band of wavelengths.

discharge 1. To remove or reduce an electric charge from a body.

2. The passage of an electric current or charge through a *gas-discharge tube or dielectric, usually accompanied by luminous effects. (*See* conduction in gases.)

3. The use of a cell, particularly an *accumulator. This involves chemical changes such that the cell eventually ceases to operate. It is then said to be *discharged*.

discharge coefficient The ratio of the actual discharge of fluid from an orifice to the discharge calculated from the velocity given by *Torricelli's law. The discharge coefficient is given by $Q/A\sqrt{2gh}$, where Q is the actual discharge through an orifice of an area A under a static head h.

discharge tube *See* gas-discharge tube.

discomposition effect *See* Wigner effect.

discriminator 1. An electronic circuit for changing a frequency-modulated or phase-modulated signal into an amplitude-modulated signal.

2. A circuit that delivers output pulses only for input pulses of greater than a certain chosen amplitude.

disc winding A type of winding used in transformers. It consists of a number of flat coils, each wound in the form of a disc. Disc windings are usually employed for the high-voltage windings of power transformers. *Compare* cylindrical winding.

dish A type of aerial used in radio astronomy, satellite communications, etc., and consisting of a sheet-metal or mesh reflector, spherical or paraboloid in shape. The reflected radio waves or microwaves are brought to a focus

above the dish, where they are collected by a secondary aerial, called a feed.

disintegration Any process in which a nucleus emits one or more particles, such as *beta particles, *alpha particles, and *gamma rays, either spontaneously or following a collision.

disintegration constant *See* decay constant.

disk *Syn.* magnetic disk. A *storage device, used in computing systems, that consists of a circular plate coated on one or more usually both sides with a magnetic film. Data is stored on a series of concentric tracks in the film; there are several thousand tracks. The disk substrate may either be rigid (in which case it is known as a *hard disk*) or it may be flexible (a *floppy disk*). The amount of data that can be stored on a disk depends mainly on its type and size, the number of tracks per disk, and the recording density along the tracks.

Data is stored on and retrieved from disks by means of a device called a *disk drive*. A single disk may be used, or a stack of disks mounted on a common spindle. The disk or disk pack is rotated in the disk drive at constant high speed. A *read-write head* can be instructed to move radially over each coated surface to select a particular track. The disk's rotation brings a particular location on the track to the read-write head. Items of data can thus be found directly and within a very short time, usually some tens of milliseconds.

dislocation *See* defect.

dispersion The decomposition of a beam of white light into coloured beams that spread out to produce spectra, or *chromatic aberration. More precisely, it is concerned with descriptions of the variation of refractive index (n) with wavelength (λ). When n is written as a function of λ a *dispersion equation* is formed. (*See* Sellmeier's equation; Cauchy dispersion formula.) *Mean dispersion* is the difference of refractive index for light of the F and C lines of hydrogen, i.e. ($n_F - n_C$). *Dispersive power* (ω) is the ratio ($n_F - n_C)/(n_D - 1$). For intermediate differences, *partial* dispersions are quoted. The dispersion of a narrow-angle prism refers to its chromatic aberration (ωP), where P is deviating power.

As most transparent substances show increasing refractive index with decreasing wavelength, the variation being more rapid at shorter wavelengths, this type of variation is called *normal dispersion*. Near an absorption band this normality apparently ceases (*see* anomalous dispersion) although the anomaly is quite general.

The rate at which an angle of refraction or diffraction varies with wavelength ($d\theta/d\lambda$) is also referred to as *dispersive power* or *angular dispersion*, while the linear separation of two lines in a spectrum per unit difference of wavelength ($dl/d\lambda$) is the *linear dispersion*.

Dispersion, in general, is a manifestation of the dependence of wave velocity on the frequency of the wave motion, and is a property of the medium in which the wave is travelling. It is not only light that is dispersed. Radio waves for example are slowed down when they travel through an ionized medium: the lower the frequency the greater the delay. *See also* dispersion of sound.

dispersion of sound At audible frequencies the speed of sound in a gas is given by the *Laplacian equation $c = \sqrt{\gamma p/\rho}$, where γ is the ratio between the two specific heat capacities of the gas, p is the pressure, and ρ is the density. This equation does not involve any variation with frequency. Air and all gases exhibit the adiabatic changes of pressure and

temperature when audible sound passes through as implied by this formula.

At higher frequencies, however, the speed of sound in certain cases, notably carbon dioxide, varies with frequency: with CO_2 the velocity increases with frequency but above 200 kHz the gas is opaque to sound waves. These variations of velocity with frequency (or dispersion) are always accompanied by abnormal values of absorption. According to the *relaxation theory*, at higher frequencies there is a lag in the interchange of translational and vibrational energy of the gas molecules. In other words the oscillatory degrees of freedom no longer have time in these rapid acoustic vibrations to adapt themselves completely to the adiabatic changes of temperature.

dispersive power *See* dispersion.

displacement 1. A vector representing in magnitude and direction the difference in position of two points. It is the basic vector used in physics. Velocity is defined as rate of change of displacement, acceleration is rate of change of velocity. Thus mechanical quantities such as *momentum and force, and electrical quantities such as *electric field depend upon the concept of displacement.

2. The quantity of fluid displaced by a submerged or partially submerged body.

3. *See* electric displacement.

displacement current The rate of change of electric flux through a dielectric when the applied electric field is varying. When a capacitor is charged, the conduction current flowing into it is considered to be continued through the dielectric as a displacement current so that the current is, in effect, flowing in a closed circuit. Displacement current does not involve motion of the current carriers (as in a conductor) but rather the formation of electric dipoles (i.e. *dielectric polarization), thus setting up the electric stress. The recognition by Maxwell that a displacement current in a dielectric gives rise to magnetic effects equivalent to those produced by an ordinary conduction current is the basis of his electromagnetic theory of light.

disruptive discharge The passage of an electric current through an insulating material when the latter breaks down under the influence of a dielectric stress equal to or greater than the *dielectric strength of the particular insulating material. *See* spark.

disruptive strength *See* dielectric strength.

dissociation **1.** The breakdown of molecules into smaller molecules or atoms. Some compounds dissociate at room temperature. For example, dinitrogen tetroxide (N_2O_4) exists in equilibrium with nitrogen dioxide (NO_2) at normal temperature. All molecules can be broken down into atoms at sufficiently high temperatures. Most dissociations of simple molecules are reversible. **2.** The breakdown of molecules into ions in solution. *See* electrolytic dissociation.

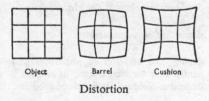

Object   Barrel   Cushion

Distortion

distortion The extent to which a system, or part, fails to reproduce accurately at its output the characteristics of the input.
(1) (electrical) The modification of a *waveform of voltage, current, etc., by a transmission system or network. It in-

volves the introduction of features that do not appear in the original or the suppression or modification of features that are present in the original. *Attenuation distortion* (or *frequency distortion*) is produced when the gain or loss of a transmission system varies with frequency. *Phase distortion* is produced when the phase change introduced by the transmission system is not a linear function of frequency. *Harmonic distortion* is due to the production of harmonics that are not present in the original. *Amplitude distortion* occurs when the ratio of the rms value of the output of the system to the rms value of the input varies with the amplitude of the input, both waveforms being sinusoidal. *Nonlinear distortion* results from transmission properties varying with the instantaneous magnitude of the input, and gives rise to harmonic and amplitude distortion.

(2) (sound) In its application to sound transmission or reproduction, the percentage of first, second, third, and higher harmonics produced in the output for a given single sine-wave input is frequently quoted. The main types of distortion are (i) nonuniform transmission of amplitude at different frequencies for a constant amplitude input (*frequency distortion*); (ii) nonlinear relation between input and output at a given frequency at different amplitudes of the input (*amplitude distortion*); (iii) a phase shift between different components on transmission, which is of great importance when considering transients; (iv) *transient distortion*, giving a duration of certain components of a note in excess of the duration of the input.

(3) (light) An *aberration of an image when it is not geometrically similar to its object. The lateral magnification (y'/y) is not constant but depends on the size of the object (y). When the magnification decreases with object size, a square object is imaged with *barrel*

distortion; the reverse case is *cushion distortion*. In general, a front stop yields barrel-shaped distortion; a rear stop yields cushion type; symmetrical doublets with a central stop for which position the lens is spherically corrected, are free from distortion.

distribution function The mathematical expression of a *frequency distribution.

distribution of velocities Maxwell's law of the distribution of velocities, based on classical statistics, states that for a gas in equilibrium the number of molecules whose total velocity lies in the range $c \rightarrow (c + dc)$ is given by the expression:

$$dN_c = 4\pi N \left\{\frac{hm}{\pi}\right\}^{3/2} (e^{-hmc^2})c^2 \, dc,$$

where N is the total number of molecules, m is the mass of a molecule, h is a constant and equal to $1/(2kT)$, where k is the *Boltzmann constant, and T the thermodynamic temperature. This relation is called the *Maxwell distribution* or *Maxwell–Boltzmann distribution*.

The distribution yields values as follows:

C, root mean square

velocity = $\sqrt{(3/2mh)}$,

$\bar{C}$, the mean velocity = $\sqrt{(3/\pi hm)}$,

C_p, the most probable

velocity = $\sqrt{(1/hm)}$.

The higher the temperature the more scattered is the distribution, but at all temperatures there is theoretically a small number of molecules whose velocity approaches infinity. *See also* canonical distribution.

diurnal motion The apparent motion of celestial bodies across the sky from east to west, caused by the rotation of the earth.

divergence Symbol: div. The *flux per unit volume leaving an infinitesimal element of volume at a point in a vector field; for example, in an electrostatic field, the divergence of the field is zero unless the volume element contains an electrostatic charge and the vector field is therefore *solenoidal. The divergence of a vector field F is the scalar product (*see* vector) $\nabla . F$, where ∇ is the differential operator *del.

diverging lens A lens that causes a beam of parallel light to diverge. *Compare* converging lens.

D-layer or region *See* ionosphere.

D-lines of sodium Two yellow lines very close together in the *emission spectrum of sodium. D_1 has a wavelength of 589.6 nm and D_2 is 589.0 nm. Because these lines are bright and easily produced, sodium light is used as a reference line in spectrometry.

Dolby A system for reducing *noise in magnetic and photographic sound recording and reproduction.

domain *See* ferromagnetism.

dominant wavelength *See* chromaticity.

donor *See* semiconductor.

doping The addition of impurities (*dopants*) to a *semiconductor to achieve a desired n-conductivity or p-conductivity.

doping level The number of impurity atoms added to a semiconductor to achieve the desired characteristic. Low doping levels yield a high-resistivity material: high doping levels yield a low-resistivity material. *See also* semiconductor.

Doppler broadening An effect observed in line spectra when radiation forming a particular spectral line has a spread of frequencies because of the Doppler effect caused usually by the thermal motion of molecules, atoms, or nuclei. The apparent frequency of radiation from a single molecule of gas depends upon the component of the velocity of the molecule in the line of sight with respect to the observer. As the molecules have a Maxwellian *distribution of velocities, the spread of frequencies will be observed to have a similar distribution, giving the line a *Doppler width*.

The motion of the atoms, etc., producing Doppler broadening may also arise from turbulence or rapid rotation or expansion of matter.

Doppler effect

Doppler effect The change in apparent frequency of a source (of light, sound, or other wave motion) due to relative motion of source and observer.

It is illustrated simply by assuming the velocities are in the line joining the source and observer; suppose C is the speed of sound, u_s the velocity of the source, u_o the velocity of observer, n the true frequency of the source, and W the velocity of the medium. If S is the initial position of source and S′ its position one second later, the waves emitted by the source in one second occupy the distance $S'A = C + W - u_s$, which contains n waves. Similarly let O be position of observer; in one second the waves received occupy the distance O′B $= C + W - u_o$, which contains n waves. Then, the apparent frequency is:

$$n' = n \times \frac{C + W - u_0}{C + W - u_s}.$$

When the medium is still:

$$n' = n \times \frac{C - u_0}{C - u_s}.$$

The principle is applicable to all types of wave motion. Thus, for sound waves it is noticed that the pitch of a whistling locomotive drops suddenly as the locomotive passes an observer. For electromagnetic radiation *relativity gives the formula:

$$\lambda = \lambda_0 \ (1 + V_r/c)(1 - V^2/c^2)^{-1/2},$$

where λ_0 is the wavelength measured by an observer at rest with respect to the source and λ is that measured by an observer with relative velocity V. V_r is the component of the velocity of the source away from the observer in the line of observation; c is the speed of light.

Doppler radar See radar.

Doppler shift The magnitude of the change in frequency or wavelength of waves that results from the *Doppler effect. See redshift.

dose A quantity of radiation or absorbed energy.

1. *Absorbed dose* (symbol: D) is the energy absorbed per unit mass in an irradiated medium. The unit is the *gray.

2. *Exposure dose* (symbol: X) is a measure of X- or gamma-radiation to which a body is exposed. It is equal to the total charge collected on ions of one sign produced in unit mass of dry air by all *secondary electrons liberated in a volume element by incident photons stopped in that element. The unit is the coulomb per kilogram, which replaces the roentgen.

3. *Dose equivalent* (symbol: H) is used for protection purposes. The unit is the *sievert. It is defined by:

1 sievert = 1 gray × QF,

where QF is the *quality factor* for a particular type of radiation and is a means of relating absorbed doses of different radiations to give the same biological effect. For X-rays, gamma-rays, and high-energy beta-rays, QF = 1; low-energy beta-rays, QF ≃ 1.8; neutrons, QF = 10.

4. *Maximum permissible dose* is the recommended maximum dose that a person, exposed to *ionizing radiation, should receive during a specified period. *See also* dosimetry; dosemeter.

dose equivalent *See* dose.

dosemeter Any instrument or material used for measuring radiation *dose. *See* dosimetry.

dosimetry The measurement of radiation *dose, the choice of method being determined by the quantity and quality of radiation delivered, the rate of delivery (*dose rate*), and the convenience. The most common method is to measure the *ionization caused by the radiation, as in an *ionization chamber.

Film dosimetry is a means of measuring dose using photographic film. The degree of blackening on the film after exposure to radiation and development under controlled conditions gives a measure of the dose received. A small piece of film is used in *film badges* to measure the dose received by personnel exposed to radiation.

High-energy radiation induces changes in the mechanical, electrical, and optical properties of polymers, such as perspex and PVC. In *perspex dosimetry* a piece of perspex is irradiated producing an increase in *transmission density, which is proportional to dose over a certain dose range.

Lithium fluoride dosimetry involves the measurement of the *thermoluminescence from the irradiated phosphor, lithium fluoride. Following irradiation the lithium fluoride is heated and the thermoluminescent output (light) is determined by using a *photomultiplier. This output is proportional to integrated dose.

double-base diode *See* unijunction transistor.

double bridge *See* Kelvin double bridge.

double refraction *Syn.* birefringence. When near objects are viewed through certain crystals, such as calcite or quartz, they appear doubled. The light is split into two parts: an *ordinary ray* (o-ray), which obeys the ordinary laws of refraction and an *extraordinary ray* (e-ray), which follows a different law. The light in the ordinary ray is polarized at right angles to the light in the extraordinary ray. Because of the crystalline nature of the medium, two groups of Huygens wavelets (*see* Huygens principle) progress; the ordinary wavefront is developed by spherical wavelets and the extraordinary wavefront is developed by wavelets that are ellipsoids of revolution. Along an optic axis these travel with the same velocity. The measurement of the double refraction of any crystal is given by the difference of its greatest and least refractive indices. Some crystals are uniaxial (calcite, quartz, ice, tourmaline); others are biaxial (mica, selenite, aragonite).

double-sideband transmission *See* single-sideband transmission.

doublet A pair of closely spaced lines in a *spectrum. *See* spin.

drag coefficient When a body and fluid are in relative motion the body experiences a *drag force* (*D*) parallel to the

direction of relative motion but in the opposite direction, and given by:
$$D = k_0 \rho l^2 V^2,$$
where ρ is the fluid density, V the relative velocity, and l some characteristic length of the body; k_0 is the drag coefficient, a dimensionless number that is a function of the *Reynolds number, lV/ν, where ν is the coefficient of kinematic viscosity. This is not a unique definition, the term $\frac{1}{2}\rho V^2$ being used sometimes instead of ρV^2.

drain The electrode in a *field-effect transistor through which *carriers leave the interelectrode region.

drift mobility Symbol: μ. In a *semiconductor, the average velocity of excess *minority carriers per unit electric field. In general, the mobilities of *holes and *electrons are different.

drift transistor *Syn.* graded-base transistor. A transistor in which the impurity concentration in the *base varies across the base region. A high *doping level at the emitter-base junction reduces across the base to a low doping level at the collector-base junction. Drift transistors have a good high-frequency response.

driver A circuit that provides the input of another circuit or controls the operation of that circuit.

driving point impedance The ratio of the rms value of the sinusoidal voltage applied to two terminals of an electrical network to the rms value of the current that flows between the terminals as a result of the applied voltage.

drum winding A type of *winding used in electrical machines. It consists of coils, usually former-wound, which are housed in slots either on the outer periphery of a cylindrical core or on the inner periphery of a core having a cylin-drical bore. Modern machines usually have this type of winding.

dry adiabatic lapse rate (DALR) *See* lapse rate.

dry cell A primary *cell in which the active constituents are absorbed in some porous material so that the cell is unspillable. The usual form consists of a zinc container (forming the negative electrode) lined with a paste of ammonium chloride and plaster of Paris, and having in the centre a carbon rod surrounded by a mixture of ammonium chloride, powdered carbon, zinc sulphate, and manganese dioxide made into a stiff paste with glycerine. Its action is the same as that of the *Leclanché cell; e.m.f. about $1\frac{1}{2}$ volts.

dry ice Solid carbon dioxide, used as a refrigerant.

dual in-line package (DIP) A standard form of package for *integrated circuits in which the circuit is encapsulated in a rectangular plastic or ceramic package with a row of metal legs down both of the longer sides. The legs are terminating pins. The legs can be soldered into holes on a printed circuit board or inserted into a chip socket.

Duane–Hunt relation The shortest wavelength (λ_{min}) generated in an X-ray tube is inversely proportional to the potential difference (V) applied to the tube. If e, h, and c are, respectively, the electronic charge, the Planck constant, and the speed of light, then:
$$Ve = hc/\lambda_{min}.$$

ductility A combination of properties of a material that enables it to be drawn out into wires.

Dulong and Petit's law The product of the mass per mole of a solid element

127

and its specific heat capacity is constant. This product is now known as the *molar heat capacity. According to Dulong and Petit's law the molar heat capacity is approximately 25 J K^{-1} mol^{-1}. This value may be deduced from the principle of *equipartition of energy: the movement of the lattice units, involving both kinetic and potential energies, requires an energy of RT per mole per degree of freedom (where R is the *molar gas constant and T the thermodynamic temperature). Thus the molar heat capacity for three degrees of freedom will be $3R$ (25 J K^{-1} mol^{-1}).

This value holds only for simple substances that crystallize in the regular or other simple systems and at high temperatures. At lower temperatures the value falls below $3R$, tending to zero as T tends to zero.

duplexer A two-channel multiplexer commonly used in *radar, in which a transmit-receive switch functions during the finite time between transmitting a pulse and receiving the return echo, so that the transmitter and receiver are connected in turn to the same aerial system.

duplex operation The operation of a communications channel between two points in both directions simultaneously. When the operation is limited to either direction but not both at once, it is called *half-duplex operation*.

duralumin A hard lightweight alloy of aluminium containing 4% copper, 5% manganese, 5% magnesium.

Dushman's equation *See* Richardson's equation.

dust core A core for magnetic devices consisting of a powdered magnetic material, such as *ferrite, sintered or cemented into a compact block. It is used

for minimizing *eddy-current loss in high-frequency equipment.

DVM Abbreviation for *digital voltmeter.

dynamic Changing, capable of being changed, or taking place over a period of time, usually while a device or system (electrical, electronic, or computing) is in operation.

dynamic equilibrium 1. *See* force.
  2. A balanced state of constant change, e.g. if water is sealed in an exhausted vessel and kept at constant temperature, although molecules are constantly being exchanged between the ice, water, and water vapour phases, it will be in equilibrium in so far as the pressure and volume of the phases are concerned.

dynamic friction *See* friction.

dynamic impedance *See* rejector.

dynamic range The range over which a useful output is obtained from a device. For an electronic device, it is often expressed as the difference in *decibels between the noise level of the system and the overload level.

dynamics The branch of *mechanics concerned with forces that change the motions of bodies. *Statics is sometimes considered as a separate branch of dynamics.

dynamic similarity *Syn.* similarity principle. The dimensions of all mechanical quantities (velocity, acceleration, force, etc.) can be expressed uniquely in terms of the fundamental dimensions of mass (M), length (L), and time (T), certain combinations of the dynamical quantities producing nondimensional numbers. Two systems in motion possess *dynamic similarity* when, for equal values of some dimensionless grouping of the dynamic quantities, they pass through geometrically similar configurations.

In the motion of fluids two systems are dynamically similar when the body boundaries and the corresponding flow patterns are geometrically similar, the nondimensional groupings consisting of a combination of one or more of the dimensionless *Reynolds, *Froude, and *Mach numbers.

In hydrodynamics and aerodynamics, the principle of similarity is used extensively in calculating the effect of flow on a system by the observation of the effect of similar flow upon a scale model.

dynamic stability Of a floating body. The amount of work performed in tilting the body over to a given angle from its position of equilibrium.

dynamic viscosity *See* kinematic viscosity; viscosity.

dynamo A machine that drives an electric current round a circuit when it is itself driven by mechanical means such as a motor. *See* generator.

dynamometer 1. *See* torquemeter. **2.** *See* electrodynamometer.

dynamotor An electrical machine having a single magnetic field system and a single armature, the latter carrying two independent windings connected to two independent *commutators. The machine operates as a motor with one of the windings and simultaneously as a generator with the other. The armature windings are usually different so that the voltage on the generator side is different from the voltage on the motor side and the machine acts as a rotary transformer.

dyne Symbol: dyn. A *CGS unit of force. 1 dyne = 10^{-5} newton.

dynode An electrode in an electron tube, whose primary function is to provide *secondary emission of electrons. *See* photomultiplier.

E

earth *Syn.* ground. **1.** A large conductor, such as the earth, that is taken as the arbitrary zero in the scale of electric potential.

2. A connection, which may be accidental, between a conductor and the earth. A point on a body that is in good conducting contact with the earth is said to be *earthed*, or at *earth potential*. A wire soldered to a cold water pipe makes an effective earth, or connection can be made to an *earth electrode*, i.e. a large copper plate buried in moist soil.

3. The point or portion in an electric circuit or device that is at zero potential with respect to earth.

earth current 1. A current that flows to earth, particularly as a result of a fault in a system.

2. A current that flows in the earth, especially one associated with disturbances in the *ionosphere.

earth plane *Syn.* ground plane. A sheet of conducting material that is adjacent to an electrical circuit and is at earth potential. It provides a low-impedance earth at any point in the circuit. For example, one side of a double-sided printed circuit board may be used as the earth plane for the circuit on the other side, contacts being made through the board at any desired point in the circuit.

earth's magnetic field *See* geomagnetism.

earth's mean density Using Newton's law of *gravitation, the mean density, ρ, is given by:
$$g = 4\pi GR\rho/3,$$
where g is the acceleration of free fall, G the gravitational constant, and R the earth's radius. Measurements of these quantities give the value:
$$\rho = 5.515 \times 10^3 \text{ kg m}^{-3}.$$
Because the density of solid matter near the earth's surface averages about 2.7 times that of water, the relative density of the core must exceed 5.5. It is calculated to be between 10 and 12, i.e. approximately that of iron-nickel under great pressure.

ebullition The process of boiling.

echelon grating A *diffraction grating capable of high resolution (100 000 to 1 000 000) for a small portion of a spectrum, e.g. for studying hyperfine structure of lines, Zeeman effect, etc. Some twenty to forty accurately parallel plates of equal thickness (to within a small fraction of a wavelength) are mounted in optical contact and staggered to form a series of steps of width about 1 mm. The grating is used either as a reflection or transmission instrument, the former giving higher resolution.

echo 1. When a sound pulse is incident upon a surface of large area, some part of the sound energy is reflected. If the time interval between the emission of the sound and the return of the reflected wave is more than about one-tenth of a second, the reflected sound is heard after a silent interval and is called an echo; the minimum path difference is about 30 metres. A reflector having a large surface area relative to the wavelength of the sound gives the best echoes. Thus a high-pitched sound usually gives a better echo than one of low

frequency. The distance of the reflector from the source can be calculated from the time taken between the emission of the sound and the return of the echo. This principle has been extensively used in *echo sounding to find the depth of the sea bed beneath a ship. Echoes are very troublesome when they occur in large buildings since the interference they cause prevents the original sound from being heard distinctly. This can be overcome by the use of sound-absorbing materials and by avoiding curved surfaces, which act as concave mirrors and thus focus the echoes. (*See* acoustics.)

2. In communications, a wave returned to the transmitter with sufficient magnitude and delay to be distinguished from the transmitted wave.

3. In *radar, the portion of the transmitted pulse that is reflected back to the receiver.

echo chamber *See* reverberation chamber.

echo sounding Any technique whereby a sound beam can be used for estimation of the depth of the sea as well as the distance from the source to the nearest solid surface at which it is reflected. By measuring the time taken between the production of sound and the receipt of its echo, the distance can be evaluated if the speed of sound is known in the medium through which it is sent.

The same principle is used in *radar.

ECL Abbreviation for *emitter-coupled logic.

eclipse Any of a number of astronomical phenomena resulting from the alignment of heavenly bodies. A planet, star, or satellite may pass behind the moon or a planet and so not be visible from earth. This is called an *occultation*. Venus and Mercury pass across the disc of the sun at irregular intervals, and satellites or the shadows of satellites pass across the disc of a planet. These phenomena are called *transits*.

The sun is eclipsed if the new moon passes directly between it and the earth. The full moon is eclipsed if the earth passes directly between it and the sun. The orbit of the moon is in a plane that makes an angle of about 5° with the ecliptic (*see* celestial sphere) hence usually the earth, moon, and sun are not in the same line at full and new moon. At times, however, the moon passes through the ecliptic at such points as to cause eclipses.

A solar eclipse occurs when the *shadow of the moon passes over the surface of the earth. A narrow strip of the surface may be in the umbra so the eclipse is here *total*. Over a wider area there is a *partial* eclipse, in the penumbra. The apparent sizes of sun and moon as seen from earth are very nearly equal but sometimes the moon does not completely cover the surface of the sun, causing an *annular* eclipse.

Lunar eclipses occur when the moon passes through the shadow of the earth.

Total eclipses have enabled astrophysicists to make valuable observations of the outer regions of the sun, the chromosphere and corona.

ecliptic *See* celestial sphere.

eddy current A current induced in a conductor when subject to a varying magnetic field. Such currents are a source of energy dissipation (*eddy-current loss*) in alternating-current machinery. The reaction between the eddy currents in a moving conductor and the magnetic field in which it is moving is such as to retard the motion, and can be used to produce *electromagnetic damping*. *See also* induction heating.

eddy viscosity In the turbulent flow (*see* turbulence) of an incompressible fluid, the formation of eddies has the effect of increasing the rate of change of momentum of any portion of the fluid. This may be considered as an increased resistance or as an apparent viscosity of the fluid greater than that pertinent to nonturbulent motion. This apparent viscosity is called eddy viscosity.

edge connector On a *printed-circuit board, tracks are taken to one edge of the board and form the edge connector. This may be plugged into a suitable socket allowing input and output to the circuit on the board, and interconnections to other circuits to be made.

edge dislocation *See* defect.

edge tones The sound made when a blade-shaped sheet of gas issues from a linear slit and meets an edge that may or may not be sharp. The distance between the slit and the edge apparently acts as a resonator and stabilizes the *jet tones produced without an edge. The frequency of the sound emitted is related to the number of vortices arriving at the edge per second.

Edison accumulator *Syn.* nickél-iron or Ni-Fe accumulator. A storage battery having steel grid plates: the positive plate is filled with a mixture of metallic nickel and nickel hydrate, and the negative plate is filled with iron oxide paste. Modern forms use cadmium or cadmium-iron alloy for the negative state. The electrolyte is a solution of potassium hydroxide of relative density about 1.2. The cells are strong, will deliver heavy currents and even withstand short circuits for a limited time, and they do not deteriorate on standing in the discharged state. They are lighter than the lead accumulator, but the voltage is lower – 1.3 to 1.4 volts per cell.

Edison–Butler bands Dark bands that appear in a continuous emission spectrum when a thin transparent plate is placed in the path of the light. They are caused by interference between waves reflected at the two surfaces, and are used in calibrating a spectrometer.

effective energy Of *heterogeneous radiation. The quantum energy of the beam of *homogeneous radiation that, under the same conditions, is absorbed or scattered to the same extent as the given beam of *heterogeneous radiation.

effective mass A parameter used in the theory of conductivity of solids to describe the behaviour of the charge carriers. When a potential difference is applied to a conductor, electrons are accelerated by the field produced. They have a mobility that depends on their position in the *energy band. This is described by an effective mass, which is a function of energy and can differ from the true mass.

effective resistance The resistance of a conductor or other element of an electric circuit when used with alternating current. It is measured in ohms and is the power in watts dissipated divided by the square of the current in amperes. It may differ from the normal value of resistance as measured with direct current, since it includes the effects of *eddy currents within the conducting material, *skin effect, etc.

effective temperature *See* luminosity.

effective value *See* root-mean-square value.

effective wavelength Of *heterogeneous radiation. The wavelength of the beam of homogeneous radiation that, under the same conditions, is absorbed to the same extent as the given beam of heterogeneous radiation.

efficiency 1. For a *machine the efficiency η is the work done by the machine divided by the work done on it. For steady operation this is equal to the output power divided by the input power. It is usually expressed as a percentage.
2. For a *heat engine the efficiency is the work done by the engine divided by the heat input. For an ideal *reversible heat engine in which all heat input is at temperature T_1 and all waste heat is discharged at a lower temperature T_2, the efficiency is $(T_1 - T_2)/T_1$, the temperatures being on the *thermodynamic scale.
See Carnot cycle; Carnot's theorem.

effusion The leakage of gas through a fine orifice. *Graham's law applies at ordinary pressures when the mean free path is small compared with the dimensions of the orifice, the flow being governed by the laws of hydrodynamics and being analogous to a fluid jet forced out by pressure. At low pressures, when the mean free path is large compared with the dimensions of the orifice, the volume of gas escaping per second is still inversely proportional to the square root of the density, but the mechanism is quite different. On the kinetic theory the volume diffusing per second into a vacuum is $s\sqrt{(kT/2\pi m)}$, where s is the area of the orifice, m the mass of a molecule, and k is the Boltzmann constant. The phenomenon plays an important part in high-vacuum techniques (*see* molecular flow) and is used in isotope separation, as well as for the measurement of vapour pressures.

EHF Abbreviation for extremely high frequency. *See* frequency bands.

Ehrenfest's rule The principle that if a system is described by quantized variables and subjected to an adiabatic change, then the *quantum numbers of the system must either change suddenly to new values or remain the same. Furthermore, if the change occurs very slowly, then the quantum numbers must remain constant; the variables are then said to be adiabatically invariant. This implies that the converse may be true, that only quantities that are adiabatically invariant can be quantized.

eigenfunction *See* wave function.

eigenvalue *See* wave function.

Einstein and de Haas effect The reverse of the *Barnett effect. When an iron cylinder that is free to move is suddenly magnetized, it rotates slightly.

Einstein coefficients Coefficients representing the probability of radiative transitions between electronic states of atoms or molecules. If atoms in a level n are subjected to a beam of electromagnetic radiation of frequency v, they may make a transition to a level of higher energy m by absorbing a photon of energy hv. The number of atoms making this transition is given by $B_{nm}N_{n}u(v)$, where $u(v)$ is the energy density of radiation of frequency v and N_{n} the number of atoms in level n. B_{nm} is the Einstein coefficient for absorption, giving the transition probability for this process. Similarly, atoms in level m can interact with the radiation and undergo *stimulated emission of photons in changing to level n. The number of atoms making this change is given by $B_{mn}N_{m}u(v)$. Atoms in level m can also undergo spontaneous emission to level n with emission of a photon, the number of atoms making this transition being given by $A_{nm}N_{m}$. The Einstein coefficients are given by the equations:

$$B_{nm}/B_{mn} = g_{m}/g_{n},$$

where g_{m} and g_{n} are the *statistical weights of level m and n respectively, and

$$A_{nm} = 8\pi hv^{3}/c^{3},$$

where h is the Planck constant.

For any system in equilibrium the probability of occupation of a state decreases with increasing energy. Thus if the equilibrium is disturbed by admitting a beam of radiation of frequency v into the system, the probability of absorption is greater than the probability of stimulated emission so the beam is attenuated. It is possible to produce nonequilibrium conditions in which states of higher energy are more densely populated. In such cases the beam is intensified as the probability of stimulated emission exceeds that of absorption. This is the basis of the action of the *laser and *maser and of experiments to study the *Lamb shift. It is seen that, other things being equal, the coefficient of spontaneous emission is proportional to the cube of the transition energy. For this reason it is difficult to operate a laser for short wavelengths. Also the lifetime of highly excited states are generally very short.

Einstein shift *Syn.* gravitational redshift. A small *redshift in the lines of a *stellar spectrum caused by the gravitational *potential at the level in the star at which the radiation is emitted (for a bright line) or absorbed (for a dark line). This shift can be explained in terms of either the special or general theory of *relativity. In the simplest terms, a quantum of energy hv has mass hv/c^{2}. On moving between two points with gravitational potential difference ϕ, the work done is $\phi hv/c^{2}$ so the change of frequency δv is $\phi v/c^{2}$.

Einstein's law The law of equivalence of mass and energy. Mass m and energy E are related by the equation $E = mc^{2}$, where c is the speed of light in vacuum. Thus, a quantity of energy E has a mass m, and a mass m has intrinsic energy E. *See also* relativity.

Einstein's photoelectric equation *See* photoelectric effect.

Einthoven galvanometer *Syn.* string galvanometer. A type of *galvanometer consisting of a single conducting thread strung tightly between the poles of a powerful electromagnet. On passing a current through the thread, it is deflected at right angles to the direction of the magnetic field. The deflection is observed with a high-power microscope. The instrument is highly sensitive. A current of 10^{-11} ampere can be detected.

elastance The reciprocal of *capacitance. It is measured in $farad^{-1}$, sometimes called a *daraf* although this term is deprecated.

elastic collision *See* collision.

elastic constants Constants, such as the *Young modulus (E) and the *Poisson ratio (μ), relating *stress to *strain in a homogeneous medium. For an isotropic material, two constants are required to specify the behaviour and these are related by linear equations. The components of normal stress (σ) and linear strain (e) in the x direction are related by the equation:

$$e_x = (1/E)[\sigma_x - \mu(\sigma_y + \sigma_z)],$$

where σ_y and σ_z are the components of stress in the y and z directions. Similar equations apply to the components of strain in the y and z directions. In general, an anisotropic solid is described by 21 elastic constants.

elastic deformation A change in the relative positions of points in a solid body that disappears when the deforming stress is removed.

elastic hysteresis A phenomenon occurring when a stress is applied in steps to an elastic body, and then removed in equal steps: it is found that the strain on unloading is greater than at the corresponding stress when loading. This hysteresis is very small for substances such as steel but is very large for imperfectly elastic materials such as rubbers. On a graph of stress against strain, the area within the *hysteresis loop* represents the energy dissipation per unit volume in a cycle of loading and unloading.

Elastic hysteresis is important in the damping of oscillations, and in animal movements.

elasticity 1. The property of a body or substance by which it tends to resume its original size and shape after being subject to deforming stresses.

2. Short for *modulus of elasticity.

elastic limit The smallest stress that leaves a detectable permanent strain after removal (a necessarily vague term). *See* Hooke's law; yield point.

elastic modulus Of an elastic material. The ratio of stress to strain, within the limit of proportionality. There are several moduli but these are not all independent. *See* elastic constants; modulus of elasticity.

elastic scattering *See* scattering.

elastoresistance The change in electrical resistance of materials when they are stressed within their elastic limits. *See* magnetoresistance.

E-layer or region *Syn.* Heaviside layer; Kennelly–Heaviside layer. *See* ionosphere.

electret A permanently electrified substance exhibiting electric charges of opposite sign at its extremities. Electrets in many ways resemble permanent magnets. If, for example, they are cut, they separate into two complete electrets, each with positive and negative charges.

electric axis The direction in a crystal of maximum electrical conductivity. It is the X-axis of a piezoelectric crystal.

electric braking A method of braking an electric motor by causing it to act as a generator, its output as a generator being either dissipated in a rheostat (*rheostatic braking*) or returned to the supply system (*regenerative braking*). It has particular application in electric traction.

electric charge *See* charge.

electric constant Symbol: ε_0. The *permittivity of *free space, with the formally defined value:
$$\varepsilon_0 = 10^7/4\pi c^2 \text{ F m}^{-1}$$
$$= 8.854\,187\,817 \times 10^{-12} \text{ F m}^{-1}$$
(c is the speed of light in vacuum).

electric current *See* current.

electric degree One-360th part of an alternating-current cycle. Currents or voltages arising in different parts of a circuit may be represented as vectors; the phase difference is the angle, expressed in electric degrees, between the vectors.

electric dipole *See* dipole.

electric dipole moment *See* dipole.

electric discharge *See* conduction in gases.

electric displacement Symbol: D. If an electric field exists in *free space with magnitude E and a dielectric is introduced into the field, the electric flux per unit area (*electric flux density*) in the medium is D, the electric displacement. The permittivity of the medium (ε) is given by D/E. The divergence of electric displacement equals the surface density of charge. Displacement is measured in coulombs per metre squared.

electric double layer *See* Helmholtz electric double layer.

electric energy 1. The potential energy of a charged particle in an *electric field. If the particle with a charge Q is at a point where the electric potential is V, then the electric potential energy is QV.
2. The potential energy of a body arising from the distribution of charges within it. If a *capacitor of *capacitance C has a potential difference V between the plates, the positive plate has charge $+Q$ and the negative plate $-Q$. $Q = CV$. The electric potential energy of the capacitor $= \frac{1}{2}QV = \frac{1}{2}CV^2 = \frac{1}{2}Q^2/C$. When the capacitor is discharged, this is equal to the work done on the circuit.
3. The energy per unit volume of an electric field is equal to $\frac{1}{2}\varepsilon E^2$, where ε is the *permittivity and E is the *electric field strength.
4. The energy of an electric current, stored in the magnetic field. For a current I in a circuit of self-inductance L, this energy is $\frac{1}{2}LI^2$. (*See* electromagnetic induction.) On switching off the current the energy is dissipated by generating a spark.
5. The energy per unit volume of a magnetic field is equal to $\frac{1}{2}\mu H^2$, where μ is the *permeability and H is the *magnetic field strength.

electric field The space surrounding an electric charge within which it is capable of exerting a perceptible force on another electric charge.

electric field strength Symbol: E. The strength of an electric *field at a given point in terms of the force exerted by the field on unit charge at that point. It is measured in volts per metre.

electric flux Symbol: Ψ. The quantity of electricity (i.e. charge) displaced across a given area in a dielectric. It is defined as the scalar product of the *electric displacement and the area; it is measured in coulombs. A line drawn in the field so that its direction at any point is the direction of the electric flux at that point is known as a *line of flux*. A space bounded by lines of flux forms a tube of electric flux. If the tube is so drawn that the flux across any cross section of it is unity, it is known as a unit tube of flux, or *Faraday tube*. The flux density at any point in a dielectric is equal to the number of unit tubes of flux crossing a unit of area drawn at right angles to the direction of the flux at that point. *See* electric displacement.

electric flux density *See* electric displacement.

electric hysteresis *See* dielectric hysteresis.

electric image A means of solving electrostatic problems arising when a point charge is in the neighbourhood of a conducting surface. It can be shown that, in certain cases, the electrical effects due to the induced charges on the plane are identical with those that would be produced by a point charge situated at some particular point relative to the surface. This imaginary charge is known as the electric image of the first charge in the surface. In the case of an infinite conducting plane the electric image is equal and opposite in sign to the actual charge, and situated as far behind the plane as the original charge is in front of it.

electric intensity Former name for electric field strength.

electric polarization *Syn.* dielectric polarization. Symbol: P. The *electric displacement D minus the product of the electric field strength E and the permittivity of free space (*electric constant), ε_0, i.e. $P = D - \varepsilon_0 E$. It is measured in $C\,m^{-2}$.

electric potential Symbol: V. The electric potential at a point in an electric field is the work required to bring unit positive electric charge from infinity to the point. It is measured in volts. If work of 1 joule is required to move a charge of 1 coulomb to the point, its potential is 1 volt. *See also* potential.

electric susceptibility *See* susceptibility.

electrocardiograph (ECG) A sensitive instrument that records the voltage and current waveforms associated with the action of the heart. The trace obtained is known as an electrocardiogram.

electrochemical equivalent The mass of any ion deposited from solution by a current of 1 ampere flowing for 1 second.

electrochemistry The study of chemical reactions caused by the passage of an electric current (*see* electrolysis), the generation of currents by *cells and related phenomena.

electrode In general, a device for emitting, collecting, or deflecting electric charge carriers, especially a solid plate, grid, or wire for leading current into or out of an electrolyte, gas, vacuum, dielectric, or semiconductor. In certain electrolytic cells a liquid-mercury electrode is used.

electrode efficiency The ratio of the actual yield of metal deposited in an electrolytic cell to the theoretical yield.

electrode potential The difference of potential between an electrode and the electrolyte with which it is in contact.

electrodisintegration The disintegration of a nucleus by electron bombardment.

electrodynamic instrument An instrument in which the operating torque is produced by interaction of the magnetic fields produced by currents in a system of movable and fixed coils. The moving system consists of one or more coils that are pivoted so as to move in the magnetic field of the fixed coils and the magnetic circuit is devoid of ferromagnetic material. The currents in all the coils are obtained from a common source. Instruments of this type will operate with either direct or alternating current.

electrodynamics The branch of science that studies the mechanical forces generated between neighbouring circuits when carrying electric currents.

electrodynamometer A measuring instrument actuated by the mechanical couple between a moving coil and one or more fixed coils when an electrical current passes through them. The moving coil may be controlled either by a torsional or by a bifilar suspension, and is mounted so that its plane is at right angles to that of the fixed coil when no current is flowing. If the coils are in series, so that the same current passes through all of them, the couple is proportional to the square of the current, and the instrument can be used to measure either a.c. or d.c. With a high resistance in series, it acts as an a.c. voltmeter.

electroencephalograph (EEG) A sensitive instrument that records the voltage *waveforms associated with the brain. The trace obtained is known as an electroencephalogram.

electroendosmosis *See* electrosmosis.

electrokinetic phenomena Phenomena in which electrically charged particles move in a liquid under the influence of an electric field and in which, if the particles are restricted, the liquid moves instead. If the particles are ions moving in a liquid, the motion is called *electrolytic migration* and is studied under *electrochemistry. The other phenomena are dependent on a *Helmholtz electric double layer, which is set up at almost every phase boundary. *See* electrophoresis; electrosmosis; streaming potential.

electroluminescence *Syn.* Destriau effect. The emission of light by certain phosphorescent substances when subjected to a fluctuating electric field. The phenomenon can be used for illumination by applying voltages of 400–500 volts across a dielectric coating, about 2.5 μm thick, in which the phosphor is dispersed.

electrolysis The production of chemical changes in a chemical compound or solution by causing its oppositely charged constituents or ions to move in opposite direction under a potential difference. *See* Faraday's laws of electrolysis.

electrolyte A substance that conducts electricity in solution or in the molten state because of the presence of ions. *See* electrolytic dissociation.

electrolytic capacitor Any *capacitor in which the dielectric layer is formed by an electrolytic method. The capacitor does not necessarily contain an electrolyte. When a metal electrode, such as an aluminium or tantalum one, is operated as the anode in an electrolytic cell, a very thin dielectric layer of the metal oxide is deposited. The capacitor is formed using either an electrolyte as the second electrode or a semiconductor, such as manganese dioxide. The electrolyte used is either in liquid form or in the form of a paste, which saturates a paper or gauze. Electrolytic capacitors have a high capacitance per unit volume but suffer from high *leakage currents.

electrolytic conductivity Symbol: κ. *See* conductivity.

electrolytic dissociation The reversible separation of certain substances into oppositely charged ions (*anions and *cations) as a result of solution. For example, sulphuric acid (H_2SO_4) molecules break down in water to give hydrogen ions and sulphate ions:

$$H_2SO_4 \rightarrow 2H^+ + SO_4{}^{2-}.$$

Sulphuric acid is in fact totally dissociated into ions. Certain compounds, such as acetic acid, are only partially dissociated in water.

electrolytic migration *See* electrokinetic phenomena.

electrolytic photocell A type of cell consisting of a metal electrode coated with metallic selenium, together with an electrode of platinum (or alternatively a similar electrode of selenium-coated metal), immersed in an aqueous solution of selenium dioxide. When a small external d.c. voltage is applied, the cell immediately becomes sensitive to light. It has a linear sensitivity of about 1 milliampere per lumen.

electrolytic polarization The tendency for the products of electrolysis to recombine. It is measured by the minimum potential difference required to cause a permanent current to pass through the electrolyte. In a primary cell, electrolytic polarization causes a decrease in the effective e.m.f. of the cell, and is counteracted by various depolarizing agents. *See* depolarizer.

electrolytic rectifier A *rectifier consisting of two electrodes of dissimilar metals immersed in an electrolyte. With certain combinations of metals and electrolytes, the current passes very much more readily in one direction than in the other.

electromagnet An electric circuit wound in a helix or solenoid so that the passage of an electric current through the circuit produces a magnetic field. The space within the windings is almost invariably filled with a core of ferromagnetic substance, to enhance the magnetic effect of the current. The strength of the magnet depends greatly on the design and continuity of the core.

electromagnetic damping *See* eddy current.

electromagnetic deflection A method of deflecting an electron beam using *electromagnets. It is most often applied to the beam in a *cathode-ray tube using two pairs of deflection coils.

electromagnetic focusing *See* focusing.

electromagnetic induction As the result of a series of experiments, Faraday came to the following conclusions about the phenomenon now known as electromagnetic induction:

(a) When a conductor is moved so as to cut the flux of a *magnetic field an *electromotive force (e.m.f.) is induced in the conductor. The original experiment was carried out in the field due to a permanent magnet, but later, the same effects were noted in the magnetic field associated with a current-carrying solenoid.

(b) The size of the induced e.m.f. depends on the size of the relative motion, reverting to zero when the motion ceases.

(c) The direction of the induced e.m.f. depends on the orientation of the magnetic field.

Further experiments showed that an e.m.f. is induced in a conductor when placed in a region of varying flux, and that it is particularly noticeable in the flux due to an applied current at the instants of switching the current on and off.

Conclusion (b) above was put in a quantitative form by Neumann, and is known as the *Faraday–Neumann law*: when a conductor cuts a magnetic flux Φ the induced e.m.f. E is proportional to the rate at which the flux is changing.

Conclusion (c) is generally expressed in the form of *Lenz's law*: The induced e.m.f. is in such a direction as to oppose the change that produces it.

The two laws may be put in the form of an equation:
$$E = -d\Phi/dt,$$
the minus sign indicating the significance of Lenz's law. When the conductor is made part of a circuit, Ohm's law may be applied to show that the charge Q flowing when the flux changes by $\Delta\Phi$ is given by:
$$Q = \Delta\Phi/R,$$
R being the total resistance of the circuit.

When the current I in the circuit varies, the associated flux varies in proportion:
$$\Phi = LI.$$
Applying the Faraday–Neumann law,
$$E = -LdI/dt.$$
The back e.m.f. induced in a circuit when the current in that circuit varies is described as *self-inductance*. L is the coefficient of self-inductance, and is defined as numerically equal to the e.m.f. induced when the current changes at unit rate. L is measured in *henrys. Similarly, the change of current in one circuit can cause an e.m.f. to be induced in a neighbouring circuit due to the flux linkage. The relationships are identical:
$$\Phi_1 = MI_2$$
$$E_1 = -MdI_2/dt$$
M is called the coefficient of *mutual inductance*; it is measured in henrys. For an ideal mutual inductance:
$$M^2 = L_1L_2,$$
where L_1, L_2 are the self-inductances of the component inductors. Mutual inductance is the principle behind the action of the *transformer. *Tuned circuits rely on the appropriate values of self-inductances.

In order to establish a current I in a circuit with self-inductance L, work must be done. The rate of doing work is EI, so the work done is given by:
$$W = \int EIdt = \tfrac{1}{2}IL^2.$$
This amount of energy is stored in the magnetic field around the circuit and must be dissipated on switching off, usually by generating a spark.

electromagnetic interaction The interaction between *elementary particles arising as a consequence of their associated electric and magnetic fields. The electrostatic force between charged particles is an example. This force may be described in terms of the exchange of virtual photons (*see* virtual particle). Because its strength lies between *strong and *weak interactions, particles decaying by electromagnetic interaction do so with a lifetime shorter than those decaying by weak interaction but longer than those decaying by strong interaction. An example of electromagnetic decay is:
$$\pi^0 \rightarrow \gamma + \gamma.$$
This decay process (mean lifetime 8.4 $\times$ 10^{-17} seconds) may be understood

as the *annihilation of the *quark and the antiquark making up the π^0, into a pair of photons.

The following *quantum numbers have to be conserved in electromagnetic interactions: angular momentum, charge, baryon number, isospin quantum number I_3, strangeness, charm, parity, and charge conjugation parity. *See also* electroweak interaction; quantum electrodynamics.

electromagnetic lens *See* magnetic lens.

electromagnetic mass That part of the total inertia of a charged body that arises from its electric charge. Providing that the speed of the body is small compared with that of light, the electromagnetic mass of a charge e carried by a sphere of radius a is equal to $\frac{2}{3}\mu(e^2/a)$, where μ is the magnetic permeability of the medium. The effect is due to the fact that the motion of the charge gives rise to a magnetic field, the energy of which is proportional to the square of the speed of the charge. For speeds approaching that of light, the mass increases according to the equation:

$$m = m_0(1 - \beta^2)^{-1/2},$$

where m_0 is the rest mass, and β is the ratio of the actual speed of the charge to that of light.

electromagnetic moment *See* magnetic moment.

electromagnetic pump A pump with no moving parts for use with conducting liquids, such as liquid metals. The liquid is contained in a flattened pipe between two poles of an *electromagnet, a strong magnetic field being applied across the pipe. If an electric current is also passed through the liquid, the liquid will experience a force along the axis of the pipe.

electromagnetic radiation Radiation in which associated electric-and magnetic-field oscillations are propagated through space. The electric and magnetic fields are at right angles to each other and to the direction of propagation. Propagation through space is fully described in terms of wave theory but interaction with matter depends on *quantum theory.

In free space the *phase speed of waves of all frequencies has the same value (c = 2.99 792 458 m s^{-1}) as seen by all observers (*see* relativity). Since there is no *dispersion the *group speed is identical with the phase speed. In any material medium the phase speed for a particular frequency has the value $v = c/n$, where n is the *refractive index of the medium at the frequency. Usually n is greater than one, but for X- and gamma-radiation it is very slightly less than one so the phase speed is greater than in a vacuum. The material properties of the radiation (energy, mass, momentum, and angular momentum) are propagated at the group speed, which is never greater than c. As for other forms of waves the phase speed v, frequency f, and wavelength λ are related by $v = f\lambda$. For a particular radiation f is constant, while λ is inversely proportional to n. Values of wavelength are normally stated for free space.

The range of frequencies over which electromagnetic radiation has been studied is called the *electromagnetic spectrum* (*see* Appendix, Table 6). The methods of generating radiations and their interactions depend very much upon frequency. It can be shown that the rate of radiation of energy caused by the acceleration of a given charge is proportional to the square of the acceleration. In most cases it is apparent that electromagnetic radiation is generated in this way, but in the decay of certain elementary particles (for example, neutral *pions) intermediate states of the system involving oscillations of hypothetical virtual particles must be assumed. Interactions can usually be seen to be caused by the electric field of the radiation. Although half of the energy of the wave is associated with the magnetic field, interaction with matter is rarely caused directly by this.

The lowest frequencies (*radio waves) are generated artificially by electric oscillations in circuits. Some astronomical bodies emit radiations of this type (*see* radio astronomy). Detection uses an *aerial in which the incident wave generates electric oscillations. Hot bodies, electric discharges, and luminescent bodies give *infrared radiation, *light, and *ultraviolet. These radiations may be affected or caused by molecular oscillations or rotations (especially for long-wavelength infrared) but usually both emission and absorption involve electrons moving between states of different energy in the outer structure of atoms, molecules, or ions. Characteristic *X-rays are generated by electrons entering states in the inner shells of atoms from which electrons have been ejected by high-energy electrons or other means. *Bremsstrahlung is caused by the acceleration of high-energy electrons in collision with atoms. *Gamma radiation is emitted by atomic nuclei. *See also* annihilation.

Electromagnetic radiations interact with matter as quanta. A quantum of radiation of frequency v transfers energy hv, mass hv/c^2, and momentum hv/c, where h is the *Planck constant. The transfer of momentum causes *radiation pressure. Changes of angular momentum follow complicated rules, but in the simplest cases emission or absorption changes a component of angular momentum by an amount $h/2\pi$.

By whatever mechanism electromagnetic radiation is absorbed, the energy is usually degraded, ultimately raising the internal energy and hence the temperature. Hence the intensity of such radiations can often be measured absolutely using suitably calibrated instruments such as the *thermopile.

electromagnetic spectrum *See* electromagnetic radiation.

electromagnetic units (e.m.u.) *See* CGS system of units.

electrometer A device for measuring potential difference. Modern instruments consist of a very high input impedance amplifier known as an electrometer amplifier, which draws a negligible amount of current.

electromotive force (e.m.f.) Symbol: E. The rate at which work is done electrically upon a circuit (power) divided by the current. It is measured in volts. Work may be done by *electromagnetic induction, *thermoelectric effects, or chemical reactions (*see* cell). If the process is driven in reverse, the circuit does work upon the source of e.m.f. Electromotive force (which is not a force in the normal sense) must be distinguished from potential difference. In a circuit, the latter is rate of energy dissipation divided by current, and is inherently irreversible.

If a battery of e.m.f. E and resistance b maintains a current I in an external resistance R, the rate of doing work is IE, which equals the rate of dissipation $(R + b)I^2$. The p.d. between the terminals is $V = RI$. If R tends to infinity, the value of E tends to that of V, so the e.m.f. is equal to the p.d. between the terminals on open circuit, although its nature is different.

electron A stable *elementary particle having a negative charge, e, equal to
$$1.602\ 189\ 25 \times 10^{-19}\ C$$
and a rest mass m_0 equal to
$$9.109\ 389\ 7 \times 10^{-31}\ kg$$

(equivalent to 0.511 0034 MeV/c^2).
It has a *spin of $\frac{1}{2}$ and obeys *Fermi–Dirac statistics. As it does not have *strong interactions, it is classified as a *lepton.

The discovery of the electron was reported in 1897 by Sir J.J. Thomson following his work on the rays from the *cold cathode of a *gas-discharge tube. It was soon established that particles with the same charge and mass were obtained from numerous substances by the *photoelectric effect, *thermionic emission, and *beta decay. Thus the electron was found to be part of all atoms, molecules, and crystals.

Free electrons are studied in a vacuum or a gas at low pressure. Beams are emitted from hot filaments or cold cathodes and are subject to *focusing by electric or magnetic fields. The force F_E on an electron in an electric field of strength E is given by $F_E = Ee$ and is in the direction of the field. On moving through a *potential difference V, the electron acquires a kinetic energy eV, hence it is possible to obtain beams of electrons of accurately known kinetic energy. In a magnetic field of *magnetic flux density B an electron with speed v is subject to a force $F_B = Bev \sin\theta$, where θ is the angle between B and v. This force acts at right angles to the plane containing B and v.

The mass of any particle increases with speed according to the theory of *relativity. If an electron is accelerated from rest through 5 kV, the mass is 1% greater than at rest. Thus account must be taken of relativity for calculations on electrons with quite moderate energies.

According to *wave mechanics a particle with momentum mv exhibits diffraction and interference phenomena similar to a wave with wavelength $\lambda = h/mv$, where h is the Planck constant. For electrons accelerated through a few hundred volts, this gives wavelengths rather less than typical interatomic spacing in crystals. Hence a crystal can act as a diffraction grating for electron beams. (*See* Davisson–Germer experiment; de Broglie waves; electron diffraction.)

At kinetic energies less than a few *electronvolt, electrons undergo elastic *collision with atoms and molecules. Because of the large ratio of the masses and the conservation of momentum, only an extremely small transfer of kinetic energy occurs, thus the electrons are deflected but not slowed down appreciably. At slightly higher energies collisions are inelastic. Molecules may be dissociated, atoms and molecules may be excited or ionized (*see* ionization potential). The excited particles or recombining ions emit *electromagnetic radiation mostly in the visible or ultraviolet. For electron energies of the order of several keV upwards, *X-rays are generated. Electrons of high kinetic energy travel considerable distances through matter, leaving a trail of positive ions and free electrons. The energy is mostly lost in small increments (about 30 eV) with only an occasional major interaction causing X-ray emission. The range increases at higher energies. *See also* positron; atom.

electron affinity Many atoms, molecules, and free radicals form stable negative ions by capturing electrons (*see* electron attachment). The electron affinity is the least amount of work that must be done to separate the electron from the ion. It is usually expressed in electronvolt.

electron attachment The attachment of a free electron to an atom or molecule to form a negative ion. The process is sometimes called *electron capture* but this term is more usually applied to nuclear processes. *See also* electron affinity.

electron capture 1. *See* capture. **2.** *See* electron attachment.

electron density 1. The number of electrons per unit mass of a given material. Most light elements (except hydrogen) have an electron density of about 3×10^{26} electrons per kilogram.
2. The number of electrons per unit volume.

electron diffraction Owing to the fact that electrons are associated with a wavelength λ given by $\lambda = h/mv$, where h is the Planck constant and (mv) the momentum of the electron, a beam of electrons suffers diffraction in its passage through crystalline material, similar to that experienced by a beam of X-rays. The diffraction pattern depends on the spacing of the crystal planes, and the phenomenon can be employed to investigate the structure of surface and other films. *See* de Broglie waves; electron.

electron gas The concept of the free electrons in the solid or liquid state as a gas whose state may be compared with that of an actual gas dissolved in a solid or liquid. This model has found application in theories of electric and thermal conduction, thermionic emission, etc. Applying Fermi–Dirac statistics (*see* quantum statistics) to the electron gas, it is shown to be completely degenerate at ordinary temperatures, i.e. it obeys a totally different distribution law from that of an ideal gas.

electron gun A device, consisting of a series of electrodes, that produces an electron beam. The beam produced is usually a narrow beam of high-velocity electrons whose intensity is controlled by electrodes in the gun.

Electrons are released from the indirectly heated cathode (*see* diagram). The control grid is a cylinder surround-

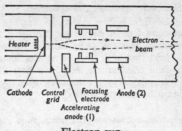

Electron gun

ing the cathode with a hole in front to allow passage of the electron beam. It controls the electron beam when its negative potential is varied. The electron beam is accelerated by the positively charged accelerating anode and passes through the focusing electrode before being further accelerated by the second anode.

electron–hole pair *See* hole.

electronic device A device in which conduction is mainly by the movement of *electrons in a vacuum, gas, or *semiconductor.

electronic flash A source of very brief but bright illumination, provided by a high-voltage discharge between electrodes in a gas-discharge tube containing xenon or neon. Electronic flash is used, for example, in photography and stroboscopy. In photography it takes the form either of battery-operated *flashguns* that may be mounted on a camera or mains-powered *studio flash*.

electronics The study, design, and use of devices based on the conduction of electricity in a vacuum, a gas, or a semiconductor. Modern electronics is principally concerned with semiconductor devices; vacuum and gas-filled devices are rapidly becoming obsolete, apart from a few specialized uses. *See* Appendix, Table 10 for symbols used in electronics.

143

electronic spectrum *See* spectrum.

electron lens A device for *focusing an electron beam by using either a magnetic field (*magnetic lens*) or an electrostatic field (*electrostatic lens*) in a way that is analogous to the focusing of a light beam by an optical lens.

electron microscope 1. *Transmission electron microscope.* An instrument, closely resembling the optical microscope but using, instead of light, a beam of energetic electrons. The *resolution, and consequently the magnification, is about one thousand times that of the optical microscope. The beam is usually focused by a magnetic lens and has an energy of 50–100 keV. It is directed onto the entire area of the sample under investigation and the electrons emerging are focused by a second magnetic lens onto a fluorescent screen, thus producing a visible image. A sharply focused image in one plane can only be obtained by using electrons of a single energy. To avoid energy losses in the incident beam, the sample must be extremely thin (usually < 50 nm) so that the scattered electrons that form the image are not changed in energy. The thinness of the sample greatly limits the *depth of field and the image is therefore two-dimensional. Electrons accelerated to an energy of 100 keV have a wavelength of about 0.04 nm (*see* de Broglie equation) so that a resolution of between 0.2–0.5 nm is possible. The maximum magnification is approximately one million diameters.

2. The *scanning electron microscope* has lower resolution and magnification but produces a seemingly three-dimensional image from a sample of any convenient size or thickness. The beam of energetic electrons originates from a heated tungsten cathode with a diameter of between 20–50 μm (Fig. *a*). It is accelerated and focused by electric and

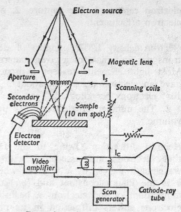

a Scanning electron microscope

magnetic fields forming a spot, with a diameter of about 10 nm, on the surface of an electrically conducting sample. (If the sample is nonconducting, a thin metallic layer, 5–50 nm thick, is evaporated onto the surface.) The electron beam is deflected by scanning coils so that the sample is scanned, point by point, in a *raster pattern. Electrons striking the sample give rise to *secondary electrons whose number is dependent on the geometry and other properties of the sample.

The secondary electrons are collected by a positively charged electron detector in which they are accelerated to about 10 kV before striking a *scintillator and so producing a large number of photons. The photons, falling upon a *photomultiplier are converted into a highly amplified electric signal. This signal modulates the intensity of an electron beam inside a *cathode-ray tube whose screen is scanned in step with the electron beam that falls on the sample. Back-scattered electrons and photons emitted by the sample are also used to produce an image. Other types of image are produced by electrons transmitted through the sample and by currents induced in the sample.

The magnification of the microscope is determined by the ratio of the variable current I_s in the scanning coils to the current I_c in the deflection coil of the cathode-ray tube. It can vary continuously from 15 diameters to 100 000 diameters. The resolution is about 10–20 nm.

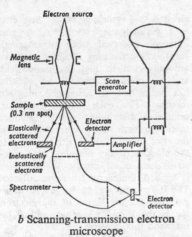

Electron source

Magnetic lens

Scan generator

Sample → (0.3 nm spot)

Elastically scattered electrons

Electron detector

Amplifier

Inelastically scattered electrons

Spectrometer →

Electron detector

b Scanning-transmission electron microscope

3. The *scanning-transmission electron microscope* combines the high resolution of the transmission instrument with the three-dimensional image of the scanning type. The electron beam, produced by *field emission from a tiny (about 10 nm) source, is accelerated by electric fields and focused magnetically to a spot, about 0.3–0.5 nm in diameter, which is made to scan the sample (Fig. *b*). Two signals are obtained from the elastically and the inelastically scattered electrons transmitted through the sample. Dividing the elastic signal by the inelastic signal gives an output proportional to *atomic number. This output is fed into a cathode-ray tube whose electron beam scans in synchrony with the primary electron beam. The output,

controlling the brightness of the final visual image, responds to changes in atomic number rather than thickness of the sample. The contrast can be enhanced electronically. Resolution varies depending on the instrument but the highest resolution will be in the region of 0.3 nm. *See also* proton microscope; scanning tunnelling microscope.

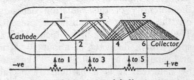

Cathode

Collector

−ve   to 1   to 3   to 5   +ve

Electron multiplier

electron multiplier An electron tube in which current amplification is secured by *secondary emission of electrons. Primary electrons (released by photoelectric effect, or otherwise) are accelerated by application of a high potential and made to strike a good secondary emitter (a *dynode), where they produce a greater number of electrons by impact (*see* diagram). These are then accelerated on to a further secondary emitter, the process being repeated several times within the same envelope. The voltage of each anode must progressively increase above that of the preceding anode, so that the final plate has to be operated at a high potential, typically 1000 V. *See* photomultiplier.

electronographic camera *See* image intensifier.

electron optics The study of the behaviour and control of an electron beam in magnetic and electrostatic fields. An analogy is drawn with the passage of a beam of light through refracting media. A limited magnetic or electric field is regarded as forming an *electron lens for *focusing or defocusing an electron beam.

145

electron paramagnetic resonance *See* electron-spin resonance.

electron-probe microanalysis A technique for analysing small quantities of solids by examining the *X-ray spectrum emitted when the sample is bombarded by a fine beam of electrons. The elements in the sample are detected by the *characteristic X-radiation that they emit and the intensity of this radiation depends on the amount of substance present. The electron beam can be focused to a spot with a diameter of about 10^{-6} m and quantities as small as 10^{-16} kg can be detected. It is a particularly useful method for determining variations in composition over a solid surface.

electron shell A group of electrons in an atom having a given total quantum number n. The innermost or K shell, with $n = 1$, cannot contain more than 2 electrons. The other shells are denoted in sequence by the letters L ($n = 2$), M (3), N (4), ... The electrons in these shells may be grouped into subshells, for which the letters s, p, d, f, g, h, ..., are used. The L shell of an atom is filled if it contains $2s$ electrons and $6p$ electrons, making 8 in all; the M shell may have a total of 18 ($2s$, $6p$, and $10d$). The stability of completed shells and of subshells provides the key to the explanation of the periodic table.

According to modern ideas, the shells do not represent precise locations in space in that electrons "belonging" to one shell may well interpenetrate the orbits or tracks of electrons from another shell and even these orbits are no longer regarded as definite paths. *See* atomic orbital.

electron spectroscopy The measurement of the distribution of electron kinetic energies in a flux of electrons, such as those resulting from beta decay, or following inelastic scattering of an electron

beam by a solid or gas, or emitted from molecules as a result of some applied stimulus such as irradiation with photons, ions, metastable atoms, or other electrons. This information is used to determine energy levels in atoms, molecules, solids, nuclei, etc. The technique depends on accurate measurement of electron energies by some form of *electron spectrometer*.

electron-spin resonance (ESR) *Syn.* electron paramagnetic resonance. A phenomenon observed when paramagnetic substances containing unpaired electrons are subjected to high magnetic fields and microwave radiation. An electron has *spin and an associated magnetic moment. For each set of *quantum numbers describing spatial location there are two states, corresponding to the two possible spin orientations. In the absence of any magnetic field, these states have equal energy. When a field is applied, the energies became different. The difference is given by $gm_B B$, where g is the *Landé factor of the electron, m_B is the Bohr *magneton, and B the magnetic flux density.

If a large number of atoms are considered, each having an unpaired electron, then at normal temperatures there is a statistical probability that slightly more atoms will be in the lower energy state than the higher one. If electromagnetic radiation is applied, they can be raised in energy to the higher state by absorption of a photon of frequency v, where $hv = gm_B B$. For the fields normally used, v lies in the microwave region of the spectrum. In electron-spin resonance spectroscopy the sample is exposed to microwave radiation of variable frequency and at the frequency v, resonance occurs and the radiation is absorbed.

Free electrons have a Landé factor of about 2 but in most compounds this is modified by contributions from the orbital and nuclear magnetic moments. Consequently, shifts in the resonant frequency and *hyperfine structure are observed and information is thus obtained on the chemical bonds in the molecule. *See also* nuclear magnetic resonance.

electron stains Substances such as phosphotungstic acid, osmic acid, silicotungstic acid, and phosphomolybdic acid that have high electron-scattering power and can be used with the *electron microscope in the same way as staining media in optical microscopy.

electron synchrotron *See* synchrotron.

electron temperature Electrons in the plasma of a discharge tube have a distribution of speeds that is approximately Maxwellian. The electron temperature of the plasma is that temperature at which gas molecules would have the same average kinetic energy as the electrons of the plasma.

electron tube Any electronic device in which the movement of electrons between two or more electrodes, through a gas or a vacuum, takes place in a sealed or continuously exhausted envelope. Examples include *cathode-ray tubes, *gas-discharge tubes, and *thermionic valves (now obsolete).

electronvolt Symbol: eV. A unit of energy extensively employed in atomic, nuclear, and particle physics. It is the work done on an electron that is displaced through a potential difference of one volt. It is equal to

$$1.602\,189\,25 \times 10^{-19} \text{ joule}$$

electro-optics The study of the changes in the optical properties of a dielectric produced by the application of an electric field. *See* Kerr effects.

electrophoresis The migration of fine particles of solid suspended in liquid to the anode (*anaphoresis*) or cathode (*catophoresis*) when an electric field is applied to the suspension.

electroplating A practical application of *electrolysis, in which the surface of one metal is covered with another, either for protection, or for decoration, or both.

electropolishing The production of a specularly reflecting surface by anodic etching of an initially rough metal.

electrorheological fluid *Syn.* smart fluid. A fluid that sets to a jelly-like solid when a high voltage is applied across it (about 3 MV m^{-1}). The stiffening is proportional to the electric field strength and is reversible. These fluids are essentially a suspension of particles in a nonconducting liquid.

electroscope An electrostatic instrument for the detection of potential differences. The commonest form consists of a pair of gold leaves hanging side by side from an insulated metal support, and enclosed in a draught-proof case. If the support is given a charge, the leaves separate, owing to their mutual repulsion. One of the leaves may be replaced by a vertical metal plate. An instrument capable of accurate quantitative indication is usually called an *electrometer.

electrosmosis *Syn.* electroendosmosis. The passage of an electrolyte through a membrane or porous partition under the influence of an electric field.

electrostatic deflection A method of deflecting an electron beam using the electrostatic fields produced between two metal electrodes – most often applied to the beam in *cathode-ray tubes using two pairs of deflection plates.

electrostatic focusing *See* focusing.

electrostatic generator A machine for producing electric charge by electrostatic action, e.g. friction or (more usually) electrostatic induction. *See* Wimshurst machine; Van de Graaff generator.

electrostatic induction The production of electric charge on a conductor under the influence of an *electric field. Thus, if an uncharged conductor is placed near a positively charged body, the portion of the conductor nearest to the body becomes negatively charged, the more remote portions being positively charged. If the conductor is insulated, the induced charges are equal in magnitude. If the conductor completely surrounds the charged body, each of the induced charges is numerically equal to the inducing charge on the body.

The separation of charges in a dielectric by an electric field is also described as electrostatic induction.

electrostatic lens *See* electron lens.

electrostatics The study of the phenomena associated with electric charge at rest.

electrostatic units (e.s.u.) *See* CGS system of units.

electrostriction The stress of extension or compression that any body in a medium of relative permittivity different from its own experiences in an electric field. If the field is not homogeneous, a body of higher relative permittivity than its surroundings will experience a force towards the area of greater field strength and vice versa.

electroweak theory A *gauge theory (also called *quantum flavourdynamics*) that provides a unified description of both the *electromagnetic and *weak interactions. In the Glashow–Weinberg–Salam (GWS) theory, also known as the *standard model*, electroweak interactions arise from the exchange of *photons and of massive charged *$W^\pm$ and neutral Z^0 bosons of spin 1 between *quarks and *leptons. The interaction strengths of the gauge bosons to quarks and leptons and the masses of the W and Z bosons themselves are predicted by the theory in terms of a single new parameter, the Weinberg angle θ_w, which must be determined by experiment. The GWS theory successfully describes all existing data from a wide variety of electroweak processes, such as neutrino–nucleon, neutrino–electron and electron–nucleon scattering. A major success of the model was the direct observation in 1983–84 of the $W^\pm$ and Z^0 bosons with the predicted masses of 80 and 91 GeV/c^2 in high energy proton–antiproton interactions. The decay modes of the $W^\pm$ and Z^0 bosons have been studied in very high energy $p\bar{p}$ and e^+e^- interactions and found to be in good agreement with the standard model.

The six known types (or *flavours*) of quarks and the six known leptons are grouped into three separate *generations* of particles as follows:

1st gen.:	e^-	ν_e	u	d
2nd gen.:	μ^-	ν_μ	c	s
3rd gen.:	τ^-	ν_τ	t	b

The second and third generations are essentially copies of the first generation (which contains the electron and the up and down quarks making up the proton and neutron) but involve particles of higher mass. Communication between the different generations occurs only in the quark sector and only for interactions involving $W^\pm$ bosons. Studies of Z^0 boson production in very high energy electron–positron interactions have shown that no further generations of quarks and leptons can exist in nature (an arbitrary number of generations is *a priori* possible within the standard model) provided only that any new neutrinos are approximately massless.

The GWS model also predicts the existence of a heavy spin 0 particle, not yet observed experimentally, known as the *Higgs boson*. This particle results from the so-called *spontaneous symmetry breaking* mechanism used to generate nonzero masses for the $W^\pm$ and Z^0 bosons and is presumably too massive to have been produced in existing particle *accelerators.

elementary particle A particle that, as far as is known, is not composed of other simpler particles. Elementary particles represent the most basic constituents of matter and are also the carriers of the fundamental forces between particles, namely the electromagnetic, weak, strong, and gravitational forces. The known elementary particles can be grouped into three classes: *leptons, *quarks, and *gauge bosons. *Hadrons, such strongly interacting particles as the *proton and *neutron, which are bound states of quarks and/or antiquarks, are also sometimes called elementary particles.

Leptons undergo electromagnetic and weak interactions but not strong interactions. Six leptons are known, the negatively charged *electron, *muon, and *tau leptons plus three associated *neutrinos: ν_e, ν_μ, and ν_τ. The electron is a stable particle but the muon and tau leptons decay through the weak interactions with lifetimes of about 10^{-6} and 10^{-13} seconds. Neutrinos are stable neutral leptons, which interact only through the *weak interaction.

Corresponding to the leptons are six quarks, namely the up (u), charm (c), and top (t) quarks with electric charge equal to $+\frac{2}{3}$ that of the proton and the down (d), strange (s), and bottom (b) quarks of charge $-\frac{1}{3}$ the proton charge. Quarks have not been observed experimentally as free particles but reveal their existence only indirectly in high-energy scattering experiments and through patterns observed in the properties of hadrons. They are believed to be permanently confined within hadrons, either in *baryons, half integer spin hadrons containing three quarks, or in *mesons, integer spin hadrons containing a quark and an antiquark. The proton, for example, is a baryon containing two up quarks and a down quark, while the π^+ is a positively charged meson containing an up quark and an anti-down ($\bar{d}$) antiquark. The only hadron that is stable as a free particle is the proton. The neutron is unstable when free. Within a nucleus, protons and neutrons are generally both stable but either particle may transform into the other by *beta decay or *capture.

Interactions between quarks and leptons are mediated by the exchange of particles known as *gauge bosons, specifically the *photon for electromagnetic interactions, the $W^\pm$ and Z^0 bosons for the *weak interaction, and eight massless *gluons in the case of the *strong interactions. The long-lived particles are given in Appendix, Table 7.

ellipsoid mirror *See* mirror.

ellipsometer An instrument for studying thin films on solid surfaces. It depends for its action on the fact that if *plane-polarized light is incident on a surface, it is reflected as elliptically polarized light. (*See* polarization.) The degree of ellipticity in the reflected beam depends on the thickness of the film.

elliptically polarized radiation *See* polarization.

e/m The ratio of the charge to the mass of an electron; sometimes called the *specific charge* of an electron. Its value decreases (owing to increase in mass) as the velocity of the electron approaches that of light. For slow-moving electrons its value is

$$1.758\ 8047 \times 10^{11}\ C\ kg^{-1}.$$

emanating power The rate of emission of radioactive inert-gas atoms (radon and thoron) from a given material expressed as a fraction of the rate of their production within the solid.

e.m.f. Abbreviation for electromotive force.

emission 1. The liberation of electrons from the surface of a solid or liquid. The types are: (i) *Thermionic emission – emission resulting from the temperature of the substance. (ii) Photoelectric emission – emission resulting from the irradiation of the substance. (*See* photoelectric effect.) (iii) *Secondary emission – emission resulting from bombardment of the substance by electrons or ions. (iv) *Field emission – emission resulting from intense electric fields at the surface of the substance. (v) Emission resulting from the disintegration of radioactive substances.

2. The release of electromagnetic radiation from an excited atom, molecule, etc. *See also* emission spectrum.

emission spectrum The *spectrum of radiation coming directly from a source as distinct from the absorption spectrum when some absorbing medium has been interposed in the path of the radiation from a source emitting a continuous spectrum. Emission spectra may be continuous, line, or band.

emissivity Symbol: ε The ratio of the power per unit area radiated from a surface to that radiated from a black body at the same temperature. Alternatively it can be defined as the ratio of *radiant exitances, $\varepsilon = M_e/M'_e$, where M_e is the radiant exitance of the body and M'_e that of the black body. The

emissivity is restricted to radiation produced by the thermal agitation of atoms, molecules, etc.

emittance Former name for *luminous or *radiant exitance.

emitter The region in a bipolar junction *transistor from which *carriers flow, through the emitter junction, into the *base. The electrode attached to this region is the *emitter electrode* (usually shortened to *emitter*).

emitter-coupled logic (ECL) A family of integrated *logic circuits. The input stage consists of an emitter-coupled transistor pair (known as a *long-tail pair), which forms an excellent differential amplifier. The output is via an *emitter-follower buffer. ECL circuits are inherently the fastest logic circuits as the transistors are operated in nonsaturated mode and the *delay time is therefore exceedingly short (approximately 1 ns). The basic ECL gate has simultaneously both the function required and its complement.

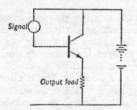

Simple emitter follower

emitter follower An amplifier consisting of a bipolar junction *transistor with *common-collector connection, the output being taken from the *emitter (*see* diagram).

A signal is applied to the *base of the transistor, which is suitably biased so that it is nonsaturated and conducting. Since the transistor is conducting, the emitter will be one *diode voltage from the base at all times, and the emitter follows the signal applied to the base. Since the emitter voltage has a constant value relative to the base voltage, the voltage gain of the amplifier is almost unity, but the current gain is high. The amplifier is characterized by high input impedance and low output impedance and is often used as a *buffer.

e.m.u. Abbreviation for electromagnetic units.

enable To activate a particular electronic circuit or device, selected from a group, in order to effect its operation. An *enable pulse* is often used to select the desired circuit or device: this pulse must be present to allow other signals to be effective.

enantiomorphy A relationship that exists between a left and a right hand, or between any two bodies that can only be brought into coincidence by means of reflection across a plane.

end correction In the elementary theory of the vibrations in a *column of air, it is supposed that the open end of the pipe is a true displacement antinode. This is not true, for some sound energy escapes at each reflection from the open end and is radiated to the atmosphere in the form of spherical waves. The air beyond the open end of the pipe is in vibration and the effective length of the pipe is greater than the actual length, the difference being called the end correction. Experiments give $0.58r$ for a flangeless cylindrical tube, where r is the tube radius. The correction is approximately independent of the wavelength of the sound. The magnitude of the correction for other than cylindrical ends depends upon the degree of openness or conductivity of the end.

endfire array *See* aerial array.

endoergic process *See* endothermic process.

endosmosis *See* electrosmosis.

endothermic process A process during which heat is absorbed by the system from outside. When a nuclear process results in the absorption of energy, it is often called an *endoergic process*.

energy Symbol: E. The quantity that is the measure of the capacity of a body or a system for doing *work. When a body does work, W, its energy decreases by an amount equal to W. The energy of the body upon which it does work increases by exactly the same amount so the total energy of the system does not change. This is the *principle of conservation of energy*. The interaction of two bodies causes the transfer of energy between them. The kind of energy may be unchanged, or it may be partially or wholly changed. If two bodies are at different temperatures, it is possible for energy, Q, to be transferred from that at higher temperatures to that at lower without any apparent forces and displacements by which work could be done. This is the process *heat, which involves work on the molecular scale.

Kinetic energy, symbol: T or E_k, is the energy possessed because of motion and is equal to the work that a body would do if brought to rest with respect to a certain observer. In classical physics it can be shown that a particle of mass m with speed v has translational kinetic energy $T = \frac{1}{2}mv^2$, while a rotating body with moment of inertia I about its axis of rotation and angular velocity ω has rotational kinetic energy $T = \frac{1}{2}I\omega^2$. In *relativity the kinetic energy of a particle is $(m - m_0)c^2$, where m is the mass as determined by the observer with respect to whom it will be brought to rest, m_0 is the rest mass, and c the speed of light. If v is very small compared to c, this formula tends to the classical form.

Potential energy, symbol V or E_p, is energy possessed by a system because of the position of a body with respect to a standard. For example, if a body of mass m is raised to a height h above the ground, the potential energy is mgh, where g is the acceleration of free fall. That is, if the body returns to ground level, the gravitational field of the earth will do this amount of work on it.

Internal energy, symbol: U, is the sum of the potential energies of the molecular interactions and the kinetic energies of the molecular motions within a body. For a solid, the vibrations of the molecules (or atoms, or ions) are nearly simple harmonic and U comprises equal amounts of molecular kinetic and potential energies. For a monatomic gas, U is almost entirely molecular kinetic energy. The value of U depends primarily upon temperature.

The internal energy of a body may be changed by both of the processes of work or heat:
$$\delta U = W + Q.$$
Thus it is misleading to use the popular term "heat energy" for U, since it is not uniquely related to heat. Moreover, it is possible to transfer internal energy from a colder to a hotter body by doing work.

See also electric energy.

energy bands According to quantum theory a single atom has a number of quantum states, each defined by a set of *quantum numbers. If the atom is isolated, each state has a characteristic energy. Normally the electrons occupy the states of lowest energy, not more than one electron going into each state according to the *Pauli exclusion principle. The energy associated with any state can be changed by external fields.

If a large number of atoms are combined in a condensed substance, the energies become spread over certain bands by the interactions with neighbouring atoms. For pure crystals the system is as shown in Fig. *a*. Each state of each atom becomes a state of the crystal. Because of the very large number of states and the effects of the *uncertainty principle, the energies form a continuous distribution in the *allowed bands*. Between these bands are ranges of energy called *forbidden bands* in which there are no quantum states in a pure substance.

The energies of the electrons in the inner states of the atoms are affected very little by the neighbouring atoms. These electrons remain tightly bound to their nuclei and play no part in electrical conduction.

The theory of these energy bands in solids depends on *quantum mechanics. In general, the *Schrödinger equation is solved for an electron moving in a varying electric potential, the periodicity of which is created by the spacing of the ions in the crystal lattice. The allowed solutions give the allowed bands of energy and the energies for which there are no solutions are the forbidden bands.

For a continuum the number of particles with energy between E and $E + dE$ is $N_E dE$, where $N_E = f_E g_E$. The quantity $g_E dE$ is the number of states with energy between E and dE. The *Fermi function* f_E expresses the probability that a state at energy E is occupied (Fig. *b*). At the *Fermi level* (or *Fermi energy*) E_F the value of f_E is exactly one half. Thus for a system in equilibrium one half of the states with energies very nearly equal to E (if any) will be occupied. The value of E_F varies very slowly with temperature, tending to E_0 as T tends to absolute zero.

The quantity g_E is zero in the forbidden bands. Fig. *c* shows schematically

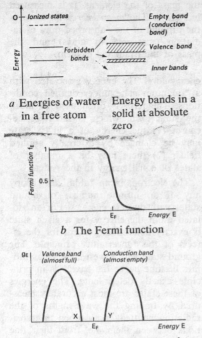

a Energies of water in a free atom   Energy bands in a solid at absolute zero

b The Fermi function

c Bands for a pure non-metal

the form of g_E for a typical pure non-metal. The quantum states of the valence electrons in the separate atoms form the *valence band*, which is separated by an *energy gap* E_G from the *conduction band*. If the value of E_G is greater than about 2 electronvolts, the substance is called an *insulator, if it is less, the substance is called a *semiconductor.

At absolute zero the valence band would be full while the conduction band would be empty. From the form of the Fermi function it can be seen that at all attainable temperatures a small number of the states near the top of the valence band (region X in Fig. *c*) will be vacant, while a few of those near the bottom of the conduction band (region Y in Fig. *c*) will be occupied. The numbers in-

crease rapidly with temperature, especially if E_G is small.

Metals do not have separate valence and conduction bands but a single band containing many more states than electrons to occupy them.

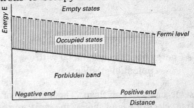

d Energy bands in a metal with potential gradient

If a potential difference is applied between the ends of a wire, the energies of the electron states are changed as illustrated in Figs. *d* and *e*. In the case of a metal there are, near the level of the Fermi energy, very many occupied states and empty states in the same locality. Thus electrons can move freely from occupied states into empty states of the same energy, giving a flow of negative charge towards the positive end. This process is initially conservative as there is a gain of kinetic energy equal to the loss of potential energy. As they move further, however, electrons undergo collisions with irregularities in the crystal structure (*see* defect; phonon) and give up energy to the vibrating ions. Thus the equilibrium energy distribution is maintained (very nearly) at all points along the wire.

In the case of a nonmetal there are relatively few electrons in the conduction band (*free electrons*). There are also a few empty states in the valence band (*holes*). The latter contribute to the conduction, as electrons can move into these states from adjacent occupied states at the same energy. The great majority of the electrons in the valence band cannot take part in conduction since there are no empty states of the

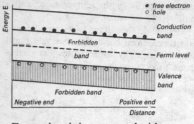

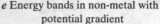

e Energy bands in non-metal with potential gradient

same energy near them. The number of free electrons and holes in most pure nonmetals is incomparably less than the number of electrons and empty states near to the Fermi level of a metal, hence the electrical conductivities of nonmetals are relatively very low. *See* semiconductor.

energy density The amount of (e.g. radiant) energy in unit volume.

energy equipartition *See* equipartition of energy.

energy fluence Symbol: Ψ. A quantity associated with nuclear reactions and ionizing radiation. It is the sum of the energies (excluding rest energies) of all particles that, within a time interval, are incident on a small sphere at a given point, divided by the cross-sectional area of the sphere. The *energy fluence rate* (or *energy flux density*), symbol ψ, is equal to $d\Psi/dt$.

energy flux density *See* energy fluence.

energy gap *See* energy bands.

energy imparted Symbol: ε. The energy delivered to a particular volume of matter by all the directly and indirectly ionizing particles (i.e. charged and uncharged) entering that volume. It is measured in joules. The *specific energy imparted* is the energy imparted to an element of irradiated matter divided by

the mass of the element. It is measured in *grays. The *mean energy imparted* is the product of the *absorbed does of radiation (measured in grays) and the mass of the irradiated element.

energy level The energy associated with a *quantum state under defined conditions. The term is often used to mean the state itself, which is incorrect because: (1) the energy of a given state may be changed by externally applied fields; (2) there may be a number of states of equal energy in the system.

The electrons in an *atom can occupy any of an infinite number of bound states with discrete energies. For an isolated atom the energy for a given state is exactly determinate except for the effects of the *uncertainty principle. The ground state with lowest energy has infinite lifetime hence the energy is in principle exactly determinate. The energies of these states are most accurately measured by finding the wavelengths of the radiation emitted or absorbed in transitions between them, i.e. from their line *spectra. Theories of the atom have been developed to predict these energies by calculation (*see* atom; wave mechanics). The energies of unbound states of positive total energy form a continuum. This gives rise to the continuous background to an atomic spectrum as electrons are captured from unbound states. The energy of an atomic state can be changed by the *Stark effect or the *Zeeman effect.

The vibrational energies of molecules also have discrete values. For example, in a diatomic molecule the atoms oscillate in the line joining them. There is an equilibrium distance at which the force is zero, the atoms repel when closer and attract when further apart. The restraining force is very nearly proportional to the displacement hence the oscillations are simple harmonic. Solution of the

*Schrödinger equation gives the energies of a harmonic oscillation as:
$$E_n = (n + \tfrac{1}{2})hf,$$
where h is the *Planck constant, f is the frequency, and n is the *vibrational quantum number*, which can be zero or any positive integer. The lowest possible vibrational energy of an oscillator is thus not zero but $\tfrac{1}{2}hf$. This is the cause of *zero point energy. The potential energy of interaction of atoms is described more exactly by the *Morse equation, which shows that the oscillations are slightly *anharmonic (*see also* phonon). The vibration of molecules are investigated by the study of band *spectra.

The rotational energy of a molecule is quantized also. According to the Schrödinger equation a body with moment of inertia I about the axis of rotation has energies given by:
$$E_J = h^2 J(J + 1)/8\pi^2 I,$$
where J is the *rotational quantum number*, which can be zero or a positive integer. Rotational energies are found from band spectra.

The energies of the states of the *nucleus can be determined from the spectra of *gamma radiations and from various *nuclear reactions. Theory has been less successful in predicting these energies than those of electrons in atoms because the interactions of *nucleons are very complicated. The energies are very little affected by external influences but the *Mössbauer effect has permitted the observation of some minute changes. *See also* energy bands; X-ray spectrum.

enhancement mode An operating mode of *field-effect transistors in which increasing the magnitude of the *gate bias increases the current.

enrich To increase the *abundance of a particular isotope in a mixture of the isotopes of an element. Applied to a nuclear fuel it means, specifically, to increase the abundance of *fissile isotopes.

The *enrichment* is the proportion of atoms of a specified isotope present in a mixture of isotopes of the same element, where this proportion is greater than that in the natural mixture. This is often expressed as a percentage.

enthalpy Symbol: H. A thermodynamic function of a system equal to the sum of its energy (U) and the product of its pressure (p) and volume (V), i.e.
$$H = U + pV.$$
For a *reversible process at constant pressure the work done by the system is equal to the product of pressure times the change of volume. The heat absorbed in such a process is thus equal to the increase of enthalpy of the system. For example, many chemical processes are carried out in the atmosphere and occur at constant pressure. The heat change involved is thus an enthalpy of reaction − for *exothermic processes the enthalpy change is taken to be negative. *See also* heat capacity; Joule−Kelvin effect.

entrance port, pupil *See* apertures and stops in optical systems.

entropy Symbol: S. A property of a system that changes, when the system undergoes a *reversible change, by an amount equal to the energy absorbed by the system (dq) divided by the thermodynamic temperature T, i.e. $dS = dq/T$.

Entropy, like other thermodynamic properties, such as temperature and pressure, depends only on the state of the system and not on the path by which that state is reached. It is a quantity with an arbitrary zero, only changes in its value being of significance.

The concept of entropy follows from the application of the second law of *thermodynamics to the *Carnot cycle. The *thermodynamic temperature is so defined that the entropy given out at the lower temperature is equal to that taken in at the higher value. As for any completed cycle the entropy of the working substance is unchanged, hence there is no overall change in the entropy of the complete system. In any real cycle there will be some degree of irreversibility and the *efficiency will be less than for the Carnot cycle. Hence more heat is discharged at the lower temperature and the entropy output is greater than the input. Thus the entropy of the whole system is increased. All real changes are to some extent irreversible so all changes within a closed system cause an increase in entropy. In applying this principle it is essential to consider every part of a system. It is quite usual for the entropy of some body to decrease, but this is always accompanied by a greater increase elsewhere.

The entropy of a system is a measure of the unavailability of its internal *energy to do work in a cyclic process. Thus if two bodies at unequal temperatures have the same internal energy, that at the higher temperature has lower entropy. More of this body's internal energy is available to do work in a heat engine than that of the cooler body.

The *Boltzmann entropy theory* relates entropy to the thermodynamic "probability", W, where W is the number of microscopically distinct states of a system that give the same macroscopically determinable state. In quantum terms W is the number of solutions of the *Schrödinger equation for the system giving the same distribution of energy. The relationship is:

$$S - S_0 = k \ln(W/W_0),$$

where k is the *Boltzmann constant and S_0 and W_0 are the entropy and probability for a standard condition. According to the third law of *thermodynamics the entropy of a perfect crystal at *absolute zero is zero. This corresponds to the idea that there is only one way in which this condition of lowest energy can be realized, i.e. taking this as a standard, W_0 is unity and as $\ln 1 = 0$ so $S_0 = 0$. A solid that is not perfect has a certain amount of disorder that gives it a *configurational entropy*. In general, entropy can be thought of as a measure of the molecular disorder of a system.

Consider N molecules arranged in N places to form a condensed substance. If the body is solid, the placing of the first molecule leaves only $N-1$ places for the second, then there are $N-2$ available for the next and so on. Thus the solid can be assembled in $N!$ ways, so $W_S = N!$ For the liquid, $W_L = N^N$ as placing one molecule does not restrict the placing of subsequent ones. The entropy increase on melting is thus:

$$k(\ln W_L - \ln W_S) = k(N \ln N - \ln N!).$$

Since $\ln N! = N(\ln N - 1)$ for very large numbers, this change is kN. If N is the *Avogadro constant, the entropy of melting is equal to R, the *molar gas constant. In practice melting involves other changes (e.g. expansion) and experimental values of the molar *latent heat of fusion divided by the thermodynamic temperature of the melting point are typically 30% above this value.

See also Carnot–Clausius equation; statistical mechanics.

epicentre *See* seismology.

epitaxy A method of growing a thin layer of material upon a single-crystal substrate so that the lattice structure is identical to that of the substrate. The technique is extensively used in the manufacture of *semiconductors when a layer (known as the *epitaxial layer*) of different conductivity is required on the substrate.

epithermal neutron A neutron with an energy just above thermal energies, often taken to be in the range from 10^{-2} to

10^2 eV (1.6×10^{-21} to 1.6×10^{-17} joule). Neutrons in the middle of this range (or a logarithmic scale) have energies of the same order of magnitude as the energies of chemical bonds.

epoxy resins Synthetic polymers that are widely used as structural plastics, surface coatings, and adhesives and for encapsulating and embedding electronic components. They are characterized by low shrinkage on polymerization, high strength, and good adhesion and chemical resistance.

EPROM Abbreviation for erasable programmable read-only memory, i.e. erasable *PROM. A type of semiconductor computer memory that is fabricated in a similar way to *ROM (read-only memory). The contents, however, are added after rather than during manufacture and then if necessary can be erased and rewritten, possibly several times. The contents are normally erased (i.e. reset to their nonprogrammed state) by exposure to ultraviolet radiation. The EPROM can then be reprogrammed using an electronic device known as a PROM programmer.

equalization In electronics, the introduction of networks that compensate for a particular type of *distortion over the frequency band required and hence reduce distortion in a system.

equal temperament *See* temperament.

equation of time *See* time; sundial.

equations of state *Syn.* characteristic equations. Equations showing the relationship between the pressure (p), volume (V), and thermodynamic temperature (T) of a substance.

(1) For a homogeneous fluid, the most familiar equation is:
$$pV = nRT,$$

where n is amount of substance and R the molar gas constant. This equation only holds good for an ideal gas, i.e. a gas that is made up of massive point particles that exert no forces on each other. This is shown by the fact that if the equation holds good over all ranges of pressure, when the pressure becomes infinitely great the volume is zero. To allow for the finite volume occupied by the particles, the equation should be written:
$$p(V - b) = nRT,$$

where b is the least volume into which the particles can be forced by an indefinitely large pressure. Further, the attractive forces between the particles result in a decreased pressure exerted on the walls of the containing vessel and the equation must be altered to:
$$(p + k)(V - b) = nRT.$$

Many equations of state have been proposed. The *van der Waals equation has $k = a/V^2$, so that the equation becomes:
$$(p + a/V^2)(V - b) = nRT.$$

The *Dieterici equation* is:
$$p(V - b) = nRT \exp(-a/RTV).$$

It gives the van der Waals equation as a first approximation. Certain tests may be applied to these characteristic equations. For example, the critical specific volume v_c, is equal to about four times the liquid specific volume (or volume of unit mass). Now the constant, b, is approximately equal to the liquid specific volume, so that $v_c = 4b$. Again, the quantity $RT_c/p_c v_c$ is roughly constant for unassociated fluids and is equal to $15/4$. Hence if the critical constants of the fluids under test are known, they should satisfy the values $v_c = 4b$ and $RT_c/p_c v_c = 15/4$. It can be shown that the van der Waals equation gives $v_c = 3b$ and $RT_c/p_c v_c = 8/3$. *See also* reduced equation of state.

(2) For a solid, using the *virial law of Clausius, the equation of state for a solid may be written:

$pV + G(U) = -E[\mathrm{d}(\log v)/\mathrm{d}(\log V)]$,
where $G(U) = V(\mathrm{d}/\mathrm{d}V)W(U)$, and $W(U)$ is the potential energy per mole of the crystal when the atoms are at rest in their mean positions; v is the frequency of oscillation of an atom about its mean position; and E is the total energy of the oscillations given by:

$$E = \int_0^T C_m \mathrm{d}T,$$

where C_m is the molar heat capacity at constant volume. Debye deduced an equation of state for solids based on thermodynamics and statistical mechanics modified to include the quantum theory. This equation is essentially the same as that of Clausius.

equatorial mounting *See* telescope.

equilibrant A single force (if one exists) capable of balancing a given system of forces.

equilibrium 1. The condition existing in a system of coplanar forces when the algebraic sums of the resolved parts of the forces in any two directions are both zero and the algebraic sum of the moments of the forces about any point in their plane is zero. If the system of forces is not coplanar, then the same results must hold between the components of the forces lying in any plane and also for the components lying in two other different planes.

As any system of forces can be reduced to a single force and a single couple, the condition of equilibrium is also that these shall both vanish.

2. A body is in stable, unstable, or neutral equilibrium according to whether the forces brought into play following a slight displacement tend to decrease, increase, or not affect the displacement respectively. These concepts of equilibrium and of stability can be generalized to apply to other physical systems, e.g. a system of electric charges in a potential field; a soap bubble on the end of a tube connected to a reservoir of air.

In general, the potential energy of a system is a minimum or a maximum if the equilibrium is stable or unstable respectively. Neutral equilibrium may turn out to be either stable or unstable equilibrium if a large enough displacement be applied. (*See* least energy principle.)

equilibrium constant Symbol: K. *See* mass action, law of.

equinoxes 1. The two points at which the ecliptic intersects the celestial equator (*see* celestial sphere).

2. The two days of the year when the sun is at these points, day and night being of equal length. They occur about March 21 and September 23.

equipartition of energy The principle that the mean energy of the molecules of a gas is equally divided among the various *degrees of freedom of the molecules. The average energy of each degree of freedom is equal to $\frac{1}{2}kT$, where k is the Boltzmann constant and T is the thermodynamic temperature.

The principle was later extended to the vibrations of atoms in crystals and to electromagnetic radiation in a cavity (*see* black-body radiation). Some of the results were consistent with experiment within certain conditions; for example, the principle predicts *Dulong and Petit's law for the specific heat capacities of solids, which was verified for most substances at the temperatures that were then attainable. In the case of radiation the principle led to difficulties and *Planck proposed the *quantum theory to overcome these. In general the equipartition principle is considered untenable, although it is an admissible approximation in certain cases, especially at high temperatures.

equipotential Having the same electric or gravitational *potential. An equipotential surface is a surface drawn so that all points on it are at the same potential. Hence no work is done in moving a small charge or mass in any direction in the surface.

equitempered scale *See* temperament.

equivalent circuit An arrangement of simple circuit elements that has the same electrical characteristics as a more complicated circuit or device under specified conditions.

equivalent focal length The distance from a principal point to its corresponding principal focal point (*see* centred optical systems). With a zoom lens or variable focal length lens, the equivalent focal length is a variable. It can be considered as the ratio of the size of an image of a small distant object near the axis to the angular distance of the object in radians.

equivalent network An electrical network that may replace another network without materially affecting the conditions obtaining in the other parts of the system, but usually only at one particular frequency.

equivalent resistance The value of total *resistance that, if concentrated at a point in an electrical circuit, would dissipate the same power as the total of various smaller resistances at different points in the circuit.

equivalent sine wave A sine wave having the same *root-mean-square value as the given wave and also the same fundamental frequency.

erecting system *Syn.* erector. An optical system, such as a pair of lenses or prisms, that may be used in a telescope, binoculars, periscope, or other viewing instrument to provide an image that is the right way up – *erect* – rather than inverted.

erg The *CGS unit of energy. 1 erg = 10^{-7} joule.

ergon A quantum of energy of an oscillator equal to the product of the frequency of oscillation and the *Planck constant.

error equation *See* frequency distribution.

errors of measurement 1. Accidental errors. In all physical measurements small errors, due to instrumental imperfections and inaccurate human judgments, always occur. It is often possible and always desirable to estimate the magnitudes of the errors in each part of the experiment and to combine these to find the likely error in the final result. The actual estimated errors in quantities that are added or subtracted should be added in finding the error in the result; the percentage errors should be added if the quantities arc combined by multiplication or division. Graphical or arithmetical methods are used to combine observations of a similar type, on the grounds that a better result can be obtained than from one single observation. (*See* probable error.)

Heisenberg's *uncertainty principle shows that there is an irremovable minimum uncertainty in all physical measurements no matter how perfect the instruments or how accurate the observer.

2. Systematic errors. The foregoing comments do not apply to these errors, which must be removed by suitable design of apparatus and technique. *See* personal equation.

Esaki diode *See* tunnel diode.

escape speed The speed that a projectile, space probe, etc., must reach in order to escape the *gravitational field of a planet or satellite. It depends on the mass and diameter of the planet or satellite. For the earth, the escape speed is about $11\,200\ \mathrm{m\,s^{-1}}$; to escape the moon's gravitational field a speed of $2370\ \mathrm{m\,s^{-1}}$ must be attained.

The escape speed is equal to the lowest speed with which a body from remote space (such as a meteorite or returning space vehicle) approaches the planet.

ESR Abbreviation for *electron-spin resonance.

e.s.u. Abbreviation for electrostatic units.

etalon *See* Fabry–Perot interferometer.

etched figures Minute pits, bounded by small faces, that are formed on crystal surfaces by treatment with solvents. These are extremely useful in helping to determine the symmetry of crystals.

ether A now-discarded hypothetical medium once thought to fill all space and to be responsible for carrying light waves and other electromagnetic waves. In order to facilitate description and to provide a physical explanation of various phenomena involving action at a distance, electricity, magnetism, transmission of light, and other radiations, a medium was postulated with mechanical properties adjusted to provide a consistent theory. For the transmission of electromagnetic radiation, it was assumed to pervade all space and matter, to be extremely elastic yet extremely light, to transmit transverse waves with the speed of light, to have a greater density in matter than in free space. *See also* Michelson–Morley experiment.

Ettinghausen effect The establishment of a difference of temperature between the edges of a plate along which an electric current is flowing when a magnetic field is applied at right angles to the plane of the plate. The effect is very small, and for copper, platinum, and silver is unappreciable.

Euler equations The three differential equations of motion of a rigid body (*a*) relative to the centre of mass, using the principal axes of the body through the centre of mass as coordinate axes, or (*b*) about a fixed point using the principal axes through this point as coordinate axes. They are:

$$A(\mathrm{d}\omega_x/\mathrm{d}t) - \omega_y\omega_z(B - C) = G_x$$
$$B(\mathrm{d}\omega_y/\mathrm{d}t) - \omega_z\omega_x(C - A) = G_y$$
$$C(\mathrm{d}\omega_z/\mathrm{d}t) - \omega_x\omega_y(A - B) = G_z$$

A, B, and C are the principal moments of inertia, ω_x, ω_y, and ω_z are the components of angular velocity, and G_x, G_y, and G_z are the components of the applied torque about the principal axes OXYZ.

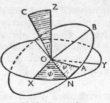

Euler's angles

Eulerian angles A set of three angles (ω, ϕ, ψ) particularly useful in describing the position of a body moving about a fixed point O (*see* diagram). Cartesian axes, OABC, are fixed in the body (OC usually being an axis of symmetry, e.g. the axis of a top) and the motion is described relative to fixed Cartesian axes OXYZ (OZ is usually vertical). θ is the angle between the axis of the body OC and the axis OZ. The plane OAB in the body intersects the plane XOY (usually horizontal) in the *nodal line* ON. The angle $\phi = X\hat{O}N$ measures the *precession* of the axis OC around the vertical *precession axis* OZ. $\psi = A\hat{O}N$ measures the rotation of the body about its own axis OC. Variations in θ are referred to as *nutation*.

eutectic A mixture of two substances that solidifies as a whole when cooled, without change in composition. The *eutectic point* is the temperature at which the eutectic mixture solidifies.

evaporation 1. The conversion of a liquid to a vapour at a temperature below the boiling point. The process involves cooling of the liquid because molecules in the liquid state have negative potential energies resulting from intermolecular interactions.
   2. The conversion of a substance, usually a metal, into a vapour at high temperatures, either from the liquid state or by sublimation from the solid metal. It is used for producing thin films of metal, used in *transistors and in studies of surface properties.

even-even nucleus A nucleus that contains an even number of protons and an even number of neutrons. Well over a half of all stable nuclides have even-even nuclei.

even-odd nucleus A nucleus that contains an even number of protons and an odd number of neutrons. About a fifth of stable nuclides have even-odd nuclei.

even parity *See* wave function.

event A point in a *four-dimensional continuum, defined by three coordinates of space and one coordinate proportional to time.

event horizon *See* Schwarzschild radius.

Ewing's theory of magnetism The theory that the individual atoms or molecules of ferromagnetic substances act as small magnets. In the unmagnetized state of the substance, these elementary magnets arrange themselves in closed chains so that the net effect of their poles externally is zero. Magnetization is produced by a realignment of the elementary magnets with their magnetic axes in the direction of magnetization, and saturation is reached when all are so aligned. The substance is prevented from following the changes in the magnetizing field owing to the force necessary to break up the molecular chains, thus explaining the phenomenon of magnetic hysteresis. This theory has been partially confirmed by modern investigations. *See* ferromagnetism.

exa- Symbol: E. A prefix denoting 10^{18}; for example, one exasecond (1 Es) equals 10^{18} seconds.

excess conduction In a *semiconductor, conduction due to electrons that are not required to complete the chemical bonding of the semiconductor and are therefore available to conduct charge. These electrons usually come from a donor impurity.

exchange force A force acting between particles due to the exchange of some property. In *quantum mechanics such forces can arise when two interacting particles can share some property: for example, the covalent bond responsible for the binding of the molecular hydrogen ion results from the two protons sharing an orbital electron. Alternatively, this electron may be regarded as being continually exchanged between the protons. In particle physics an exchange of particles, known as *gauge bosons, is now considered responsible for the four fundamental interactions – *strong, *electromagnetic, *weak, and *gravitational interactions. *See also* ferromagnetism.

exchange relation The statement
$$(pq - qp) = (h/2\pi i),$$
in which p and q are matrices replacing momentum and positional coordinates in *matrix mechanics. It replaces the old Wilson–Sommerfeld type of quantum condition. *See* quantum theory; quantum mechanics.

excitation 1. The addition of sufficient energy to an atom, molecule, etc., to change it to a state of higher energy. *See* excitation energy.
   2. The production of magnetic flux in an electromagnet by means of a current in a winding. The current is referred to as the *exciting current*.
   3. The application of an electrical signal to drive a device such as an amplifier, tuned circuit, or piezoelectric oscillator.

excitation energy *Syn.* critical potential. The energy required to change an atom or molecule from one quantum state to another of higher energy. It is equal to the difference in energy of the states and is usually the difference in energy between the *ground state of the atom and a specified *excited state.

excitation purity *See* chromaticity.

excited state The state of a system, such as an atom or molecule, when it has a higher energy than its *ground state. *See also* excitation energy.

exciton An electron in combination with a *hole in a crystalline solid. The electron has gained sufficient energy to be in an *excited state and is bound by electrostatic attraction to the positive hole. The exciton may migrate through the solid and eventually the hole and electron recombine with emission of a photon.

exclusion principle *See* Pauli exclusion principle.

exclusive OR gate *See* logic circuit.

exitance 1. *See* luminous exitance. **2.** *See* radiant exitance.

exit port, pupil *See* apertures and stops in optical systems.

exoergic process *See* exothermic process.

exosphere The outermost *atmospheric layer of the earth, extending from about 400 km.

exothermic process A process during which heat is evolved from the system. When a nuclear process results in the production of heat, it is often called an *exoergic process*.

exotic atom An unstable atom in which an electron has been replaced artificially by another negatively charged particle, such as a muon, pion, or kaon. Following capture the particle drops through the atomic energy states, causing X-ray photons to be emitted, before colliding with the nucleus. Exotic atoms are studied by means of these X-rays.

expanded sweep A technique whereby the electron beam in a *cathode-ray oscilloscope is made to move at greater speed during part of its horizontal traverse across the screen.

expander *See* volume compressors (and expanders).

expanding universe Lines in the spectrum of the light from remote galaxies are shifted towards the long wavelength end by an amount that is greatest for those galaxies known to be farthest away. If this *redshift is interpreted as due to a velocity away from the earth in the line of sight, then all galaxies (beyond the Local Group) are receding from us and those galaxies that are farthest away are moving fastest. This leads to the conclusion that the distance between clusters of galaxies is continuously increasing. Thus the universe is expanding. *See* big-bang theory.

expansion *See* coefficient of expansion; adiabatic process; isothermal process.

exploring coil *Syn.* search coil. A coil used for measuring magnetic flux. It is commonly used in conjunction with a *ballistic galvanometer or a *fluxmeter.

exponential decay The decrease of some physical quantity, usually with time, according to a negative exponential law, represented by an equation of the type $y = y_0 \, e^{-at}$. Examples occur in many diverse branches of physics, e.g. the fall of amplitude in damped harmonic oscillations (*see* damped vibrations), the fall in voltage of a charged capacitor leaking through a high resistance, and the fall in activity of a pure radioactive substance with an inactive daughter product.

exponential functions Any expression of the form $y = A \, e^{ax}$ is an exponential function but *the* exponential function is e^x, where $e = 2.718\,281\,8\ldots$, the sum of the infinite series:

$$1 + 1/1! + 1/2! + 1/3! + \ldots + 1/n! \ldots.$$

The *negative exponential function* refers to e^{-x}. For convenience of printing, the exponential functions may be written: $\exp(ax)$, especially when the index or exponent (whence the name) is a complicated expression. These functions are solutions of the differential equation $dy/dx = ay$, where a is a constant.

exposure 1. Symbol: H. The product of the *illuminance, or the *irradiance, and the time, Δt, for which the material in question is illuminated or irradiated. (*See* light exposure; radiant exposure.) The quantity Δt is the *exposure time* and should not be confused with the term exposure.

2. *See* dose.

exposure meter *Syn.* light meter. A photographic instrument that measures light intensity by means of a *cadmium sulphide cell or *selenium cell. For a particular type of film it indicates the *f-number required for a given shutter speed, or vice versa, to give the correct *exposure. A similar instrument giving a measurement in terms of the actual light intensity rather than as the required f-number for a given shutter speed is called a *light meter*.

extensive air shower *Syn.* Auger shower. *See* cosmic rays.

extensometer A device for measuring the small change in length of an arbitrary length of a sample undergoing strain.

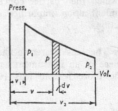

External work done on reversible expansion

external work Work done by a substance expanding against an external resistance. For a *reversible expansion the work done is $\int_{v_1}^{v_2} p\,dv$, where v_1 and v_2 are the initial and final volumes and p is the pressure (*see* diagram). For a cyclic reversible process the external work done per cycle is given by the area in the diagram representing the cycle.

For real processes the system departs from equilibrium and pressure is no longer uniform in the substance or the immediate surroundings. The work done cannot then always be calculated accurately from the formula.

extinction coefficient *See* linear attenuation coefficient.

extraordinary ray *See* double refraction.

extrapolation The estimation of the value of a function for a value of the variable lying outside the range of those for which the function is known. This may be done graphically, by extending the graph of the function beyond the actual plotted points, or by calculation as for *interpolation. Extrapolation is of necessity less accurate than interpolation.

extremely high frequency (EHF) *See* frequency band.

extrinsic semiconductor A *semiconductor in which the *carrier concentration is dependent upon impurities or other imperfections.

eye lens The lens nearer to the eye in an eyepiece of an instrument as distinct from the more remote lens (the field lens).

eyepiece *Syn.* ocular. The single lens, doublet, or combination of lenses, acting virtually as a magnifying lens to examine the image formed by an objective. (*See under* Ramsden, Huygens, Kellner, Fraunhofer eyepieces.) It is usual to arrange that the image from the objective lies in the focal plane of the eyepiece, which thus delivers parallel rays out of the instrument (infinity adjustment). *See*

microscope; refracting telescope; reflecting telescope.

F

Fabry–Perot interferometer An interferometer that is a spectroscopic device of extremely high resolution and also serves as a laser resonant cavity. In its simplest form it consists of two parallel optically flat semisilvered or semialuminized glass plates. The plates are separated by an air gap of a few millimetres or centimetres, or of much greater length when serving as a resonant cavity. If the gap can be varied, the device is called an interferometer. If the separation is fixed, and the plates are only adjustable for parallelism, it is called an *etalon*.

A ray of light from a particular point on the source enters through the first partially reflecting plate and is multiply reflected within the gap. The rays transmitted through the second plate are focused on a screen, where they interfere to form either a bright or dark spot. All rays incident on the gap at a given angle will produce a single circular fringe. With a broad diffuse source, the interference pattern will be narrow concentric rings.

On account of its sharp fringes and high resolving power the device is used for accurate comparison of wavelengths and the study of the *hyperfine structure of spectral lines.

face-centred The form of crystal structure in which the atoms occupy the centres of the faces of the lattice as well as the vertices.

facsimile transmission A system that provides electronic transmission of documents, including pictorial matter. The original document is scanned at the sending station, so providing a successive analysis from which an electrical representation (either analogue or digital) is produced. These electrical signals are sent over a communications channel to the receiving station, which produces a duplicate image on paper; this image is called a *facsimile*. The commercial system known as *Fax* uses the telephone network to transmit the information.

fading In communications, variations in the signal strength at the receiver caused by variations in the transmission medium. It is usually caused by destructive interference between two waves travelling to the receiver by two different paths. If all frequencies in the transmitted signal are attenuated approximately equally, the fading is known as *amplitude fading* and results in a smaller received signal. If different frequencies are attenuated unequally, the fading is known as *selective fading* and results in a distorted signal at the receiver.

Fahrenheit scale The temperature scale on which the ice point is defined as 32 °F and the steam point as 212 °F. It is no longer in use for scientific purposes.

fail-safe device An automatic device that causes a system to cease operation when a failure occurs in the supply or control of power, or the overall structure is found defective.

fallout 1. Radioactive materials that fall to earth following a nuclear-bomb explosion. *Local fallout* is observed downwind of the explosion after a few hours, no more than about 500 km from the source; it consists of large particles. During the month or so that follows, a *tropospheric fallout* of fine particles is observed in various locations at roughly the same latitude as the explosion. The

particles that are drawn up to high altitudes often take many years before being deposited all over the surface of the earth, and are referred to as *stratospheric fallout*.

2. A substance that enters the atmosphere from a source on the earth's surface (e.g. a volcano, nuclear reactor, car exhaust, etc.) that is later deposited as particles either in the vicinity of the source or elsewhere.

fall time A measure of the rate of decay of a periodic quantity. The time required for the ratio of the amplitude at a particular instant and its peak value to fall from 0.9 to 0.1.

fan-in The maximum number of inputs to a *logic circuit.

fan-out The maximum number of inputs to other circuits that can be driven by the output of a given *logic circuit.

farad Symbol: F. The *SI unit of *capacitance, defined as the capacitance of a capacitor that acquires a charge of one coulomb when a potential difference of one volt is applied. The farad is far too large a unit for ordinary use and the submultiples microfarad (μF), nanofarad (nF), and picofarad (pF) are generally employed.

Faraday cage An arrangement of conductors, or conducting sheet or mesh, surrounding electrical equipment and so shielding it from external electrical disturbances: no electric field can be produced inside a hollow conductor by an external charge (Faraday, 1836).

Faraday constant Symbol: F. The quantity of electricity equivalent to one mole of electrons, i.e. the product of the *Avogadro constant and the charge on an electron in coulombs. It is, therefore, the quantity of electricity required to liberate or deposit 1 mole of a univalent ion. Its value is

$$9.648\ 453\ 1 \times 10^4 \text{ C mol}^{-1}$$

Faraday cylinder 1. A closed or nearly closed hollow conductor, usually earthed, placed round electrical apparatus to shield it from the external electric fields.
  2. A similar structure for the collection of a stream of charged particles (electrons or gaseous ions), usually shielded by an earthed cylinder. The inner conductor is insulated and connected to suitable detecting apparatus.

Faraday dark space *See illustration*, gas-discharge tube.

Faraday effect *Syn.* Faraday rotation. The rotation of the plane of polarization experienced by a beam of plane-polarized light when it passes through certain substances exposed to a strong magnetic field. (There must be a component of the field in the direction of propagation.) The effect occurs, for example, in heavy flint glass (as discovered by Faraday), quartz, and water. Faraday rotation is also observed with other plane-polarized radiation, such as radio waves passing through a plasma in which a magnetic field is present. The angle of rotation is directly proportional to the magnetic field strength and to the pathlength in the substance. *See also* Kerr effects.

Faraday–Neumann law *See* electromagnetic induction.

Faraday rotation *See* Faraday effect.

Faraday's disc An early model of an electromagnetic d.c. generator. It consists of a copper disc that can be rotated (usually by hand) about a horizontal axis through its centre at right angles to the plane of the disc, between the poles of a permanent horseshoe magnet. When the disc is in rotation, an e.m.f. is induced between the axis and the circumference, and can be tapped off by sliding contacts or brushes.

Faraday's laws of electrolysis The mass of any substance liberated from an electrolyte by the passage of current is proportional to the product of the current and the time for which it flows. If the same current passes for the same time through a series of different electrolytes, the masses of the different substances liberated are directly proportional to their relative atomic masses (atomic weights) divided by the charge of the ion carrying the current.

Faraday's laws of induction 1. Whenever the number of lines of magnetic induction linked with a conducting circuit is changing, an induced current flows in the circuit, which continues only so long as the change is actually taking place.
  2. The direction of the induced current in the circuit is such that its magnetic field tends to keep the number of lines linked with the circuit constant.
  3. The total quantity of electricity passing round the circuit is directly proportional to the total change in the lines of induction divided by the resistance of the circuit. *See* electromagnetic induction.

Faraday tube *See* electric flux.

far infrared or ultraviolet *See* near infrared or ultraviolet.

fast axis In negative crystals (e.g. calcite), the electric vibrations of the extraordinary ray (which travels faster) are parallel to the optic axis, which is then referred to as the fast axis. In positive crystals (e.g. quartz), the fast axis is at right angles to the optic axis.

fast breeder reactor A fast *nuclear reactor that breeds more *fissile material than it consumes (*see also* breeder reactor). These reactors are much more economical in the use of fuel than thermal reactors, being able to utilize some 75% of the uranium ore as it comes from the earth, compared to less than 1% in thermal reactors. After the first fuelling of a fast breeder reactor, which requires some 3000 kilograms of plutonium per 1000 megawatts of electricity produced, the net fuel requirement is a very small quantity of natural uranium.

Fuel must be arranged in a compact *core, resulting in a large heat flow from a small surface area. A liquid metal (usually sodium) is used as the *coolant. Usually two sodium circuits are used: in the first the sodium passes through narrow channels round the fuel elements, becoming radioactive in the process. Heat exchangers are used to transfer heat from this radioactive sodium to another sodium circuit from which the heat is in turn extracted by a second heat exchanger in which steam is raised. The extra complexity of fast breeder reactors compared to *thermal reactors means that they are unlikely to be economic until the early decades of the next century.

fast fission Nuclear fission that is induced by *fast neutrons.

fast neutron A neutron with a kinetic energy greater than some specified value, usually above 0.1 MeV (1.6 × 10^{-14} joule). However, is is also applied to neutrons that have an energy greater than the fission threshold in ^{238}U, which is about 1.5 MeV. Neutrons of this energy are capable of initiating fast fission.

fast reactor *See* nuclear reactor; fast breeder reactor.

fatigue The progressive decrease of a property due to repeated stress, e.g. the elasticity of a metal under continuous vibration.

fault A defect in any apparatus that interferes with or prevents normal operation.

Fax *See* facsimile transmission.

feedback The process of returning a fraction of the output signal of a signal device to the input. It usually applies to *amplifiers, the gain of the amplifier being either increased or reduced according to the relative phase of the returned signal.

Feedback may occur through one electrical path or through several paths. Capacitive feedback employs a *capacitor as the feedback device, and inductive feedback employs an *inductor or *inductive coupling. If the phase of the feedback is such that the input signal (voltage or current) is increased, the feedback is known as *positive feedback*. If sufficient positive feedback is applied, the amplifier will oscillate. If the phase is such that the input signal is decreased, the feedback is known as *negative feedback*. This is the type of feedback most commonly employed as it tends to stabilize the amplifier or reduce noise and distortion in the circuit: feedback used for this purpose may also be called stabilized feedback.

feedback control loop *See* loop.

feeder **1.** A system of wires or waveguides that conveys radio frequency power between a radio aerial and a transmitter or receiver, with minimum loss.

2. An electric line that conveys electric power from a generating station to a point of a distributing network. It is not tapped at any intermediate points.

See transmission line.

feedthrough A contact between one layer of interconnections on a *printed circuit board and the next layer, passing through the insulating materials separating them. Usually only a double-sided board is used, but up to 12 layers have been mounted on a single board. Likewise, in an *integrated circuit with multilayer interconnections, feedthroughs can be used to make contact between one layer of interconnections and the next.

Felici balance A method of determining the mutual inductance between the windings of an inductor by means of an alternating-current bridge circuit.

femto- Symbol: f. A prefix denoting 10^{-15}. For example, one femtometre (1 fm) equals 10^{-15} metre.

Fermat's principle The path of a ray in passing between two points during reflection or refraction is the path of least time (*principle of least time*). It is now more usually expressed as the principle of *stationary* time: that the path of the ray is the path of least *or* greatest time. If a reflecting or refracting surface has smaller curvature than the aplanatic surface tangential to it at the point of incidence, the path is a minimum; if its curvature is greater than the aplanatic surface, its path is a maximum.

fermi A unit of length used in nuclear physics, equal to 10^{-15} metre.

Fermi age theory An approximate method of calculating the *slow-down density of neutrons in a *nuclear reactor, based on the assumption that they lose energy continuously rather than in discrete amounts. The *age equation*:
$$\nabla^2 q - dq/d\tau = 0$$

relates the slowing-down density q to *neutron age, τ. Because of the assumptions made, the theory is least applicable to media containing light elements.

Fermi–Dirac distribution function Symbol: f_E. For any system of identical *fermions in equilibrium, the probability that a quantum state of energy E is occupied. It is given by:
$$f_E = 1/[\exp(\alpha + E/kT) + 1],$$
where k is the *Boltzmann constant, T is the *thermodynamic temperature, and α depends on the temperature and the concentration of particles. Since the exponential function cannot be less than zero, f_E cannot be greater than one, in agreement with the *Pauli exclusion principle. *See* quantum statistics.

Fermi–Dirac statistics *Syn.* Fermi statistics. *See* quantum statistics.

Fermi function *See* energy bands.

Fermi gas model A model of the nucleus in which the neutrons and protons are regarded as independent particles obeying Fermi–Dirac statistics (*see* quantum statistics) but confined within a cube having a volume equal to that of the nucleus. The model is similar to the theory of electrons in solids. It is useful in describing collisions in high-energy nuclear processes.

Fermi level *See* energy bands.

fermion Any *elementary particle having half-integer *spin. Fermions obey Fermi–Dirac statistics (*see* quantum statistics). All particles are either fermions or *bosons; *leptons, *quarks, and *baryons are fermions.

Ferranti effect An effect occurring in *transmission lines when the load is suddenly reduced. The charging current through the line inductance causes a sharp rise in the voltage at the end of the line.

ferrimagnetism The property of certain solid substances, such as ferrites, that show both ferromagnetic and antiferromagnetic properties. (*See* ferromagnetism, antiferromagnetism.) It is characterized by a small positive magnetic *susceptibility that increases with temperature. It is caused by the presence of two types of ion in the crystal with unequal electron *spins – arranged so that the magnetic moments of adjacent ions are antiparallel. Thus the situation is similar to that in antiferromagnetic materials with the difference that the magnetic moments are unequal.

ferrite A low-density ceramic oxide of iron to which another oxide has been added. The formula for a typical ferrite is $Fe_2O_3.XO$, where X is a divalent metal such as cobalt, nickel, zinc, or manganese. Ferrites possess insulating properties and exhibit *ferrimagnetism or *ferromagnetism according to the nature of X.

ferroelectric materials Dielectric materials, usually ceramics such as barium titanate, that in an alternating electric field develop very large values of the dielectric constant, generally in one particular direction, within a certain temperature range. They exhibit dielectric hysteresis and in most cases the piezoelectric effect. These properties are in many ways analogous to *ferromagnetism.

ferromagnetism A property of certain solid substances that, having a large positive magnetic *susceptibility, are capable of being magnetized by weak magnetic fields. The chief ferromagnetic elements are iron, cobalt, and nickel,

and many ferromagnetic alloys based on these metals also exist.

Ferromagnetic materials exhibit magnetic *hysteresis. Their relative *permeability is much greater than unity and they achieve a maximum magnetization (magnetic saturation) at fairly low external magnetic field strengths. At a certain temperature, the Curie temperature, there is a change from ferromagnetism to *paramagnetism. The magnetic *susceptibility then varies according to the *Curie–Weiss law.

Ferromagnetics are able to retain a certain amount of magnetization when the magnetizing field is removed. Those materials that retain a high percentage of their magnetization are said to be hard, and those that lose most of their magnetization are said to be soft. Typical examples of hard ferromagnetics are cobalt steel and various alloys of nickel, aluminium, and cobalt. Typical soft magnetic materials are silicon steel and soft iron. (*See* coercive force.)

The characteristic features of ferromagnetism are explained by the presence of *domains*. A ferromagnetic domain is a region of crystalline matter, whose volume may be between 10^{-12} and 10^{-8} m^3, which contains atoms whose magnetic moments are aligned in the same direction. The domain is thus magnetically saturated and behaves like a magnet with its own magnetic axis and moment. The magnetic moment of a ferromagnetic atom results from the spin of the electrons in an unfilled inner shell of the atom. The formation of a domain depends upon the strong interatomic forces (*exchange forces*) that are effective in a crystal lattice containing ferromagnetic atoms.

In an unmagnetized volume of a specimen, the domains are arranged in a random fashion with their magnetic axes pointing in all directions so that the specimen has no resultant magnetic moment. Under the influence of a weak magnetic field, those domains whose magnetic axes have directions near to that of the field grow at the expense of their neighbours. In this process the atoms of neighbouring domains tend to be aligned in the direction of the field but the strong influence of the growing domain causes their axes to align parallel to its magnetic axis. The growth of these domains leads to a resultant magnetic moment and hence magnetization of the specimen in the direction of the field. With increasing field strength the growth of domains proceeds until there is, effectively, only one domain whose magnetic axis approximates to the field direction. The specimen now exhibits strong magnetization. Further increases in field strength cause the final alignment and magnetic saturation in the field direction. This explains the characteristic variation of magnetization with applied field strength.

The presence of domains in ferromagnetic materials can be demonstrated by the use of *Bitter patterns or by the *Barkhausen effect.

fertile Of a nuclide. Capable of being transformed into a *fissile material in a *nuclear reactor. Uranium-238 is an example of a fertile nuclide.

Féry total radiation pyrometer A *pyrometer used for the direct measurement of temperature up to 1400 °C by measuring the total energy of radiation of all wavelengths from the source.

Fessenden oscillator An efficient form of electromagnetic or electrodynamic underwater sound generator and receiver. It is commonly used for signalling through sea when a large range is required.

FET Abbreviation for *field-effect transistor.

Feynman diagram and propagator *See* quantum electrodynamics.

fibre-optics system An optical system in which a single glass or plastic fibre or an array of fibres transmits light between two points. Such systems have many uses, including the direct transmission of images and illumination and medium- and long-distance communications. An optical fibre consists of a single flexible rod of high refractive index, less than 1 mm in diameter, having polished surfaces coated with transparent material of lower refractive index. This coating, known as *cladding*, prevents light from leaking between fibres in close proximity. In a *stepped-index fibre*, the indexes of both cladding and core are constant throughout. Light falling on one end within a certain solid angle will undergo *total internal reflection at the cylindrical surface of the glass core. The light is trapped within the core and travels in zig-zag paths down the length of the fibre with little or no absorption. The fibre can continue to reflect light when it is considerably curved, as long as the reflection angle remains greater than the critical angle.

In a *graded-index fibre*, the refractive index of the core decreases radially outwards. Light rays then spiral smoothly around the central axis rather than zig-zagging. This reduces the difference in time delays of the transmitted light rays. There may be many hundreds of ray paths, or modes, by which energy can propagate down the core. In a *single-mode fibre*, the core is very narrow relative to the cladding and rays travel parallel to the central axis; it may be stepped- or graded-index. These are the most efficient fibres for communications. A laser is used as the light source and information is carried by modulating the light. In general, all optical fibres have a much greater data-handling-capacity than, say, telephone systems and electric cables as the frequency of the transmitted signal is so much greater.

Images may be transmitted by using a bundle of optical fibres, between 0.01–0.5 mm in diameter, in a fixed array. If a pattern of light is displayed at one end, each fibre will transmit light from a small area of the pattern to the other end, and the image will be reassembled on that surface. If an *objective is placed at one end, and an *eyepiece or camera at the other end to view the image, the fibre bundle can be used in both industry and medicine to view inaccessible locations, an external layer of fibres transmitting light into the location.

Screens on *cathode-ray tubes, *image intensifiers, etc., can be made from parallel glass fibres, fused together. There is therefore no loss in definition, which would normally occur when light from the inner phosphor layer passes through the glass envelope.

Fick's law A law expressing the process of *diffusion of liquids and solids in mathematical form. The mass of dissolved substance crossing unit area of a plane of equal concentration in unit time is proportional to the concentration gradient. The constant of proportionality is called the *coefficient of diffusion*, symbol: D.

field 1. A region under the influence of some physical agency. Typical examples are the *electric, *magnetic, and *gravitational fields that result from the presence of charge, magnetic dipole, and mass respectively; these are *vector fields. A field can be pictorially represented by a set of curves, often referred to as *lines of flux* (or *force*); the density of these lines at any given point represents the strength of the field at that point, and their direction represents the direction conventionally associated

with the agency. Thus, electric fields run from positive to negative, magnetic fields from north-seeking to south-seeking, and gravitational fields from lighter to heavier.

A field is also used to describe the region inhabited by *nucleons, in which *exchange forces are set up. In addition, it has been used in connection with scalar quantities to describe distributions of temperature, electric potential, etc. *See also* quantum field theory.
2. *See* field of view.

field coil A coil that, when carrying a current, magnetizes a *field magnet of an electrical machine (dynamo or motor).

field curvature *See* curvature of field.

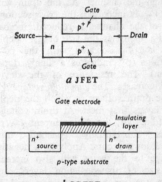

a JFET

b IGFET

field-effect transistor (FET) A *semiconductor device, a *transistor, in which current flow depends on the movement of *majority carriers only; it is thus a unipolar rather than a bipolar device. Current flows through a narrow conducting *channel* between two regions, the *source* and *drain*, to which electrodes are attached. Application of a suitable bias across the transistor causes the charge carriers to flow from source to drain. The current is modulated by an electric field applied to a third electrode attached to the *gate* region. Devices with n-type source and drain regions and hence an n-type channel are called *n-channel devices*; those with p-type regions are called *p-channel devices*. (*See* n-type conductivity, p-type conductivity.)

There are two main types of FET. In the *junction FET* (JFET), the conducting channel forms part of the structure of the device. In the *insulated-gate FET* (IGFET), the channel is formed in use by the action of the gate voltage. The basic structures are shown in Figs. *a* and *b*.

In the case of the JFET, when a positive voltage is applied to the drain of an n-channel device, electrons in the source are attracted into the drain, flowing through the channel. As the drain voltage, V_D, is increased, the cross-sectional area of the channel is reduced: increasing V_D increases the size of the *depletion layers associated with the p^+-n junctions (p^+ indicates a highly doped region). The resistance of the device thus increases. The *pinch-off voltage V_P*, is the drain voltage at which the depletion layers first meet. At drain voltages exceeding V_P, the drain current I_D, remains substantially constant until *breakdown occurs. Applying a negative voltage to the gates will also increase the size of the depletion layers, and hence the pinch-off condition will be reached at a lower drain voltage. A family of characteristics will thus be generated for different values of gate bias (Fig. *c*). The gate voltage is therefore used to modulate the channel conductivity.

The IGFET has an insulating layer that is formed on the surface between the highly doped source and drain regions (Fig. *b*). A conductor deposited on top of this layer forms the gate electrode. The most widely used insulated-gate FET is the MOSFET (metal-oxide-

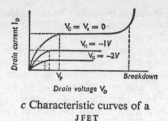

c Characteristic curves of a
J F E T

silicon FET), in which silicon dioxide is the insulator.

In the case of an n-channel IGFET, the channel only forms when a positive voltage of sufficient magnitude is applied to the gate electrode. At this voltage, known as the *threshold voltage, V_T*, *inversion occurs in the semiconductor just below the insulating layer, i.e. a layer forms of opposite conductivity-type to the p-type semiconductor. The inverted layer constitutes the narrow channel that connects source and drain.

Applying a small positive voltage to the drain attracts electrons from the source, and current then flows through the channel. As with JFETs, depletion layers determine channel size and shape. The channel acts as a resistor, and the drain current, I_D, increases approximately linearly with the drain voltage, V_D. As V_D is increased the channel depth near the drain is reduced to zero. I_D then remains substantially constant with any further increase of V_D (Fig. *d*). Again as with JFETs, the gate voltage is used to modulate the channel conductivity; the I_D characteristics are similar to those of JFETs (Fig. *c*).

A p-channel device operates in a similar manner to an n-channel device but negative gate and drain voltages are applied.

Depletion-mode devices are those in which conduction takes place with zero gate bias, and *enhancement-mode devices* are those in which a voltage must be applied to the gate before conduction can occur. All junction FETs are deple-

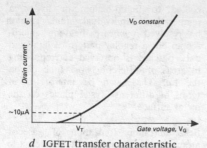

d IGFET transfer characteristic

tion-mode devices. Ideally, all insulated-gate FETs, as described above, are enhancement-mode devices, but in n-channel devices a spontaneous inversion layer may exist, even with zero gate bias, causing them to be depletion-mode devices. Enhancement-mode devices are "off" at zero gate bias and hence simpler to use as switches. n-channel devices are preferred to p-channel devices, however, due to the greater mobility of electrons compared to holes, which leads to a higher gain.

FETs may be broadly described as square-law devices, i.e. the output current I_{DS} varies with the square of the input voltage V_{GS}. (In a bipolar junction transistor there is an exponential dependence.) The input impedance of an FET is always very high. In the junction devices the input is through a reverse-biased diode, and in the insulated-gate version the input impedance is purely capacitive (the gate is separated from the channel by an insulator).

Junction FETs are invariably used as discrete devices in circuits in which their square-law and/or high input impedance characteristics are required. These include high input impedance *amplifiers and square-law *mixers. They are also used as bidirectional switches. Insulated-gate FETs are also used in similar discrete device applications, but their main use is in MOS *integrated circuits.

field emission Emission of electrons from a solid that is subjected to a high electric field. In a metal, the outer electrons of the atoms move through the metal as an *electron gas and their energies lie within allowed bands. The diagram shows one such band in an idealized metal; the electron energy is plotted against distance from the metal surface, both within the metal and outside it.

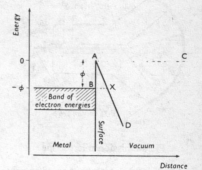

Energy band at the surface of a metal

If the metal is at the same potential as its surroundings, there is no external electric field and the potential energy of an electron outside the metal does not vary with distance, as represented by the line AC. However, if the metal has a high negative potential with respect to an external electrode, there is a corresponding high electric field at its surface; the potential energy therefore falls off with distance, as represented by the line AD. If the field is very high, there is a barrier (represented by the area BAX) keeping the electrons inside the solid. Nevertheless, electrons can escape without having to surmount the barrier as a consequence of the *tunnel effect, and it is possible for electrons to appear at point X outside the metal.

This is the process of field emission, and its probability increases as the barrier width BX decreases. Consequently, the field-emission electron flux goes up if the electric field strength is increased and is greater for solids of low work function. High electric fields of the order of 10^{10} volts per metre are necessary for the effect to be observed and these are usually obtained by subjecting very sharp points to high potentials.

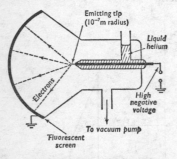

Emitting tip (10^{-7}m radius)

Liquid helium

Electrons

High negative voltage

To vacuum pump

Fluorescent screen

Field-emission microscope

field-emission microscope An instrument for observing the surface structure of a solid by causing it to undergo *field emission. In the simple instrument shown, a sharp metal tip is maintained at a high negative potential with respect to a conducting fluorescent screen. The instrument is evacuated to a very low pressure to prevent *gas discharge. Ideally the tip has a regular shape and is composed of a single crystal of material. Field emission occurs under the influence of the high local electric field and the resulting electrons are accelerated to the screen where they cause fluorescence. Field-emitted electrons leave a surface at right angles and if r_t is the tip radius and r_s its distance from the screen, an image of the surface of the tip is projected on the screen with a linear magnification of r_s/r_t. The resolution is limited by vibrations of the metal atoms and the tip is therefore usually

cooled to liquid-helium or hydrogen temperatures. Individual atoms cannot be resolved (*compare* field-ion microscope) but a regular pattern of light and dark patches is observed corresponding to areas of different *work function on the tip. These can be interpreted in terms of different crystal planes on the metal surface.

field ionization Ionization of gaseous atoms and molecules by a high electric field at a solid surface. Electrons can normally only escape from an atom if they gain energy equal to the *ionization potential of the atom. If the atom is close to, say, a metal and there is a high electric field near the surface, an electron can tunnel (*see* tunnel effect) through the *potential barrier from the atom into the metal. The process is similar to *field emission, with the difference that electrons tunnel from atoms or molecules into a metal rather than out of a metal. The fields required are of the order of 10^7 volts per metre and are produced by subjecting very sharp points to very high positive potentials (~ 10–20 kV). Ions formed at the metal surface are accelerated away by the field.

field-ion microscope An instrument for observing the surface structure of a metal using *field ionization. It is identical in form to the *field-emission microscope with the difference that a positive voltage is applied to the tip rather than a negative one and the image is formed by positive ions from gas atoms rather than by electrons from the metal itself. Helium is allowed into the microscope at low pressure and helium ions form at the surface of the tip and are accelerated to the screen where they cause fluorescence. The field-ion current from a point depends on the magnitude of the electric field, and on the atomic scale there is a local intensification of the field in the region of a surface atom. Consequently, a magnified image of the atomic structure of the surface is projected on the screen. The tip is cooled to liquid hydrogen or helium temperatures to reduce the vibrations of the metal atoms and improve the resolution.

field lens Of an eyepiece. The front lens of a two-lens eyepiece that serves to bend the chief rays towards the optical centre of the eye lens.

field magnet The magnet that provides the magnetic field in an electrical machine. Usually it is an electromagnet but may be a permanent magnet in small machines.

field of view *Syn.* field. The angular extent of object space that can be observed or embraced by an optical instrument. *See* apertures and stops in optical systems.

field stop *See* apertures and stops in optical systems.

filament A threadlike body, particularly the conductor of metal or carbon in an incandescent lamp.

film dosimetry *See* dosimetry.

filter 1. A device for removing solid matter suspended in a liquid by forcing the suspension through a material (e.g. sand, filter paper) that retains the solid matter while allowing the liquid to pass. **2.** An electrical network designed to transmit signals with frequencies that lie within one or more designated ranges (*pass bands*) and to suppress signals of other frequencies (in one or more *attenuation bands*). The cut-off frequencies are those that separate the several pass and attenuation bands (symbols f_1, f_2,

etc. or f_c if there is only one). The four main types of filter are *low-pass*, *high-pass*, *band-pass*, and *band-stop filters*. Their frequency limits are given in the table.

Type	Pass band(s)	Attenuation band(s)
Low-pass	0 to f_c	f_c to ∞
High-pass	f_c to ∞	0 to f_c
Band-pass	f_1 to f_2	0 to f_1, f_2 to ∞
Band-stop	0 to f_1, f_2 to ∞	f_1 to f_2

3. A device for transmitting light (and also infrared and ultraviolet) with restricted ranges of wavelength. Commonly, transparent substances that absorb selectively are used (coloured glasses or films; *see also* Polaroid). *Interference filters* use a different principle and can be produced to yield a narrower band of wavelength 1–10 nm for half transmission.

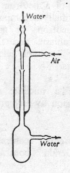

Filter pump

filter pump *Syn.* water aspirator. A fast-working vacuum pump in which a jet of water is used to trap and remove air (*see* diagram). The lowest pressure attainable is approximately the vapour pressure of water at the prevailing temperature.

finder *See* collimator.

fine-structure constant Symbol: α. A dimensionless quantity formed from the four basic physical constants, electronic charge e, speed of light in vacuum c, the *Planck constant h, and the *permittivity of free space ε_0 (electric constant):
$$\alpha = e^2/2hc\varepsilon_0 = 7.297\ 353\ 1 \times 10^{-3}$$
$$\simeq 1/137$$
It is a convenient measure of the strength of the *electromagnetic interaction. The equivalent pure number measuring the strength of the *strong interaction is about 1. On this scale the *weak interaction has a strength of 10^{-13} and the gravitational interaction of 10^{-38}.

fissile Of a nuclide. Capable of undergoing *fission by interaction with *slow neutrons. *Compare* fissionable.

fission The splitting of a heavy nucleus of an atom into two or more fragments of comparable size usually as the result of the impact of a neutron on the nucleus. It is normally accompanied by the emission of neutrons or gamma rays. Plutonium, uranium, and thorium are the principal fissionable elements. *See* nuclear reactors.

fissionable Of a nuclide. Capable of undergoing fission by any process. *Compare* fissile.

FitzGerald–Lorentz contraction *See* Lorentz–FitzGerald contraction. *See also* relativity.

five-fourths power law The law of cooling applicable to free convection. The rate of loss of heat is proportional to the five-fourths power of the excess temperature of the body over the temperature of its surroundings.

fixed points Reproducible invariant temperatures used to define a temperature scale. The melting point and boiling point of water were used to define nearly all temperature scales for two centuries. The *thermodynamic temperature scale now has one fixed point, the *triple point of pure air-free water. The *International Practical Temperature Scale of 1968 uses eleven fixed points all defined in terms of the equilibrium between phases of pure substances under specified conditions.

flash barrier A structure of fireproof material designed to minimize the formation of an electric arc between conductors or to minimize the damage caused by such an arc in an electrical machine.

flashover An abnormal formation of an arc or spark (called *arcover* and *sparkover* respectively) between two electrical conductors or between a conductor and earth.

flashover voltage The dry flashover voltage is the voltage at which the air surrounding a clean dry insulator (especially one supporting electric lines) breaks down completely and *flashover between the conductors occurs. The wet flashover voltage is the voltage at which flashover occurs when the clean insulator is wet (to simulate rain).

flash point The lowest temperature at which a substance will provide sufficient inflammable vapour (under special conditions) to ignite upon the application of a small flame.

flavour *See* quark.

F-layer (or **region**), F_1-**layer**, F_2-**layer** *Syn*. Appleton layer. *See* ionosphere.

Fleming's rules Rules for the relation between current, motion, and field in the dynamo and electromotor. *Right-hand rule* (dynamo principle). Hold the thumb, first finger, and middle finger of the right hand at right angles to one another. Point the thumb in the direction of motion, the first finger along the lines of the field; the middle finger will then point in the direction of the induced current. *Left-hand rule* (motor principle). Hold the thumb, first finger, and middle finger of the left hand at right angles to each other. Point the first finger in the direction of the field, the middle finger in the direction of the current; then the thumb points in the direction in which the force acts on the conductor.

flicker photometer A photometer that presents alternately to the eye two surfaces illuminated by the sources to be compared. If both sources are white and the frequency of alternation is not too high, disappearance of flicker signifies equality of brightness. Flicker photometry is useful when there are colour differences between the lights to be compared, as there occurs a speed of alternation at which brightness flicker exists only while colour flicker is absent.

F-line A green-blue line in the *emission spectrum of hydrogen, wavelength 486.133 nm. It is used as a reference line for specifying the refractive index and dispersion of optical glass, etc.

flip-flop *Syn.* bistable. An electronic circuit element that is capable of exhibiting either of two stable states and of switching between these states in a reproducible manner. Flip-flops are widely used in computers, especially in *logic circuits: the two states are made to correspond to logic 1 and logic 0, so that the flip-flop is a one-bit memory element. Various types of flip-flop have been developed, the simplest being the R-S flip-flop and the most useful being the J-K and D flip-flops. With a *clocked flip-flop*, the device cannot change state

until triggered by the application of a clock pulse.

floating Of a circuit or device. Not connected to a source of potential.

fluence *See* energy fluence; particle fluence.

fluid A collective term embracing liquids and gases. A "perfect fluid" offers no resistance to change of shape (i.e. has zero viscosity).

fluidity Symbol: ϕ. The reciprocal of dynamic *viscosity.

fluorescence A type of *luminescence in which the emission of electromagnetic radiation ceases as soon as excitation ceases (*compare* phosphorescence). The radiation emitted is usually, but not necessarily, light. Excitation is commonly by *ionizing radiation or by electromagnetic radiation of different wavelength from that which is emitted. Normally the emitted radiation is of longer wavelength than the incident electromagnetic radiation (Stokes' law), but it can be shorter if the substance is not initially in the ground state.

fluorescent lamp A lamp in which light is generated by *fluorescence. The common forms of fluorescent lamp consist of a *gas-discharge tube containing a gas, such as mercury vapour, at a low pressure. The inner surface of the lamp is coated with a *phosphor. When an electric current is passed through the vapour, ultraviolet radiation is produced and this, in striking the phosphor, produces visible radiation. The phosphors used are such that the combination of colours emitted by the gas and by the phosphor gives white. Lamps of this type are more efficient than filament lamps because less of the radiation is in the infrared.

The common sodium-vapour and mercury-vapour street lamps do not have a fluorescent coating. The light is emitted directly from atoms of vapour that have been raised to an *excited state by electrons in the discharge.

fluorescent screen A surface coated with a luminescent material that fluoresces when excited by electrons, X-rays, etc., and hence displays visual information.

flutter An undesirable form of *frequency modulation heard in the reproduction of high-fidelity sound and characterized by variations in pitch above about 10 Hz. *See also* wow.

flux 1. A measure of the strength of a *field of force through a specified area. (*See* electric flux; magnetic flux.)
 2. A measure of the rate of flow of a scalar quantity. (*See* luminous flux; sound flux.)
 3. *See* neutron flux.

flux density 1. *See* magnetic flux density.
 2. *See* electric displacement.
 3. *See* particle fluence.

fluxmeter An instrument for measuring changes in magnetic flux. The *Grassot fluxmeter* is essentially a moving-coil galvanometer in which the restoring couple on the moving coil is negligibly small and the electromagnetic damping is large. The galvanometer is used in conjunction with an exploring coil of known area. Any change in the magnetic flux through the exploring coil causes an induced charge to circulate round the suspended coil, and the latter is deflected ballistically through an angle that is directly proportional to the change in flux through the exploring coil. The instrument is calibrated empirically, using a standard magnetic flux.

flux refraction An abrupt change of direction of lines or tubes of magnetic flux when they pass from one region to another of different *permeability. The ratio of the tangents of the angles of incidence and refraction are constant for any pair of media. The same effect is observed when lines of electric flux pass across a boundary between media of different *permittivity.

flyback *See* time base.

flying-spot microscope A microscope in which the lens system is used to produce a minute spot of light that, after passing through the object, falls on a photocell for subsequent amplification and display. The spot is made to scan the object and the image is produced on a *cathode-ray tube scanned in synchronization. The contrast of the display is partly determined by the amplifier gain and can thus be varied. Further advantages are the greater *quantum efficiency of the photocell over the photographic plate and increased resolution.

f-number *Syn.* relative aperture; stop number. The number, used especially in photography, equal to the ratio of the focal length of a particular lens to the effective diameter of the lens aperture for a parallel beam of incident light. For a ratio of, say, 4 it is written f/4, f4, f:4, etc. For a given shutter speed, it always corresponds to the same *exposure. The variable diaphragm used to change the effective diameter of a camera lens can be set to different f-numbers.

focal length The distance from the *pole of a curved mirror, from the centre of a thin lens, or from the principal point of a system to the principal *focal point. In general, there are two focal lengths, anterior or first (f), and posterior or second (f'), and $n/f = -n'/f'$ where n and *n'* are the refractive indexes of the medium on the two sides of the system. When the focal length is measured from the last vertex of a lens, it is referred to as the *back focal length. *See also* centred optical systems.

focal plane *See* focal point.

focal point *Syn.* focus. A point to which parallel light rays, incident on a lens or curved mirror, converge or from which they appear to diverge. The plane perpendicular to the axis through a focal point is the *focal plane*. The *principal focal point* is the focus for a light beam parallel to the principal axis of the system. (*See also* centred optical systems.)

The concept of focal point (and *focal length) has been extended to electron lenses, acoustic lenses, and lenses and mirrors designed for use with infrared and ultraviolet radiation and radio waves. The focal point is then the point on the axis of the system to which a parallel beam of incident radiation converges.

focus 1. *See* conic sections.
2. *See* focal points.
3. *See* seismology.

focusing 1. Of charged particles in an accelerator. In *accelerators magnets are used to focus the beam of particles. In cyclic accelerators a narrow ring of magnets is used. The magnets usually perform two functions. They bend the path of the particles into a circle, the particles being accelerated by an electric field in the gaps between the magnets. They also focus the beam into a central orbit of small cross section. *Strong focusing* or *alternating gradient* (*AG*) focusing is achieved by using pairs of magnets. One member focuses the beam in one plane and defocuses it in the plane at right angles; the other member focuses the defocused plane and

defocuses the orthogonal plane. The net effect is a strongly focused beam. The accelerating radio frequency field in both linear and cyclic accelerators tends to keep the particles in phase with the r.f. field, so that they orbit in bunches.

2. Of an electron beam in, for example, a *cathode-ray tube. The principal methods are: (i) *Electrostatic focusing*. The beam is made to converge by the action of electrostatic fields between two or more electrodes at different potentials. The electrodes are commonly cylinders coaxial with the electron tube, and the whole assembly forms an electrostatic *electron lens. The focusing effect is usually controlled by varying the potential of one of the electrodes (called the focusing electrode). (ii) *Electromagnetic focusing*. The beam is made to converge by the action of a magnetic field that is produced by the passage of direct current through a focusing coil. The latter is commonly a coil of short axial length mounted so as to surround the electron tube and to be coaxial with it.

folded dipole *See* dipole aerial.

foot-pound-second system of units (f.p.s. system) The system of units formerly used in English-speaking countries, based on the foot, pound, and second. It has now been replaced for scientific and technical purposes by *SI units. Appendix, Table 1 gives interconversions between SI, c.g.s., and f.p.s. units.

forbidden band *See* energy bands.

forbidden transition A transition between two states of a system that violates certain *selection rules. Such transitions are not necessarily impossible but have much lower probability than allowed transitions of similar energy.

force Symbol: *F*. Unit: newton (N). In classical physics real forces are defined

179

by a set of axioms, *Newton's laws of motion, with reference to an *inertial reference frame. By Newton's second law the resultant force F acting on a body of constant mass m is equal to ma, where a is the *acceleration of the body, and has the direction of a. Force is a *vector quantity.

Forces are either *long range* or *short range*. Long-range forces, such as *gravitation and the *Coulomb force, fall off less rapidly than the inverse fourth power of the distance. Short-range forces, such as those inside the atomic nucleus (*see* strong interaction; weak interaction) and those between molecules, fall off more rapidly than the inverse fourth power. It can be shown that in a condensed body, short-range forces are small at distances not much greater than those of near neighbours.

Some workers find it convenient to use fictitious forces in analyses. There are two kinds – *inertia forces* and *inertial forces*.

An inertia force is a fictitious force that is supposed to act on a body, being equal and opposite to the resultant of the real forces. Since it does not represent any actual interaction, an inertia force does not obey Newton's third law. According to the *principle of d'Alembert* (1742) any accelerated body can be treated as if it were in equilibrium under the action of the real forces and the fictitious one.

When a problem is considered from the point of view of an observer who is accelerated with respect to an inertial reference frame, Newton's laws are not applicable to real interactions. It is possible to apply these laws in such a case by introducing a fictitious force, in this case called an inertial force. In particular, an observer on a rotating body may use an inertial force called a *Coriolis force*, which is supposed to act at right angles to the path of a body that moves towards or away from the axis of rotation (*see* Coriolis theorem). Problems on projectiles and movements of the atmosphere and oceans are often treated in this way.

When a particle of mass m moves in a circular arc of radius r with uniform angular velocity ω, there is an acceleration $r\omega^2$ towards the centre, so by Newton's second law the resultant force on the particle, called the *centripetal force*, is $mr\omega^2$, acting radially inward. In the system of d'Alembert there is a fictitious inertia force equal to this, supposedly acting radially outwards. Also an observer orbiting with the particle could introduce an inertial force that would be, in this case, equal and opposite to the centripetal force. Both these fictitious forces may be called *centrifugal forces* since they are directed away from the centre. Now the centripetal force is real, so there must be an equal and opposite real force acting on another body, by Newton's third law. This real force is also often called a centrifugal force.

forced convection Ventilated cooling in a strong draught. For a body cooling by this means, *Newton's law applies. *See also* convection (of heat).

forced oscillations *Oscillations produced in an electric circuit when it is acted on by an external driving force, as when a resonant circuit is coupled to a fixed-frequency oscillator. The resultant oscillations have two components: a transient component whose frequency is determined by the natural frequency of the circuit and decays rapidly, and a steady component whose frequency equals that of the external driving force. *See also* resonance; forced vibrations.

forced vibrations Motion produced when a system, capable of vibrating, is acted upon by an external driving force. The resulting vibrations consist of two components, namely a transient component of frequency given by the natural frequency of the system and a steady component of frequency equal to that of the driving force.

The amplitude of the steady vibration is a maximum when the frequency of the driving force is equal to the natural frequency of the system without the effect of the damping. This condition is called *resonance*, and the system is said to resonate with the driving force. Then the forced vibrations lag 90° behind the driving force. *See also* forced oscillations.

force ratio *Syn.* mechanical advantage. *See* machine.

form factor Of a periodic function (e.g. an alternating current or e.m.f.). The ratio of the *root-mean-square value of the function to the mean value taken over a half period beginning at a zero point. For a simple sine wave, the form factor is $\pi/2\sqrt{2}$, i.e. 1.111.

Fortin's barometer *See* barometer.

forward bias *Syn.* forward voltage. A voltage applied to a circuit or device in such a direction as to produce the larger current (known as the *forward current*). The term commonly refers to *semiconductor devices.

Foster–Seeley discriminator *See* frequency discriminator.

Foucault pendulum A simple pendulum that demonstrates the earth's rotation. The original, set up by Foucault (1851), consisted of a lead ball weighing about 28 kg suspended by a fine steel wire 67 m long. The plane of swing is invariable but, owing to the rotation of the earth, it appears to rotate through 360° in T hours, where T is equal to $24/\sin\lambda$. (λ is the latitude.)

four-dimensional continuum *Syn.* space-time continuum. In certain formulations of the theory of *relativity, use is made of a four-dimensional coordinate system in which three dimensions represent the space coordinates x, y, z and the *fourth dimension* is ict, where t is time, c is the speed of light, and i is $\sqrt{-1}$. Points in this space are called *events*. The equivalent to the distance between two points is the *interval* between two events. The distance between two points is not invariant under a *Lorentz transformation, because the measurements of the positions of the points that are simultaneous according to one observer are not simultaneous according to an observer in uniform motion with respect to the first. By contrast, the interval between two events is invariant.

The equivalent to a vector in the four-dimensional space is a *four vector*, which has three space components and one time component. For example: the four-vector momentum has a time component proportional to the energy of a particle; the four-vector potential has the space coordinates of the magnetic vector potential, while the time coordinate corresponds to the electric potential.

Fourier analysis It is possible to express any single-valued periodic function as a summation of sinusoidal components, of frequencies that are multiples of the frequency of the function. Such a summation is called a *Fourier series*, and the analysis of a periodic function into its simple harmonic components is a Fourier analysis.

A function of time, $x = f(t)$, may thus be expressed as follows:

$$x = a_0 + a_1 \cos \omega t + a_2 \cos 2\omega t$$
$$+ a_3 \cos 3\omega t + \dots$$
$$+ b_1 \sin \omega t + b_2 \sin 2\omega t$$
$$+ b_3 \sin 3\omega t + \dots$$

The values of the coefficients a_0, a_1, a_2, $a_3, \dots, b_1, b_2, b_3, \dots$, are as follows:

$$a_0 = \omega/2\pi \int_0^{2\pi/\omega} x \, dt$$
$$a_n = \omega/\pi \int_0^{2\pi/\omega} x \cos n\omega t \, dt$$
$$b_n = \omega/\pi \int_0^{2\pi/\omega} x \sin n\omega t \, dt$$

Fourier integral The limiting form of the Fourier series (*see* Fourier analysis), when the period is made indefinitely great. This representation is very useful for pulses and for limited trains of waves.

Fourier number Symbol: *Fo*. A dimensionless quantity used in the study of heat transfer. It is defined by the function $\lambda t/c_p \rho l^2$, where λ = thermal conductivity, t = time, c_p = specific heat capacity at constant pressure, ρ = density, and l = a characteristic length.

Fourier pair *See* Fourier transform.

Fourier series *See* Fourier analysis.

Fourier transform A mathematical operation by which a function expressed in terms of one variable, x, may be related to a function of a different variable, s, in a manner that finds wide application in physics. The Fourier transform, $F(s)$, of the function $f(x)$, is given by:
$$F(s) = \int_{-\infty}^{\infty} f(x) \exp(-2\pi i x s) \, dx.$$
An analogous formula gives $f(x)$ in terms of $F(s)$. The variables x and s are called *Fourier pairs*. Many such pairs are useful, for example time and frequency.

fourth dimension *See* four-dimensional continuum.

four vector *See* four-dimensional continuum.

frame aerial *See* loop aerial.

frame of reference 1. A rigid framework relative to which positions and movements may be measured; e.g. latitude and longitude define position on the earth's surface, the earth being used as a frame of reference.

2. A Galilean frame of reference is a rigid framework isotropic with respect to mechanical and optical experiments (used in the special theory of *relativity). *See* inertial frame of reference; Newtonian system.

Fraunhofer diffraction *See* diffraction of light.

Fraunhofer eyepiece A terrestrial *eyepiece that has a lenticular erecting system in addition to the optical system of the *Huygens eyepiece or *Ramsden eyepiece.

Fraunhofer lines Absorption lines that occur in the spectrum of the photosphere (the sun's visible surface layer) and arise mainly as a result of absorption in the higher levels of the photosphere. A few lines are due to absorption in the terrestrial atmosphere. First studied in detail by Fraunhofer (1814), over 25 000 lines have now been identified. The most prominent at visible wavelengths are due to the presence of singly ionized calcium, neutral hydrogen, sodium, and magnesium. Many weaker lines are due to iron.

free convection *Syn.* natural convection. Loss of heat vertically that occurs in the absence of draughts when the surrounding fluid circulates freely. The rate of cooling under these conditions obeys the *five-fourths power law although for a small temperature difference between the body and its surroundings *Newton's law of cooling may be applied. *See* convection (of heat).

free electron An electron that is not permanently attached to a specific atom or molecule and is free to move under the influence of an applied electric field. *See also* energy bands; semiconductor.

free-electron paramagnetism *See* paramagnetism.

free energy A thermodynamic function that gives the amount of work available when a system undergoes some specified change. *See* Gibbs function; Helmholtz function.

free fall The acceleration of a body under the action of a *gravitational field only, there being no air resistance or buoyancy. Near the surface of the earth the *acceleration of free fall, g,* is measured with respect to a nearby point on the surface. Because of the axial rotation the reference point is accelerated to the centre of the circle of its latitude, hence *g* is not quite equal in magnitude or direction to the acceleration towards the centre of the earth given by the theory of *gravitation. It varies slightly in magnitude with position on the surface because of the greater centripetal acceleration at lower latitudes, variation of distance from the centre, and local variations of density or level. For some purposes the standard value, g_n = 9.806 65 m s^{-2}, is used.

The value of *g* can be measured by electronically timing the fall of a small sphere *in vacuo* between two levels determined by laser beams. Formerly, the most accurate methods for absolute measurements used the *pendulum. Variations of *g* are found using a *gravity meter* (*syn.* gravimeter). Moderately accurate measurements of variations due to local abnormalities are made with *gravity balances.* These are, in effect, sensitive spring balances in which the change in weight of a fixed mass is measured.

A body in a state of free fall is said to be *weightless.* All living organisms are normally subject to stress as the force of gravity acts throughout the volume while support forces act on the surface. In the case of man the stress is large because the support usually acts over a small area. An astronaut in a freely falling space vehicle is not supported, so there is no stress (although there is still a gravitational force). Physiological changes occur, involving weakening of the bones.

free-piston gauge An absolute device for measuring high fluid pressures. The pressure is applied to one side of a small piston working in a cylinder and the force necessary to keep the piston stationary is a measure of the pressure.

free space A space that contains no particles (such as gas molecules) and no *fields of force. Formally it is distinguished from a *vacuum,* which contains no particles but may contain fields. For many purposes the two terms may be regarded as equivalent. The values of the properties possessed by free space fall into one of the following classes:

(a) zero (e.g. temperature);

(b) unity (e.g. *refractive index);

(c) the maximum possible (e.g. the *speed of light);

(d) a particular, formally defined value (e.g. *permeability and *permittivity).

free surface energy *See* surface tension.

free vibrations (or **oscillations**) A vibrating system when displaced from its neutral position oscillates about this position with a frequency characteristic of the system – the *natural frequency* of the system. The amplitude decays gradually, depending on the resistance of the medium to the motion and on the inertia of the system, until the energy supplied by the initial displacement has been expended into the medium. These vibrations are called free vibrations.

The expression for the displacement, *x,* in a free undamped vibration is given by

$$x = a \sin (\kappa/m)^{1/2} t$$

where a is the amplitude, κ and m are the elastic and inertia factors respectively, and t is the time.

The corresponding expression for free vibrations with damping is:

$$x = a\, e^{-\alpha t} \sin(\omega t - \delta),$$

where the damping factor α is equal to $\mu/2m$ (μ being the resistive term), and the angular frequency ω is given by $\sqrt{[(\kappa/m) - \alpha^2]}$. δ is the angular displacement at $t = 0$ and is called the epoch.

Free oscillations occur not only in mechanical systems. They can also arise in electric circuits, as when a capacitor discharges through a resistance and inductance. The oscillations decay gradually, the frequency depending on the circuit parameters.

Compare forced vibrations.

freezing mixture A mixture of two or more substances that absorb internal energy when they mix and thus produce a lower temperature than that of the original constituents.

freezing point *Syn.* melting point. The temperature at which the solid and liquid phases of a substance can exist in equilibrium together at a defined pressure, normally standard pressure of 101 325 Pa.

F-region *See* ionosphere.

Frenkel defect *See* defect.

frequency 1. Symbol: ν or f. The number of complete oscillations or cycles in unit time of a vibrating system. It is measured in *hertz. The frequency, ν, is related to the *angular frequency, ω, by the formula $\omega = 2\pi\nu$.

In the case of an alternating current, the frequency is the number of times the current passes through its zero value in the same direction in unit time. The frequency of a wave is, ideally, the number of complete oscillations in unit time, the oscillations being propagated through a medium. The frequency ν is then related to the wavelength λ and *phase speed v by: $v = \nu\lambda$. In practice ν and λ do not have exactly determinate values unless the wave has infinite duration.

2. The number of values of a statistical variable lying in a given range. *See* frequency distribution.

Wavelength	Band	Frequency
1 mm–1 cm	Extremely high frequency; EHF	300–30 GHz
1 cm–10 cm	Super-high frequency; SHF	30–3 GHz
10 cm–1 m	Ultra-high frequency; UHF	3–0·3 GHz
1 m–10 m	Very high frequency; VHF	300–30 MHz
10 m–100 m	High frequency; HF	30–3 MHz
100 m–1000 m	Medium frequency; MF	3–0·3 MHz
1 km–10 km	Low frequency; LF	300–30 kHz
10 km–100 km	Very low frequency; VLF	30–3 kHz

frequency band A range of *frequencies forming part of a larger continuous series of frequencies. The internationally agreed radio frequency bands are given in the table.

frequency changer 1. Generally, an electrical machine or circuit for converting alternating current at one frequency to alternating current at another frequency.

2. *See* mixer.

frequency discriminator A device that selects input signals of constant amplitude and produces an output voltage proportional to the amount by which the input frequency differs from a fixed frequency. It is commonly employed in *automatic frequency-control systems (the output being used to correct the frequency) and in *frequency-modulation systems (the frequency-modulated signals being converted to amplitude-modulated signals). The most common type is the *Foster-Seeley discriminator*.

frequency distortion *See* distortion.

frequency distribution A table, graph, or equation describing how a particular attribute is distributed among the members of a group, e.g. the distribution of a set of measured quantities about their mean value.

When the *deviations of the members of the set from the true value are the algebraic sum of a very large number of independent small deviations, the resulting frequency distribution is said to be *normal* or *Gaussian*. The graph of a Gaussian distribution, following *normalization, has the equation (sometimes called the *error equation*):

$$y = h\pi^{-1/2} \exp [-h^2(x-a)^2],$$

where $y.dx$ is the probability of a value of x lying in a small range from x to $x + dx$; a is the arithmetic mean of all the values of x; h is a constant determining the spread of the distribution. The quantity $s = (h\sqrt{2})^{-1}$ is the standard deviation (*see* deviation) of the distribution and is small if the graph is narrow. *See also* Poisson distribution.

frequency divider An electronic device, the output frequency of which is an exact integral submultiple of the frequency of the input.

frequency-division multiplexing A method of *multiplex operation in which a different frequency band is used for each of the input signals. The transmitted sig-

nal consists of a series of carrier waves of different frequencies, each modulated with a different input signal.

frequency doubler A particular type of *frequency multiplier in which the frequency of the output is twice the frequency of the input.

frequency function *Syn.* probability density function. *See* probability.

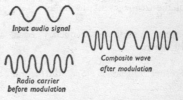

Frequency modulation

frequency modulation (FM, f.m.) A type of *modulation in which the frequency of the *carrier wave is varied above and below its unmodulated value by an amount that is proportional to the amplitude of the modulating signal and at a frequency equal to that of the modulating signal, the amplitude of the carrier wave remaining constant. (*See* diagram.) If the modulating signal is sinusoidal, then the instantaneous amplitude, e, of the frequency-modulated wave may be given as:

$$e = E_m \sin (2\pi Ft + (\Delta F/f) \sin 2\pi ft),$$

where E_m = amplitude of the carrier wave, F = frequency of the unmodulated carrier wave, ΔF = the peak variation of the carrier-wave frequency away from the frequency F, caused by the modulation, f = frequency of the modulating signal. ΔF is called the *frequency swing*.

Compared with *amplitude modulation, frequency modulation has several advantages, the most important of which is improved *signal-to-noise ratio. *Compare* phase modulation.

185

frequency multiplier An electronic device that produces an output signal whose frequency is an exact integral multiple of the frequency of the input. One type uses a nonlinear amplifier (usually *class B or *class C), so that the output is rich in harmonics of the input. The desired harmonic is selected by means of a *filter. Another type consists of a *multivibrator triggered by the input signal to produce oscillations with a frequency that is an exact integral multiple of the frequency of the input.

fresnel A unit of frequency equal to 10^{12} hertz.

Fresnel diffraction *See* diffraction of light.

Fresnel lens

Fresnel lens A lens consisting of a large number of steps, each one having a convex surface of the same curvature as the corresponding section of a normally shaped convex lens (*see* illustration). It was originally designed for use in lighthouses, to reduce the thickness and weight of the large lenses required. It is now also used as a *field lens in spotlights, camera viewfinders, etc., producing a large increase in image brightness.

Fresnel rhomb A glass rhombohedron that, by two internal reflections, changes *plane-polarized light into circularly polarized light.

Fresnel zone *See* diffraction of light.

friction 1. Forces opposing the sliding of one surface over another. For a given value of the normal forces of interaction, N, between the surfaces, no sliding occurs for applied forces up to a limiting value F_l; this is the *limiting* or *static friction* at which sliding begins. For any pair of surfaces the *coefficient of limiting* (or *static*) *friction*, μ_l, is defined by:
$$\mu_l = F_l / N.$$
When sliding takes place at constant speed, the reuniting force F_k is the *kinetic* or *dynamic friction*. The *coefficient of kinetic* (or *dynamic*) *friction*, μ_k, is defined by:
$$\mu_k = F_k / N.$$
μ_k is lower than μ_l for low speeds and is roughly independent of speed, but it may rise considerably at high speeds.

By *Coulomb's laws of friction* μ_l and μ_k are constants and are independent of the apparent area of contact for a given value of N. These laws are not quite exact but are good approximations over a wide range of values.

For typical metal-metal surfaces μ_l may have values 0.5 to 2.0, but much larger values may be given for very clean smooth surfaces. Friction between hard substances is caused by the interaction of minute surface irregularities. With softer materials, *elastic hysteresis is a major cause. Sliding friction is greatly reduced by *lubrication.

(2) *Rolling friction*. The force F_R resisting the rolling of a circular body over a plane surface. The *coefficient of rolling friction* μ_R is defined by:
$$\mu_R = F_R / N.$$
μ_R has values typically 0.1 for rubber tyres on a hard road, and 10^{-2} for steel wheels on steel rails. It is caused mostly by elastic hysteresis and is unaffected by lubrication. The values increase with increasing load and decrease with increasing radius.

(3) *Journal friction*. The resistance to rotation of an axle in its bearing. For sliding bearings lubrication is necessary to prevent metal-metal contact. For most purposes ball or roller bearings are used, reducing the friction to the low values for rolling friction of hard bodies.

frictional electricity *Syn*. triboelectricity. The electric charge produced by rubbing together two dissimilar substances, e.g. ebonite and paper or glass and silk. The charges produced are equal and opposite, one of the substances becoming positively charged, the other negatively charged.

fringes Bands, rings, or other patterns of alternate light and dark or of colour, produced by *interference or *diffraction.

front *See* air mass.

fuel cell A device for the direct use of energy from an oxidation/reduction chemical process to maintain a flow of electricity. The requisite reagents are introduced continuously from outside the cell and react together with the aid of a catalyst. A typical simple fuel cell utilizes hydrogen and oxygen, which are fed to separate porous nickel plates in an electrolyte of weak potassium hydroxide solution; the plate fed by oxygen becomes the anode. The water that is formed from the gases so dilutes the electrolyte that its concentration must be increased from time to time. Fuel cells are able to deliver currents of twenty or more amperes for long periods, but are bulky and have efficiencies of 60% (as opposed to *accumulators, with efficiencies of 75%). The essential difference between a fuel cell and an accumulator is that the former feeds on chemicals and needs no charging, whereas the latter is recharged electrically and its chemicals do not need replenishing.

fuel element The smallest unit of a *fuel assembly* containing *fissile nuclides for powering a *nuclear reactor. The assembly (consisting of fuel elements and their supporting mechanism) together with the *moderator (if any) form the *core of the reactor.

fugacity Symbol: *f*. A corrected pressure used in the thermodynamic equations of real gases to give them the same form as the equations of ideal gases.

full load The maximum output of an electrical machine or *transformer under certain specified conditions, e.g. of temperature rise.

full radiator *See* black body.

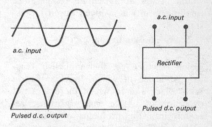

Full-wave rectifier circuit

full-wave rectifier circuit A *rectifier circuit that rectifies both the positive and negative half-cycles of the single-phase a.c. input and delivers unidirectional current to the load (*see* diagram). *Compare* half-wave rectifier circuit.

function generator A *signal generator producing specific waveforms, which may be used for test purposes, over a wide range of frequencies.

fundamental Generally that component of a complex vibration constituting a note by which the pitch of a note is described. In a given note it is generally the tone having the lowest frequency.

fundamental constants *See* Appendix, Table 5.

fundamental interaction *See* interaction.

fundamental particle *See* elementary particle; particle physics.

fuse 1. To melt or to cast.

2. A short length of easily fusible wire put into an electrical circuit for protective purposes. It is arranged to melt ("blow") at a definite current. The term includes all the parts of the complete device.

fusion 1. *Syn.* melting. The change of state of a substance from solid to liquid, which occurs at a definite temperature (melting point) at a given applied pressure.

2. *See* nuclear fusion.

fusion reactor *Syn.* thermonuclear reactor. A device in which *nuclear fusion takes place and in which there is a net evolution of usable energy. Intense research into the problems of designing such a device has occupied laboratories in many countries. The two central problems are (*a*) containing the *plasma in such conditions that it will yield more energy than is required to raise its temperature and confine it, and (*b*) extracting the energy in a usable form.

Three parameters control the first problem: temperature, *containment time, and plasma density. When fusion occurs the plasma temperature has to be high enough for the fusion energy released to exceed the energy lost by *bremsstrahlung radiation. The temperature above which this occurs is called the *ignition temperature*. The deuterium-tritium reaction

$$^2H + {}^3H \rightarrow {}^4He + n + 17.6 \text{ MeV}$$

has the lowest known ignition temperature of 40×10^6 °C. This temperature has been achieved.

The problem of confining a plasma for long enough to release fusion energy has proved more difficult. Plasma instabilities have been the main cause of plasma leakage, but a workable plasma containment has now been achieved, though not at the same time as ignition temperature or adequate particle density. (*See also* Lawson criterion.)

Several types of device have been used for fusion experiments; in most of them a strong pulse of current is passed through the gas to create the plasma. At the same time this current pulse creates a strong magnetic field that makes the charged particles in the plasma travel along helical paths around the lines of force of the field. This causes a contraction of the plasma away from the walls of the tube. This *pinch effect* partially solves the containment problem, but the confined plasma is not stable and tends to develop kinks. In *zeta pinch* devices, the current is passed axially through the plasma and the magnetic field forms round it. In the *theta pinch*, current-carrying coils run round the plasma and the magnetic field is axial. Both devices are toroidal. Since the mid-1970s most research into torus-shaped plasma confinement systems has been concentrated on the *tokamak configuration. This originated in the Soviet Union but research in many countries has confirmed that it has the greatest potential at present for achieving the Lawson criterion. The Joint European Torus (JET) experiment at Culham in England uses this principle.

Linear devices are often called *magnetic bottles, their ends being "stoppered" with *magnetic mirrors. Greater stability in these linear devices is achieved by using extra current carriers.

Another experimental device for the creation of plasma uses a pellet of fuel that is ionized instantaneously by a pulse from a high-power *laser. The use of superconducting magnets might facilitate plasma containment.

Methods of extracting the energy ·of fusion reactions fall into two classes. When most of the energy is in the form of energetic neutrons, e.g. the deuterium-tritium reaction, the neutron energy could be absorbed by a liquid lithium *coolant surrounding the reactor tube. The heat so absorbed would be transferred to a conventional water, steam, and turbogenerator cycle.

When most of the energy is carried by charged particles, as in the deuterium-deuterium reaction, it has been proposed that some of this energy might be used directly to drive an electric current, without using a steam cycle. Placing less reliance upon neutrons as the medium of energy transfer would also serve to reduce radioactive waste arising from neutron-induced radioactivity in the reactor structure. *See also* cold fusion.

G

Gaede molecular air pump *See* air pumps.

gain A measure of the advantage of using an electronic system. For an amplifier the gain is measured by the ratio of the power or voltage delivered by the amplifier to that of the input signal. For a directive aerial the gain is measured by the ratio of the voltage produced by a signal entering along the path of greatest sensitivity to that produced by the same signal entering an omnidirectional aerial. The gain is measured in *decibels, or sometimes in *nepers.

galaxy A giant assembly of stars, gas, and dust held together and organized largely by the gravitational interactions between its components. Galaxies contain most of the observable matter in the universe (*see also* dark matter). Few exist in isolation. The majority occur in groups, known as *clusters of galaxies,* which may contain up to a few thousand members.

Galaxies can be divided into three broad categories: *elliptical, spiral,* and *irregular galaxies.* Elliptical galaxies are dense spheroidal systems with no clearly defined internal structure. The stars are mainly cool and old and there is very little interstellar gas and dust. Spiral galaxies are disc-shaped systems with conspicuous spiral arms winding out from a dense central nucleus, which is sometimes bar-shaped. The arms contain mainly bright young stars and interstellar gas and dust, with older stars occurring in the nucleus. Irregular galaxies have no discernable shape or structure. They are small systems with a large amount of gas and dust. In many there is intense activity observed as emission of large amounts of radiation.

The *Galaxy* is the term used to refer to the spiral system in which the sun lies, about 10 kiloparsecs from the centre and near the edge of one of the spiral arms.

The formation of galaxies is thought to have occurred several hundred thousand years after the *big bang, and to have resulted from slight fluctuations in the density of the primordial gas. Details of the formation and subsequent evolution of galaxies are highly uncertain.

Galilean telescope *See* refracting telescope.

Galilean transformation equations The set of equations:
$$x' = x - vt$$
$$y' = y$$
$$z' = z$$

$$t' = t$$

They are used for transforming the parameters of position and motion from an observer at the point O with coordinates (x,y,z) to an observer at O′ with coordinates (x',y',z'). The x axis is chosen to pass through O and O′. The times of an event are t and t' in the *frames of reference of observers at O and O′ respectively. The zeros of the time scales are the instant that O and O′ coincided. v is the relative velocity of separation of O and O′. The equations conform to *Newtonian mechanics. *Compare* Lorentz transformation equations.

Galitzin pendulum *See* pendulum.

gallium arsenide (GaAs) devices Semiconductor devices based on the 3-5 *semiconductor gallium arsenide. The semiconducting properties of GaAs give it several advantages over silicon for certain applications. For example, it has a high *drift mobility, allowing it to be used for high-speed applications such as high-speed *logic circuits. It can be operated at microwave frequencies at which silicon devices cannot function: gallium arsenide is used for microwave devices, such as *Gunn diodes and *IMPATT diodes, and in microwave integrated circuits used, for example, in *direct broadcast by satellite and in phased-array *radar. In addition GaAs is a type of semiconductor known as a *direct-gap semiconductor*, allowing it to be used for optical components, such as *light-emitting diodes and *semiconductor lasers, and optically coupled devices. It therefore offers the potential for fabricating integrated optoelectronic circuits. GaAs devices are also more tolerant to ionizing radiation than silicon devices. It is, however, much more difficult to fabricate GaAs devices.

Galton whistle A whistle that can produce high frequencies. It consists essentially of a short cylindrical pipe blown from an annular nozzle. The distance of the pipe can be varied by turning a micrometer screw. By suitable adjustment of this distance and the pressure of the air blast, the pipe is set into resonant vibration at a frequency corresponding to its length and diameter. (*See* edge tones.) Frequencies above the human audible limit (normally above 20 000 hertz) can be produced.

galvanomagnetic effect Any of various phenomena occurring when a current is passed through an electrical conductor or semiconductor in the presence of a magnetic field. They include the *Hall effect, *magnetoresistance, and the *Nernst effect.

galvanometer An instrument for measuring or detecting small currents, usually by the mechanical reaction between the magnetic field of the current and that of a magnet. The most commonly used type is the *moving-coil galvanometer*, in which a small coil carrying the current is suspended in the field of a permanent horseshoe magnet (*see* ammeter). For high-frequency currents, use may be made of a *thermogalvanometer* in which the temperature rise in a resistance wire through which the current is passing is employed to measure the current, the temperature rise being measured by means of a thermocouple. *See also* astatic system; ballistic, Helmholtz, and Einthoven galvanometers.

gamma (γ) 1. The symbol used to denote the ratio of the principal *heat capacities C_p/C_V of a substance, where C_p is the heat capacity at constant pressure and C_V that measured at constant volume. For a gas the ratio is given by $C = \sqrt{(\gamma p/\rho)}$ or $\sqrt{(\gamma RT)}$, where C is the speed of sound in the gas and R is the gas constant per unit mass of gas. Using the law of *equipartition of energy, the classical theory predicts that the ratio should be given by $\gamma = 1 + 2/F$, where F is the number of *degrees of freedom of the molecule. For a monatomic gas for which $F = 3$, $\gamma = 1.667$, in agreement with experimental determinations on the rare gases. For polyatomic molecules *quantum theory must be applied. (*See also* specific heat capacity.)

2. A measure of the contrast obtained from a photographic material, given by the gradient of the linear part of the *characteristic curve of the material.

3. A former unit of magnetic flux density, used in *geomagnetism, equal to 10^{-9} *tesla.

gamma camera A device for visualizing the distribution of radioactive compounds in the human body during diagnosis using radioisotopes. It consists of a large thin *scintillation crystal, with an array of *photomultiplier tubes mounted above the crystal and connected with it by a section of transparent material. Usually a low-energy gamma-ray source is used. Radiation causing a scintillation at point X in the crystal will create pulses in the photomultiplier tubes, the size of the pulses being dependent on the relative position of the tubes and X. The output of the photomultiplier tubes is fed into a circuit that analyses the pulses produced. The total sum of the pulses gives the intensity of radiation from X, i.e. the energy of the incident radiation; the relative sizes of the pulses gives positional information. The output, after amplification, controls the position of the spot on a *cathode-ray tube. After a suitable time interval a picture of the area below the crystal is built up on the screen and can be photographed. More sophisticated gamma cameras may have the output fed into and analysed by a computer.

gamma rays (γ-rays) Electromagnetic radiation emitted by nuclei. A nucleus may be raised to an excited state by a nuclear interaction, such as the inelastic *scattering of a neutron or the absorption of radiation. A nucleus may be produced in an excited state by the emission of α- or β-rays, *electron capture, or the capture of a neutron. Excited states usually have lifetimes less than picoseconds and decay directly or indirectly to the ground state with the emission of one or more quanta of gamma radiation. Some states are much longer-lived because of certain *selection rules. In such cases decay by *internal conversion may compete with gamma-ray emission.

The quantum energies of gamma rays are mostly in the range 10^4 eV to 5×10^6 eV, giving wavelengths 10^{-10} m to 2×10^{-13} m. Some gamma-ray photons of cosmic origin have been detected with wavelengths down to 10^{-15} m. (*See also* X-rays.) Those of long wavelength are almost totally absorbed by thin metal foils, while short-wave radiations are detectable through many centimetres of lead. The radiations usually form a line spectrum with a few wavelengths, but in several cases just one wavelength is observed.

Gamma rays ionize matter by the electrons they eject by the *photoelectric effect, or *Compton effect, or by *pair production. They can be detected by all forms of *counter, by *ionization chambers, and photographic emulsions.

Electromagnetic radiations from sources other than nuclei are sometimes loosely called gamma rays when the quantum energy is more than a few hundred keV. Cases include *annihilation and the decay of some *elementary particles.

gamma-ray spectrum A series of wavelengths in the gamma-ray region emitted by a given gamma-ray source. *See also* spectroscopy.

gamma-ray transformation A radioactive disintegration accompanied by the emission of *gamma rays.

Gamow barrier *See* nuclear barrier.

ganged circuits Two or more circuits having variable elements mechanically coupled by a single control so that the circuits can be adjusted simultaneously.

gas A fluid that expands to fill any container, however large, without any change of phase. If the substance is below the *critical temperature it is called a *vapour*. Usually the term gas is understood to apply in this case also, that is, the vapour is a special case of a gas and not a distinct form.

In any substance molecules have translational kinetic energy (which is positive) and potential energy of intermolecular interaction (which is negative on the average). In a gas the total energy of the molecules is positive, whereas in condensed phases (liquid and solid) the total energy is negative.

gas amplification *See* gas multiplication.

gas breakdown A type of *breakdown that occurs when the voltage across a *gas-filled tube reaches a given value. Electrons in the gas are accelerated by the field to such energies that further ion-electron pairs are produced by collision but little recombination of ions occurs due to the high kinetic energies. Positive ions eject electrons from the cathode by *secondary emission. A multiplication effect is present causing breakdown of the gas. The process is analogous to *avalanche breakdown in a *semiconductor.

gas constant *See* molar gas constant.

gas-cooled reactor A type of thermal *nuclear reactor in which a gaseous *coolant is used. In the *Magnox reactors the coolant is carbon dioxide and the outlet temperature is about 350 °C: natural uranium metal fuel is used with a graphite *moderator, the fuel elements being encased in Magnox alloy. In the *advanced gas-cooled reactors* (AGRs), the fuel is ceramic uranium dioxide encased in stainless steel. The same coolant and moderator as in the Magnox type are used, but the outlet temperature is considerably higher – usually about 600 °C. As a result of these higher temperatures, special precautions have to be taken to cool the graphite to avoid chemical attack.

In the *high-temperature gas-cooled reactor* (HTR), the reactor core is composed entirely of ceramic materials, with helium as coolant. A variant of this type is the pebble-bed reactor, in which ceramic pebbles (incorporating both fuel and moderator) are loaded into a vessel to form the reactor core.

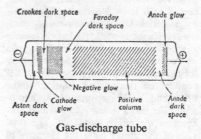

Gas-discharge tube

gas-discharge tube An *electron tube in which the presence of gaseous molecules contributes significantly to the characteristics of the tube. Normally a gas is a poor electrical conductor but if a sufficiently high electric field is applied, conduction can occur. If two plane electrodes are sealed in a tube and a potential difference applied between them, the gas can conduct as a result of an external ionizing agent, such as *ultraviolet radiation. If the ionization agent is removed, the current ceases. Under certain conditions the *discharge can be self-sustaining and independent of the external agent.

In self-sustaining discharges the ions and electrons initially formed in the tube are accelerated to the electrodes and the electrons cause further ionization along their path. Electrons are also produced by *secondary emission at the electrodes. Electrons and ions are removed at the *anode and *cathode respectively and by recombination. A stable state can be reached when the rate of production of ions and electrons is equal to the rate at which they are removed. The characteristics of the discharge depend on the gas, the pressure, the electric field, and the shapes and materials of the electrodes.

The most common type of discharge is the *glow discharge*, characterized by several luminous regions in the tube (*see* diagram). Electrons emitted from the cathode as a result of ion bombardment (*see* secondary emission) are accelerated towards the anode, and for a short distance they have not enough kinetic energy to ionize the atoms of gas or to excite them. The positive ions moving towards the cathode have, in this region, a high velocity and a low probability of recombining with electrons. Any excited ions produced further down the tube have returned to their *ground state by the time they reach this region. Consequently, it emits no radiation and is called the *Aston dark space*. The *cathode glow* is a luminous region near the cathode where positive ions that have been excited by electrons return to their ground state with emission of luminous radiation.

In the *Crookes dark space* electrons moving from the cathode have gained enough kinetic energy to ionize atoms

but the electrons thus produced do not have sufficient energy to excite atoms. Consequently, the region produces little radiation. In the *negative glow* the electrons have gained sufficient energy to cause excitation and the excited atoms return to their ground state with emission of radiation. A small amount of the radiation is also produced by recombination of ions and electrons in this region. In passing the region of the negative glow the electrons lose much of their energy and in the *Faraday dark space* they again have insufficient energy to excite or ionize the gas. Further along is a large luminous region (the *positive column*) in which the gas is excited and emits radiation. The relative sizes of the negative glow and positive column depend on the gas pressure, which determines the *mean free path of charged particles. At pressures below about 15 pascals the positive column often displays *striations*, i.e. alternate dark and light regions caused by the electrons alternately gaining and losing kinetic energy in their journey to the anode.

In a glow discharge the potential drop across the tube is independent of the current and does not vary uniformly down the length of a glow discharge. Most of the potential drop occurs between the cathode and the negative glow.

gaseous ions Positively or negatively charged systems formed in gases by the action of *ionizing radiation (e.g. X-rays); when an electric field is applied across the gas, the motion of the gaseous ions under the action of the field conveys an ionization current across the gas. They differ from electrolytic ions in the fact that they are not permanent, but recombine to form neutral molecules within a short time after the ionizing radiation has been cut off. *See* conduction in gases.

gas-filled relay *See* thyratron.

gas-filled tube An *electron tube containing a gas (or vapour, e.g. mercury vapour) in sufficient quantity to ensure that, once ionization of the gas has taken place, the electrical characteristics of the tube are determined entirely by the gas. *See also* conduction in gases.

gas laws Laws governing the variation of physical conditions (temperature, pressure, etc.) of a gas. *See* equations of state; ideal gas.

gas multiplication *Syn.* gas amplification.
1. The process by which, in a sufficiently strong electric field, ions produced in a gas by *ionizing radiation can produce additional ions.
2. The factor by which the initial ionization is multiplied as a result of this process.

gas thermometer *See* constant-pressure gas thermometer; constant-volume gas thermometer.

gas turbine *See* turbine.

gate 1. An electrode or electrodes in any of a number of devices. In a *field-effect transistor it is the electrode(s) to which a bias is applied for the purpose of modulating the conductivity of the channel.
2. Digital gate. A digital electronic circuit, with one or more inputs but only one output, frequently used in *logic circuits. The output is switched between two or more discrete voltage levels, depending on the input conditions.
3. Analogue gate. A *linear circuit or device, frequently used in radar or electronic control systems, that passes signals only for a specified fraction of the input signal. The output is a continuous function of the input signal for the period that the circuit is switched on.

gate array An integrated logic circuit comprising a two-dimensional array of digital logic *gates that can be interconnected in an arbitrary manner during manufacture. The interconnections determine the performance of the chip, according to the required application. It is thus a programmable device.

gauge boson An *elementary particle that mediates particle interactions in *gauge theories. The *photon, the *$W^\pm$ and Z^0 bosons, and the *gluon are the gauge bosons in gauge theories of the *electromagnetic, *weak, and *strong interactions, respectively.

gauge field In *gauge theories, a quantum field corresponding to a *gauge boson.

gauge theory A *quantum field theory for which all measurable quantities remain unchanged under a *gauge transformation,* in which the phases of the fields are altered by an amount that is a function of space and time. Gauge theories are now believed to provide the basis for a description of all elementary particle interactions. *Quantum electrodynamics and *quantum chromodynamics, the quantum field theories of the *electromagnetic and *strong interactions, are gauge theories, as are *electroweak and *grand unified theories. Einstein's theory of general *relativity can also be formulated as a gauge theory.

gauss Symbol: G. The *CGS-electromagnetic unit of *magnetic flux density. $1 \, G = 10^{-4}$ tesla.

Gaussian distribution *Syn.* normal distribution. *See* frequency distribution.

Gaussian optics *Syn.* paraxial theory; first-order theory. A simplified theory of geometric optics concerned only with light rays close to the optic axis (i.e. paraxial rays). *See* centred optical systems; Seidel aberrations.

Gaussian system of units *See* CGS system of units.

Gauss's theorem For any closed surface drawn in an electric field the integral $\int D.dS$ of the normal component of the *electric displacement, D, over the surface is equal to the total charge within the surface. If the surface encloses no charge the electric field strength within the space is equal to zero. Gauss's theorem applies also to surfaces drawn in a magnetic field. Analogous statements of the theorem may be made for gravitational, magnetostatic, and fluid-velocity fields. The general mathematical statement is that the total flux of a vector field through a closed surface is equal to the volume integral of the divergence of the vector taken over the enclosed volume.

Gay-Lussac's law **1.** Of volume. The volumes in which gases combine chemically bear a simple relation to one another and to that of the resulting product if this is also gaseous. The volumes must all be measured under the same conditions of temperature and pressure. *See* ideal gas.
2. *See* Charles's law.

Geiger counter *Syn.* Geiger–Müller counter. An instrument for counting ionizing particles and photons. Counting rates up to a few hundred per second are practicable. A wire anode lies on the axis of a cylindrical cathode (*see* diagram). The electrodes are contained within a glass tube, or the cathode itself serves as a container. Thin mica windows or, for more penetrating par-

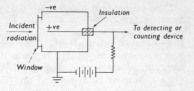

Geiger-counter circuit

ticles, light alloy windows are used. The tube contains argon at a pressure of a few tenths of an atmosphere. A potential difference slightly lower than that required to produce a *discharge through the gas is maintained between the electrodes. Any charged particle that passes through the gas will initiate a discharge. The detection efficiency is very low for electromagnetic radiation, which must eject an electron into the gas by *photoelectric effect or *Compton scattering in order to cause a response.

Whenever any free electrons, however few, appear in the sensitive volume of the tube, a discharge is initiated. The electrons move to the anode causing further ionization by collision. Ultraviolet radiation from excited atoms ejects photoelectrons from the cathode causing the discharge to grow until an *avalanche occurs. The surge of current to the anode causes a voltage pulse, which operates an electronic counter. The size of this pulse is independent of the original amount of ionization so the instrument only counts particles or photons without distinguishing them or measuring their energies.

As the discharge develops a large number of positive ions are produced, and until most of these are collected by the cathode the counter tube is insensitive to further incident radiation. This gives a *dead time* of the order of 10^{-4} s. The presence of a small quantity of *quenching gas* in the tube stops the discharge. Halogen-quenched tubes operate typically at 400 volts.

195

Compare proportional counter; scintillation counter.

Geiger–Nuttall relation (1911) The relation, discovered empirically, between the range, R, of an α-particle emitted by a given radioactive substance and the *decay constant, λ, of the substance:

$$\log \lambda = A + B \log R,$$

where B has the same value for all four radioactive series, while the constant A has a different value for each series. The law is of only approximate validity. Similar laws, in which the energy of the α-particle replaces the range, can be applied with greater accuracy to certain groups of alpha emitters, for example the isotopes of a given element.

Geissler tube A *gas-discharge tube specially designed to demonstrate the luminous effects of an electrical discharge through a rarefied gas.

generalized coordinates Symbol: q_i or q. Coordinates describing the motion of a mechanical system without specifying its exact nature. The *generalized momentum*, symbol: p_i or p, is related to q_i by the relation:

$$p_i = \partial L / \partial q_i,$$

where L is the *Langrangian function. *See also* Langrange's equations; degrees of freedom.

generalized force The quotient of the work done by all the forces acting in a system, if one of the generalized coordinates alters by an infinitesimal amount while the others remain constant, to the change in that generalized coordinate.

generalized momentum *See* generalized coordinates.

general relativity *See* relativity.

generating station *See* power station.

generator A machine that drives an electric current when it itself is driven mechanically. In the electromagnetic generator (dynamo), a coil is moved so as to cut the lines of induction in a magnetic field. In the electrostatic generator (*see* Van de Graaff generator; Wimshurst machine), work is done in separating equal and opposite electrical charges produced by electrostatic induction or by friction. *See also* alternating-current generator.

geodesic 1. The path with minimum (or maximum) length between two points in a mathematically defined space. In three dimensions it is a straight line. On the surface of a sphere it is a *great circle.

2. The equivalent to (1) in the *four-dimensional continuum in the general theory of *relativity. It is the path of electromagnetic radiation, or of a particle that is subject to no nongravitational force.

geomagnetism The study of the earth's magnetic field and its variations. At any point on the earth's surface three magnetic elements are defined:

B_0 is the horizontal component of the magnetic flux density at the location; δ is the angle of *dip (often called the *inclination*), being the angle between the vector B_0 and the resultant magnetic flux density at the location; α is the angle of *declination (often called the *variation*), being the angle between the vector B_0 and the geographic true north. The vertical component of the earth's magnetic flux density B_v is given by:

$$B_v = B_0 \tan \delta.$$

Two main kinds of variation in these elements are observed. Secular variations take place slowly and are associated with periodic and semiperiodic terrestrial and solar phenomena. Abrupt changes, termed *magnetic storms*, are the result of solar phenomena, e.g. flares.

The approximate values of the magnetic elements in the UK are:

$$B_0 = 1.88 \times 10^{-5} \text{ T}$$
$$\alpha = 9.8° \text{ W}$$
$$\delta = 66.7° \text{ N}$$
$$B_v = 4.35 \times 10^{-5} \text{ T}.$$

Due to the slow change with time of the angle of declination, the position of the magnetic poles is changing with time. At times of great magnetic disturbance (usually linked with increased solar activity) the poles can be displaced by 150 km in short periods. It is also known that roughly every 200 000 to 300 000 years the magnetic poles reverse: north becomes south and vice versa.

The geomagnetic field is similar in shape to the field of a bar magnet. It is thought, however, to result from the presence of an internal dynamo, maintained in some way by the flow of matter within the inner layers of the earth.

geometric image *See* image.

geometric mean *See* mean.

geometric optics The study of reflection and refraction of rays of light without reference to the wave or physical nature of light. The elementary studies of light are essentially geometric, graphical methods being used for determining the positions of images formed by lenses, mirrors, etc. It is still a valuable tool in the more advanced studies of technical optics.

geophysics The physics of the *earth. Studies include the following: the evolution and constitution of the planet itself, its atmosphere, and oceans; movements within the earth, such as those associated with continental drift, mountain building, and earthquakes (*see* plate tectonics, seismology); *geomagnetism; geophysical prospecting of mineral deposits; the circulation of the atmosphere and oceans.

geostationary orbit A circular orbit around the earth that lies in the plane of the equator and has a period equal to the period of the earth's rotation on its axis (nearly 24 hours). The altitude of the orbit is approximately 35 780 km. A *satellite in geostationary orbit will appear from earth to be very nearly stationary. An earth orbit with the same period but inclined to the equatorial plane is called a *geosynchronous orbit*.

geosynchronous orbit *See* geostationary orbit.

geothermal energy *See* renewable energy sources.

getter A material with a strong chemical affinity for other materials. Such materials may be used to remove unwanted atoms or molecules from an environment; for example, barium may be used in a sealed vacuum system to remove residual gases or phosphorus may be introduced into oxide layers on silicon to remove mobile impurities such as sodium.

g-factor *See* Landé factor.

giant star A highly luminous star of large dimensions in comparison with average stars like the sun. A typical giant star will have a diameter ten times that of the sun, although *supergiant stars may be as much as 500 times the sun's diameter. Giants have a dense core but a very tenuous atmosphere. They are in a late stage of stellar evolution and are situated above the main sequence in the *Hertzsprung–Russell diagram. *See also* red giant.

Gibbs function *Syn.* Gibbs free energy; thermodynamic potential. Symbol: G. A thermodynamic function of a system given by its *enthalpy (*H*) minus the product of its *entropy (*S*) and its thermodynamic temperature (*T*), i.e. $G = H - TS$. In a *reversible change occurring at constant temperature and pressure the change in the Gibbs function of a system is equal to the work done on it. If a system is considered at constant pressure and temperature and the only work done is that caused by changes in volume, it can be shown that the system is in equilibrium when *G* has a minimum value. In a chemical reaction the change in *G* (ΔG) is zero when equilibrium has been attained. If ΔG is negative for a particular reaction, it can proceed spontaneously to equilibrium, whereas if it is positive the reaction cannot occur without energy being supplied. *See also* Helmholtz function.

Gibbs–Helmholtz equation The thermodynamic expression for the internal energy (*U*) in terms of the free energy (*A*) and its variation with thermodynamic temperature:

$$U = A - T(\partial A/\partial T)_V.$$

giga- 1. Symbol: G. A prefix meaning 10^9; for example, one gigahertz (1 GHz) = 10^9 hertz. **2.** When the binary number system is in use (e.g. in computing) a prefix meaning 2^{30} (i.e. 1 073 741 824); for example, one gigabyte is 2^{30} bytes.

gilbert Symbol: Gb. The *CGS-electromagnetic unit of magnetomotive force or magnetic potential. A point has a magnetic potential of one gilbert if the work done in bringing a unit positive pole up to that point is one erg. One turn of wire carrying a current of 1 ampere produces a magnetomotive force of $4\pi/10$ gilberts.

Giorgi units Units based on the metre, kilogram, and second as fundamental mechanical units, together with one electrical unit of practical size. When first proposed (in 1900) the ohm was the fourth unit chosen, although in 1950 this was replaced by the ampere. In 1954 Giorgi's system was superseded by *SI units. As well as unifying mechanical, thermal, and electrical units, Giorgi recognized and recommended the principle of *rationalization.

Gladstone–Dale law If the density, ρ, of a substance is altered by compression or by increasing its temperature, there is a corresponding rise in the refractive index, *n*, given by:

$$(n - 1)/\rho = k,$$

where *k* is a constant.

glancing angle The complement of the angle of incidence, *i*, i.e. the angle $(90° - i)$.

Glan–Foucault polarizer *See* Nicol prism.

Glashow–Weinberg–Salam theory *See* electroweak theory.

glide The movement of one atomic plane over another in a crystal. It is the process by which a solid undergoes plastic deformation.

glide plane In metal physics, a plane upon which glide can take place upon application of a suitable shearing stress. Sometimes the glide is in a particular direction (the glide direction) in the plane.

glove box An enclosure with gloves fitted to holes in the walls, enabling a substance to be manipulated in an environment quite distinct from that of the operator. It is employed for working with sources of alpha and beta particles and for work in environments with special properties (e.g. controlled humidity, sterilized, or inert). It is usual for the pressure to be maintained slightly above atmospheric to reduce the possibility of contamination from without.

glow discharge An electric discharge through a gas, usually at a relatively low pressure, in which the gas becomes luminous (*see* gas-discharge tube). A *glow lamp* (or *glow tube*) is a gas-discharge tube operated under conditions producing a glow discharge throughout the tube. The colour is characteristic of the gas: a *neon tube emits a red glow. Glow lamps are often used as voltage regulators.

gluon Symbol: g. The elementary particle that mediates the *strong interaction between *quarks (and antiquarks). It is thus a *gauge boson. *See* quantum chromodynamics.

gnomonic projection From a point within a crystal (the pole of projection) lines are drawn normal to the crystal faces (or sets of planes in the crystal) and these produced will meet any plane in a pattern of points, which is the gnomonic projection of the crystal on that plane.

Golay cell A small transparent device containing gas, used to detect infrared radiation. A very thin film within the cell absorbs the incident radiation, causing the gas temperature and consequently the gas pressure to increase (for a fixed volume of gas). The amount of incident radiation can therefore be indicated by recording the changes in pressure.

gold-leaf electroscope *See* electroscope.

gold point The melting point of pure gold taken as a fixed point (1064.43 °C) on the *International Practical Temperature Scale.

goniometry The measurement of angles. In crystallography it is the measurement of interfacial angles for the comparison of crystals of different development.

goniophotometer *See* photometry.

G-parity A quantum number associated with *elementary particles that have zero *baryon number and *strangeness. It is conserved in *strong interactions only.

graded-base transistor *See* drift transistor.

graded-index device *See* fibre-optics system.

gradient 1. Of a graph at any point. The slope of the tangent to the graph at that point as measured by the increase of the ordinate divided by the increase of the abscissa.

2. (grad) Of a scalar field $f(x,y,z)$ at a point. The *vector pointing in the direction of the greatest increase in the scalar with distance (i.e. perpendicular to the level surface at the point in question). It has components along the coordinate axes that are the partial derivatives, f_x, f_y, f_z of the function with respect to each variable:
$$\text{grad } f = \nabla f = i f_x + j f_y + k f_z,$$
where ∇ is the differential operator *del and i, j, and k are unit vectors along the x-, y-, and z-axis. Electric field is the negative gradient of electrical potential. *See* potential gradient.

gradient-index lens (GRIN lens) An optical lens composed of an inhomogeneous medium in which the refractive index varies in a prescribed fashion. It normally varies radially, decreasing parabolically from the central axis. GRIN lenses can take the form of small-diameter parallel flat-faced rods.

These are usually grouped into large arrays.

Graetz number Symbol: *Gz*. A dimensionless coefficient of importance in the study of hydrodynamics:

$$Gz = q_m c_p / \lambda l,$$

where q_m = mass flow rate, c_p = specific heat capacity at constant pressure, λ = thermal conductivity, and l = a characteristic length.

Graham's law of diffusion (1846) The rates of efflux of different gases through a fine hole at the same temperature and pressure are inversely proportional to the square roots of their densities. Knudsen showed that the law is only true when the mean free path in the issuing gas is at least ten times the diameter of the hole. *See* effusion.

gram One-thousandth of a *kilogram.

gram-atom or -molecule The former name for a *mole.

Gramme winding *See* ring winding.

grand unified theory (GUT) A unified *quantum field theory of the *electromagnetic, *weak, and *strong interactions. In most models, the known interactions are viewed as a low-energy manifestation of a single unified interaction, the unification taking place at energies (typically 10^{15} GeV) very much higher than those currently accessible in particle accelerators. One feature of GUTs is that *baryon number and *lepton number would no longer be absolutely conserved quantum numbers, with the consequence that such processes as *proton decay* (for example, the decay of a proton into a positron and a π^0, p $\to$ e$^+\pi^0$) would be expected to be observed. Predicted lifetimes for proton decay are very long, typically 10^{35} years. Searches for proton decay are being undertaken by many groups, using large underground detectors, so far without success.

graphical symbols Symbols that represent the various types of components and devices used in electronics, electrical engineering, telecommunications, and allied subjects. A selection of the more commonly used symbols (as recommended by the British Standards Institute) is shown in the Appendix, Table 10.

graphic equalizer An electronic device in a radio, tape recorder, etc., that controls the *tone, i.e. alters the relative frequency response of the audiofrequency amplifier. The frequency range of the amplifier is divided into bands. The power of the output signal in each frequency band is adjusted by sliding contacts, the positions of which therefore indicate the frequency response in each band.

graphics *Syn.* computer graphics. A mode of computer processing and output in which a large proportion of the output is in pictorial form. The information presented may be in the form of graphs, engineering or architectural drawings, maps, models, etc. It may be in one or more colours and may be labelled. The output may be displayed on the screen of a *VDU or may be recorded by a plotter. The information is fed into the computer by various means, such as *light pen or *mouse. The computer can be made to manipulate the information, for example by straightening lines, moving or removing specified areas, or expanding or contracting details. Apparently three-dimensional images can be produced, which can sometimes be observed from different viewpoints.

Grashof number Symbol: *Gr*. The dimensionless parameter:

$$l^3 g\gamma\rho^2\theta/\eta^2,$$

which occurs in the dimensional analysis of convection in a fluid due to the presence of a hot body; l is a typical dimension of the body, g is the acceleration of free fall, γ is the cubic expansion coefficient of the fluid, ρ is the density of the fluid, η is the viscosity of the fluid, and θ is the temperature difference between the hot body and the fluid. *See* convection (of heat).

Grassot fluxmeter *See* fluxmeter.

graticule A network of fine lines set at the focal point of the *eyepiece of a microscope or telescope, and therefore in focus simultaneously with the object viewed. It acts as a field reference system and may be used for the purpose of measurement. The graticule consists either of a grid or pattern of fine wires or threads (often then referred to as a *reticle*), or of a transparent glass disc with the lines engraved on it.

grating *See* diffraction grating.

gravimeter *See* free fall.

gravitation The mutual attraction of all bodies, independent of electromagnetic, strong, or weak interactions. *Galileo studied falling bodies (late 16th century) and introduced the concept of *acceleration. He argued that in a vacuum all bodies would have the same acceleration, that of *free fall. In 1687 Newton presented his *law of universal gravitation*, according to which every particle attracts every other particle with a force F given by:

$$F = Gm_1 m_2/x^2,$$

where m_1, m_2 are the masses of two particles a distance x apart. G is the *gravitational constant*, which, according to modern measurements, has the value

$$6.672\,59 \times 10^{-11} \text{ m}^3 \text{ kg}^{-1} \text{ s}^{-2}.$$

For extended bodies the forces are found by integration. Newton showed that the external effect of a spherically symmetric body is the same as if the whole mass were concentrated at the centre. Astronomical bodies are roughly spherically symmetrical so can be treated as point particles to a very good approximation. On this assumption Newton showed that his law was consistent with *Kepler's laws. Until recently, all experiments have confirmed the accuracy of the inverse square law and the independence of the law upon the nature of the substances, but in the past few years evidence has been found against both.

The size of a *gravitational field* at any point is given by the force exerted on unit mass at that point. The field intensity at a distance x from a point mass m is therefore Gm/x^2, and acts towards m. Gravitational field strength is measured in newtons per kilogram. The *gravitational potential V at that point is the work done in moving a unit mass from infinity to the point against the field. Due to a point mass

$$V = Gm \int_\infty^x dx/x^2 = -Gm/x.$$

V is a scalar measured in joules per kilogram. The following special cases are also important: (a) Potential at a point distance x from the centre of a hollow homogeneous spherical shell of mass m and outside the shell:

$$V = -Gm/x.$$

The potential is the same as if the mass of the shell is assumed concentrated at the centre. (b) At any point inside the spherical shell the potential is equal to its value at the surface:

$$V = -Gm/r,$$

where r is the radius of the shell. Thus there is no resultant force acting at any point inside the shell (since no potential difference acts between any two points). (c) Potential at a point distance x from the centre of a homogeneous solid

201

sphere and outside the sphere is the same as that for a shell:

$$V = -Gm/x.$$

(d) At a point inside the sphere, of radius r:

$$V = -Gm(3r^2 - x^2)/2r^3.$$

The essential property of gravitation is that it causes a change in motion, in particular the acceleration of free fall (g) in the earth's gravitational field. According to the general theory of *relativity, gravitational fields change the geometry of space-time, causing it to become curved. It is this curvature of space-time, produced by the presence of matter, that controls the natural motions of bodies. General relativity may thus be considered as a theory of gravitation, differences between it and Newtonian gravitation only appearing when the gravitational fields become very strong, as with *black holes and *neutron stars, or when very accurate measurements can be made.

gravitational collapse The contraction of an astronomical body resulting from the mutual gravitational pull of all its constituents. The term is most commonly applied to the sudden collapse of the core of a star when energy can no longer be produced by nuclear-fusion reactions. There is then no outwardly directed gas pressure or radiation pressure to counterbalance the inwardly directed gravitational force, and the hydrostatic equilibrium is destroyed. The three most likely end-products of such a collapse are (in order of mass) *white dwarfs, *neutron stars, and *black holes. In the first two the collapse can be halted by a quantum mechanical effect: a *degeneracy pressure* exerted by tightly packed electrons stripped from the atomic nuclei (in white dwarfs) or by tightly packed neutrons (in neutron stars).

gravitational constant *See* gravitation.

202

gravitational field The space surrounding a massive body in which another massive body experiences a force of attraction. *See* gravitation.

gravitational lens An astronomical body (usually a galaxy or cluster of galaxies) whose gravitational field bends light and other radiation from a more distant source (usually a *quasar), so that multiple images of the latter are produced. This effect, known as *lensing*, can be explained by general *relativity and was first observed in 1979.

gravitational mass *See* mass.

gravitational potential *See* gravitation; potential.

gravitational redshift *See* Einstein shift.

gravitational unit A unit of force, pressure, work, power, etc., involving g, the acceleration of *free fall.

gravitational waves *Syn.* gravitational radiation. The propagation of a changing gravitational field at the speed of light, caused by the displacement of masses. Gravitational waves are predicted by the *general theory of relativity but have not yet been detected experimentally. Astronomical observations of *supernova explosions or the orbital motions of binary *pulsars could present indirect evidence.

graviton A hypothetical *elementary particle responsible for the effects of *gravitation; it is the quantum of the *gravitational field and is thus a *gauge boson. It is postulated to be its own *antiparticle, to have zero charge and rest mass, and a spin of 2. It is firmly predicted by theory but its direct observation is at present unlikely.

gravity 1. An alternative name for *gravitation.

2. For a body at or near the surface of a planet, the apparent force of gravitation. If this is combined vectorially with the centripetal force of axial motion it gives the real force of gravitation. *See* free fall; weight.

gravity balance *See* free fall.

gravity cell A primary electric cell in which two electrolytes are kept apart by their different densities.

gravity meter *See* free fall.

gravity wave A wave in the surface layers of a liquid, being controlled by gravity and not by surface tension. For example, if the depth of the liquid is large compared with the wavelength λ, the speed is given by $v = \sqrt{(g\lambda/2\pi)}$, where g is the acceleration of free fall. The amplitude falls off exponentially with depth, decreasing by a factor e in a distance $\lambda/2\pi$ (where e = 2.718).

gray Symbol: Gy. The derived SI unit of absorbed *dose of ionizing radiation, and of specific *energy imparted. It is equal to an absorption or delivery of one joule per kilogram of irradiated material. It replaces the *rad. 1 gray = 100 rad.

grease-spot photometer A design of *photometer head consisting of a thin white opaque paper with a translucent spot at the centre. Lights illuminate both sides. The intensity of illumination (i.e. the *illuminance) is assumed to vary inversely as the square of the distance. Any convenient auxiliary source can be fixed on one side. The two sources to be compared (of illuminance C_1 and C_2) are moved along a bench until the spot disappears for each in turn. Then

$$C_1/C_2 = (d_1/d_2)^2.$$

Alternatively the two sources C_1 and C_2 are on opposite sides and the distances d_1 and d_2 when disappearance occurs on one side, and d'_1 and d'_2 for disappearance on the other side, are measured. Then,

$$C_1/C_2 = d_1 d'_1/d_2 d'_2.$$

great circle A circle in which a plane passing through the centre of a sphere intersects the surface. The shortest distance between two points on the surface of a sphere is along the great circle joining them.

greenhouse effect A process whereby an environment is heated by the trapping of *infrared radiation. It operates on several bodies in the solar system, including Venus (where it is responsible for the high surface temperature) and to a much lesser but increasing extent on earth. In the case of earth, sunlight is absorbed by the earth's surface and is reradiated at longer (infrared) wavelengths. Although most of the infrared escapes from the atmosphere, some is absorbed by certain gases in the atmosphere, notably carbon dioxide (CO_2). The amount of CO_2 in the atmosphere is known to have risen recently, as a result of man's activities, raising fears about global warming.

Green's theorem A vector form of *Gauss's theorem.

Greenwich Mean Time (GMT) *See* time.

Gregorian calendar *See* time.

grenz rays X-rays of long wavelength produced when electrons are accelerated by voltages of 25 kV or less. They are generated in many types of electronic equipment using electron beams but have a very low penetrating power.

203

grey body A body that emits radiation of all wavelengths in constant proportion to the *black-body radiation of the same wavelengths at the same temperature.

grid 1. *See* control grid; thermionic valve.

2. The high-voltage transmission-line system that interconnects many large *generating stations. Voltages of 275 kV, or in some cases 400 kV, are commonly used, though in some countries voltages as high as 735 kV are used.

grid bias A polarizing potential difference applied between the cathode and control grid of a *thermionic valve to cause it to operate on any desired part of its characteristic curve, or to modify its cut-off values.

GRIN lens *See* gradient-index lens.

ground *See* earth.

ground state The state of a system with the lowest energy. An isolated body will remain indefinitely in it. It is possible for a system to have two or more ground states, of equal energy but with different sets of *quantum numbers. In the case of atomic hydrogen there are two states for which the quantum numbers n, l, and m are 1, 0, and 0 respectively, while the *spin may be $+\frac{1}{2}$ or $-\frac{1}{2}$ with respect to a defined direction (*see* atomic orbital). There is in fact an extremely small difference of energy according to whether the electron spin is parallel or antiparallel to the proton spin. For nearly all purposes one can assume that there is just one ground state of a unique energy, but transitions between these states do occur in interstellar atomic hydrogen, giving rise to radiation of wavelength 21 cm. *See also* excited state; zero-point energy.

ground wave An electromagnetic radio wave that is radiated from a transmitting aerial on the surface of the earth, and that travels along the surface of the earth. *Compare* ionospheric wave.

group (mathematics) A set of elements or operations a, b, c, ... for which a law of "combination" may be defined so that the "product" ab of any two elements is well defined and satisfies the following conditions:
(1) If a and b belong to the set, so does ab.
(2) "Combination" is associative; that is $a(bc) = (ab)c$.
(3) The set contains an element e, called the *identity*, such that $ae = ea = a$ for all elements a of the set.
(4) For every element a in the set there is an element b such that $ab = ba = e$. We denote b by a^{-1}, which is not necessarily the reciprocal of a, but depends on the combination (see below).

The law of combination referred to above is not necessarily ordinary multiplication. The set of integers
$$\ldots, -2, -1, 0, 1, 2, \ldots$$
form a group with addition as the law of combination. For example, take $a = 2$, $b = 3$ then ab means we take $2 + 3 = 5$, which is also an integer. The identity is 0 since $0 + n = n$, where n is any integer. The inverse of $a = 2$ is -2 since $aa^{-1} = 2 + (-2) = 0$.

Some of the most important groups have *matrices as elements. For example, the set of all $n \times n$ matrices that have nonzero determinants form a group called $GL(n)$ having matrix multiplication as its law of combination.

Two elements a,b of a group are said to commute if
$$ab = ba.$$
If all the elements of a group commute with each other the group is said to be commutative or an *Abelian group*. Noncommutative groups are *non-Abelian*; this distinction is important in *gauge

theories. Group theory is also important in physics in the analysis of such symmetries as the rotations and reflections of molecules, which underlie the quantum theory of angular momentum. More abstract and generalized aspects of group theory are used to describe the fundamental *interactions by gauge theories.

group speed In certain forms of wave motion the *phase speed (i.e. the speed with which the phase of an oscillation is propagated) varies with the wavelength. As a result of this, a nonsinusoidal wave appears to travel with a speed distinctly different from the phase speed.

The phenomenon is most readily seen with waves on the surface of water. Considering the group of waves resulting from a stone dropped into water, it may be observed that the waves within the group travel faster than the group itself, fresh waves appearing at the rear of the group as the existing waves vanish at the leading edge. The speed of the group is called the group speed, while that of the waves within the group is the phase speed. Since wave speeds are usually measured by the arrival of the disturbance caused by the wave at different points in its path, it is seen that measurements usually give the group speed of the wave and not the phase speed.

An expression for group speed may be obtained by considering the propagation of two sinusoidal waves of slightly different wavelengths, λ and $\lambda - \delta\lambda$, say, the corresponding phase speeds being c and $c - \delta c$. The superposition of these waves produces *beats that travel with the group speed of the waves. This speed, U, is given by:

$$U = (c\delta t - \lambda)/\delta t = c - \lambda(\delta c/\delta\lambda).$$

In the limit, for $\delta\lambda = 0$, this equation becomes

$$U = c - \lambda dc/d\lambda.$$

Otherwise expressed $U = d\nu/d\nu'$, where ν is the frequency and $\nu' = 1/\lambda$, the reciprocal wavelength or *wavenumber.

If the phase speed does not vary with wavelength, then the group speed and the phase speed become identical. There is then absence of dispersion of speed, as in the case of light in a vacuum. The direct measurement of light speed in dispersing media measures group speed; light energy flows with the speed of the group.

Grove cell A two-fluid primary *cell in which the negative element consists of a zinc rod in dilute sulphuric acid, and the positive element, which is separated from the negative by a porous partition, consists of a platinum plate immersed in fuming nitric acid. The e.m.f. is 1.93 volts.

grown junction A *p-n junction formed in a single crystal of semiconductor by varying, in a precise manner, the types and amounts of impurities added to the semiconductor, while the crystal is being grown from the melt.

Grüneisen's law A law derived from the *equation of state for solids that states that the ratio of the coefficient of linear expansion of a metal to its specific heat capacity is a constant independent of the temperature at which the measurements are made.

guard band In allocating the bands of frequencies within the frequency spectrum to various communication channels, it is desirable that the frequency bands of adjacent channels be separated by narrow bands of frequencies (called guard bands) to reduce the possibility of mutual interference between the bands.

guard ring 1. (electrical) A large metal plate surrounding and coplanar with a small metal plate from which it is separated by a narrow air gap. The device is used in standard capacitors to ensure a uniform and calculable field over the area of the smaller plate, which can be treated as an infinite plane – the variations in the field that occur as the edge of the plane is approached affect only the guard ring. An extra electrode, equivalent to a guard ring, is commonly used in semiconductor devices and vacuum tubes.

2. (heat) A similar device, used in experiments on heat flow, that produces a temperature gradient in the region all round the specimen identical with that down the specimen so that heat losses from the latter are eliminated.

guard wires Earthed conductors placed beneath overhead-line conductors so that if the latter break, they will be earthed before they reach the ground. A series of guard wires arranged to form a net is known as a *cradle guard* and is used where a high-voltage line crosses a telephone wire or a thoroughfare.

Gudden–Pohl effect A form of *electroluminescence from a phosphor following exposure to ultraviolet radiation, which excites the phosphor into a metastable state.

Guillemin effect A type of *magnetostriction in which a bent bar of *ferromagnetic material tends to straighten out under the influence of a magnetic field applied along its length.

Guillemin line An electrical network designed to produce pulses with very sharp rise and fall times so that they are almost square.

Gunn diode A two-terminal device consisting of a sample of n-type gallium arsenide operated under such conditions that microwave oscillations are produced due to the *Gunn effect.

Gunn effect An effect in which coherent microwave oscillations are generated when a large d.c. electric field is applied across a short n type sample of gallium arsenide; the value must be above a threshold value of several thousand volts per cm. The effect is caused by charge *carriers of different mobilities forming bunches, known as *domains*, under the influence of the electric field. Some of the conduction electrons move from a low-energy, high-mobility state into a higher-energy, low-mobility state causing domains of low mobility to be set up. It is these domains that produce the microwave output.

GUT Abbreviation for *grand unified theory.

GWS theory *See* electroweak theory.

gyrator A component, usually used at microwave frequencies, that reverses the phase of signals transmitted in one direction but has no effect on the phase of signals transmitted in the opposite direction. The gyrator may be entirely passive, or contain active components.

gyrocompass A nonmagnetic compass using a *gyroscope fitted with a pendulous weight or some equivalent to induce precession due to gravity. The subsequent damped motion of the gyro aligns this with the true N-S direction.

gyrodynamics The study of rotating bodies, particularly when subject to *precession.

gyromagnetic effects The relationships between the magnetization of a body and its rotation. *See* Barnett effect; Einstein and de Haas effect.

gyromagnetic ratio Symbol: γ. The ratio of the *magnetic moment of a system to its *angular momentum. An orbiting electron has a value of $e/2m$, where e is the electron charge and m its mass. The gyromagnetic ratio of an electron due to its *spin is twice this value.

gyroscope A device in which a suitably mounted flywheel or rotor is spun at high speed. The mounting, usually of gimbal type, allows the axis of rotation to be in any direction in space.

If a couple is applied to the frame of the gyroscope, the resulting motion of *precession tends to align the gyro with the axis of the couple. The rate of turning or precession is proportional to the moment of the applied couple and inversely proportional to the angular momentum of the gyro.

In the absence of disturbing couples, the direction in space of the spin axis stays constant; hence, gyroscopic devices are useful for guidance of aircraft during turns, etc., and they are widely used in automatic-guidance devices. Large gyros are used on some ships to achieve stability against rolling. *See* gyrostat; gyrocompass.

gyrostat A *gyroscope, especially a version intended primarily to indicate or to use the constancy of direction of axis of a fast-running gyroscope.

H

habit The set of natural faces that appear on crystal.

hadron An elementary particle composed of *quarks and/or antiquarks that can take part in *strong interactions. Hadrons with zero or integer spin are known as *mesons (consisting of a quark–antiquark pair), and those with half integer spin as *baryons (consisting

of three quarks). *See also* Appendix, Table 7.

Hagen–Poiseuille law *See* Poiseuille flow.

Haidinger fringes Interference fringes formed by rays reflected practically normally from two plane and parallel surfaces relatively widely separated. The observing eye or telescope must be focused for infinity.

hail Roughly spherical ice particles, usually a few millimetres in radius, produced in very turbulent clouds. A hailstone may be carried upwards by an eddy several times, each time falling through a region in which it grows by impact with the supercooled water droplets of which clouds below 0 °C are generally composed. On hitting the hailstone the droplet freezes onto the surface and minute fragments of ice are thrown off. According to one theory, these fragments have opposite electric charge to that of the hailstone and cause the electrification of thunderclouds. *See* snow.

hair hygrometer A *hygrometer that depends for its action on the increase in length of a hair occurring when the relative *humidity of the surrounding air increases.

halation 1. An exposed ring surrounding a strongly illuminated spot on a photographic emulsion. It is caused by light scattered so as to strike the opposite face of the plastic or glass base at angles greater than the *critical angle, giving total reflection. It is prevented by an *antihalation backing* to the base, consisting of a film of refracting material containing a light-absorbing dye.

2. A similar phenomenon when a fluorescent screen is coated on a sheet of transparent material, for example, the screen of a *cathode-ray tube. In this case the incident radiation is fully absorbed but the light is emitted at all angles, giving an illuminated ring around a bright spot.

half-cell One electrode of an electrolytic cell and the electrolyte with which it is in contact.

half-duplex operation *See* duplex operation.

half-life Symbol: $T_{1/2}$, $t_{1/2}$. The time in which the amount of a radioactive nuclide decays to half its original value. It is given by:
$$T_{1/2} = (\log_e 2)/\lambda = 0.693\,15/\lambda,$$
where λ is the *decay constant, or by:
$$T_{1/2} = \tau \times 0.693\,15,$$
where τ is the *mean life.

half-value thickness The thickness of a uniform sheet of material that, when interposed in a beam of radiation, will reduce the intensity or some other specified property of the radiation passing through it to one half. The half-value thickness is often used as a means of defining the quality of the radiation.

half-wave dipole An *aerial consisting of a straight conductor that is approximately half a wavelength the dipole is excited, it has a voltage node and current antinode at its centre, and a voltage antinode and current node at each end. The feeder is commonly, but not always, connected across a small gap in the centre of the dipole. *See also* dipole aerial.

half-wave plate A thin double-refracting optical element, often of quartz or mica, that can be used to change the *polarization of an incident wave. It is cut parallel to the optic axis of such thickness as to introduce a half-wavelength path difference, i.e. a relative phase difference of 180° between the ordinary and extraordinary rays. *Plane-polarized light incident normally on the plate has its plane of polarization rotated through twice the angle between the axis and the incident vibrations.

half-wave rectifier circuit A rectifier circuit in which only alternate half waves of the single-phase a.c. input wave are effective in delivering unidirectional current to the load.

half-width Half the width of a spectrum line measured at half its height. In some branches of spectroscopy the half-width is used for the full width of the line at half its height.

Hall coefficient *See* Hall effect.

Hall effect An effect occurring when a current-carrying conductor is placed in a magnetic field and orientated so that the field is at right angles to the direction of the current: an electric field is produced in the conductor at right angles to both the current and the magnetic field. The field produced is related to the vector product of the current density j and magnetic flux density B by the relation:
$$E_H = -R_H(j \times B).$$
The constant R_H is the *Hall coefficient*. The electric field results in a small transverse potential difference, the *Hall voltage*, V_H, being set up across the conductor.

In metals and *degenerate semiconductors, R_H is independent of B and is given by $1/ne$, where n = carrier density and e = electronic charge. In nondegenerate semiconductors additional factors are introduced due to the energy distribution of the current carriers.

The Hall effect is a consequence of the *Lorentz force acting on the charge-carrying electrons. A sideways drift is imposed on the motion of the electrons by their passage through the magnetic field. For materials in which the current is carried by positive charge carriers (*holes), the direction of the Hall field, E_H, is reversed.

Under certain conditions the *quantum Hall effect* is observed. The motion of the electrons must be constrained so that they can only move in a two-dimensional "flatland"; this can be achieved by confining the electrons to an extremely thin layer of semiconductor. In addition, the temperature must be very low (around 4.2 K or below) and a very strong magnetic flux density (of the order of 10 T) must be used. The magnetic field, applied normal to the semiconductor layer, produces the transverse Hall voltage as in the ordinary Hall effect. The ratio of the Hall voltage to the current is the *Hall resistance*. At certain values of flux density, both the conductivity and the resistivity of the solid become zero, rather like in a superconductor. A graph of Hall resistance against flux density shows steplike regions, which correspond to the values at which the conductivity is zero. At these points, then, the Hall resistance is quantized; calculations show that

$$(V_H/I)n = h/e^2,$$

where n is an integer, h is the Planck constant, and e is the electron charge. The Hall resistance can be measured very accurately: it is equal to 25.8128 kΩ. Hence the quantum Hall effect can be used to calibrate a conventional resistance standard, and can also be used in the determination of h and e.

Hall mobility Of a *semiconductor or conductor. Symbol: μ_H. The product of the Hall coefficient R_H (*see* Hall effect) and the electrical conductivity κ.

Hall resistance *See* Hall effect.

Hall voltage *See* Hall effect.

Hamiltonian function *Syn.* Hamiltonian. Symbol: *H*. A function that expresses the energy of a system in terms of *generalized momenta, p_i, and positional coordinates, q_i; for example,

$$(p_i{}^2/2m + \mu q_i{}^2/2)$$

expresses the energy of a body in *simple harmonic motion. The Hamiltonian function may also involve the time. It is much used in *wave mechanics. *See* Hamilton's equations; Hamilton's principle.

Hamilton's equations A restatement of *Lagrange's equations with emphasis on momenta rather than forces. Much used in advanced mechanics including *quantum mechanics. There are twice as many Hamiltonian equations as Lagrangian equations but they are only first-order instead of second-order differential equations. They involve the *Hamiltonian function *H*, which in ordinary cases is the total energy expressed as a function of the *generalized coordinates q_i and momenta p_i:

$$dq_i/dt = \partial H/\partial p_i,$$
$$dp_i/dt = -(\partial H/\partial q_i).$$

Hamilton's principle If the configuration of a system is given at two instants, t_0 and t_1, then the value of the time-integral of the *Lagrangian function, $L = T - V$, is stationary (maximum or minimum) for the path described in the motion compared with any other infinitely near paths that might be described (for instance under constraints) in the same time between the same configurations. That is,

$$\delta \int_{t_0}^{t_1} (T - V)dt = 0,$$

where T = total kinetic energy, V = total potential energy. It has been more freely stated as "Nature tends to equalize the mean potential and kinetic energies during a motion". The principle is important for it contains in itself all the $3n$ equations of motion of the n particles comprising the system. The form given here is that for a conservative system, but the principle is of general application.

handset A telephone transmitter and receiver mounted in a single holder.

hard disk *See* disk.

hardness Of a crystal. The resistance that a face of the crystal offers to scratching, which may differ in different directions. For many substances, hardness is an inverse measure of plasticity, and the hardness may be measured by the *Brinell test*, which consists in determining the load necessary to produce an indent of measured dimensions on the material under test. *Mohs' scale* of comparative hardness uses ten selected solids arranged in such an order that a substance can scratch all substances below it in the scale, and cannot scratch those above it: (1) talc, (2) rock salt, (3) calcite, (4) fluorite, (5) apatite, (6) felspar, (7) quartz, (8) topaz, (9) corundum, (10) diamond. The scale is not quantitative.

hard radiation *Ionizing radiation with a high degree of penetration, especially X-rays of relatively short wavelengths. *Compare* soft radiation.

hard-vacuum tube *Syn.* high-vacuum tube. A *vacuum tube in which the degree of the vacuum is such that ionization of the residual gas has a negligible effect upon the electrical characteristics.

hardware The physical components of a *computer system, such as VDUs, disk drives, printers, and the electronic circuitry making up semiconductor memory and logic circuits. *See also* software.

Hare hydrometer *See* hydrometer.

harmonic 1. An oscillation of a periodic quantity whose frequency is an integral multiple of the fundamental frequency.
2. A tone of a series constituting a *note, and having a frequency that is an integral multiple of the fundamental frequency of the note.

harmonic analyser A device that evaluates the coefficients of the Fourier series corresponding to a particular function. *See* Fourier analysis.

harmonic distortion *See* distortion.

harmonic generator A *signal generator that produces a large number of odd and even *harmonics of the fundamental frequency of the input.

Hartmann formula A formula giving the variation of refractive index n of a medium with the wavelength of light:
$$n = n_\infty + c/(\lambda + \lambda_0)^a,$$
where n_∞, λ_0, and a are constants. For common forms of glass it is usual to take $a = 1$.

Hartmann generator An apparatus for producing ultrasonic edge tones on the principle of the *Galton whistle. It differs from this mainly in the greater blast velocity employed whereby the energy of the output is markedly increased. Frequencies up to 100 000 hertz can be produced.

hartree *See* atomic unit of energy.

Hawking radiation The emission of particles by a *black hole. A black hole's gravitational field causes particle–antiparticle pairs to be produced in the vicinity of the event horizon (*see* Schwarzschild radius); one member of each pair falls into the black hole and the other escapes, the energy of the particle falling in is negative and exactly balances the positive energy of the particle that escapes. The result of this process is that the flux of emitted particles escapes with some of the mass of the black hole. This process is unlikely to be significant with black holes having a mass comparable to that of the sun, because the radiation temperature is inversely proportional to the mass of the black hole. However, a black hole of mass 10^{12} kg (as might have formed in the early universe) could be a copious source of Hawking radiation.

H–D curve *See* Hurter–Driffield curve.

head 1. A device that records, reads, or erases signals or data on a medium such as a magnetic tape or disk. **2.** *See* pressure head.

health physics A branch of *medical physics concerned with the health and safety of personncl in medical, scientific, and industrial work. It is most particularly concerned with protection from *ionizing radiation and from neutrons.

Problems involved in radiation protection include the detection and measurement of ionizing radiation, cleaning of both personnel and surfaces contaminated by radioactive substances, disposal of *radioactive waste, design of laboratories and the *shielding* of equipment for radiation work, and the supervision of tolerance *doses received by personnel in the course of their duties.

The shielding of equipment and personnel is usually achieved by using concrete (*see* loaded concrete) or lead. The conditions and the type of radiation under consideration (whether it is *hard or *soft, etc.) determine the position and thickness of the shielding.

heat Symbol: Q. *Syn.* quantity of heat. The energy transferred from a body at a higher temperature to one at a lower temperature because of the difference of the temperature only. Before the

principles of *thermodynamics were clearly established the word heat was used with various meanings, including *temperature and *internal energy. It is important to distinguish such quantities and to avoid any implication that heat now refers to any property or condition of a body, or anything but a process of transfer. The ambiguous term "heat energy" should be avoided. The processes are *conduction of heat, *convection, and *electromagnetic radiation.

Radiation is regarded as heat when the spontaneous emission from a hotter body is absorbed by one at lower temperature. All wavelengths emitted (ultraviolet, visible, infrared) are heat. Radiation can also transfer energy by doing work, for example a transmitter does work on a radio receiver, the temperatures being irrelevant.

The unit is the *joule (J); former units include the *calorie.

heat capacity Symbol: C. The quantity of heat required to raise the temperature of a body through one degree. It is measured in joules per kelvin. *See also* specific heat capacity.

heat death The condition of any isolated system when its *entropy is a maximum. The matter present is then completely disordered and at a uniform temperature, and there is therefore no internal energy available for doing work. If the universe is a closed system, it should eventually reach this state. This is called the *heat death of the universe.*

heat engine A device that takes in heat from a hot source and does work, waste heat being discharged to a colder body, which is usually the atmosphere. In most cases work is done mechanically but, for example, a *thermocouple is a heat engine that does work electrically.

Ideally heat engines work cyclically, a *working substance* being taken through a

211

sequence of operations and being returned to the initial state; in practice a fresh supply of working substance may be taken in for each cycle. A heat engine may be driven in reverse as a *refrigerator or *heat pump. *See* Carnot cycle; Diesel cycle; Otto cycle; Rankine cycle; efficiency.

heater In general, any *resistor used to provide a source of heat when carrying an electric current. Heaters are used, for example, in indirectly heated cathodes and in domestic appliances. The term is also used to indicate a complete heating device, e.g. convector heater.

heat exchanger A device for transferring heat from one fluid to another without the fluids coming in contact. Its purpose is either to regulate the temperatures of the fluids for optimum efficiency of some process, or to make use of heat that would otherwise be wasted. The simplest form consists of two coaxial pipes, the inner one finned on the outside to maximize the contact area, with the fluids moving through the pipes in opposite directions.

heat flow rate Symbol: Φ. The rate of heat flow across a surface; it is measured in watts. The *density of heat flow rate*, symbol: ϕ or u, is the heat flow rate per unit area. The *thermal conductivity is the density of heat flow rate divided by temperature gradient.

heating effect of a current *Syn.* Joule effect. When a current I is maintained in a resistor of resistance R by a potential difference V for a time t, the work done electrically is:
$$W = IVt = I^2Rt = V^2t/R.$$
By the first law of *thermodynamics the internal energy of the resistor increases by δU while heat Q is given out to the surroundings, such that $W = Q + \delta U$. In the steady state δU is zero and $Q =$

W, hence the resistor supplies heat at the rate given by the equation above.

The law in the form $Q = I^2Rt$ was originally proposed by *Joule (1840) on the basis of experiment, before the electrical quantities were fully defined or the concept of energy clearly formulated.

heat pump A device for heating buildings, in the form of a *heat engine driven in reverse. The internal energy of some part of the environment (the atmosphere, soil, a river, etc.) is used as an energy source, giving heat Q_I to the working substance. An electric motor does work W in taking the substance round a cycle in which the temperature is raised to slightly above that of the building. The heat given out Q_O is equal to $Q_I + W$. Thus the internal energy of the colder body is decreased and that of the hotter body is increased. This does not violate the laws of thermodynamics since the process involves work being done upon the system and is not overall a process of heat transfer.

heat sink 1. A device employed (especially in association with *transistors and other electronic components) when it is essential to dispose of unwanted heat and prevent an unwelcome or damaging rise in temperature. It generally consists of a set of metal plates in a finlike formation that conducts and radiates the heat away.

2. A system that is considered to absorb heat at a constant temperature. The concept is useful in *thermodynamics, as in the operation of a heat engine.

heat-transfer coefficient The heat flow per unit time through unit area divided by the temperature difference. When applied to conduction of heat through a body, it is called the *thermal conductance* and has the symbol K. When applied to the emission of heat from a surface, it has the symbol E or α. It is measured in watt metre^{-2} kelvin^{-1}.

Heaviside layer *See* ionosphere.

Heaviside–Lorentz units A *CGS system of electrostatic and electromagnetic units. They are a rationalized form of Gaussian units, in which the magnetic constant has the value 4π and the electric constant $1/4\pi$. Like Gaussian units, they are still used in particle physics and relativity.

heavy-fermion substance A substance in which electrons have an effective mass several hundred times that of a normal electron. The high effective mass occurs in f electrons in narrow *energy bands associated with many-body effects in atoms of the actinium series and in rare earths; for example, $CeCuSi_2$ is a substance in which this phenomenon occurs. These substances have unusual magnetic, thermodynamic, and superconducting properties, which are not well understood. Their *superconductivity does not comply with the BCS theory as the Cooper pairs are apparently formed by these electrons of high effective mass, but in some cases the superconducting properties occur at a much higher temperature (high-temperature superconductivity).

heavy-water reactor (HWR) A type of thermal *nuclear reactor in which heavy water (deuterium oxide) is used as the *moderator because of its much smaller capture cross section for neutrons, relative to water. It is sometimes also used as the *coolant.

hecto- Symbol: h. A prefix denoting 100; for example, one hectometre (1 hm) = 100 metres.

Heisenberg uncertainty principle *See* uncertainty principle.

Helmholtz coils A pair of identical cylindrical coils of wire mounted coaxially and separated by a distance equal to the radius of the coils. When a current is passed through the coils, connected in series, a uniform magnetic field is produced over a considerable volume on either side of the midpoint between the coils.

Helmholtz electric double layer When a body is brought into contact with another body composed of a different material, the two bodies become oppositely charged, the substance with the higher relative *permittivity, ε_r, becoming positive. Helmholtz postulated that a film one molecule thick forming a double layer of positive and negative charges is set up and maintained by the inherent electrical forces of matter. Lenard extended the theory by suggesting that at the surface of any solid or liquid the molecules show orientation of the dipoles, negative charge outwards, forming an electric double layer. In materials of high ε_r, the attraction between the opposite charges is smaller, and a substance of small ε_r can thus remove free negative charges from one of greater ε_r. Contact needs to be close for this to happen (*see* frictional electricity). On separation of the charges, which initially were at molecular diameters apart, the lines of force are considerable extended, and the potential difference thus produced may be made very large.

Helmholtz function *Syn.* Helmholtz free energy. Symbol: A, F. A thermodynamic function of a system given by its internal *energy (U) minus the product of its *entropy (S), and its thermodynamic temperature (T), i.e.

$$A = U - TS.$$

If a system undergoes a *reversible change at a constant temperature, the Helmholtz function increases by an amount equal to the work done on it.

The change in A (ΔA) between any two states of a system gives the maximum work that could be obtained from the system during this change if the optimum pathway were to be followed. If this change is negative, work is obtained from the system. *See also* Gibbs function.

Helmholtz resonator An acoustic resonator in the form of an air-filled cylindrical or spherical cavity connecting with the atmosphere through a neck, whose internal capacity is smaller than that of the cavity. These resonators in general have more selective resonance than the columns of air in pipes, since a very small proportion of energy is radiated into the atmosphere and therefore the damping is very small.

It is assumed that the air in the neighbourhood of the neck acts like a piston, alternately compressing and rarefying the air in the cavity, and that the wavelength of the vibrations in the free air is large compared with the dimensions of the cavity. The system in its simplest form is equivalent to a mass attached to a spring, the air piston in the neck being regarded as the mass and the air in the cavity as the spring. In acoustic terms the neck is considered as the inertance and the cavity as the capacitance (*see* acoustic impedance). The dissipation is mainly due to the energy radiated and this is equivalent to the acoustic resistance.

The resonance frequency v is that corresponding to the value of angular frequency for which the reactance term disappears, i.e. for which
$$2\pi v = c\sqrt{(S/lV)},$$
where c is the speed of sound, l is the length of the neck, S its cross-sectional area, and V is the volume of the cavity. Cylindrical Helmholtz resonators can be adjusted in volume by sliding one part over the other, so changing the resonant frequency.

Helmholtz resonators are used as extremely sensitive detectors of sound at a particular frequency. The use of two connected resonators – a *double resonator* – produces further increase of sensitivity.

henry Symbol: H. The *SI unit of self- and mutual inductance, defined as the inductance of a closed loop that gives rise to a magnetic flux of one weber for each ampere of current that flows. *See* electromagnetic induction.

heptode A *thermionic valve having five grids between the cathode and anode (i.e. a total of seven electrodes).

Herschel–Quincke tube An apparatus to demonstrate the *interference of sound. It consists of a tube that divides into two tubes of different lengths, the ends of which join together to form one tube again. One of the tubes can usually be varied in length by means of an arrangement similar to the slide of a trombone. A source of sound is placed at one end of the apparatus and the sound travels along the two different paths, the resultant being heard at the other end. When the path difference is a whole number of wavelengths, the sounds arrive at the ear in phase and reinforce each other. When the path difference is an odd number of half-wavelengths, however, they are out of phase and no sound is heard.

hertz Symbol: Hz. The *SI unit of *frequency, defined as the frequency of a periodic phenomenon that has a period of one second.

Hertzian oscillator An electrical system for the production of electromagnetic waves of radio frequency. It was first used by Heinrich Hertz to demonstrate the existence and properties of such waves (originally called *Hertzian waves*).

It consists of two capacitors, e.g. two plates or spheres joined by a conducting rod in which there is a small spark gap. If the two halves of the oscillator are raised to a sufficiently high potential difference, a spark passes across the gap, rendering it temporarily a conductor; an oscillatory discharge takes place, the period of the oscillations being equal to $2\pi\sqrt{LC}$, where L is the self-inductance and C the capacitance of the system. Electromagnetic waves of the same period are given off during the discharge. The oscillator is usually activated by a small induction coil, and a group of waves is emitted at each discharge. Owing to the resistance of the spark gap, the waves are highly damped. Their wavelength is usually of the order of a few metres.

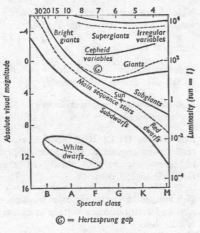

Temperature/kK

© = Hertzsprung gap

Hertzsprung–Russell diagram

Hertzsprung-Russell diagram (H–R diagram) A diagram showing the variation of absolute *magnitude (i.e. intrinsic brightness) in stars against their spectral type, and hence temperature (*see* stellar spectra). Instead of a uniform distribu-

tion, the stars occupy well-defined regions on the diagrams (*see* diagram). Some 90% lie along a diagonal band, known as the *main sequence*. The brighter *giant stars form another group, as do the brightest and relatively rare supergiants. Other groups can also be distinguished.

The H–R diagram is of great importance in studying *stellar evolution*. Diagrams determined from a theoretical basis can be tested against those obtained by astronomical observations. Diagrams can be drawn, for example, for the brightest stars, for stars in a particular locality (e.g. in the sun's neighbourhood), for pulsating *variable stars, and for globular clusters of stars (which contain some of the oldest stars in our Galaxy). These diagrams show different distributions of stars. For example, globular-cluster stars appear mainly in the giant branches of the H–R diagram.

A star spends most of its life on the main sequence, appearing there once nuclear-fusion reactions have begun in its core. Main-sequence stars obtain their energy from these reactions, in which hydrogen is converted to helium (*see* stellar energy). As a star burns up the last of its hydrogen, its radius increases and the star "moves" across from the main sequence to the giant region. This movement occupies only a relatively short length of time, and thus there is a space (known as the *Hertzsprung gap*) between the two regions. After eventually becoming *red giants, the stars tend to develop a variation in magnitude (*see* variable star), and they occupy the lower left-hand part of the giant region. In due course, after a sequence of as yet imperfectly understood events, the stars become *white dwarfs, *neutron stars, or possibly *black holes, depending on the final stellar mass. *See also* gravitational collapse.

heterodyne reception *See* beat reception.

heterogeneous radiation A particular type of radiation, such as X-rays or gamma rays, having a variety of wavelengths or quantum energies.

heterogeneous reactor A type of *nuclear reactor in which the fuel is separated from the *moderator.

heterogeneous strain *See* homogeneous strain.

heterojunction A junction between two dissimilar *semiconductors of opposite polarity (i.e. p-type and n-type) and with different energy gaps between the valence and conduction bands (*see* energy bands). Such junctions have several advantages over the usual *homojunction. They are used for example, in *semiconductor lasers, *light-emitting diodes, and in the *heterojunction bipolar transistor* (HJBT). *See also* heterostructure.

heteropolar generator A *generator in which the active conductors pass through magnetic fields of opposite sense in succession. Most modern generators are of this type. An alternating e.m.f. is induced in each conductor so that a direct-current generator of this type must be fitted with a *commutator.

heterostructure The composite structure resulting when a layer of one semiconductor is deposited on a layer of a different semiconductor. There may be several layers of the two dissimilar materials. The two semiconductors are selected to have different energy gaps between the valence and conduction bands (*see* energy bands). Heterostructure semiconductors, involving ultrathin layers, can now be produced by advanced crystal-growing techniques. The choice of materials, and the thickness and number of layers, may be varied by design over wide limits. The electronic properties of such a device can therefore be designed for a particular application.

hexagonal system *See* crystal systems.

hexode A *thermionic valve having four grids between the cathode and anode (i.e. a total of six electrodes).

HF Abbreviation for high frequency. *See* frequency bands.

hidden matter *See* missing mass.

Higgs boson *See* electroweak theory.

high elasticity A property that enables some substances to obey *Hooke's law with fair exactitude up to enormously greater strains than are normally met (e.g. cellulose hydrate and some other organic polymers).

high frequency (HF) *See* frequency bands.

high-pass filter *See* filter.

high-temperature gas-cooled reactor (HTR) *See* gas-cooled reactor.

high-temperature superconductivity *See* superconductivity.

high tension (H.T.) High voltage, especially when applied to the anode supply of thermionic valves; usually in the range 60 to 250 volts.

high voltage In electrical-power transmission and distribution, a voltage in excess of 650 volts.

HII region *See* interstellar matter.

Hilbert space A multidimensional space in which the proper (eigen) functions of *wave mechanics are represented by orthogonal unit vectors.

HI region *See* interstellar matter.

histogram A graphical representation of a frequency distribution in which rectangular areas standing on each interval into which the observations are grouped, show the frequency of observations in that interval.

hoar frost A layer of ice deposited on surfaces whose temperature is below 0 °C by the condensation (strictly speaking, sublimation) of moisture from the atmosphere.

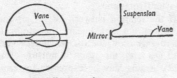

Hoffmann electrometer

Hoffmann electrometer A sensitive *electrometer using only a half-vane (*see* diagram) moving in two half-segments of a closed metal box, but otherwise working on the same principle as the *quadrant electrometer. Very heavy copper shields are used to minimize thermal variations, which might affect the needle, and the instrument is usually operated with the pressure reduced to a few hundred pascals.

hole In a solid, an empty state near the top of the valence band (*see* energy bands). Electrons can move into such empty states from adjacent occupied states, permitting the conduction of a current – *hole conduction*. In effect the hole travels through the material, acting like a positive charge (as shown by the *Hall effect) with a positive mass of the same order of magnitude as the mass of an electron but not identical with it (*see* effective mass). Generally, the *mobility of holes is less than that of the electrons in the conduction band.

A hole – an empty state in the valence band – may result from an elec-

tron being thermally excited from the valence to the conduction band, generating an *electron-hole pair*, or from an electron being trapped by an acceptor impurity (*see* semiconductor). In an intrinsic semiconductor the numbers of holes and free electrons are equal, with concentration n_i per unit volume. In an extrinsic semiconductor the concentration of holes n_+ and electrons n_- is given by $n_+ \times n_- = n_i^2$. A p-type semiconductor is one in which n_+ greatly exceeds n_- so electrical conduction is mostly by the holes, which are the *majority carriers.

hole conduction *See* hole.

hologram *See* holography.

holographic interferometry A technique that involves the superposition on the same photographic plate of two or more holograms (*see* holography) of an object under study. Any movement of the object between the holographic exposures shows up as *interference fringes distributed across the reconstructured image. The movement may result from vibration, heating, strain, etc. Analysis of the fringes provides information about the properties of the object.

holography A technique for the reproduction of a stereoscopic image without cameras or lenses. A monochromatic, coherent, and highly collimated beam of light from a *laser is separated into two beams, one of which is directed to a photographic plate coated with a film of high resolution. The second beam hits the subject whose image is to be reproduced, and is diffracted to the plate where a *hologram* is formed, consisting of an interference pattern (rather than a collection of light and dark areas as with a conventional photographic negative). The original subject may be recreated by placing the hologram in a beam of coherent light, generally from the same laser; the hologram behaves as a *diffraction grating, producing two beams of diffracted radiation, one giving a real image that may be recorded on a photograph, and the other a sterereoscopic virtual image.

It is also possible now to produce full-colour holograms: three laser beams are used (instead of a single beam), the three wavelengths corresponding to three *primary colours, and a thick emulsion is used on the holographic plate. It is also possible to view such holograms with ordinary (reflected) sunlight or tungsten light and see a stereoscopic full-colour image.

homocentric Converging to or diverging from a common point.

homogeneous radiation Radiation that has only one constant wavelength or quantum energy.

homogeneous reactor A type of *nuclear reactor in which the fuel and moderator present a uniform medium to the neutrons; for example, the fuel, in the form of a salt of uranium, may be dissolved in the moderator.

homogeneous solids Those in which the physical and chemical properties are the same about every point; they may be amorphous or crystalline.

homogeneous strain (1) When a body is strained (*see* strain), a particle whose Cartesian coordinates with respect to axes fixed outside the body are (x, y, z) is displaced to a new position (x', y', z'). If the following relations (in which the as, bs, and cs are nine constants) exist, the strain is said to be homogeneous and uniform:
$$x' = a_1x + a_2y + a_3z$$
$$y' = b_1x + b_2y + b_3z$$
$$z' = c_1x + c_2y + c_3z.$$

(No constant terms are included on the right-hand side as these would merely indicate a superimposed translation of the body.)

In a uniform homogeneous strain, a plane in the body remains a plane but changes its position; a parallelogram becomes a parallelogram in a different plane and with different angles. A sphere becomes an ellipsoid whose three mutually perpendicular axes are derived from three mutually perpendicular diameters of the sphere by their elongation and rotation.

If the relation between (x', y', z') and (x, y, z) are not linear, the strain is *heterogeneous*; if the as, bs, and cs vary with (x, y, z), the strain is not uniform.

(2) Strains may be expressed in terms of displacements, or shifts, of the particles. These are the quantities:
$$u = x' - x, v = y'-y, \text{ and } w = z'-z.$$

(3) If the squares and products of the displacements of a homogeneous strain are negligible in comparison with the displacements themselves, it can be shown that such strains can be combined by adding corresponding displacements. For example, two small homogeneous strains whose displacements are u_1, v_1, w_1, and u_2, v_2, w_2 are equivalent to a single strain with displacements $u_1 + u_2, v_1 + v_2, w_1 + w_2$. Using this theorem, it is found that any small uniform homogeneous strain can be resolved into two parts:

(*a*) a pure uniform homogeneous strain in which (see (1) above) $c_2 = b_3$, $a_3 = c_1$, and $b_1 = a_2$. This may be written in terms of displacements as:
$$u = s_xx + g_zy + g_yz$$
$$v = g_zx + s_yy + g_xz$$
$$w = g_yx + g_xy + s_zz$$
(there being now only six different constants $s_x = (a_1 - 1)$, $g_z = a_2$, $g_y = a_3$, etc.);

(*b*) a set of displacements representing a rotation of the body as a whole. (This brings the three mutually perpendicular diameters of the sphere (see (1) above) into the same directions as the axes of the ellipsoid derived from them.)

Thus, a pure uniform homogeneous strain, alone, changes a sphere into an ellipsoid and the three mutually perpendicular diameters of the sphere become the axes of the ellipsoid by changing their lengths without rotating. This ellipsoid is known as the *strain ellipsoid* and its axes are called the *principal axes of strain*. If the axes OX, OY, OZ are parallel to the principal axes of strain, the constants g_x, g_y, and g_z vanish, and s_x, s_y, and s_z are known as the *principal strains*.

homojunction A *p-n junction between two regions of opposite polarity (p-type and n-type) within a semiconductor. *Compare* heterostructure.

Hooke's law (1676) The law that forms the basis of the theory of elasticity and in its most general form states that, for a certain range of stresses, the strain produced is proportional to the stress applied, is independent of the time, and disappears completely on removal of the stress. (*See* strain; stress.) The point on the graph of stress versus strain for a real material, where it ceases to be linear, is known as the *limit of proportionality*. *See* yield point; elastic limit.

hook-up A temporary connection between electrical or electronic circuits or a temporary communications channel.

horizontal polarization 1. *Polarization of electromagnetic radiation in which the electric-field vector is horizontal, and hence the magnetic-field vector is vertical.

2. The transmission of radio waves so that their electric-field vectors are horizontal. *Dipole aerials are arranged horizontally for the transmission and reception of horizontally polarized signals.

In both cases, when the electric-field vector is vertical, then there is *vertical polarization*.

horn A tube of which the cross-sectional area increases progressively from the small end (throat) to the large end (mouth). It is used as an acoustic transmission line to couple the acoustic impedance as seen looking back into the diaphragm at the throat as efficiently as possible to the load as seen looking out of its mouth. Exponential horns are most generally used, the cross-sectional area increasing exponentially from throat to mouth. Horns can amplify sounds and also make them directional. They are used, for example, in musical instruments and loudspeaker systems.

horn antenna A microwave *aerial consisting of a metal device that flares out from the end of a *waveguide to a large circular, square, or rectangular aperture. The central axis may be straight, curved, folded, or bifurcated. Maximum radiation is along the centre of a straight horn.

horsepower (h.p.) A unit of power in the f.p.s. system equal to 745.7 watts. The *indicated* h.p. of an engine is the theoretical power developed in the cylinder by the steam, gas, or oil (according to the type of engine). The *effective* or *brake* h.p. is the rate of doing external work and is equal to the indicated h.p. minus the rate of working against friction in the engine itself.

horseshoe magnet A permanent magnet or electromagnet shaped so that the two poles are close together.

hot Highly radioactive. A *hot atom* is one in an excited state or having kinetic energy above the thermal level of the surroundings, usually as a result of nuclear processes.

hot cathode A cathode, used for example in *electron guns, that is operated at high temperatures in order to provide a source of electrons by *thermionic emission.

hot-wire ammeter An instrument in which the thermal expansion of a wire or strip due to the temperature rise caused by a current passing through it is employed to measure the current. Some mechanical device is used to magnify the actual increase in length of the wire, and to cause it to rotate a pointer over a circular scale. The scale is not uniform and the instrument must be calibrated empirically; but since the heating effect varies as the square of the current, the hot-wire ammeter can be used both for direct and for alternating current. It can be used as a voltmeter by adding suitable resistances in series.

hot-wire anemometer An instrument that measures the speed of a fluid in motion by virtue of the convective cooling it experiences when exposed to the fluid. It has the great advantage over other *anemometers that it can be made to occupy a very small space. It may, for example, take the form of a few centimetres of very fine nickel wire mounted on a fork and heated to 50 °C above its surroundings (in still air or water). The change of temperature of the wire when exposed to the flow is measured in terms of the change in its electrical resistance.

hot-wire gauge A pressure gauge depending on the cooling by the gas of a hot filament. *See* Pirani gauge.

hot-wire microphone An instrument used for comparative (or sometimes absolute) measurements of sound intensity and distribution of sound, for example in buildings. It is based on the change of the steady resistance of an electrically heated fine wire when subjected to sound waves. The resistance drop is the same as would result from a steady draught of speed equal to the maximum speed of the SHM of the particles set into motion.

hour-angle *See* celestial sphere.

howl A high-pitched audiofrequency tone heard in receivers and generally caused by acoustic feedback: the sound output of the loudspeaker can be detected and amplified by electronic circuits in the sound-reproduction system; above a critical level, oscillations occur and produce the howl in the loudspeaker. The process is similar to electrical *feedback, which can also produce howl. An oscillator designed to produce such a tone is called a *howler*.

H–R diagram *See* Hertzsprung–Russell diagram.

HTR Abbreviation for high-temperature *gas-cooled reactor.

Hubble constant *Syn.* Hubble parameter. Symbol: H_0. According to the theory of the *expanding universe, the *redshifts observed in the spectra of galaxies represent recessional velocities. The Hubble constant is defined as the ratio of recessional velocity to distance, and is commonly measured in kilometres per second per megaparsec. Galaxy redshifts, and hence recessional velocities, can be accurately determined. Distances are known with less certainty, especially for the more distant galaxies, and this leads to uncertainty in the value of H_0. Current estimates place H_0 as either about 55 or 80–100 km s^{-1} Mpc^{-1}.

The reciprocal of H_0 has the dimensions of time. It is a measure of the age of the universe, but only if the rate of expansion has always been constant. Since gravitation tends to diminish the expansion rate, $1/H_0$ can only give an upper limit: using the value 55 $\text{km s}^{-1} \text{Mpc}^{-1}$ gives an age of 18 $\times$ 10^9 years.

hue *See* colour.

hum An extraneous low-pitched droning noise heard in sound-reproduction systems. It originates in an associated or nearby electric power circuit. The most common hum is that caused by mains 50 hertz alternating current.

humidity A measure of the degree of wetness of the atmosphere. *Absolute humidity* is the mass of water vapour present in unit volume of moist air and is measured in kg m^{-3}. In meteorological studies this is often called the *vapour concentration*.

The absolute humidity of a volume of air is strongly dependent on temperature. A more useful measure is the *relative humidity* (symbol: U), defined as the ratio e/e' of the actual *vapour pressure to the *saturation vapour pressure over a plane liquid water surface at the same temperature, expressed as a percentage. This is the quantity normally referred to by the single word "humidity". As the temperature of the air is reduced, the water-vapour concentration increases until, at a particular temperature known as the dew point, θ_d, the air becomes saturated. Because only temperature has been altered, the actual vapour pressure at a temperature θ must equal the saturation vapour pressure at the dew point. Hence:
$$U = 100e'(\theta_d)/e'(\theta)$$
Because values of e' for water vapour are well known, the relative humidity can easily be calculated. The *mixing ratio* (symbol: r) is defined as the ratio of the mass of water vapour to the mass of dry air that contains it. If p is the total atmospheric pressure (in pascals),
$$r = 0.622 \, e/(p - e)$$

Humidity, either relative or absolute, is measured with a *hygrometer. A rough indication of the humidity is given by a *hygroscope.

hunting Variation of a controlled quantity such as the temperature of a thermostat above and below the desired value. Time-lags or bad positioning of regulating devices may lead to violent hunting. It can be eliminated by damping.

Hurter–Driffield curve (H–D curve) The *characteristic curve of a photographic material, showing the relation between the transmission density and the exposure. *See also* reciprocity law.

Huygens's eyepiece An *eyepiece that consists of two planoconvex lenses, having the convex faces towards the incident light and a field stop between them to reduce the somewhat large *spherical aberration. The focal length of the field lens is usually from two to three times that of the eye lens. The distance between the lenses is half the sum of the focal lengths, which minimizes *chromatic aberration.

Light from the objective converges towards a virtual focus inside the eyepiece so an external *graticule or cross-wires cannot be used. Because the eye lens by itself is uncorrected for aberrations, a graticule placed in the field stop would be unsuitable. Thus the Huygens eyepiece is not used for measuring instruments, although it is satisfactory for those used purely for observation.

Huygens's principle (published 1690) Every point on a primary *wavefront acts as a source of spherical *wavelets* or *secondary waves*, such that the primary wavefront at some later time is the envelope of these wavelets. The wavelets advance with a speed and frequency equal to those of the primary wave at each point in space. A back wave does not exist in practice, a fact explained by later modifications to the principle (Fresnel, Kirchhoff), giving it a firmer mathematical basis. If the medium is homogeneous, the wavelets can be constructed with finite radii. The amplitude in a wavelet falls off in proportion to $(1 + \cos\theta)$, where θ is the angle with the forward direction. The concept of wavelets was useful in explaining refraction (regular and double) and diffraction.

hybrid integrated circuit *See* integrated circuit.

hydrated electron *Syn.* aqueous electron. When an aqueous solution is irradiated with *ionizing radiation, the water molecules become ionized and secondary electrons are released. Such an electron rapidly loses its energy by ionizing and exciting neighbouring water molecules. After about 10^{-11} seconds from its formation, the water molecules have sufficient time to orientate themselves around the electron without it escaping, and a region of radial polarization results, trapping the electron in the centre. (The water molecule is polarized such that the hydrogen atom is slightly more positive than the oxygen atoms.) This species is called the hydrated electron, the electron existing in a *potential well. It is an extremely reactive species.

hydraulic press A device for producing large forces, consisting basically of two cylinders, one much wider than the other, fitted with pistons A and B and connected by a pipe C, the whole being filled with water. A force f applied downwards at A develops a pressure equal to f/A, where A is the cross-sectional area of A. This pressure is transmitted throughout the fluid (*see* Pascal's principle) and is thus applied at B. The

total force upwards on B equals the pressure times the area of B $= fB/A$. Hence a much larger force appears at B if B is much larger than A. On the other hand, if A moves downwards, B moves upwards by a much smaller distance and the work done on B never exceeds that done on A.

hydrodynamics A branch of the science of deformable bodies, being a study of the motion of fluids (liquid and gaseous). The classical theory of hydrodynamics is concerned with the mathematical treatment of perfect fluids. This theory is subject to modification to include the effect of the viscosity of a real fluid. Aerodynamics is essentially a specialized branch of hydrodynamics.

hydroelectric power station An electricity-generating station powered by water falling under gravity through a water turbine. Usually a dam is built across a river to provide a large reservoir of water, which permits continued operation during a dry season.

hydrogen bomb *See* nuclear weapons.

hydrogen electrode An electrode system in which hydrogen is in contact with a solution of hydrogen ions. It consists of a *half-cell in which a platinum foil is immersed in a dilute acid solution. Hydrogen gas is bubbled over the foil, which is usually coated with finely divided platinum to increase absorption of the hydrogen. Hydrogen electrodes are used in cells to measure standard *electrode potentials.

hydrogen spectrum The *emission spectrum of a *gas-discharge tube containing hydrogen has a faint continuum caused by the recombination of ions with electrons, band *spectra of hydrogen molecules (H_2) and molecular ions (H_2^+), and lines in the visible, ultraviolet, and infrared attributed to separated hydrogen atoms. Both emission and absorption lines of H atoms are observed in stellar spectra.

The study of the spectrum of atomic hydrogen has been of outstanding importance in the understanding of spectra, atomic structure, and quantum theory. Balmer (1885) showed that the wavelengths λ of the visible lines of hydrogen were represented accurately by the empirical equation:

$$1/\lambda = R(1/4 - 1/n^2),$$

where R is the *Rydberg constant and n is an integer greater than or equal to 3. The first members of the *Balmer series* are as follows (wavelength in brackets): Hα (656.3 nm), Hβ (486.1 nm), Hγ (434.0 nm). The series limit, for $n = \infty$, has λ = 364.6 nm and is in the near ultraviolet.

Later, other series were found, following the general formula:

$$1/\lambda = R(1/n_1^2 - 1/n_2^2),$$

where n_1 and n_2 are integers. The value $n_1 = 1$ gives the *Lyman series* in the *vacuum ultraviolet; $n_1 = 3$ gives the *Paschen series*; $n_1 = 4$ gives the *Brackett series* and $n_1 = 5$ gives the *Pfund series*. The last three are all in the infrared.

The formation of the Balmer series was discussed by Bohr in his theory of the *atom (1913). He obtained the formula for the energy E_n of a quantum state of the H atom:

$$E_n = -me^4/8h^2\varepsilon_0^2 n^2,$$

where m is the *reduced mass of the electron of charge e, h is the *Planck constant, ε_0 the *electric constant, and n is a positive integer. When the atom goes to a state of lower energy, electromagnetic radiation is emitted with a *quantum of energy (*photon) $h\nu$, where ν is the frequency of the waves. The solution of *Schrödinger's wave equation gives the same formula.

hydrometer An instrument for determining relative densities, usually of liquids. The *Hare hydrometer* consists of two vertical glass tubes, one standing in a vessel of water, the other in the liquid under test. The tubes are connected at their upper ends by a glass T-piece, and by applying suction, the liquids may be raised in the tubes; the relative densities are inversely proportional to the height to which the liquids are raised in the tubes.

In hydrometers of variable immersion, the hydrometer consists of a glass tube blown out to two bulbs at its lower end. The lower bulb contains mercury, so that the hydrometer may float vertically in whatever it is immersed, and the graduated scale, which is fixed to the tube, will indicate the relative density of the liquid in which it floats.

In hydrometers of constant immersion, the instrument carries a pan at its upper end and is adjusted by means of weights put into the pan until the hydrometer is sunk to a fixed mark on its neck; the relative density of the liquid may be obtained from a knowledge of the weights required, first in water, and then in the liquid under test.

hydrophone An instrument for detecting sounds under water. It usually consists of a carbon microphone or electromagnetic detector fixed to a diaphragm in contact with the water.

hydrostatic equation An equation giving the relation between atmospheric pressure, p, and altitude z:

$$dp/dz = -g\rho,$$

where g is the acceleration of free fall and ρ is the density. The rate of fall of pressure with altitude is usually sufficiently regular to allow pressure readings to be used to determine altitude. Strictly, the hydrostatic equation is only valid when the atmosphere has no accelerations in the vertical; these are usually very small compared with g.

hydrostatics *See* statics.

hygristor An electronic component with an electrical *resistance that varies with *humidity. It is used as the basis of some recording *hygrometers.

hygrometer An instrument for the measurement of the *humidity of air. The main types are (1) the *chemical hygrometer, in which the mass of water vapour actually present is measured and can be compared with the mass of vapour to saturate the same volume at the same temperature; (2) the dew-point instruments, such as the *Regnault hygrometer, in which the *dew point is measured and the relative humidity obtained from the ratio of the saturation pressure of water vapour at the dew point to that at the temperature of the air; (3) the *wet-and-dry bulb hygrometers; (4) the recording instruments, in which an indicator gives the relative humidity directly, the instrument having been previously calibrated.

hygroscope A device for giving a rough guide to the *humidity of the atmosphere. A very simple form of hygroscope is a card impregnated with chemicals that change colour when the moisture content increases above a certain level, e.g. cobalt chloride (changing from blue to pink).

hyperboloid mirror *See* mirror.

hypercharge Symbol: Y. A *quantum number associated with an elementary particle. It is the sum of its *baryon number and *strangeness and is conserved in strong and electromagnetic interactions.

hyperfine structure of spectral lines Some spectral lines (apparently single lines at ordinary resolution) consist of a multiple system with differences of wavelength of the order of a thousandth of a nanometre. These can only be resolved by using high-resolution (10^6) gratings, etc., for example the *Fabry–Perot interferometer and *Lummer–Gehrcke plate. This structure is caused by interactions between the magnetic moments of the electrons and the very small magnetic moments of the nuclei.

hypermetropia *Syn.* long sight. *See* refraction.

hyperon The collective name given to long-lived *baryons other than the proton and neutron. Long-lived in this context is taken to mean not decaying by *strong interaction, i.e. particles with lifetimes much greater than 10^{-24} seconds. The lamda, sigma, xi, and omega-minus particles are hyperons. *See* elementary particle; strangeness; charm.

hypersonic speed A speed not less than five times the speed of sound through a medium at the same level and under the same physical conditions, i.e. not less than *Mach 5.

hypothesis A provisional supposition that, if true, would account for known facts and serves as a starting point for further investigation by which it may be proved or disproved.

hypsometer An apparatus for the calibration of a thermometer at the steam point. The thermometer T is placed in the steam above the water boiling under a known pressure, usually atmospheric, applied at P, the manometer M being used to ensure that the water is not being boiled too vigorously. The central space is enclosed by a steam jacket from which any condensed liquid flows

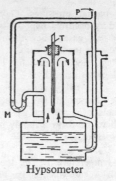

Hypsometer

back to the boiler. The boiling point under the measured pressure is deduced from tables and the reading of the thermometer T corrected.

hysteresis A delay in the change of an observed effect in response to a change in the mechanism producing the effect.
   1. (magnetic) A phenomenon shown by ferromagnetic substances, whereby the magnetic flux through the medium depends not only on the existing magnetizing field, but also on the previous state or states of the substance. The existence of permanent magnets is due to hysteresis. The phenomenon necessitates a dissipation of energy when the substance is subjected to a cycle of magnetic changes. This is known as the magnetic *hysteresis loss. See hysteresis loop.
   2. (dielectric) *See* dielectric hysteresis.
   3. (elastic) *See* elastic hysteresis.

hysteresis loop (magnetic) A closed curve obtained by plotting the magnetic flux density, B, of a ferromagnetic material against the corresponding value of the magnetizing field H. The area enclosed by the loop is equal to the *hysteresis loss per unit volume in taking the specimen through the prescribed magnetizing cycle. The general form of the hysteresis loop for a symmetrical cycle

between H and $-H$ is shown in the diagram, but any complete magnetizing cycle, say between the limits $H + h$ and $H - h$, will give rise to a hysteresis loop. OC is the *coercive force and OR the *remanence. The area enclosed by the loop varies with the nature and heat treatment of the magnetic substance, being a minimum for electrolytic iron and reaching a value some twenty times greater for tungsten steel. *See* ferromagnetism.

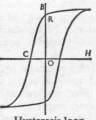

Hysteresis loop

hysteresis loss 1. (magnetic) The dissipation of energy that occurs, due to magnetic *hysteresis, when the magnetic material is subjected to changes (particularly cyclic changes) of magnetization. *See* hysteresis loop.
   2. (dielectric) The dissipation of energy that occurs, due to *dielectric hysteresis, when the dielectric is subjected to a varying (in particular, an alternating) electric field.
   3. (elastic) The dissipation of energy through *elastic hysteresis.

I

IAT Abbreviation for *International Atomic Time.

IC Abbreviation for *integrated circuit.

ice The solid form of water, the transition point at the *standard atmosphere being defined as 0 °C (*see* ice point). The specific latent heat of fusion is 0.3337 MJ kg⁻¹. Its density at 0 °C is 916.0 kg m⁻³, compared to water at 0 °C with a density of 999.8 kg m⁻³. There are several allotropic forms of ice, mostly stable only under high pressure.

ice calorimeter *See* Bunsen ice calorimeter.

ice line *Syn.* solidification curve. A curve expressing the relation between the melting point of ice and the applied pressure. It may be calculated by using the *Clausius–Clapeyron equation.

ice point The temperature of equilibrium of ice and water at standard pressure, the water being saturated with dissolved air. Its former importance was as the lower fixed point on the Celsius scale of temperature. Now, however, *thermodynamic temperature and the kelvin are based on the *triple point of water, and its value (273.16 K) has been chosen so as to make the ice point equal to 0 °C within the limits of experimental measurement. *Compare* steam point.

ideal crystal The crystal structure considered as perfect and infinite, that is, ignoring all questions of *crystal texture.

ideal gas *Syn.* perfect gas. A gas defined for the purposes of *thermodynamics as one that obeys *Boyle's law and that, in addition, has an internal energy independent of the volume occupied, i.e. it obeys *Joule's law of internal energy. These two requirements are, from the point of view of the kinetic theory, both equivalent to saying that the intermolecular attractions are to be negligible, but the first requires also that the molecules shall be of negligible volume. An ideal gas in fact obeys Boyle's law, Joule's law of internal energy, Dalton's

law of partial pressures, Gay-Lussac's law, and Avogadro's hypothesis exactly, whereas real gases obey them only as their pressure tends to zero.

The *equation of state for 1 mole of an ideal gas is given by:
$$pV = RT,$$
R being the molar gas constant. The isothermals of a perfect gas on a p/V graph therefore form a family of rectangular hyperbolas.

ignition temperature 1. The temperature to which a substance must be heated before it will burn in air (or some other specified oxidant).

2. *See* fusion reactor.

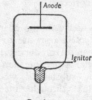

Ignitron

ignitron A type of mercury-arc rectifying tube in which the discharge is initiated by a subsidiary *ignitor electrode*. The tube has an anode and a mercury-pool cathode in which the ignitor is immersed (*see* diagram). The ignitor usually consists of a rod of semiconductor, such as silicon or boron carbide. The voltage applied to the anode is insufficient to strike an arc, but if a current is passed between the ignitor and the mercury, a hot spot forms that is sufficient to enable the arc to strike.

illuminance *Syn.* illumination; intensity of illumination. Symbol: E_v, E. The *luminous flux, Φ_v, incident on a given surface per unit area. At a point on a surface, of area dS, the illuminance is given by:
$$\Phi_v = \int E_v\, dS.$$

It is measured in lux. *See also* cosine law. *Compare* irradiance.

illumination 1. *See* illuminance.

2. The extent to which a surface is illuminated or the application of visible radiation to a surface. The brightness of an object depends on its *illuminance and its *reflectance.

image From the geometric optics point of view an image point is the point to which rays are converged (*real image*) or from which they appear to diverge (*virtual image*) after reflection or refraction. The object point from which rays have diverged and the corresponding image point, are said to be *conjugate. If the pencils of rays do not reunite to foci in the image, the latter suffers from *aberration and may be more or less blurred; the greatest concentration of ray intersections is taken as the *geometric image*. If the blur circles are sufficiently small (say 0.1 mm for viewing at 25 cm), the image formation may be practically regarded as being sharp. This means that there is an allowable *depth of focus of the image.

From the physical optics point of view, the distribution of light in the image is considered in relation to phase and path differences, which gives the image a focal depth, throughout which there is little deterioration of quality. The *image space* is a convenient mathematical conception to describe where images may lie; it may be real or virtual, i.e. beyond or in front of the second principal point. A convenient device to represent image-side quantities is to use the accented (dash) letters, e.g. l' the image distance, y' the size of the image. The lateral magnification of the image is y'/y where y is the size of the object.

image converter An evacuated electron tube that is similar to the *image intensifier but operates with infrared, ultraviolet, X-ray, or electron images rather than faint optical images; such images occur, for example, in astronomy, microscopy, and medical diagnosis. The focused image falls on a suitable surface from which electrons can be liberated by, for instance, the *photoelectric or *photovoltaic effect. The electrons are accelerated and focused on a detector or recorder, such as a positively charged fluorescent screen, so that a visible image is produced.

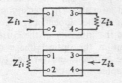

Image impedances

image impedances Of a *quadripole. The two impedances Z_{i1} and Z_{i2} that satisfy both the following conditions: (i) When Z_{i2} is connected across one pair of terminals, the impedance between the other pair is Z_{i1}. (ii) When Z_{i1} is connected across the other pair of terminals, the impedance between the first pair is Z_{i2}.

image intensifier An evacuated electron tube used to intensify a faint optical image. The image falls on a *photocathode so that electrons are emitted by the *photoelectric effect. The electrons are accelerated by an electric field and may be detected and recorded by a variety of methods. They can be focused (by an *electron-lens system) on a positively charged *fluorescent screen; the resulting optical image is many times brighter than the original image, and can be photographed.

In some devices the image on the screen can be made to fall on a second photocathode so that the intensification process is repeated; several image intensifiers can be linked in this way to form a multistage device. The image produced can be photographed or it can be recorded by a special TV camera. The TV signal is fed to and stored in a computer. The image of an extremely faint object can hence be slowly built up. The stored image (with the noise removed) can be displayed, analysed, and/or manipulated electronically.

In the *electronographic camera*, the liberated electrons are focused on a very sensitive high-resolution photographic emulsion in which the electrons are recorded directly. The density at every point on the developed image is proportional to the intensity at the corresponding point in the optical image over almost the entire intensity range (unlike a photographic image).

image orthicon *See* camera tube.

image potential If a charged particle (electron or ion) is a distance r from a metal surface, it experiences an electrostatic force. The interaction is equivalent to the interaction between the particle and an image of the particle a distance r below the surface. The potential energy of the particle is then $e^2/16\pi\varepsilon_0 r^2$, where e is its charge and ε_0 the *electric constant.

image processing The analysis and manipulation, by computer, of information contained in images, such as those obtained from satellites and spacecraft, medical diagnostic equipment, or electron microscopes. The original may be an actual object or scene, or may be a photograph, drawing, etc. This is converted into a digitized form by spatial *sampling, i.e. it is converted into a two-dimensional array of tiny elements, and a set of numbers produced corresponding to the brightness and possibly the colour of each element. The sampling is often done by a form of TV

camera. The numerical version of the image is stored in a computer, and manipulated in various ways to highlight different aspects of the original, compare or superimpose slightly different images, correct over- or underexposure or blur, etc.

image space *See* image.

image-transfer coefficient Of a *quadripole. The quantity

$$\theta = \tfrac{1}{2}\log_e[(E_1 I_1)/(E_2 I_2)],$$

where E_1 and I_1 are the voltage and current, respectively, at the input terminals, and E_2 and I_2 are the corresponding quantities at the output terminals under steady-state conditions when the network is terminated in its *image impedance. The voltages and currents are to be expressed in vector (e.g. complex) form and θ is in general complex.

image tube An *image intensifier or *image converter.

Immersion objective

immersion objective Microscope objectives use the principle of *aplanatic refraction, and to reduce the refraction at the front lens of the objective for higher powers, cedar-wood oil (refractive index 1.517) is placed between the cover glass (index 1.51, say) and the plane surface of the front lens of higher index (*see* diagram). Besides aiding aplanatism the *numerical aperture is increased by this process, which therefore increases resolving power.

IMPATT diode Abbreviation for *imp*act ionization *a*valanche *t*ransit *t*ime. A diode that provides a very powerful source of microwave power. It consists essentially of a p-n junction that is reverse biased into *avalanche breakdown. It then exhibits negative resistance at microwave frequencies and may be used as an *oscillator. The current is delayed, usually by half a cycle, with respect to the voltage. This delay is due to (and characteristic of) the avalanche and to the transit time during which the charge carriers are collected by the electrodes.

impedance In general the ratio of one sinusoidally varying quantity (e.g. force or e.m.f.) to a second quantity (e.g. acceleration or current) that measures the response of the system to the first quantity.

1. (electrical) Symbol: Z. If an alternating e.m.f. is applied to an electric circuit, the *alternating current produced is affected by the capacitance and inductance of the circuit as well as its resistance. The consequent opposition is the *reactance of the circuit and the total opposition to current flow is the impedance. Impedance is measured in ohms. It is the ratio of *root-mean-square voltage to root-mean-square current and is equal to $\sqrt{(R^2 + X^2)}$, where R is the resistance and X the reactance.

The magnitude of a sinusoidal alternating current varies with time according to the equation

$$I = I_0 \cos (2\pi ft),$$

where f is the frequency and I_0 the maximum current. More generally, such a quantity can be represented by a rotating *vector, the magnitude of the current being the projection of the vector onto a line (*see* simple harmonic motion; phase; phase angle). The e.m.f.

in the circuit will not be in phase with the current if reactance is present. It follows an equation of the form

$$V = V_0 \cos (2\pi ft + \phi),$$

where ϕ is the phase angle. It is often convenient to represent such quantities by complex numbers on an *Argand diagram. Thus

$$I = I_0 e^{i\omega t}$$
$$V = V_0 e^{i(\omega t + \phi)},$$

where ω is $2\pi f$. The real parts of these are the instantaneous current and voltage. The impedance is then the complex voltage divided by the complex current, i.e. $Z = V/I$, and is thus equal to $|Z| e^{i\phi}$, where $|Z|$ is $\sqrt{(R^2 + X^2)}$. This quantity is sometimes called the *complex impedance*. It is given by:

$$Z = R + iX,$$

the real part being the resistance and the imaginary part the reactance.

2. (acoustic) Symbol: Z_a. The complex ratio of the alternating *sound pressure to the rate of volume displacement of the surface that is vibrating to produce the sound. Defined for a surface producing a simple sinusoidal source of sound, it is related to *acoustic resistance* (R_a) and *reactance* (X_a) by:

$$Z_a = R_a + iX_a.$$

3. (mechanical) Symbol: Z_m; ω. The complex ratio of the force acting in the direction of motion to the velocity. It is related to *mechanical resistance* (R_m) and *reactance* (X_m) by:

$$Z_m = R_m + iX_m,$$

and is measured in newton second per metre.

impedance drop (or rise) *See* voltage drop.

impedance magnetometer An instrument for measuring local variations of the magnetic field of the earth (e.g. in a building) by measurement of the change in impedance of a nickel-iron wire of high permeability caused by the axial component of the field in which the wire is placed.

impedance matching In a system in which power is transferred, the matching of the *impedances of parts of the system to ensure optimum conditions for transfer of power. The system is usually electrical in nature, but it may be an acoustic system such as a *horn.

If electrical power is transferred from an *amplifier to a *load, the load impedance is made the *conjugate of the amplifier output impedance to effect the transfer of maximum power. In *transmission lines, the line impedance is made equal to the generator output impedance, and also to the load impedance, to ensure no reflection of wave power in the transmission line. If transmission lines of differing line impedance are joined in a system, a section of line, one quarter of a wavelength long, is used to couple the lines and effect matching. Such a section is termed a *quarter-wavelength line*.

impedance voltage *See* voltage drop.

impulse Of a constant force F, the product, Ft, of the force and the time t for which it acts. If the force varies with time, the impulse is the integral of the force with respect to the time during which the force acts. In either case, impulse of force equals the change of momentum produced by it. An *impulsive force* is one that is very large but acts only for a very short time; it can be represented by a *Dirac function.

impulse current *See* impulse voltage (or current).

impulse noise *See* noise.

impulse voltage (or current) A unidirectional voltage (or current) that rises rapidly to a maximum value without ap-

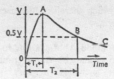

Waveshape of impulse voltage

preciable superimposed oscillations and then falls to zero more or less rapidly. The diagram shows a typical waveshape, which is described as a T_1/T_2 wave.

Relevant terms are: *Peak value*: maximum value, V. *Wavefront*: rising portion, OA. *Wavetail*: falling portion, ABC etc. *Duration of the wavefront*: time interval (T_1) for the voltage (or current) to rise from zero to its peak value (usually measured in microseconds). *Time to half value of the wavetail*: time interval (T_2) for the voltage (or current) to rise from zero, pass through its peak value (V), and then fall to half its peak value (0.5 V) on the wavetail (usually measured in microseconds).

An impulse voltage having $T_1 = 1$ μs and $T_2 = 50$ μs (described as a 1/50 microsecond wave) has been found to be representative of those produced by surges that are propagated along a transmission line as a *travelling wave as a result of lightning.

impulsive force *See* impulse.

impulsive sound A sharp sound that is completed in a small period of time and is extended throughout the aural spectrum.

impurities In a semiconductor. Foreign atoms, either naturally occurring or deliberately introduced into the semiconductor. They have a fundamental effect on the amount and type of conductivity. *See* semiconductor.

incandescence The emission of visible radiation from a substance at a high temperature. The term also refers to the radiation itself. *Compare* luminescence.

incandescent lamp An electric lamp in which light is produced by the heating effect of a filament of carbon, osmium, tantalum, or (more usually) tungsten. Inert-gas fillings are often used to suppress disintegration of the filament at the high temperature (>2600 °C), and efficiency is often increased thermally by winding the filament into a close spiral, and then this into a second close spiral (coiled coil) to reduce heat loss by conduction through the gas.

incidence (angle of) The angle between the ray striking a reflecting or refracting surface (i.e. the incident ray) and the normal to the surface at the point of incidence.

inclination, magnetic *See* dip.

inclined plane A rigid plane, inclined at an angle to the horizon and used to facilitate the raising of heavy bodies. The lifting force may be applied along the plane or at an angle to it.

inclinometer An instrument for measuring the magnetic inclination or dip. *See* dip circle.

incoherent Denoting radiation that is not *coherent. If two sources of optical radiation are independent, there is normally no regular relationship between the phases of waves from them arriving at any point.

indeterminacy principle *See* uncertainty principle.

index error *Syn.* zero error. A scale error on a measuring instrument such that the instrument shows a reading x when

it should show zero reading. Provided there is no other error, all readings on the instrument require a correction of −x.

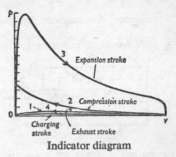

Indicator diagram

indicator diagram The cycle traced out during the motion of a piston in the cylinder of an engine. Vertical displacements represent the pressure and horizontal displacements the volume of the working substance. The area enclosed by the diagram gives the work done per cycle and is used in estimating the efficiency of the engine.

indicator tube A minute *cathode-ray tube, often with a screen diameter measured in millimetres, in which the shape or size of the image on the screen varies with the input voltage V in such a manner that it is used to measure V, and hence can indicate the value of a varying signal.

indirect stroke *See* lightning stroke.

induced e.m.f. *See* electromagnetic induction.

inductance A property of an electric circuit that results from the magnetic field set up when a current flows. Inductance relates the *magnetic flux through the circuit to the current flowing in that circuit – self-inductance – or in a nearby circuit – mutual inductance. *See* electromagnetic induction.

induction 1. *See* electromagnetic induction.

  2. (magnetic) *See* magnetic flux density.

  3. *See* electrostatic induction.

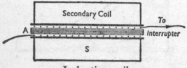

Induction coil

induction coil A device for producing a series of pulses of high potential and approximately unidirectional current by *electromagnetic induction. It consists of a primary circuit of a few turns of wire, wound on an iron core and insulated from a secondary coil of many turns, which surrounds it coaxially. The current in the primary is continuously and suddenly interrupted. Owing to the high resistance introduced into the primary circuit by each break of the circuit, the *time constant of the primary is much smaller at break than when the contact is remade, and the induced e.m.f. in the secondary is consequently much higher. The efficiency of the coil thus depends on the sharpness of the break. The output from the secondary consists of a succession of sharp pulses, corresponding to the breaks in the primary circuit, with much smaller inverse pulses produced when the current is remade.

induction flowmeter A device whereby the rate of flow of a conducting liquid passing through a tube T (see diagram) in a magnetic field can be measured by the e.m.f. induced across a diameter, between electrodes E. The relationship is $e = BLv$, where e is the e.m.f. in volts, B is the flux density in tesla, L is the

tube diameter in metres, v is the velocity in m s^{-1}, and B, L, and v are mutually perpendicular.

induction heating The heating effect of induced *eddy currents in a conducting material subjected to a varying magnetic field. The field can be produced by an alternating current flowing in a coil surrounding the material. Induction heating can be employed for metal melting. The advantage is that the heat is generated in the metal itself, and after melting the eddy currents set up circulatory movements that stir the melt.

induction instrument An instrument in which the deflecting force or torque is produced by the interaction of *eddy currents induced in a movable conducting mass and the magnetic field of an a.c. electromagnet.

induction motor An a.c. motor consisting of a *stator and a *rotor in which the current in one member (usually the rotor) is generated by *electromagnetic induction when alternating current is supplied to a winding on the other member (usually the stator). The torque is produced by interaction between the rotor current and the magnetic field produced by the current in the stator. For motors used industrially, there are two main types of rotor: (i) *cage rotor* (originally called *squirrel-cage rotor*) in which all the rotor conductors are permanently short-circuited at both ends of the rotor by means of end-rings; (ii) *slip-ring rotor* (or *wound rotor*), which carries a *polyphase winding connected to *slip rings. The object of the slip rings is to enable resistances to be connected temporarily in the rotor circuits so that the torque and current at starting may be controlled. The brushes on the slip rings are short-circuited when the motor is running normally.

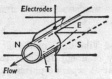

Induction flowmeter

inductive coupling *See* coupling.

inductive load *See* lagging load.

inductive reactance *See* reactance.

inductive tuning *See* tuned circuit.

inductor *Syn.* reactance coil. A coil, or other piece of apparatus, possessing *inductance and selected for use because of that property. *See also* choke.

inelastic collision *See* collision.

inelastic scattering *See* scattering.

inertance *See* acoustic inertance.

inert cell A primary *cell that is inert until water is added to produce an electrolyte. It contains the chemicals and other necessary ingredients in solid form.

inertia The property of a body by virtue of which it tends to persist in a state of rest or uniform motion in a straight line. *See* Newton's laws of motion.

inertia force *See* force.

inertial force *See* force.

inertial observer 1. In classical physics, an observer who finds that *Newton's laws of motion are valid.
2. In *relativity, an observer who finds that the special theory of relativity is valid.
See inertial reference frame.

inertial reference frame *Syn.* inertial coordinate system. A frame of reference used by an *inertial observer in which any body that is not subject to a resultant force has constant *velocity, i.e. the speed and direction of motion are unchanging.

inflationary universe A possible phase in the very early universe when its size increased by an immense factor. It is postulated (Guth et al) that at an age of 10^{-35} second the state of the universe changed in such a way that additional energy was released, allowing the existing expansion of the universe to accelerate rather than decelerate. The phenomena occurring during this rapid expansion are very speculative. The existence of such an inflationary phase can lead, for example, to an explanation of the origin of galaxies and stars. After a period of time the inflationary expansion changed to that of the standard *big-bang theory.

information technology (IT) Any form of technology, primarily electronic equipment and techniques, used by people to handle and distribute information. It incorporates the technology of both computing and of telephony, television, and other forms of telecommunication.

information theory An analytical technique for determining the optimum (generally minimum but sufficient) amount of information required to solve a specified problem in communication or control.

infrared astronomy The study of astronomical sources emitting *infrared radiation. These include the dust clouds around many stars and all newly forming stars, and extragalactic objects such as *quasars and *Seyfert galaxies. Observations can be made from the ground through several *atmospheric windows

233

up to a wavelength of about 20 μm. Satellites, rockets, and balloons are required for longer-wavelength studies. The equipment used includes reflecting telescopes to collect the radiation plus detectors such as *photovoltaic and *photoconductive cells and *bolometers. The detectors, and sometimes the telescope optics, must be cooled to very low temperatures to minimize thermal emission and hence noise. Siting ground-based telescopes at high altitudes minimizes atmospheric absorption of the infrared radiation, which is due mainly to water vapour and carbon dioxide.

infrared radiation (IR) Long-wave radiation emitted by hot bodies, with wavelengths ranging from the limit of the red end of the spectrum, about 730 nm, to about 1 mm. The longest wavelengths are adjacent to the microwave and radio-wave regions of the electromagnetic spectrum, where the radiation is produced electronically. The shorter wave or *near infrared (first detected by Herschel in 1800 by its heating effect) is examined by a spectroscopic method using fluorite or other material prisms in place of glass, and concave reflectors in place of lenses. This is because most glasses are absorbent at wavelength 2 μm. The quartz limit is 4 μm, fluorite 10 μm, rocksalt 15 μm, sylvin 23 μm. Detectors operating in the near-infrared region, and in some cases at longer wavelengths, include the *bolometer, *thermopile, and semiconductor devices such as *photoconductive and *photovoltaic cells. Photographic methods can be only used to about 1 μm. To examine the far infrared (to about 75 μm) the method of *selective reflection, using the residual rays, is employed.

infrared spectroscopy *See* spectroscopy.

infrared windows *See* atmospheric windows.

infrasound Vibrations of the air below a frequency of about 16 hertz, recognized by the ear as separate pulses rather than as sound. Infrasonic waves are generated at the muzzles of guns when they are fired and also accompany many industrial noises.

injection 1. In general, the application of a signal to an electronic circuit or device.
2. In a *semiconductor, the process of introducing carriers (electrons or holes) into the semiconductor so that the total number of carriers exceeds the number present at thermal equilibrium. The carriers may be introduced in various ways, e.g. across a forward-biased junction or by irradiation. There is *high-level injection* when the number of excess carriers is comparable to the thermal-equilibrium numbers.

injection efficiency The efficiency of a *p-n junction when forward bias is applied, defined as the ratio of the injected minority *carrier current to the total current across the junction.

injection laser *See* semiconductor laser.

in parallel *See* parallel.

in phase *See* phase.

in-phase component *See* active current; active voltage; active volt-amperes.

input The signal or driving force applied to a circuit, device, machine, or other plant. Also the terminals to which this is applied.

input impedance The *impedance presented by any circuit or device at its input terminals.

in series *See* series.

insertion loss (or gain) The reduction (or increase) of power in a load that occurs when a network is interposed between the load and the generator supplying the load. It is usually expressed in *nepers or *decibels and, in general, it is a function, not only of the network parameters, but also of the load and generator impedances.

instantaneous axis That straight line in a rigid body about which it may be regarded as rotating at any instant. If a rigid body is constrained to rotate about a fixed point O (*compare* Poinsot motion), the instantaneous axis will occupy different positions in the body but will always pass through O.

instantaneous frequency The rate of change of phase of an electric oscillation in radians per second, divided by 2π. It has particular applications in connection with *frequency and *phase modulation.

instantaneous value The value of any varying quantity at a particular instant of time, or (strictly) the average value over an infinitesimal period of time. The symbol for an instantaneous value is usually the lower-case form of the capital letter used for the quantity itself; e.g. p, i, and v (or u) are the symbols for instantaneous power, current, and potential difference.

insulate To surround or support an electrical (or thermal) conductor by insulating material so that the flow of electricity (or heat) is confined to the desired path.

insulated-gate field-effect transistor (IGFET) *See* field-effect transistor.

insulating resistance The resistance between two electrical conductors or systems of conductors that are normally separated by an insulating material. It is usually expressed in megaohms or, in the case of cables, in megaohms per mile (or kilometre).

insulation 1. Material that insulates an electrical conductor.
2. Material that reduces the transmission of heat, sound, etc., from a body or region.

insulator A substance that provides very high resistance to the passage of an electric current; an appliance made of insulating material used to prevent the loss of electric charge or current from a conductor. *See also* energy bands; dielectric.

integrated circuit (IC) A complete electronic circuit manufactured in a single package and implementing a particular function. Both *digital and *linear circuits are produced. Normally, all the circuit components – transistors, etc. – are manufactured in or on top of a single *chip of *semiconductor, usually silicon. Interconnections between the various parts of the circuit are made by means of a pattern of conducting material on the surface of the IC. The individual parts are not separable from the complete circuit. This type of circuit is often called a *monolithic integrated circuit*.

In contrast, a *hybrid integrated circuit* has the individual circuit components attached to an insulating substrate and interconnected by conducting tracks laid down on the substrate. The individual devices are unencapsulated, and may be diodes, transistors, monolithic ICs, or thick-film resistors and capacitors; the complete circuit is very small.

In *MOS integrated circuits*, the active devices are MOS *field-effect transistors, which operate at low currents and high frequencies. They have a high *packing

235

density and consume very little power. The development of MOS technology has allowed extremely complex MOS ICs to be fabricated. In *bipolar integrated circuits*, the components are bipolar junction *transistors and other devices that are fabricated using the p-n junction properties of semiconductors. They have higher operating speeds than MOS circuits but have a high power consumption, low packing density, and are less simple to manufacture. Both MOS and bipolar integrated circuits are monolithic.

The complexity of digital circuits that may be produced on a single chip is usually described in terms of the number of transistors or the number of logic gates involved. This leads to the following groupings, in order of complexity:

SSI — small-scale integration
MSI — medium-scale integration
LSI — large-scale integration
VLSI — very large-scale integration

LSI and VLSI technologies produce at least 10 000 and 100 000 transistors respectively on a single chip.

integrating meter A measuring instrument that integrates the measured quantity with reference to time.

integration time The period over which a noisy signal is averaged in order to improve the *signal-to-noise ratio in an electronic system.

integrator A mechanical or electrical device for performing the mathematical operation of integration. For example, the *capacitance integrator* uses a capacitor, usually in series with a resistor, to perform integration; the direct current i, flowing into the capacitor C gradually builds up a voltage on the capacitor equal to $(1/C)\int i\,dt$, i.e. an integration of i with respect to time is performed.

intensifying screen A screen coated with a fluorescent material, such as calcium tungstate crystals, that emits light under the action of X-rays; each X-ray photon produces several hundred light photons. Such screens are used in medical X-ray diagnosis, the emitted light being recorded on photographic film adjacent to the screen. They reduce the exposures needed in radiography by a factor of about 60. There is, however, some reduction of definition as the screen impresses its own grain on the film.

intensity A measure of the concentration of some factor, such as sound or light, usually over a given area or volume. (*See* sound intensity; luminous intensity; radiant intensity.) The term *illuminance is replacing *intensity of illumination*, and the terms *magnetic and *electric field strength have replaced *magnetic* and *electric intensity*.

intensity modulation *Syn.* z-modulation. The variation of the brilliance of the spot on the screen of a *cathode-ray tube in accordance with the magnitude of a signal.

intensity of magnetization For a uniformly magnetized body, the *magnetic dipole moment per unit volume.

interaction A process in which bodies exert forces on each other, or in which electromagnetic radiation and particles exert mutual forces. Usually one is concerned with those processes in which one or more bodies undergo some change of structure.

Every particle of matter in the universe feels the influence of the force of *gravitation, which derives from the existence of mass. *Electromagnetic interactions bind matter on a smaller scale, holding atoms and molecules together. In nuclear science *strong interactions occur between *quarks, and between *hadrons and systems of hadrons (nuclei). Such interactions are attributed to *exchange forces involving virtual bosons. The forces are short-range (*see* force) and do not operate significantly at separations much greater than 10^{-15} m, taking place in times typically 10^{-23} s. At a deeper level, strong interactions can be described in terms of the exchange of virtual *gluons between quarks or antiquarks (*see* quantum chromodynamics; *see also* electroweak theory; gauge theory). *Weak interactions are typically 10^{-12} times as powerful as strong ones. They are most commonly observed in decay processes (such as *beta decay) when there is some principle (e.g. a conservation law) that prevents the operation of a strong or electromagnetic interaction.

interactive Allowing continuous two-way transfer of information between user and *computer.

interelectrode capacitance The capacitance between specified electrodes of an electronic device, in which the electrodes form a small capacitor, e.g. between emitter and base of a *transistor. These capacitances may have a significant effect on the operation of such devices.

interface A common boundary between two parts, devices, or systems. It may be a surface separating two fluids of different densities or velocities. Another example is the electronic circuitry and its associated software that allows communication between two computer systems or two components of a computer system.

interfacial angles Angles between the normals to crystal faces.

interference 1. Of light. The phenomenon arising when two beams of *coherent light, which have travelled different

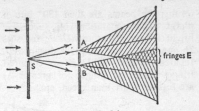

a Young's fringes

distances, are superimposed. (In fact interference occurs when any two beams of coherent radiation, such as radio waves, are superimposed.) The beams should be of approximately equal intensity. A *laser is normally used now as the source of the two beams. The beams should be of approximately equal intensity and should not intersect at too great an angle. *Interference fringes* are then observed in the region of overlap.

The phenomenon was discovered by Thomas Young (1801). Fig. *a* shows the apparatus schematically, the distance between the pinholes A and B being actually about a thousandth of their distance from the primary pinhole S. The latter was illuminated by white light, causing coloured fringes in the region E. The beams going through A and B *interfere*, since each alone gives a patch of light on the screen without fringes. Later experiments have used *monochromatic radiation giving light and dark fringes, and slits (perpendicular to the plane of the figure) have been used instead of pinholes.

These and similar fringes are readily explicable on wave theory and were used by Fresnel and Young as evidence to establish wave theory. A bright fringe will be observed at P (Fig. *b*) if the path difference BP–AP is equal to an integral number of wavelengths, $n\lambda$. The beams reinforce each other at P and *constructive interference* occurs, the beams being in *phase. If BP–AP is equal to $\frac{1}{2}n\lambda$, where n is an odd integer, *destructive interference* occurs and a dark fringe results; the beams are then 180° out of phase. The separation of successive bright fringes comes to $\lambda D/2b$, for slits $2b$ apart, with the screen at distance D. Incident white light gives coloured fringes since red fringes (of greater λ) are further apart than green, etc.

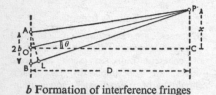

b Formation of interference fringes

Thin films (e.g. soap bubbles; oil on water) often display brilliant colorations when reflecting white light and show fringes when in monochromatic light. Here, light reflected from the front and back surfaces may be out of step by various amounts, destructive interference and consequent darkness occurring in some directions while constructive interference or reinforcement occurs in others. The irregular appearance of the coloured bands results from the uneven thickness of the film. Parallel-sided layers give fringes of equal inclination, visible by eye or telescope focused on infinity; thin films give fringes of equal thickness, effectively contours, located in or near the film. *Newton's rings*, formed between a convex-lens face and a plane glass slab, are of the latter type. Fringes become much sharper when formed by multiple reflections between surfaces of high reflecting power, e.g. if the film is between surfaces thinly silvered so that light can enter and leave the interspace.

An *interferometer* is any instrument or device designed to produce interference fringes. In some, including the *Fabry–Perot interferometer, highly accurate measurements of wavelength can be made. Measurements of very small distances can also be made, such as the angular separation of stars or the appar-

ent diameters of stars. Observation of interference fringes can be used, for example, to test the flatness of optical surfaces and to study very small movements such as those produced by elastic strain or thermal expansion.

2. Of sound. The phenomenon arising as a result of the superposition of two or more waves originating from a common source, but traversing different paths. At certain regions in the transmitting medium there is a minimum intensity and in other regions there is a maximum intensity; the resulting pattern in the field of radiation is called an *interference pattern*. Interference is produced, for example, by the interaction of a wave reflected from a wall and the corresponding incident wave; this contributes, in particular, to the poor audibility at certain positions in some theatres. Interference can also be produced by waves originating from two or more different sources. A ready demonstration of interference in sound is provided by the *Herschel–Quinke tube.

3. In a communications system. A disturbance to a signal caused by undesired signals, which may be man-made or may arise from natural causes such as changes in the atmosphere. In radio reception, electrical machines and apparatus commonly give rise to interference. Motor-car ignition systems sometimes produce serious interference with the reception of TV signals. *See* crosstalk; hum.

interference filter *See* filter.

interference fringes *See* interference (of light).

interference microscope *See* microscope.

interference pattern *See* interference (of sound).

interferometer Any device or instrument designed to produce and study optical or acoustic *interference, or radio-wave interference (*see* radio telescope).

intermediate frequency *See* superheterodyne receiver.

intermediate vector boson Either of the two forms of particle, the *W particle and Z particle, that mediate the *weak interaction.

intermodulation Modulation of the component sinusoidal waves of a complex wave by each other. The resulting wave contains, in particular, frequencies that are equal to the sum of and also the difference between the frequencies, taken in pairs, of all the components of the original complex wave. *See also* distortion.

internal absorptance Symbol: α_i. A measure of the ability of a substance to absorb radiation as expressed by the ratio of flux absorbed between the entry and exit surfaces of the substance to the flux leaving the entry surface. Internal absorptance does not apply to loss of intensity by scattering or to reflection of radiation at the surface of the substance. (*Compare* absorptance.) It is related to the *internal transmittance (τ_i) by:

$$\alpha_i + \tau_i = 1.$$

internal conversion The process in which a nucleus in an *excited state decays to a lower state and gives up energy to one of its orbital electrons, usually a K-electron. If this energy is large enough to overcome the *binding energy of the electron, then it is ejected from the atom as a *conversion electron*. The process is independent of *gamma-ray emission; it is not the production of a gamma-ray photon, which then knocks the electron out by the photoelectric effect. Usually excited states of nuclei have very short lifetimes and decay by gamma emission within picoseconds so

internal conversion is not observed. Sometimes, however, the *selection rules prevent rapid decay by gamma rays and the excited state is relatively long-lived. In these circumstances internal conversion becomes significant and may be the principal mode of decay. The conversion electrons have a line spectrum so are easily distinguished from beta rays. Electrons from outer shells fall into the vacant states caused by conversion, hence the substance emits its characteristic X-rays.

internal energy *See* energy.

internal friction The effect that causes a damping of elastic vibrations in a solid and similar effects. It is analogous to viscosity in liquids and results from the *anelasticity of the material.

internal resistance Of a cell, accumulator, or dynamo. The resistance obtained by dividing the difference between the generated e.m.f. and the potential difference between the terminals of the device by the current.

internal transmission density *Syn*. absorbance. Symbol: D_i. A measure of the ability of a body to absorb radiation as expressed by the logarithm to base ten of the reciprocal of the *internal transmittance, i.e.

$$D_i = \log_{10}(1/\tau_i).$$

internal transmittance Symbol: τ_i. A measure of the ability of a material to transmit radiation as expressed by the ratio of the flux reaching the exit surface of the body to the flux leaving the entry surface. The internal transmittance only applies to regular transmission and not to substances that scatter light or to reflection at the surfaces of the body. (*Compare* transmittance.) The internal transmittance is related to the *internal absorptance (α_i) by:

239

$$\tau_i + \alpha_i = 1.$$

internal work The work done in separating the molecules of a system against their forces of attraction. Its value is zero for an *ideal gas.

international ampere *See* ampere.

International Atomic Time (IAT, TAI in France) The most precisely determined timescale now available, set up by the Bureau Internationale de l'Heure in Paris and adopted in 1972. Atomic time is measured by means of atomic *clocks, the fundamental unit being the SI *second. Civil timekeeping is based on IAT.

international candle A standardized unit of luminous intensity superseded (in 1948) by the *candela.

international ohm *See* ohm.

International Practical Temperature Scale (IPTS) An easily and accurately reproducible temperature scale introduced (1927) for the purpose of practical measurements. It is based on the meaning of *thermodynamic temperature, and employs experimentally determined values of particular temperatures (known as the *primary fixed points*) and particular experimental methods for measuring temperature between and beyond these fixed points. The 1968 version, IPTS-68, uses both thermodynamic and Celsius temperatures, and the following eleven fixed points are defined (temperatures given in °C):

triple point of equilibrium hydrogen	−259.34
boiling point of equilibrium hydrogen at a pressure of $^{25}/_{76}$ atmosphere	−256.108
boiling point of equilibrium hydrogen	−252.87

boiling point of neon	−246.048
triple point of oxygen	−218.789
boiling point of oxygen	−182.962
triple point of water	0.01
boiling point of water	100.0
freezing point of zinc	419.58
freezing point of silver	961.93
freezing point of gold	1064.43

(Equilibrium hydrogen is an equilibrium mixture of ortho-hydrogen and para-hydrogen.)

Temperatures below 630 °C are measured by means of a platinum resistance thermometer; between 630 °C and 1064 °C a platinum-platinum/rhodium thermocouple is used; above 1064 °C a *radiation pyrometer is used based on Planck's law of radiation.

The lower and upper limits of the scale were extended in 1990. A provisional scale, EPT-76, extends the lower limit to −272.68 °C.

international table calorie (IT calorie) A standardized heat unit now replaced by the joule; 1 cal_{IT} = 4.1868 joules. *See* calorie.

interpolation Estimation of the value of a function, $f(x)$, for a value of the variable, x, which lies between those for which the function is known. This may be done by graphing $f(x)$ against x using the known values and reading off the value of $f(x)$ for the x required. Alternatively, various interpolation formulae (due to Newton, Bessel, etc.) can be used. *Compare* extrapolation.

interstage coupling In a multistage *amplifier, employing several amplifying stages in *cascade, the system that effects the transfer from the output of one stage to the input of the next. Common types of *coupling are direct, resistive, capacitive, etc.

interstellar matter *Syn.* interstellar medium (ISM). The matter – both gas and dust – that occurs in the regions between the stars of our Galaxy and tends to be concentrated in the spiral arms. The gas is mainly hydrogen, gathered into immense clouds. Roughly spherical clouds of predominantly ionized hydrogen (*HII regions*) are known to exist, usually less than 200 parsecs across. There are also smaller, more diffuse, and relatively cool (about 70 K) clouds of neutral predominantly atomic hydrogen (*HI regions*). Between the HI regions there is more tenuous neutral hydrogen gas at temperatures of several thousand kelvin. In addition, very cool (10–20 K) very dense *molecular clouds* exist, consisting principally of molecular hydrogen but with a large variety of other molecules; these are major sites of star formation. These different regions have been detected and studied through their radio, X-ray, ultraviolet, and infrared emissions.

Interstellar dust is found throughout the interstellar region. It causes dimming and reddening of starlight by absorption and *scattering; the effect is greatest for observations directed towards the centre of the Galaxy, where the extent and density of the dust is greatest. The dust also produces partial *polarization of starlight. The dust consists of solid grains, mainly of carbon, between about 0.01 to 0.1 μm in size.

interstitial structures Crystalline arrangements in which small atoms occupy some of the interstices between large atoms, which themselves form a regular crystalline pattern. They are of considerable importance in connection with the structure of steels and other alloys as well as *semiconductors. *See* defect.

interval 1. The relationship in frequency between two notes of a scale, the relationship being expressed either as a ratio or logarithmically. If the ratio form is used, it is conventionally written as a value greater than one. The most common units of interval expressed in a logarithmic form are the *millioctave* and the *cent*. An interval I in millioctaves between frequencies f_1, f_2 is given by:
$$I = (10^3/\log_{10}2)\log_{10}(f_1/f_2).$$
In cents the interval is given by:
$$I = (1200/\log_{10}2)\log_{10}(f_1/f_2).$$
(The octave is an interval of size 2/1 = 1000 millioctaves = 1200 cents.) *See also* temperament.

2. The separation of two events in a *four-dimensional continuum.

intrinsic conductivity The *conductivity of a *semiconductor that is associated with the semiconductor itself and is not contributed by impurities. At any given temperature equal numbers of charge carriers – electrons and holes – are thermally generated, and it is these that give rise to the intrinsic conductivity.

intrinsic mobility The mobility of *carriers in an *intrinsic semiconductor. Electrons are approximately three times as mobile as *holes.

intrinsic pressure A term in the *equation of state of a liquid resulting from intermolecular attractions. It has the form a/V^2 in the van der Waals equation of state.

intrinsic semiconductor *Syn.* i-type semiconductor. A pure *semiconductor in which the *electron and *hole densities are equal under conditions of thermal equilibrium. In practice absolute purity is unattainable and the term is applied to nearly pure materials.

Invar An alloy of iron with 36% of nickel, which has a very small coefficient of thermal expansion. It is used in instruments, pendulums, and accurate standards of length.

inverse gain The *gain of a bipolar junction *transistor when it is connected in reverse, i.e. with the *emitter acting as the *collector, and the collector as the emitter. It is usually less than the gain normally observed, as the emitter has a higher *doping level than the collector and therefore a higher *injection efficiency into the *base than the collector.

inverse-speed motor *See* series-characteristic motor.

inverse-square law A law relating the intensity of an effect to the reciprocal of the square of the distance from the cause. The law of *gravitation is an inverse-square law, as is *Coulomb's law relating the force associated with static electric charges. Another case is the fall-off in *illuminance of a screen placed normal to the light direction, assuming a point source.

inversion 1. A reversal in the usual direction of a process or variation, as in the change of density of water at 4 °C or, in meteorology, an increase in temperature with altitude as opposed to the normal decrease.

2. The production of a layer of opposite type in the surface of a *semiconductor, usually under the influence of an applied electric field. The presence of mobile minority carriers is necessary for inversion to take place, otherwise a *depletion layer forms. The phenomenon is utilized in the formation of the channel in an insulated-gate *field-effect transistor. A spontaneous inversion layer is often found in the surface of p-type semiconductor material in contact with an insulating layer even when no external electric field is applied.

inversion temperature 1. If one junction of a *thermocouple is kept at a constant low temperature, the temperature to which the other junction must be raised in order that the thermoelectric e.m.f. in the whole circuit shall be zero is known as the inversion temperature. For the same thermocouple, the sum of the temperatures of the two junctions is a constant. If the inversion temperature is exceeded, the direction of the e.m.f. in the thermocouple is reversed.

2. *See* Joule–Kelvin effect.

inverter 1. Any device that converts d.c. into a.c., particularly a rotating machine designed for the purpose.

2. *Linear inverter.* An amplifier that inverts the polarity of a signal, i.e. introduces a 180° phase shift.

3. *Digital inverter.* A *logic circuit whose output is low when the input is high and vice versa.

Ioffe bars Heavy current-carrying bars used in experimental fusion devices to increase *plasma stability.

ion An electrically charged atom, molecule, or group of atoms or molecules. A negative ion (or *anion) contains more electrons than are necessary for the atom or group to be neutral, a positive ion (or *cation) contains less. *See also* gaseous ions.

ion-beam analysis The analysis of materials on microscopic and macroscopic scales using a beam of positively charged light ions – protons, deuterons, or alpha particles. The ions generally have energies of a few MeV, the beams being produced in particle accelerators. Various analytical techniques have been developed, including *Rutherford back scattering* (RBS), *nuclear reactions analysis* (NRA), and *particle-induced X-ray emission* (PIXE). The ions are allowed to impinge on the surface of the sample being analysed, and the radiations subsequently emitted from the sample – particles from RBS and NRA, gamma rays from NRA, and X-rays from PIXE – are detected and processed using conventional devices. Information is obtained on where sample atoms are sited within the crystal lattice.

ion exchange (IX) A reversible process in which a liquid runs through or over a suitable solid and in so doing exchanges cations or anions. The solid employed is often an ion-exchange resin (a synthetic or natural polymer), zeolite (a synthetic or natural alumino-silicate of sodium, calcium, etc.), or a specially prepared carbonaceous mineral. The process is employed for water softening, desalination of brine, isotope separation, and the extraction of metals from their ores.

ionic atmosphere In an electrolyte, the accumulation of anions around cations, and vice versa. When an electric field is applied, the ions migrate in the reverse direction to their ionic atmosphere, and the symmetry of the atmosphere with respect to the ion is disturbed in such a manner that the ion is retarded. If, however, rapidly alternating (or short-duration direct) current is applied, there is insufficient time to disturb the symmetry and a higher value of conductivity is found. The conductivity depends, under these conditions, on the applied voltage gradient. *See* Wien effect.

ionic conduction In a *semiconductor, the movement of charges within the semiconductor due to the displacement of ions within the crystal lattice. An external contribution of energy is required to maintain such movement.

ionic crystals Crystals in which the interatomic forces are of the coulomb type, in which positively and negatively ionized atoms attract each other, the attraction being balanced by the repulsive force that comes into play when the outer electronic shells approach too closely.

ionic mobility The average speed attained by an ion when acted on by an electric field of unit strength. It is usually measured in m/s per V/m, i.e. $m^2 V^{-1} s^{-1}$.

ion implantation A technique used in the manufacture of *integrated circuits and *transistors in which the *semiconductor material is bombarded by high-velocity ions under controlled conditions. The ions penetrate the surface of the semiconductor and can be made to assume lattice positions within the semiconductor crystal. The technique may be used in conjunction with diffusion or as an alternative to it.

ionization The process of forming *ions. Ionization occurs spontaneously when an electrolyte dissolves in a suitable solvent. Ionization in gases requires the action of some *ionizing radiation, e.g. X-rays, α-, β-, or γ-rays. *See* conduction in gases.

ionization chamber A chamber containing two oppositely charged electrodes so arranged that when the gas in the chamber is ionized, e.g. by X-rays, the ions formed are drawn to the electrodes, creating an ionization current. This is used as a measure of the intensity of the *ionizing radiation. The sensitivity of an ionization chamber is dependent on the mass of gas enclosed in the sensitive volume. Extremely large ionization chambers have been developed for measuring *background radiation levels, whereas extremely small chambers are used for calibrating high-output beams of X-rays or electrons.

ionization gauge A vacuum pressure gauge consisting basically of a three-electrode thermionic valve and used for

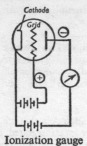

Ionization gauge

measuring small gas pressures of the order of micropascals. The tube is fused to the gas system to be measured, and is connected up as shown. Electrons are accelerated between the cathode and grid, but cannot reach the plate since it is at negative potential. Some electrons, however, pass through the grid and collide with gas molecules and ionize them, leaving them positively charged. The positively charged gas molecules then go to the plate and the plate current produced provides a measure of the number of molecules present.

ionization potential Symbol: I. The minimum energy necessary to remove an electron from a given atom or molecule to infinity. It is thus the least energy that causes an ionization:

$$A \rightarrow A^+ + e^-,$$

where the *ion and the electron are far enough apart for their electrostatic interaction to be negligible and no extra kinetic energy is produced by the ionization. The electron removed is that in the outermost orbit, i.e. the least strongly bound electron. It is also possible to consider removal of electrons from inner orbits in which their *binding energy is greater. The minimum energy required to remove the second least strongly bound electron from a neutral atom is called its *second ionization potential*. The first ionization potential (I_1) corresponds to formation of the *ground state of the singly charged ion. The second ionization potential (I_2) corresponds to formation of the ion in its first *excited state. *Compare* electron affinity.

ionizing radiation Any radiation that causes *ionization or *excitation of the medium through which it passes. It may consist of streams of energetic charged particles, such as electrons, protons, alpha particles, etc., or energetic ultraviolet, X-rays, or gamma-rays. A large number of ions, secondary electrons, and excited molecules are produced in the medium by particles; electromagnetic radiation produces a lower number by processes such as the *photoelectric effect, *Compton effect, and *pair production.

Ionizing radiation occurs naturally as *cosmic rays and the *solar wind, and is emitted by *radionuclides. It is produced artificially by X-ray machines and particle *accelerators. Its effects can be observed visually by using such apparatus as the *bubble chamber or *spark chamber, or by examination of the tracks made in photographic emulsion. More quantitative measurements are made with *counters.

Ionizing radiation can damage biological tissue as a result of the highly reactive transient species produced by its ionizing effects on water molecules in the tissue. Under controlled conditions, however, it can be used, for example, for diagnosis and therapy in medicine and for sterilization of perishable food. *See also* dosimetry.

ionosphere A spherical shell of ionized air surrounding the earth, extending from about 50 km (the top of the *stratosphere) to over 1000 km. Nitrogen and oxygen molecules are split into atoms, ions, and free electrons by *ionizing radiation from space, especially ultraviolet radiation and X-rays from the sun.

Following the first radio transmission, it was postulated by Heaviside and Kennelly in 1902 that transmission was achieved by reflection of radio waves from a layer of charged particles in the atmosphere. The particles were detected by Appleton in 1924. Long-distance radio transmission between any two points on the earth's surface is still obtained sometimes by successive reflections from the ionosphere.

The ionosphere can be divided into distinct layers or regions whose degree of ionization varies with time of day, season, latitude, and state of solar activity. There are three major layers.

The *D-layer*, altitude approximately 60–90 km, contains a relatively low concentration of free electrons and reflects low-frequency waves.

The *E-layer* (*Syn.* Kennelly–Heaviside layer), altitude approximately 90–150 km, has a higher electron concentration than the D-layer and reflects medium-frequency waves.

The *F-layer* (*Syn.* Appleton layer) lies approximately 150–1000 km above the earth. During the day it splits into the F_1-*layer* (lower) and F_2-*layer* (higher). It has the highest fractional concentration of free electrons and is the most useful region for long-range radio transmissions at frequencies up to about 30 GHz.

At night the electron concentrations in the D- and E-layers fall owing to the absence of sunlight and the consequent recombination of electrons and ions. In the higher F-layer the density is lower and collisions between electrons and ions are less frequent. The F-layer can therefore be used for radio transmission at all times.

Radio waves deflected by the electrically conducting ionospheric layers are called *ionospheric waves* (or *sky waves*). Some wavelengths, lying in the radio window (*see* atmospheric windows) between about a millimetre and 30 m,

are not reflected but transmitted through the ionosphere; long-distance television, broadcast at high frequencies, must therefore be reflected by means of artificial *satellites, usually in *geostationary orbits. *Radio astronomy is restricted to using these transmitted frequencies.

ionospheric wave *See* ionosphere.

ion pair A pair of positively and negatively charged ions generated when an electron transfers from one atom or molecule to another.

ion pump A type of vacuum pump in which the gas is ionized by a beam of electrons and the positive ions attracted to a cathode. It is only operated at very low pressures (less than about 10^{-6} Pa) and the gas is not completely removed from the system but simply trapped on the cathode. The pump thus saturates after a certain time. The capacity of ion pumps can be increased by continuously evaporating a film of metal onto the cathode during its operation, for example by *sputtering of the cathode. *See also* getter.

ion source A device that provides ions, especially for use in a particle *accelerator. A minute jet of gas, such as hydrogen or helium, is ionized by bombardment with an electron beam and the resulting protons, alpha particles, etc., are ejected into the accelerator.

ion trap In a *cathode-ray tube, a device to prevent the *ions present in the tube impinging on the phosphor coating of the screen and so causing blemishes.

IPTS Abbreviation for *International Practical Temperature Scale.

IR Abbreviation for infrared.

iridescence A display of colours on a surface, commonly as a result of the *interference of light of the various wavelengths reflected from superficial layers in the surface.

I²R loss *Syn.* copper loss. The power loss due to the flow of electric current in the windings of a machine or transformer. It is calculated by multiplying the square of the current by the resistance of the winding.

iron loss *See* core loss.

irradiance Symbol: E_e, E. The *radiant flux of electromagnetic radiation, Φ_e, incident on a given surface per unit area. At a point on the surface, of area dS, the irradiance is given by:
$$\Phi_e = \int E_e \, dS.$$
It is measured in joules per square metre. *Compare* illuminance.

irradiation The exposure of a body or substance to *ionizing radiation, either electromagnetic (X-rays and gamma rays) or corpuscular (alpha particles, electrons, etc.).

irreversible change *See* reversible change.

irrotational motion Of a fluid. Motion such that the equation of relative motion of any element of a finite portion of the fluid does not include rotational terms. The mathematical condition of irrotational motion is that
$$\text{curl } V = \nabla \times V = 0,$$
where V is the vector velocity of an element of the fluid. When the motion is irrotational, there exists a *velocity potential and conversely when a velocity potential exists, the motion is irrotational. If once irrotational, then the motion of a fluid under conservative forces is always irrotational. Any motion of a fluid, such that the component angular velocities of rotation do not vanish together, is called *rotational* and a velocity potential does not then exist.

isenthalpic process A process that takes place without any change of *enthalpy, i.e. so that the total heat energy (internal plus external) remains constant.

isentropic process A process that occurs with no change in entropy. *Compare* adiabatic process.

isobar 1. A line on a map passing through places of the same atmospheric pressure.
2. One of two or more nuclides that have the same *mass number but different *atomic numbers. Isobars are different elements and have different properties.

isobaric Taking place without change of pressure.

isochore A curve representing two variables involved in an isometric (constant volume) thermodynamic change, e.g. pressure/temperature, temperature/entropy.

isochronous Maintaining the same period of vibration or orbital time; having a regular periodicity.

isoclinal A curve drawn in such a manner that all places on the curve have the same magnetic *dip. *See* aclinic line.

isodiapheres Two or more *nuclides that have the same difference in the number of neutrons and protons. For example, in *alpha decay the parent and daughter nuclides are isodiapheres.

isogam A line on a map joining points at which the acceleration of free fall is constant. Used in geophysical prospecting.

isogonal A curve drawn in such a manner that the magnetic *declination (or variation of the compass) is the same at all places on the curve. *See* agonic line.

isolating The act of disconnecting a circuit or piece of apparatus from an electric supply system: it usually implies the opening of a circuit which, at the time, carries no current.

isolating transformer A *transformer used to isolate any circuit or device from its power supply.

isolator A device that allows microwave radiation to pass in one direction, while absorbing it in the reverse direction.

isomers 1. Compounds of the same relative molecular mass and percentage composition differing in some or all of their chemical and physical properties.
   2. *See* nuclear isomerism.

isometric change A change in a gas that takes place at constant volume.

isomorphism Similarity of crystalline form or of structure in substances that are chemically related.

isophote A line on a diagram joining points of equal flux density or intensity.

isospin *Syn.* isotopic spin; i-spin. Symbol: I. A *quantum number associated with *elementary particles. It is found experimentally that the *strong interaction between two protons and between two neutrons is the same. This suggests that the proton and neutron may be regarded as two states of the

same "particle" as far as strong interactions are concerned. Similarly, the three pions, π^+, π^0, and π^-, may be regarded as three states of a single "particle" when only strong interactions are considered. When *electromagnetic interactions are taken into account, there will be differences between interactions involving π^+ and π^0 because only the π^+ has a charge. However, as electromagnetic interactions are about 100 times weaker than strong interactions they can often be ignored.

*Hadrons with very similar masses and differing only in their charge can thus be combined into groups (called *multiplets*) that can be regarded as different states of the same object. It is found that to each hadron two quantum numbers I and I_3 may be assigned. The quantum number I is the isospin. It can take values
$$0, \tfrac{1}{2}, 1, \tfrac{3}{2}, 2, \ldots$$
and is the same for all particles in a multiplet. The *isospin quantum number* I_3 can have values
$$-I, -I + 1, \ldots, I - 1, I$$
and labels the particles in a multiplet. Examples of isospin multiplets are the nucleon doublet and the pion and sigma triplets (*see* table). In general, the charge Q of any elementary particle is related to its *hypercharge Y and the quantum number I_3 by the equation:
$$Q = I_3 + \tfrac{1}{2}Y$$
For systems of strongly interacting particles a total isospin may be defined. Two particles having isospin quantum numbers (I, I_3) and (I', I'_3) have a total I_3 quantum number given by $I_3{}^{TOT} = I_3 + I'_3$. The quantum number I^{TOT} of the combined system can have a number of different values:
$$I^{TOT} = I + I', I + I' - 1,$$
down to the larger of $|I_3{}^{TOT}|$ and $|I - I'|$.

Strong interactions only depend on the total quantum number I^{TOT} of the system and are independent of $I_3{}^{TOT}$. Both

I	I_3		I	I_3		I	I_3
n	$\tfrac{1}{2}$	$-\tfrac{1}{2}$	π^-	1	-1	Σ^-	1	-1
p	$\tfrac{1}{2}$	$\tfrac{1}{2}$	π^0	1	0	Σ^0	1	0
			π^+	1	1	Σ^+	1	1

I^{TOT} and $I_3{}^{TOT}$ are conserved in strong interactions. For electromagnetic interactions there is a dependence on the charge of the particles and I^{TOT} is no longer conserved although $I_3{}^{TOT}$ is.

isothermal 1. Occurring at constant temperature.
2. A line joining all points on a graph that correspond to the same temperature.

isothermal process A process that occurs at a constant temperature. For example, if a gas is expanded in a cylinder by a piston, its temperature can be kept constant by supplying heat from a thermostatically controlled source during the expansion. In such a process the wall separating the gas from the source has to allow them to remain in thermal equilibrium with each other. It is then called a *diathermic* wall. *Compare* adiabatic process.

isotones Nuclides having the same *neutron number.

isotopes Two or more *nuclides that have an identical nuclear charge (i.e. the same atomic number) but differ in nuclear mass; the nuclides are said to be *isotopic*. Such substances have almost identical chemical properties but differing physical properties, and each is said to be an isotope of the element of given atomic number. The difference in mass is accounted for by the differing number of *neutrons in the nucleus. For most elements several different naturally occurring isotopes have been discovered, and artificial radioactive isotopes can generally be prepared by bombardment of suitable materials by high-speed particles, or by slow neutrons.

In the case of hydrogen three isotopes are known. Ordinary hydrogen (^{1}H), with no neutrons, makes up 99.985% of

the total and deuterium (^{2}H), with one neutron, makes up the remaining 0.015%. The artificial isotope tritium (^{3}H), with two neutrons, has a *half-life of 12.26 years and decays through *beta emission.

Methods of separating isotopes use either physical or chemical processes taking place at a rate that depends on the mass of the atoms or molecules.

isotopic number *Syn.* neutron excess. The difference between the number of neutrons and the number of protons in a nuclide.

isotropic Possessing a property or properties, such as *permittivity, *susceptibility, or *elastic constants, that do not vary with direction.

iterative impedance Of a *quadripole. The impedance presented by the quadripole at one pair of terminals when the other pair is connected to an impedance of the same value. In general, a quadripole has two iterative impedances, one for each pair of terminals. Sometimes the two are equal and in this case their common value is called the *characteristic impedance* of the network. *Compare* image impedance.

i-type semiconductor *See* intrinsic semiconductor.

J

Jamin refractometer *Syn.* Jamin interferometer. *See* refractive index.

jamming Deliberate *interference in communications and radar caused by an undesired signal that is so strong that the desired signal cannot be understood.

jansky Symbol: Jy. A unit of *radiant flux density that is used in astronomy throughout the spectral range but especially for radio and infrared measurements. It refers to a particular frequency. One jansky is equal to 10^{-26} W m^{-2} Hz^{-1}.

Jansky noise High-frequency static disturbance of cosmic origin. *See* radio noise.

JET Abbreviation for Joint European Torus. *See* fusion reactor.

jet propulsion Propulsion of aircraft or other vehicles in which one or more jets of hot gases are ejected at high speed from backwardly directed nozzles. The ejected gases exert forces in the forward direction upon the system. Jet engines are devoid of reciprocating parts; air is drawn through an intake into a compressor, whence it passes to a combustion chamber, where it mixes with an oil fuel. The products of the combustion are expanded into the jet, driving the compressor on their way by means of a turbine. *See also* ramjet.

jet tones The rather unsteady tones produced when a stream of air is projected into still air from an orifice, such as a linear slit. The moving fluid tends to curl outwards into the stationary fluid forming alternate vortices on each side of the jet. Instability in jets is, however, very high and where the velocity of efflux and the fluid are suitably chosen to give sufficient vortices per second for an audible sound, the tones produced are weak, uncertain, and fluctuating. The sound, generally of high frequency, is more suitably described as a variable hiss.

jitter A short-term instability in either the amplitude or phase of a signal, particularly the signal on a *cathode-ray tube. It has the effect of causing momentary displacements of the image on the screen.

Johnson–Lark–Harowitz effect The change in resistivity of a metal or *degenerate semiconductor due to scattering of the charge carriers by impurity atoms.

Johnson noise Thermal *noise.

Johnson–Rahbeck effect If a semiconducting plate of material such as slate or agate is placed against a metal plate, the two hold strongly together during the application of a potential of about 200 volts. The plate and stone are only in actual contact at a few points through which a very small current flows to equalize a high potential difference. This potential difference is therefore applied across a very small distance and the forces of attraction are correspondingly great.

Joint European Torus (JET) *See* fusion reactor.

Josephson constant *See* Josephson effect.

Josephson effect Any of the phenomena that occur at sufficiently low temperatures when a current flows through a thin insulating layer between two superconducting substances (*see* superconductivity). The narrow insulating gap between the superconductors is known as a *Josephson junction*, and is usually in the form of a very thin film. The electrons forming the current are able to leak across the junction as a result of the *tunnel effect.

The current can flow across the junction in the absence of an applied voltage: this is the *d.c. Josephson effect*. In certain circuit configurations of Josephson junctions, the superconducting current is highly sensitive to a magnetic field. This allows it to be used as an extremely fast electronic switch with very low power dissipation. (*See also* squid.)

If a voltage is applied across a Josephson junction, then an alternating current flows through the junction: this is the *a.c. Josephson effect*. The current varies at a microwave frequency, ν, which is related to the voltage V:

$$\nu = (2e/h)V,$$

where h is the Planck constant and e the electron charge; the quantity $2e/h$ is called the *Josephson constant*. Conversely, if microwave radiation (frequency 10–100 GHz) impinges on a Josephson junction, the microwave frequency can be related to increments in the voltage developed across the junction when a superconducting current flows. The voltage increments are very precise, being equal to multiples of $(h/2e)$ times the frequency. It is now possible to connect many thousands of Josephson junctions in a long line to obtain a measurable voltage. These voltages can be used, for example, to compare laboratory voltage standards, and to standardize the volt to within a few parts in 10^8.

Josephson junction *See* Josephson effect.

joule Symbol: J. The *SI unit of all forms of *energy (mechanical, thermal, and electrical), defined as the energy equivalent to the work performed as the point of application of a force of one newton moves through one metre distance in the direction of the force. In electrical theory the relationship 1 J = 1 W s (watt second) is most useful. Since 1948 the joule has replaced the calorie as a unit of heat. As a formally defined conversion factor 1 calorie = 4.1868 joules.

Joule effect The liberation of heat by the passage of a current through an electric conductor, due to its resistance. *See* heating effect of a current.

Joule–Kelvin effect *Syn* Joule–Thomson effect. A change of temperature observed when a gas undergoes an irreversible adiabatic expansion on being pumped continuously through a porous plug or a very fine orifice (*throttle*). Provided that the change of kinetic energy of the flowing gas is small, the *enthalpy is unchanged in the process. According to the form of the *equation of state, the work done by part of the gas upon the gas in front of it on leaving the system may be greater or less than the work done on it by the pump, so the net external work may be of either sign. Generally, it is found that for each gas there is an *inversion temperature* (dependent on the pressure) above which there is a rise of temperature on expansion and below which there is a fall.

The change of temperature with pressure at constant enthalpy, H, is called the *Joule–Thomson* (or *Joule–Kelvin*) *coefficient*, symbol: μ, and is given by:

$$\mu = \left(\frac{\partial T}{\partial P}\right)_H = \frac{T\left(\frac{\partial v}{\partial T}\right)_p - v}{c_p},$$

where v is the specific volume, p the pressure, T the thermodynamic temperature, and c_p the specific heat capacity at constant pressure.

The Joule–Kelvin effect is used in *refrigerators and in the *liquefaction of gases. Hydrogen and helium have inversion temperatures far below room temperature so these gases must first be cooled to below their inversion temperatures before further cooling can be produced by this effect.

Joule magnetostriction Positive *magnetostriction.

Joule's equivalent The *mechanical equivalent of heat.

Joule's law 1. The principle that the heat produced by an electric current, I, flowing through a resistance, R, for a fixed time, t, is given by the product I^2Rt. If the current is expressed in amperes, the resistance in ohms, and the time in seconds then the heat produced is in joules.

2. The principle that the internal *energy of a gas is independent of its volume. It only applies to *ideal gases, i.e. when there are no intermolecular forces.

Joule–Thomson effect, coefficient *See* Joule–Kelvin effect.

journal friction *See* friction.

J/psi particle A massive unstable *elementary particle (3097 MeV), more precisely a meson resonance (*see* resonances). When detected (1974) by two independent US groups (hence its dual name), the width of the resonance peak was found to imply a lifetime of 10^{-20} second. This is considerably longer than the 10^{-23} second characteristic of resonance decay, and led to the concept of *charm: the J/psi is composed of a charm quark and a charm antiquark ($c\bar{c}$) but the particle itself has zero charm. The decay inhibition results because decays into final states containing charmed hadrons are kinematically forbidden, the rest masses of these hadrons being too large for these decays to occur.

junction 1. A contact between two different conducting materials, e.g. two metals, as found in a *rectifier or *thermocouple.

2. In a *semiconductor device. A transition region between semiconducting regions of differing electrical properties. *See* p-n junction.

3. A connection between two or more conductors or sections of transmission lines.

4. *See* Josephson effect.

junction field-effect transistor *See* field-effect transistor.

junction transistor *See* transistor.

just intonation *See* temperament.

Juvin's rule When a capillary tube of internal radius r stands vertical in a liquid of density ρ and surface tension γ the liquid rises a distance h up the tube, given by:

$$h = (2\gamma/rg\rho) \cos \alpha,$$

where α is the angle of contact between the liquid and the walls of the tube, and g is the acceleration of free fall. For a liquid that does not wet glass, α exceeds 90° and h will be negative: the liquid is depressed in the bore below the general level.

K

kaon *Syn.* K meson. *See* meson.

K-capture *See* capture.

keeper *Syn.* armature. A piece of iron or steel that is placed across the extremities of a permanent horseshoe magnet, or across pairs of extremities of permanent bar magnets, when the magnets are not in use, thereby completing the magnetic circuit. The regions near the ends of a magnet produce an induced flux within it opposing the original magnetizing flux, the effect being greatest for the shortest magnets. The magnetization of the keeper(s) neutralizes the demagnetizing effect.

Kellner eyepiece A type of *Ramsden eyepiece with an *achromatic eye lens that corrects chromatic aberration and

251

distortion inherent in the original design. It is commonly used as an eyepiece in *prism binoculars.

kelvin Symbol: K. The *SI unit of *thermodynamic temperature, defined as 1/273.16 of the thermodynamic temperature of the *triple point of water. The kelvin is also used as a unit of temperature difference on the thermodynamic and Celsius scales, where 1 K = 1 °C. *See also* degree Celsius.

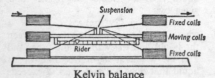

Kelvin balance

Kelvin balance A type of *current-balance instrument that consists of six coils, four fixed and two that move between them on a balanced rod. The suspension consists of two flexible multiple copper ribbons that serve to carry current to the coils in the manner shown in the diagram, so that each fixed coil tends to displace the balanced arm in the same direction when current flows. A rider, moving along an arm graduated in amperes, is used to rebalance the coil system – the scale divisions being uneven as the displacement of the weight is proportional to the square of the current. The instrument is suitable for measuring alternating and direct currents, and can also be adapted to measure wattage.

Kelvin contacts A means for testing or making measurements on electronic circuits or components. Two sets of leads are used to each test point, one set carrying the test signal and the other going to the measuring instrument. This removes the effect of the resistance of the leads on the measurement.

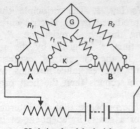

Kelvin double bridge

Kelvin double bridge A special development of the d.c. *Wheatstone bridge for precision measurement of low resistances (*see* diagram). A is the low resistance to be measured, and B is a known low resistance of the same order. The bridge is balanced by varying R_1 and r_1 with the connection K open and closed until the balance is exact; then:

$$A/B = R_1/R_2 = r_1/r_2.$$

The method eliminates possible errors due to contact resistance and the resistance of leads. A and B are usually *four-terminal resistors, as shown.

Kelvin effects *See* thermoelectric effects.

Kennelly–Heaviside layer *See* ionosphere.

Kepler's laws Three fundamental laws of planetary motion.

1. Every planet moves in an ellipse, the sun occupying one focus of the ellipse.

2. The radius vector drawn from the sun to the planet sweeps out equal areas in equal times (i.e. the areal velocity is constant).

3. The squares of the times taken to describe their orbits by two planets are proportional to the cubes of the major semiaxes of the orbits. The first two laws were published in 1609, the third in 1619.

These laws were later shown to apply also to the orbits of comets around the sun, and to natural and artificial satellites around planets. Similar laws apply to the orbits of double stars. Although the laws are followed very accurately, there are small deviations that are explained by *perturbation theory, the imperfect symmetry of the central body, or the theory of *relativity, according to the case. *See* gravitation.

Kepler telescope *Syn.* astronomical telescope. *See* refracting telescope.

kerma (*k*inetic *e*nergy *r*eleased in *ma*tter) Symbol: *K*. The sum of the initial kinetic energies of all charged particles produced by the indirect effect of *ionizing radiation in a small volume of a given substance divided by the mass of substance in that volume. The SI unit is the *gray.

Kerr cell *See* Kerr effects.

Kerr effects Two effects concerned with the optical properties of matter in electric and magnetic fields.

The *electro-optical effect* is the effect in which certain liquids and gases become double-refracting when placed in an electric field at right angles to the direction of the light. The substance acts as a *uniaxial crystal with optic axis parallel to the field. If n_1 and n_2 are the refractive indexes of light with planes of polarization respectively parallel and perpendicular to the field, then:

$$n_1 - n_2 = k\lambda E^2,$$

where k is the *Kerr constant*, λ the wavelength of the light, and E the electric field strength. The *Kerr cell* consists of two parallel plate electrodes immersed in a liquid that shows a marked electro-optical effect. Polarized light passes through the cell and can be interrupted by the application of an electric field. The device is also called an *elec-*tro-optical shutter*. (*See also* Pockel effect.)

The *magneto-optical effect* refers to the production of a slight elliptic polarization, produced when plane-polarized light is reflected from the polished pole face of an electromagent. The incident light is plane-polarized in, or normal to, the plane of incidence. *See also* Faraday effect.

Kew magnetometer A type of *magnetometer used to make accurate measurements of the earth's magnetic field and the magnetic declination. The magnetic needle is a steel tube with a graduated transparent scale on one end and a lens on the other. Its precise position can be observed with a coaxial telescope.

keyboard A manually operated device by means of which people can communicate with a computer. It consists of an array of labelled keys, operated by finger pressure as in a typewriter. Operation of a particular key (or combination of keys) produces a coded digital signal that can be fed directly into a computer. A keyboard usually has a standard QWERTY layout plus some additional keys. These can include a control key, function keys, cursor keys, and a numerical keypad.

kilo- Symbol: k. **1.** A prefix meaning 10^3, i.e. one thousand; for example one kilometre is equal to 1000 metres.

2. In computing, etc., where the binary rather than the decimal system of numbers is used, a prefix meaning 2^{10}, i.e. 1024; for example, one kilobyte is equal to 1024 bytes. The symbol K is not recommended.

kilogram Symbol: kg. The *SI unit of mass represented by the international prototype kilogram at the International Bureau of Weights and Measures at Sèvres in France. It consists of a cylinder whose height is equal to its diameter and is made from an alloy consisting of 90% platinum and 10% iridium.

Decimal multiples and submultiples of the kilogram are formed by adding the SI prefixes to the word gram, for example milligram (mg) rather than micro-kilogram. One kilogram is equal to 2.204 62 pounds.

kilowatt-hour Symbol: kWh. A unit of energy equivalent to the work done when power of 1 kilowatt is expended for 1 hour, or equal to one thousand watt-hours. It is used for electric work.

kinematics The branch of *mechanics dealing with the motion of bodies without reference to mass or force.

kinematic viscosity (coefficient of) Symbol: ν. The ratio of the coefficient of *viscosity (η) to the fluid density (ρ). It is used in modifying the equations of motion of a perfect fluid to include the terms due to a real fluid. The units of kinematic viscosity are metres squared per second. At room temperature water has a kinematic viscosity of $10^{-6} \, m^2 \, s^{-1}$. The ordinary viscosity coefficient is often called the coefficient of *dynamic viscosity* to avoid confusion.

kinetic energy *See* energy.

kinetic friction *See* friction.

kinetic potential *See* Lagrangian function. *See also* Hamilton's principle.

kinetic theory The work of Rumford, Joule, and others, led to the establishment of the concept of heat as a process of *energy transfer. The kinetic theory combines this conclusion with the molecular theory of chemistry and interprets the internal *energy of a body as being the energy of the motions and positions of the molecules of which the

body is made up. The basis of the theory is in fact that of the kinetic theory of matter as a whole, namely, that the particles of matter in all states of aggregation are in a violent state of agitation.

In gases the molecules move rapidly in all directions, being so small that they are mostly removed from one another at distances large compared with their own dimensions, being virtually free from the influence of other molecules. In a liquid the molecules are very close to one another and their mutual influence is significant. The molecules are in continuous motion, but there seems to be some semblance of a patterned structure, which is not present in a gas. In the solid state matter may be crystalline, exhibiting a very definite patterned structure, or amorphous without such a pattern. Each fundamental unit of the structure is vibrating about a mean position, these vibratory motions constituting thermal agitation, becoming more energetic with increase of temperature. The energy of these motions accounts for the greater part of the specific heat capacity of a solid substance. Evidence of molecular agitation is afforded by such phenomena as *diffusion and *Brownian movement.

During their motion gas molecules will collide with the walls of the vessel in which they are contained, thereby delivering momentum to them and giving rise to an exertion of pressure. The kinetic interpretation of an *ideal gas is one in which the molecules occupy a negligible volume and exert no influence on one another except when they actually collide, collisions being on the average perfectly elastic in equilibrium. The pressure exerted by an ideal gas is given by:

$$p = \tfrac{1}{3} m v C^2,$$

where m is the mass of the molecules, v is the number of molecules per m^3, and C is the *root-mean-square velocity of the molecules.

Maxwell's law of the *distribution of speeds gives the actual distribution of speed among the molecules in equilibrium state, and is based on the main concepts of classical statistics. These concepts lead also to the principle of *equipartition of energy, namely that the total energy of a system is equally divided between the different degrees of freedom, and that each degree of freedom possesses a mean energy $\frac{1}{2}kT$, where k is the Boltzmann constant and T is the thermodynamic temperature.

The application of kinetic theory leads to the relation for gases:

$$\gamma = 1 + 2/n,$$

where γ is the ratio of the principal *specific heat capacities and n is the number of *degrees of freedom for each molecule. In the case of solids it leads to *Dulong and Petit's law that the *molar heat capacity of a solid is constant and equal to $3R$. The simple kinetic theory of specific heat capacities is quite unable to account for the variation of specific heat capacity with temperature both in solids and in gases, and the classical concepts of equipartition had to be replaced by those of the *quantum theory.

Kirchhoff formula A formula for the variation of *vapour pressure with temperature:

$$\log p = A - B/T - C \log T,$$

where A, B, and C are constants. It is valid over limited temperature ranges.

Kirchhoff's law (for radiation) The principle that at a given temperature the spectral *emissivity of a point on the surface of a thermal radiator in a given direction is equal to the spectral *absorptance for incident radiation coming from that direction. A *thermal radiator* describes a body emitting radiation as a result of thermal vibration of the atoms or molecules. The adjective *spectral* implies that the emissivity or absorptance

is considered for monochromatic radiation.

Kirchhoff's laws (for an electric circuit) 1. The algebraic sum of the electric currents that meet at any point in a network is zero.

2. In any closed electric circuit the algebraic sum of the products of current and resistance in each part of the network is equal to the algebraic sum of the electromotive forces in the circuit.

Klein–Gordon equation A relativistic equation of *wave mechanics applicable to *bosons of zero spin, such as the *pion.

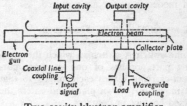

Two-cavity klystron amplifier

klystron An *electron tube that employs *velocity modulation of an electron beam, and is usually used for either the amplification or generation of *microwaves. Several varieties of the basic klystron exist.

In the simple two-cavity klystron (*see* diagram), a beam of high-energy electrons from an electron gun is passed through a *cavity resonator excited by high-frequency radio waves. The interaction between the high-frequency waves and the electron beam produces velocity modulation of the beam. After leaving the cavity, bunching of the electrons will occur; the current density of the beam thus varies and has the same frequency as the exciting radio waves.

The modulated beam then passes through a second cavity resonator, where its current-density variations pro-

255

duce a voltage wave. This is tuned to the exciting radio frequency or a harmonic of it. Voltage amplification is obtained by conversion of the energy of the original beam into r.f. energy in the output cavity, power being taken from the beam. If positive feedback to the input cavity is employed, the device can be made to oscillate. A *reflex klystron* employs only one cavity, the electron beam being reflected after velocity modulation has occurred. Reflex klystrons are most commonly used as low-power oscillators. *Multicavity klystrons* employ more than two cavities in the beam, and a higher overall gain may be achieved. They are used when extremely high-power pulses or continuous waves of moderate power are required.

K meson *Syn*. kaon. *See* meson.

knife switch A switch in which the moving part consists of one or more current-carrying blades each hinged so that it moves in its own plane and enters the fixed contact or contacts.

Knudsen flow *See* molecular flow.

Knudsen gauge

Knudsen gauge A device for the measurement of very low pressures where the mean free path of the molecules is large compared with the dimensions of the apparatus. Two cold plates, B_1, B_2, at a temperature T_2 are free to rotate in an evacuated vessel about a vertical quartz suspension. Stationary plates, A_1, A_2, are electrically heated to a temperature T_1. The gas molecules striking B from the side A have greater momentum than those striking the other side of B and so the vanes B_1, B_2 experience a force per unit area equal to F in the direction shown. F is calculated from the measured deflection of the suspended system and the torsion constant of the fibre. (This constant is not necessarily known.) The force and pressure are then related by the formula:

$$F = p\sqrt{[(T_1/T_2) - 1]}.$$

Kundt's rule The principle that the refractive index of a medium does not vary continuously with wavelength in the region of absorption bands. *See* anomalous dispersion.

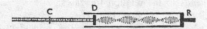

Kundt's tube

Kundt's tube An apparatus to measure the speed of sound in gases under different controllable conditions of temperature, density, and humidity. A column of gas in a tube D is closed by a reflector piston R at one end and has a source of sound at the other; in between there is a dry powder for detecting resonance. If the piston is adjusted so that the length of the gas column gives an exact number of stationary waves, the dust will be violently disturbed at the displacement antinodes and will form a series of striations. A circulation exists in the tube; it takes place from antinode to node in the neighbourhood of the walls and from node to antinode along the centre.

The absolute determination of speed of sound in a gas is made from the relation $c = f\lambda$, where c is the speed of sound, f is the frequency, and λ is the wavelength, equal to $2d$ (d being the distance between two nodes or antinodes in the tube).

The determination of the ratio of the specific heat capacities of a gas (γ), especially those supplied in small quantities as rare gases, can be found from the relation $c = \sqrt{(\gamma p/\rho)}$, where p and

ρ are the pressure and density of the gas.

L

labelled *See* radioactive tracer.

ladder filter A network consisting of a succession of series' and shunt impedances, usually acting as a transmission line with a known attenuation or delay. *See* filter.

laevorotatory *Syn.* laevorotary. Capable of rotating the plane of polarization of polarized light in an anticlockwise direction as viewed against the direction of motion of the light. *See* optical activity.

lag 1. Of a periodically varying quantity. The interval of time or the angle in *electric degrees by which a particular phase in one wave is delayed with respect to the similar phase in another wave. (*Compare* lead.)
2. The time elapsing between the transmission and reception of a signal.
3. In a control system. The delay between a correcting signal and the response to it.

lagging current An alternating current that, with respect to the applied electromotive force producing it, has a *lag. *Compare* leading current.

lagging load *Syn.* inductive load. A *reactive load in which the inductive *reactance exceeds the capacitive reactance and therefore carries a *lagging current.

Lagrange's equations A set of second-order differential equations for a system of particles that relate the kinetic energy T of the system to the generalized coor-

dinates q_i, the generalized forces Q_i, and the time t. There is one equation for each of the n *degrees of freedom possessed by the system:

$$\frac{\mathrm{d}}{\mathrm{d}t}\left(\frac{\partial T}{\partial \dot{q}_i}\right) - \frac{\partial T}{\partial q_i} = Q_i, \quad (i = 1, 2 \ldots n).$$

Here $\dot{q}_i$ denotes $\mathrm{d}q_i/\mathrm{d}t$.

Lagrange's equations provide a uniform method of approach for all dynamical problems.

Lagrangian function *Syn.* Lagrangian; kinetic potential. Symbol: L. An expression for the kinetic minus the potential energy in a conservative system:
$$L = T(q_i, \dot{q}_i) - V(q_i, \dot{q}_i),$$
where q_i are the *generalized coordinates. *See also* Lagrange's equations; Hamilton's principle.

Lagrangian points Five points in space, associated with a system of two bodies orbiting around a common centre of mass, at which a small body can maintain a stable orbit despite the gravitational influence of the two much more massive bodies. Groups of minor planets are found at two Lagrangian points in the sun-Jupiter gravitational field (60° ahead and behind Jupiter in its orbit). The other three points, along the line joining the centres of mass, are in unstable equilibrium; this is the case in any system.

Lalande cell A *primary cell with zinc and iron electrodes in caustic soda solution as electrolyte and with copper oxide as depolarizer.

lambda particle Symbol: λ. An uncharged elementary particle, a *hyperon, with spin ½ and a mass about 1.1 times that of the proton. It can replace a neutron in a nucleus to form an extremely unstable *hypernucleus*.

lambda point The temperature at which the two forms of liquid helium can exist together. *See* superfluid.

lambert A former unit of *luminance equal to 1 lumen of flux emitted per square centimetre of surface assumed to be perfectly diffusing. In terms of the *candela:

$$1 \text{ lambert} = 1/\pi \text{ cd cm}^{-2}$$

Lambert's law 1. *Syn.* cosine law of emission. The *luminous intensity of a small element of a perfectly diffusing surface in any direction is proportional to the cosine of the angle between the direction and the normal. The law is used to define the perfect diffuser.
2. *See* linear absorption coefficient.

Lamb shift A small difference in energy between the energy levels of the $^2S_{1/2}$ and $^2P_{1/2}$ states of hydrogen. These levels would have the same energy according to the *wave mechanics of Dirac. The shift can be explained by a correction to the energies on the basis of the theory of the interaction of electromagnetic fields with matter (*see* quantum electrodynamics) in which the fields themselves are quantized.

lamellar field A *field in which the *vector associated with the field is derivable from a scalar potential (by taking its *gradient). The scalar potential field may be mapped by level surfaces (or laminae), hence the name. Such a vector field has no *curl.

laminar flow Steady flow in which the fluid moves in parallel layers or laminae, the velocities of the fluid particles within each lamina not being necessarily equal. In the motion of a fluid through a straight horizontal pipe the velocities of the particles within each lamina are the same until the *critical velocity is attained and the motion changes from

laminar to *turbulent. When a velocity potential exists the flow is called *potential flow* and is essentially laminar flow, although the laminae are not necessarily plane.

lamination A thin steel or iron stamping, oxidized or lightly varnished on the surface, a number of which can be built up to form the *core of a transformer, transductor, relay, choke, or similar apparatus. The laminations reduce losses by preventing *eddy currents circulating in the core.

Lamy's theorem If a particle is in equilibrium under the action of three forces P, Q, and R, then

$$P/\sin\alpha = Q/\sin\beta = R/\sin\gamma,$$

where α is the angle between Q and R, β the angle between R and P, and γ the angle between P and Q.

LAN Abbreviation for *local area network.

Landau damping The damping of a space charge oscillation by a stream of particles moving at a speed slightly less than the phase speed of the associated wave.

Landé factor *Syn.* g-factor. Symbol: g. A constant factor used in expressions for changes in energy level in a magnetic field. It is a correction for the fact that there is not a simple relationship between the total *magnetic moment of an atom, nucleus, or particle and its angular momentum. The Landé factor is necessary to explain fine structure in spectral lines due to coupling between orbital and spin angular momenta. It is also used in the magnetic moments of particles resulting from their *spin. For example, a nucleus with a spin quantum number I has a magnetic moment given by:

$$g\sqrt{[I(I+1)]} \cdot \mu_N,$$

where μ_N is the nuclear *magneton and g is a constant for a particular nucleus.

langley A former unit of energy density used for solar radiation; it is equal to one calorie per cm^2 or 4.1868×10^4 $J\,m^{-2}$.

Langmuir–Blotchett film (LB film) An ordered layer of organic molecules that is formed on a solid surface as it is passed through the surface of a liquid on which is spread a quasi-solid mono-molecular layer of the film material. Perfect films can now be achieved. Multilayer structures are also being produced. Large areas can be fabricated. LB films have many (potential) applications, for example as insulating coatings and layers in transistors; multilayer films have even greater potential.

Langmuir effect An ionization that occurs when atoms of low *ionization potential come into contact with a hot metal of high *work function. It has been used in the production of intense beams of ions of such elements as the alkali metals.

Laplace equation 1. A linear differential equation of the second order:

$$\frac{\partial^2 V}{\partial x^2} + \frac{\partial^2 V}{\partial y^2} + \frac{\partial^2 V}{\partial z^2} = 0.$$

V may, for example, be the potential at any point in an electric field where there is no free charge.

2. For speed of sound. An equation relating the speed of sound (c) in a gas to the density (ρ), pressure (p), and ratio of heat capacities (γ) of the gas. It has the form:

$$c = \sqrt{(\gamma p/\rho)}.$$

The equation implies that there is an adiabatic change in pressure and temperature when sound passes through any gas. *See* dispersion of sound.

Laplace operator The differential operator

$$\left(\frac{\partial^2}{\partial x^2} + \frac{\partial^2}{\partial y^2} + \frac{\partial^2}{\partial z^2} \right)$$

often represented by the symbol ∇^2. *See* del.

lapse rate The rate of decrease of a quantity, usually temperature, with height in the atmosphere. The *dry adiabatic lapse rate* (DALR) is the rate of cooling for dry air that is subjected to an adiabatic ascent. Its value is equal to 9.76 °C per kilometre. This is also the lapse rate for moist air that remains unsaturated.

large-scale integration (LSI) *See* integrated circuit.

Larmor precession A uniform magnetic field applied to a plane electron orbit of an atom causes the plane to precess about the direction of the field in such a way that the normal to the plane traces a cone with its axis in the field direction. The frequency of precession is given by:

$$\nu = eB/4\pi m,$$

e being the electron charge and m its mass, and B the magnetic flux density.

laser (*l*ight *a*mplification by *s*timulated *e*mission of *r*adiation) A source of near monochromatic *coherent radiation in the visible, ultraviolet, and infrared regions of the spectrum. (X-ray lasers are under development.) The beam produced is narrow and emerges almost perfectly collimated. In consequence, laser beams can be brought to a very fine focus in which the power per unit area can attain extremely high values. Many of the applications depend on this.

Production of the laser beam depends on *stimulated emission*. The *emission of photons that occurs following *excitation of electrons in a system is usually spontaneous and cannot be controlled. Stimulated emission is a process whereby an incoming photon of energy $h\nu$ (where h is the *Planck constant and ν the frequency) can stimulate an electron in a high-energy state E_1 to jump to a lower energy state E_2, where $E_1 - E_2 = h\nu$. The photon resulting from this process has the same frequency, $\nu = (E_1 - E_2)/h$, as the stimulating photon and travels in the same direction. If there are sufficient electrons in the high-energy level, both stimulating and stimulated photons can cause further stimulated emission and a narrow beam of monochromatic radiation results, the intensity of which increases exponentially. The beam is coherent (i.e. spatially and temporally in phase) and can have a very high energy density.

A laser beam is produced by stimulated emission but can only operate efficiently if a large number of electrons are in a particular high-energy level. This condition, called *population inversion*, is a nonequilibrium condition and power must be fed into the system to maintain the inversion.

Laser action has been achieved in gaseous, solid, and liquid media. The beams may be pulsed or continuous wave (CW), and vary widely in the power generated and the overall efficiency. The first laser (1958) was the pulsed ruby laser, output wavelength 694.3 nm; it is still in use. Chromium ions in the ruby lattice are excited by an intense flash of light. Electrons, raised to a high-energy state, decay immediately to a slightly lower *metastable state, and thus population inversion occurs. The level slowly depopulates and this spontaneous emission triggers stimulated emission of the same frequency. The photons are reflected back and forth along the cylindrical ruby crystal between two parallel flat external mirrors, one only partially silvered. During

this process stimulated emission continuously builds up the number of photons, so increasing the power; a small percentage of photons emerge from the semisilvered end, forming an intense pulse of light lasting about a millisecond. The laser plus mirrors act as an optical *cavity resonator.

There are many other solid-state lasers, their outputs varying in wavelength and power. For example, trivalent rare earths undergo laser action in a variety of media, including yttrium aluminium garnet (YAG) and glass (which may both be doped with neodymium). *Semiconductor lasers are especially important, being small, robust, and cheap to produce.

Gas lasers form another large group, operating across the spectrum from far infrared to far ultraviolet. Examples include the helium-neon, argon, krypton, carbon dioxide, and hydrogen fluoride lasers. The helium-neon laser usually consists of a *gas-discharge tube containing a mixture of helium and neon at low pressure. Helium atoms, excited by a continuous electrical discharge, collide inelastically with neon atoms and an energy transfer occurs. The helium atom relaxes to its ground state and the neon atom becomes excited, a large population inversion occurring in certain high-energy levels. Once stimulated emission has been triggered, a continuous laser beam is obtained at wavelengths including 1.152 μm, 3.391 μm (both infrared) and 632.8 nm (visible). The continuous output is usually a few tens of milliwatts.

As with the solid-state laser, two reflecting surfaces are used with the gas-laser tube to form a cavity resonator.

Liquid lasers are frequently liquid organic dyes. These dye lasers have been made to lase at frequencies from the IR to the UV. They have the advantage that they can be tuned continuously over a range of wavelengths.

The laser is not truly monochromatic but has a linewidth of the order of 10^4 hertz. This is much narrower than other radiation sources, the best "monochromatic" beams having a linewidth between $10^8 - 10^9$ Hz.

latch An electronic device in which a single bit of data is stored temporarily. The storage is under the control of a clock signal, a given transition of which fixes the contents of the latch at the current value of its input. The contents remain fixed until the next transition. The latch is an extension of a simple *flip-flop.

latent heat Symbol: *L*. The quantity of heat absorbed or released in an isothermal transformation of phase. The quantity of heat released or absorbed per unit mass is called the specific latent heat. (*See* specific latent heat of fusion; specific latent heat of vaporization; specific latent heat of sublimation.) The quantity of heat absorbed or released per unit amount of substance (per mole) is called the *molar latent heat.*

latent image *See* photography.

latent magnetization The property possessed by certain metals, such as manganese and chromium, that are weakly magnetic in themselves but form strongly magnetic alloys or compounds.

lateral aberration *See* chromatic aberration.

lateral magnification The inverse ratio of the size (*y*) of an object perpendicular to the axis of a reflecting or refracting system, to the size (*y'*) of the image, i.e. $m = y'/y$. According to the sign convention adopted, the algebraic sign of *m* will determine whether the image is erect or inverted.

lattice 1. A regular periodic repeated three-dimensional array of points that specify the positions of atoms, molecules, or ions in a crystal. *See* Bravais lattice; crystal systems.

2. The internal structure of the *core of a *nuclear reactor, consisting of a regular array of *fissile material and nonfissile material, especially a *moderator.

lattice constant The length of edge or the angle between the axes of the *unit cell of a crystal. It is usually the edge length of a cubic unit cell.

lattice dynamics The study of the excitations that a crystal *lattice can experience and their effects on the thermal, optical, and electrical properties of solids.

Laue diagram A symmetrical pattern of spots produced on a photographic plate by a collimated beam of X-rays that have passed through a stationary single crystal. There is a range of wavelengths present in the X-ray beam, which usually comes from a tungsten source operated at a voltage too low to excite characteristic K radiation (*see* X-rays). A heterogeneous beam of electrons or neutrons could also be used. The different atomic planes diffract the beam and give rise to the series of symmetrically arranged spots on the Laue diagram. From the diagram it is possible to determine the type of crystal and calculate its crystal structure.

law A theoretical principle deduced from particular facts expressed by the statement that a particular phenomenon always occurs if certain conditions are present.

Lawson criterion The product of the particle density of a *plasma (in particles per cm^3) and the *containment time (in seconds) at or above its *ignition temperature, such that the fusion energy released equals the energy required to produce and confine the plasma. For the deuterium-tritium reaction the value of the Lawson criterion is 10^{14} s/cm^3. *See* fusion reactor.

LC A prefix to an electronic device, circuit, or effect (as with LC filter, LC network, LC coupling) whose action is based on the properties and arrangement of one or more inductors and capacitors.

LCD Abbreviation for *liquid-crystal display.

lead 1. An electrical conductor.
 2. Of a periodically varying quantity. The interval of time or the angle in *electric degrees by which a particular phase in one wave is in advance of the similar phase in another wave. *Compare* lag.

lead accumulator *See* accumulator.

lead equivalent A measure of the absorbing power of a radiation screen expressed as the thickness of metallic lead (usually in millimetres) that could give the same protection as the given material under the same conditions.

leader stroke The initial discharge that establishes the track of a *lightning flash. It may develop from the cloud towards earth or from the earth towards the cloud. Its development may be continuous, or may develop in a series of definite steps of relatively short length. A high-current discharge that flows upwards through a lightning path as soon as a downward leader stroke has made contact with earth is called a *return stroke*.

leading current An alternating current that, with respect to the applied electromotive force producing it, has a *lead. *Compare* lagging current.

leading load *Syn.* capacitive load. A *reactive load that takes a *leading current.

leadless chip carrier (LCC) A form of package commonly used for integrated circuits in which connections to the device are made by means of small metallic contacts arranged around the outer periphery of the package, flush with the edges. This allows the package to be inserted in, for example, printed circuit boards.

leakage 1. The flow of an electric current, due to imperfect insulation, in a path other than that intended. This *leakage current* is small compared to that of a short circuit.
 2. *See* magnetic leakage.
 3. A net loss of particles from a region or across a boundary in a nuclear reactor.

leakage flux In any electrical machine or transformer in which there is a magnetic circuit. The flux that is outside the useful portion of the flux circuit.

leakage reactance Reactance caused in a transformer by the leakage inductance associated with losses due to some of the magnetic flux cutting one coil but not the other.

leap second *See* time.

least-action principle In a conservative dynamical system, the *action has a stationary value for the actual path, as compared with various paths between the same points for which the total energy has the same constant value. *Compare* Hamilton's principle.

least-energy principle A dynamic system is in stable equilibrium only if the potential energy of the system considered as a whole is a minimum. *See* Le Chatelier's rule.

least squares, method of A technique for finding the equation that gives the best fit for a set of experimental data. A simple example is in fitting a linear equation $y = ax$ to a set of measurements of the dependent variable y in terms of the independent variable x. A value of a is chosen and the *deviation of experimental values can be obtained. The best fit is considered to occur for a value of a at which the sum of the squares of these deviations is a minimum.

least-time principle *See* Fermat's principle.

Le Chatelier's rule (or principle) *Syn.* Le Chatelier–Braun principle. When a constraint is applied to a dynamic system in equilibrium, a change takes place within the system, opposing the constraint and tending to restore equilibrium.

Leclanché cell A primary *cell in which the anode is a rod of carbon, and the cathode a zinc rod, which may be amalgamated. The electrolyte is 10–20% NH_4Cl solution. The depolarizer consists of manganese dioxide mixed with graphite or crushed carbon, contained in a fabric bag or porous pot. The e.m.f. is about 1.5 volts, but falls off fairly rapidly on closed circuit as the depolarizer is slow in action. The cell is particularly useful for intermittent applications. In another form, the *agglomerate cell*, an attempt is made to reduce the internal resistance by having the depolarizer made into solid blocks held to a carbon plate by rubber bands. In this variant

the cathode is usually a large zinc cylinder surrounding the blocks. *See* dry cell.

LED Abbreviation for *light-emitting diode.

Leduc effect *Syn.* Righi effect. If heat is flowing through a metal strip, a magnetic field set up perpendicular to the plane of the strip causes a temperature difference to appear across the strip. The disposition of the higher and lower temperature regions depends on the metal of which the strip is composed. The effect is related to the *Nernst effect.

Lees' rule A formula for calculating moments of inertia, I:
$$I = \text{mass} \times \{a^2/(3 + n) + b^2/(3 + n')\},$$
where n and n' are the numbers of principal curvatures of the surface that terminates the semiaxes in question and a and b are the lengths of the semiaxes. Thus, if the body is a rectangular parallelepiped, $n = n' = 0$, and
$$I = \text{mass} \times (a^2/3 + b^2/3).$$
If the body is a cylinder then, for an axis through its centre, perpendicular to the cylinder axis, $n = 0$ and $n' = 1$ and
$$I = \text{mass} \times (a^2/3 + b^2/4).$$
If I is desired about the axis of the cylinder, then $n = n' = 1$ and $a = b = r$ (the cylinder radius) and
$$I = \text{mass} \times (r^2/2).$$
Compare Routh's rule.

left-hand rule For electric motor. *See* Fleming's rule.

Legendre equation The differential equation of the form

$$\frac{d}{dx}\left((1 - x^2)\frac{dy}{dx}\right) + ay = 0.$$

The solutions of this equation are known as *Legendre polynomials*.

Lennard–Jones potential *See* van der Waals forces.

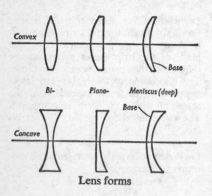

Convex

Base

Bi-   Plano-   Meniscus (deep)

Base

Concave

Lens forms

lens 1. A piece of transparent substance (commonly glass, plastic, quartz, etc.) bounded by two surfaces of regular curvature. Most commonly the surfaces are parts of spheres but they may be cylindrical, parabolic, toroidal, plane, etc. The general function of a lens is to change the curvature of *wavefronts so that light may be focused to a desired position. Lenses are classed as *convergent* or *divergent*. Most lenses are intended to be *stigmatic*, uniting rays to point foci; others (*cylindrical and *toric) have a different converging effect in meridians at right angles, producing two focal lines instead of one point focus – such lenses are *astigmatic*.

Lens shape refers to the shape of the periphery while *lens form* refers to the relative allocation of surface curvature (*see* diagram). A *thin lens* is one whose thickness is small compared with the focal lengths of its surfaces or of the focal length of the lens: the *powers of the individual surfaces add up to the power of the lens. This is not true for thick lenses. *See also* lens formula; aberration;

achromatic lens; blooming of lenses; convex; concave.

2. *See* electron lens.

lens antenna A microwave *aerial with an electronic focusing arrangement placed in front of the radiator in order to produce a desired shape and direction in the radiated beam. This is achieved by introducing selected phase shifts over different paths through the lens, which could, for example, be a system of metal slats or shaped insulator segments.

lens formula The relationship between object distance, u, and image distance v, from a thin lens:

$$1/v + 1/u = 1/f,$$

where f is the *focal length. For a *real object or image, v and u are positive; for a virtual object or image, v and u are negative. For a convex lens, f is positive; it is negative for a concave lens. For thick lenses, refraction must be taken into account. The relation for refraction at a single curved surface is:

$$n_2/v + n_1/u = (n_2 - n_1)/2f,$$

where n_2 and n_1 are the refractive indexes of the lens material and the medium.

Lenz's law *See* electromagnetic induction.

lepton The class of *elementary particles that do not take part in *strong interactions. They have no substructure of *quarks and are considered indivisible. They are all *fermions. There are six distinct types: the *electron, *muon, and *tauon (which all carry an identical charge but differ in mass) and the three *neutrinos (which are all neutral and thought to be massless or nearly so). In their interactions the leptons appear to observe boundaries that define three families, each composed of a charged lepton and its neutrino.

The families are distinguished mathematically by three *quantum numbers, l_e, l_μ, and l_τ, called *lepton numbers*:

Particle	l_e	l_μ	l_τ
e^-, ν_e	1	0	0
$e^+, \bar{\nu}_e$	-1	0	0
μ^-, ν_μ	0	1	0
$\mu^+, \bar{\nu}_\mu$	0	-1	0
τ^-, ν_τ	0	0	1
$\tau^+, \bar{\nu}_\tau$	0	0	-1
others	0	0	0

In *weak interactions the total lepton numbers $l_e{}^{TOT}$, $l_\mu{}^{TOT}$, and $l_\tau{}^{TOT}$ (obtained by adding up the values of l_e, l_μ, and l_τ for the individual particles) are conserved.

LET Abbreviation for *linear energy transfer.

lethargy Symbol: u. The negative natural logarithm of the ratio of the energy (E) of a neutron to a specified reference energy (E_0).

lever A simple *machine that can be regarded as a rigid bar that can turn about a pivot. The relative positions of load, effort, and pivot determine the type of lever.

levitation Any process by which a body is supported in a vacuum or in a gas by means of radiation or a field of force.

Laser levitation can be produced in a vacuum by projecting a continuous laser beam upwards and converging it with a lens. Minute spheres can be supported just above the focus using a beam of power about one watt. The suspended body is stable against vertical displacements because the support force falls off with height. To give stability against lateral displacements the intensity of the beam must vary across its area of cross section. The support depends upon the momentum of radiation, which is equal to the energy transfer divided by the speed of light.

Acoustic levitation is analogous to the above, being produced by focusing an ultrasonic beam in a gas. *Magnetic levitation* is the repulsion of a magnetized small body by a suitably magnetized surface. One form of this levitates a small magnet in the space over a dish-shaped piece of superconducting material, using the Meissner effect (*see* superconductivity.)

Lewis number Symbol: *Le*. A dimensionless number used in problems involving both heat and mass transfer. It is equal to $\lambda/\rho Dc$, where λ is the thermal conductivity, ρ is the density, D is the *diffusion coefficient, and c is the *specific heat capacity.

Leyden jar A historic form of *capacitor consisting of a glass jar, with metal foil on the outside and inside surfaces. The connector to the inside foil usually terminated in a small brass knob.

LF Abbreviation for low frequency. *See* frequency bands.

Lichtenberg figures The surface of a solid dielectric is affected by proximity of a potential high enough to produce ionization of the surrounding air. If fine powder is dusted over the surface and then blown by a current of air, some still remains adherent in a symmetrical star-shaped pattern, often of great complexity. This is called a Lichtenberg figure and is of use in the study of the properties of insulators under high fields. The figures can also be obtained on photographic emulsions and can be developed by ordinary photographic chemicals. In some dielectrics they can be "developed" directly on the dielectric by subsequent heating.

lidar (*light detection and ranging*) A radar-like technique employing pulsed or continuous-wave *laser beams for remote sensing. It is used, for example, in atmospheric physics to study the distribution of clouds, dust particles, and pollutants. The incident laser beam is backscattered by a target of interest and detected by a receiver. The photons in the beam interact with the target species by elastic or inelastic *scattering.

lifetime 1. In a *semiconductor. The mean time interval between generation and recombination of a charge *carrier.
   2. *See* mean life.

lift coefficient For a body and a fluid in relative motion. The ratio of the component of resistance force perpendicular to the direction of relative motion (lift) to the quantity $\rho l^2 V^2$, where ρ is the fluid density, V the relative velocity, and l is some characteristic length of the body. The quantity $\frac{1}{2}\rho V^2$ is sometimes used in place of ρV^2. The coefficient is a function of the *Reynolds number and is dependent on the circulation round the body.

light The agency that when it strikes the retina of the eye stimulates the sensation of the same name.
   Light may be regarded as a form of *electromagnetic radiation, consisting of interdependent mutually perpendicular transverse oscillations of an electric and magnetic field. It forms a narrow section of the electromagnetic spectrum, the wavelength range (for normal vision) being approximately 390 nanometres (violet) to 740 nanometres (red). According to the *quantum theory, light is absorbed in packets of light quanta or *photons. When light frequencies are multiplied by the *Planck constant, h, the energy of the photon for the frequency concerned is obtained.
   The electromagnetic-wave theory and the quantum theory are regarded as complementary; phenomena involving the propagation of light can be interpreted adequately on the wave basis, but when interactions of light (e.g. the *photoelectric effect) are under consideration, the quantum theory must be employed.
   See also colour; speed of light.

light-emitting diode (LED) A small cheap p-n junction *diode formed from certain *semiconductor materials, e.g. gallium arsenide, in which direct radiative recombination of excess electron-hole pairs is possible: a photon of radiation is emitted as an electron in the conduction band recombines with a hole in the valence band (*see* energy bands). When the p-n junction is *forward biased, light emission can take place; the intensity is proportional to the bias current, i.e. to the numbers of excess minority carriers. The useful light obtained will be dependent on the optical quality of the crystal surfaces, and the colour will depend on the particular material used. Such diodes are used as small indicator or warning lights and in self-luminous display devices.

light exposure Symbol: H_v, H. **1.** The surface density of the total quantity of light received by a material.
   2. A measure of the total amount of light energy incident on a surface per unit area, expressed as the product of the *illuminance and the time for which it is illuminated. When the intensity of the light varies over the period of illumination the exposure is given by the integral $\int E_v dt$, where E_v is illuminance and t is time. It is measured in lux second. *Compare* radiant exposure.

light guide An informal name for an optical fibre. Large-diameter (about 5 mm) fibres are generally called *light pipes*. *See* fibre-optics system.

light meter *See* exposure meter.

lightning conductor A device used to protect a building from damage by lightning. It consists of a metal strip earthed at its lower end and with an air terminal attached to the highest point of the building. It provides a path along which a *lightning stroke can pass.

lightning flash A complete lightning discharge along a single discharge path. It may be made up of more than one *lightning stroke, in which case it is described as a multiple stroke. *See also* leader stroke.

lightning stroke A discharge that is a component of a complete *lightning flash. It is the discharge of one of the charge regions of a thunder cloud. The polarity of a lightning stroke is the polarity of the electric charge that is brought to earth. A lightning stroke to any part of a power or communication system is described as a *direct stroke*. It is an *indirect stroke* If it induces a voltage in that system without actually striking it.

light pen A penlike device attached to an *on-line *VDU by which information may be input to a *computer. This is usually achieved by pointing the pen at small areas of the screen, for example indicating a selection from a displayed list, or sometimes by drawing shapes on the screen. The pen is connected by cable to the computer, which is able to identify the position of the pen when light from the screen is detected by an electronic photodetector in the pen.

light pipe *See* light guide.

light-year (l.y.) A unit of distance employed particularly in popular works on astronomy and equal to the distance travelled by light through a vacuum in one year. It equals $9.460\,528 \times 10^{15}$ metres or $0.306\,5949$ *parsec.

limiter In general, an electronic or electrical device that automatically sets a boundary value or values on some output characteristic. In particular, a device that for inputs below a specified instantaneous value gives an output proportional to the input, but for inputs above that value gives a constant peak output. The output of a *base limiter* comprises that part of an input signal that exceeds a predetermined value.

limiting current *Syn.* diffusion current. In electrolysis the rate of diffusion of ions towards the electrodes may not keep pace with their deposition. There is thus a limiting value of current that can be passed under the particular conditions of ionic concentration and characteristics of the electrolyte.

limiting friction *See* friction.

linac *See* linear accelerator.

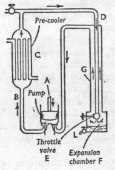

Linde process for liquefying air

Linde process For the liquefaction of air. Air is compressed by the pump A, and passes through B to the cooler C where the heat of compression is removed (*see* diagram). The compressed air passes through D and expands through a throttle valve E into the vessel F. The cooled air then returns

267

through the heat exchanger G to the pump chamber A, cooling the oncoming air in DE, until finally liquid air is formed in F and may be withdrawn at L.

linear 1. Having components arranged in a line, as in a *linear accelerator.

2. Measured along or having one dimension.

3. Having an output that is directly proportional to its input, as in a linear *amplifier.

linear absorption coefficient *Syn.* absorption coefficient. Symbol: a; unit: m^{-1}. During its passage through a medium radiation is absorbed to an extent that depends on the wavelength of the radiation and the thickness and nature of the medium. If $d\Phi$ is the change in *radiant or *luminous flux of a parallel beam of monochromatic radiation on passing through a small thickness dl of an absorbing medium, the linear absorption coefficient is defined by the equation:
$$a = \Phi^{-1}(d\Phi/dl),$$
which integrates to
$$\Phi_x/\Phi_0 = \exp(-ax),$$
where Φ_0 is the initial flux and Φ_x the flux after a distance x.

This equation is known as *Bouguer's law* or *Lambert's law* of absorption. It only applies in practice if factors such as reflection and scattering are negligible or can be corrected for. (*See also* linear attenuation coefficient; internal absorptance.)

For X-ray absorption it is often more convenient to consider the mass per unit area, rather than the thickness of the absorbing radiation. The corresponding coefficient is known as the *mass absorption coefficient*. It is equal to a/ρ, where ρ is the density of the material.

linear accelerator (linac) A particle *accelerator in which electrons or protons are accelerated along a straight evacuated chamber by an electric field of radio frequency produced by a *klystron or *magnetron. In older low-energy machines, cylindrical electrodes (or *drift tubes*) of the RF supply are aligned coaxially with the chamber. Keeping in phase with the RF supply, the charged particles are accelerated in the gaps between the electrodes.

Modern high-energy linacs are usually travelling-wave accelerators in which particles are accelerated by the electric component of a *travelling wave set up in a *waveguide. The RF is boosted at regular intervals along the chamber length by means of klystrons. Only a small magnetic field, supplied by magnetic lenses between the RF cavities, is required to focus the particles and maintain them in a straight line. Typical rates of energy gain in a linac are 7 MeV per metre (electrons) and 1.5 MeV m^{-1} (protons).

linear amplifier *See* amplifier.

linear attenuation coefficient *Syn.* (linear) extinction coefficient. Symbol: μ; unit: m^{-1}. A measure of the ability of a medium to diffuse and absorb *radiation. If a collimated beam of radiation is passing through the medium it loses intensity due to absorption and scattering. The linear attenuation coefficient is defined by the equation:
$$\mu = -\Phi^{-1}(d\Phi/dl),$$
where $d\Phi$ is the decrease in *luminous or *radiant flux Φ passing through a section dl of the material perpendicular to its face. The linear attenuation coefficient is more general than the *linear absorption coefficient, which only applies to an absorbing medium.

linear circuit *Syn.* analogue circuit. A circuit in which the output varies continuously as a given function of the input. *Compare* digital circuit.

linear energy transfer (LET) The average energy locally imparted to a medium by a charged particle, of specified energy, in traversing a short distance.

linear expansion coefficient *See* coefficient of expansion.

linear extinction coefficient *See* linear attenuation coefficient.

linear inverter *See* inverter.

linear momentum *See* momentum.

linear motor A type of *induction motor in which stator and rotor are linear and parallel rather than cylindrical and coaxial.

linear network *See* network.

linear stopping power *See* stopping power.

linear strain *See* strain.

line broadening *See* line profile.

line defect *See* defect.

line frequency *See* television.

line of force *Syn.* line of flux. An imaginary line whose direction at all points along its length is that of the electric, gravitational, or magnetic field at those points. *See* field.

line printer *See* printer.

line profile A plot of intensity against wavelength or frequency for a spectral line, showing the fine structure of the line. The natural width of a spectral line is determined by quantum-mechanical uncertainty. Other factors, however, can lead to additional *line broadening*. These include the *Doppler effect, *Zeeman

effect, and high density and hence high pressure of the emitting or absorbing material (*pressure broadening*). Analysis of the line profile yields information about the physical conditions at the source.

line spectrum *See* spectrum.

line voltage *See* voltage between lines.

linkage A measure of the amount of magnetic flux embraced by an electric circuit.

Linke–Fuessner actinometer *See* pyrheliometer.

lin-log receiver A radio receiver that has a linear response for small input signals and a logarithmic response for large signals.

liquefaction of gases The following methods for the liquefaction of gas are available: (1) The application of high pressure after cooling the gas below its critical temperature, as in *cascade liquefaction. (2) The Joule–Kelvin effect in which gas at high pressure is cooled by expansion through a porous plug or throttle valve, as in the *Linde process. (3) Adiabatic expansion in which a compressed gas is cooled by performing external work as in the *Claude process. (4) Adiabatic *desorption in which the gas is adsorbed in cooled charcoal and further cooling produced when the gas is removed adiabatically.

liquid A phase of matter, intermediate between a gas and a solid, that is characterized by ease of flow and near incompressibility. It takes the shape of its container but, unlike a gas, does not expand to fill the container. The intermolecular forces are intermediate in strength between those in gases and solids. Liquids possess a short-range structural regularity and these bundles of atoms, molecules, or ions can move relative to each other.

liquid-column manometers Open liquid-column *manometers are based on the U-tube of liquid, which measures the pressure difference between the two sides. If the difference in vertical level of the surfaces of the liquid is h and its density is ρ, the applied pressure difference is $h\rho g$ (g being the acceleration of free fall).

The mercury *barometer is a special case in which one pressure is zero. When measuring very high pressures, which would need a prohibitively great difference in height, several U-tubes of mercury are used in tandem with either compressed air or a light liquid in between.

liquid crystal An arrangement of certain kinds of long molecules in a liquid with the result that the liquid does not suffer an entire loss of fluidity. In *nematic* liquid crystals, the molecules are aligned in the same direction but randomly so. In *cholesteric* and *smectic* liquid crystals, the molecules are still aligned but in distinct layers; their long axes are parallel to (cholesteric) or usually perpendicular to (smectic) the plane of the layers. Liquid crystals have characteristic optical properties. Cholesteric and nematic types can have very large rotatory powers (*see* optical activity). Smectic and nematic structures both exhibit *double refraction and are optically positive. Cholesteric types exhibit *iridescent colours as a result of *dichroism, but often only in a certain temperature range.

liquid-crystal display (LCD) A low-power device used to display numbers, letters, and other characters (black on white) in, for example, digital watches, calculators, measuring instruments, and some computer displays. The LCD depends for its action on the change produced by an electric field on the optical properties of *liquid crystals.

In one form, a thin film of liquid crystal is sandwiched between thin transparent electrodes. The upper electrode is etched into a pattern of segments making up the characters of the display. Two crossed *polarizers (i.e. with their vibration planes at right angles) lie on either side of the film. Light falling on an unenergized liquid crystal segment is largely reflected back and it appears transparent: the plane-polarized light entering the segment has its plane of polarization rotated through 90° so that it can then pass through the second polarizer, and is reflected back. When an electric field is applied across selected segments, they appear dark: the rotatory power of the liquid-crystal segments is affected by the field and the light cannot pass through the second polarizer.

liquid-drop model A model of the nucleus in which the nuclear matter is regarded as being continuous. The interactions between *nucleons are thought of as being analogous to those between molecules in a liquid. As in a liquid drop, the net effect of the interaction of particles near the surface of the nucleus is interpreted as a *surface tension, which maintains the shape of the nucleus. The liquid-drop model is most applicable to heavy nuclei and is used in the theory of *nuclear fission.

Lissajous' figure The displacement pattern traced out by the superposition of two vibrations in directions at right angles to each other. These figures can be constructed graphically; or they may be obtained practically using either a mechanical device or a cathode-ray oscilloscope.

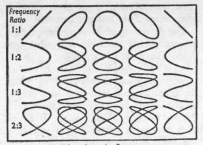

Lissajous's figures

Examples are given in the illustration for various frequency ratios and phase differences between 0 and π. The figures are useful, in particular, in identifying the phase relationship of two vibrations of the same frequency, and in verifying that two given vibrations are of the same frequency.

lithium fluoride dosimetry *See* dosimetry.

lithography A group of techniques, including *photolithography, used in the manufacture of *integrated circuits, *thin-film circuits, *printed circuits, etc. Advantages, especially increased resolution, are obtained if the light or ultraviolet radiation used in photolithography to expose the resist is replaced by a beam of X-rays or high-energy electrons or ions.

litre Symbol: l or L. A unit of volume formerly defined as the volume occupied by a mass of 1 kg of pure water at its maximum density and under standard atmospheric pressure. This volume is equal to 1.000 028 decimetres cubed. Subsequently the litre was defined as a special name for the decimetre cubed, but owing to confusion between the two definitions the unit is not recommended for scientific purposes, although the ml (millilitre) is used synonymously with the cc where great accuracy is not implied.

litzendraht wire (litz) A multistranded wire composed of many fine conducting filaments, used, for example, in low-loss coils for filters and radios to reduce their high-frequency resistance. *See* skin effect.

live 1. Connected to a voltage supply; not at earth potential.

2. Reverberant; having a reverberation time that is normal or above normal.

3. Involving the direct transmission of a TV or radio signal without prior recording.

Lloyd's mirror A glass or metal mirror for producing interference fringes in overlapping beams. Light from a slit parallel to and near the mirror is reflected from the surface at a near *glancing angle, undergoing a 180° phase shift in the process. This beam interferes with the direct beam to produce the fringes.

load 1. A device or material in which electrical signal power is dissipated or received, i.e. a device that absorbs power from a source of electrical signals. Examples include loudspeakers, TV and radio receivers, logic circuits, and the material to be heated by dielectric and induction heating.

2. The power delivered by a machine, generator, transducer, or electronic circuit or device.

3. The mechanical force applied to a body.

4. The weight supported by a structure.

loaded concrete Concrete containing material of high atomic number or capture cross section, such as barium, iron, or lead, to increase its effectiveness as a radiation shield in *nuclear reactors.

load impedance The *impedance presented by a *load to the driver circuit that supplies power to it.

load line A line drawn on the graph of a family of *characteristics of an electronic device, showing the graphical relationship between voltage and current for the particular *load of the circuit under consideration.

lobe A region on a *radiation pattern representing enhanced response of an aerial. The *main* or *major lobe* corresponds to the direction of best transmission and reception. All the other lobes are called *side lobes* and are usually unwanted.

local area network (LAN) A simple communications system linking a number of computers within a small and defined locality, such as an office building, industrial site, or university. This allows the computers (usually microcomputers) to share single and/or expensive resources, such as a line printer or hard disks, and to share data files and databases. Messages can be sent by electronic mail. The transmission medium is usually electric cables or optical fibres.

local oscillator *See* superheterodyne receiver.

logarithmic resistor A form of variable resistor designed so that the fractional change in resistance is directly or inversely proportional to the movement of the contact.

logic circuit A circuit designed to perform a particular logical function based on the concepts of "and", "either-or", "neither-nor", etc. Normally these circuits operate between two discrete voltage levels, i.e. high and low logic levels, and are described as binary logic cir-

cuits. Logic using three or more logic levels is possible but not common.

The devices used to implement the elementary logic functions are called *logic gates*. The basic gates are:

(a) *AND gate*. A circuit with two or more inputs and one output in which the output signal is high if and only if all the inputs are high simultaneously.

(b) *Inverter (NOT gate)*. A circuit with one input whose output is high if the input is low and vice versa.

(c) *NAND gate*. A circuit with two or more inputs and one output, whose output is high if any one or more of the inputs is low, and low if all the inputs are high.

(d) *NOR gate*. A circuit with two or more inputs and one output, whose output is high if and only if all the inputs are low.

(e) *OR gate*. A circuit with two or more inputs and one output whose output is high if any one or more of the inputs are high.

(f) *Exclusive OR gate*. A circuit with two or more inputs and one output whose output is high if any one of the inputs is high.

These circuits are for use with *positive logic*: that is the high voltage level represents a logical 1 and low a logical 0. *Negative logic* has high level representing a logical 0 and low a logical 1. The same circuits may be used in negative logic, but become the complements of the positive logic circuits, i.e. a positive OR circuit becomes a negative AND circuit.

Any logical procedure may be effected by using a suitable combination of the basic gates. Binary logic circuits are extensively used in *computers to carry out instructions and arithmetical processes. They may be formed from discrete components or, more commonly, from *integrated circuits. Families of integrated logic circuits exist based on bipolar transistors. (*See* emitter-coupled logic (ECL); transistor-transistor logic (TTL).) *MOS logic circuits are based on *field-effect transistors.

logic gate *See* logic circuit.

longitudinal aberrations Aberration distances measured along the principal axis. In *chromatic aberration it is the distance between the foci for the two standard colours. In *spherical aberration, it is the distance from the paraxial focus to the intersection of a zonal ray with the axis.

longitudinal mass In special *relativity theory. The ratio of force to acceleration in the direction of the existing velocity of a particle. It is given by:
$$m_1 = m_0 / \sqrt{(1 - \beta^2)^3},$$
where m_0 is the rest mass of the particle and $\beta = v/c$, i.e. its velocity expressed as a fraction of the speed of light. *Compare* transverse mass.

longitudinal strain *See* strain.

longitudinal vibrations Vibrations in which the displacement is along the main axis or direction of the vibrating body or system.

longitudinal waves Waves in which the particles of the transmitting medium are displaced along the direction of propagation. The speed, c, of longitudinal waves in a bar is given by $c = \sqrt{(E/\rho)}$, where E is the Young modulus and ρ is the density; the corresponding equation for longitudinal waves in a fluid is $c = \sqrt{(k/\rho)}$, where k is the bulk modulus. Sound waves in a gas form the chief example of longitudinal wave motion. *See* speed of sound.

long-range force *See* force.

long-tailed pair Two matched bipolar *transistors that have their *emitters coupled together, with a common-emitter bias resistor acting as a constant-current source. The name is derived from the physical resemblance of the bias resistor to a tail: the larger (i.e. longer) the bias resistor, the more nearly it resembles a constant-current source (because of the relatively large voltage developed across it). The long-tailed pair forms the basis of most *differential amplifiers.

long-wave Designating radio waves with wavelengths exceeding 1000 metres, i.e. with frequencies below 300 kHz.

loop For feedback and control. *Syn.* feedback control loop. A means of control used in many types of control system, in which part of the output derived from the control system is fed back to the input circuit in order to control the output in a desired manner. The portion of the signal fed back to the input circuit is the feedback signal. This is mixed with the external signal applied to the loop (the loop input signal) to produce a loop actuating signal, which is used to produce the controlled output.

loop aerial *Syn.* frame aerial. An *aerial that is essentially a coil having one or more turns of wire wound on a frame and having an axial length that is usually small compared with its other linear dimensions. The plane of the coil is the direction of maximum sensitivity and transmission. It is commonly employed in radio direction finders and in small portable radio receivers.

Lorentz–FitzGerald contraction The theory that a material body moving through the *ether with a velocity v, contracts by a factor of $\sqrt{(1 - v^2/c^2)}$ in the direction of motion, where c is the speed of light in vacuum. The theory was advanced to account for the failure of the *Michelson–Morley experiment to detect the earth's motion through the ether, but has been superseded by the theory of *relativity. According to relativity theory the length of a body as measured by an observer in uniform relative motion is less than that measured by an observer at rest with respect to the body by the factor given above. This is not a physical change in the body so should not be confused with the hypothetical change postulated in the older theory.

Lorentz force The force acting on a moving charge q in magnetic and electric fields. It is given by:

$$F = q(E + v \times B),$$

where F is the force, E the electric field, and $v \times B$ the *vector product of the particle's velocity and the *magnetic flux density.

The magnetic contribution to this force is often called the Lorentz force and this, in nonvector notation, is given by:

$$F = qvB \sin \theta,$$

where θ is the angle that the direction of motion of the particle makes with the magnetic field. The force acts in a direction that is perpendicular to both the direction of motion and the magnetic field.

Lorentz transformation equations A set of equations for transforming the position-motion parameter from an observer at a point $O(x, y, z)$ to an observer at $O'(x', y', z')$, moving relative to one another. The equations replace the *Galilean transformation equations of Newtonian mechanics in *relativity problems. If the x-axis is chosen to pass through OO' and the time of an event is t and t' in the *frame of reference of the observers at O and O' respectively (where the zeros of their time scales were the instant that O and O' coincided) the equations are:

$$x' = \beta(x - vt)$$
$$y' = y$$
$$z' = z$$
$$t' = \beta(t - vx/c^2),$$

where v is the relative velocity of separation of O, O', c is the speed of light, and β is the function $(1 - v^2/c^2)^{-1/2}$.

Loschmidt constant Symbol: n_0. A constant that gives the number of molecules in one cubic metre of an ideal gas at STP. It is equal to

$$2.686\,763 \times 10^{25} \text{ m}^{-3}.$$

It is the ratio of the *Avogadro constant to the molar volume of an ideal gas.

loss *See* core, copper, dielectric, and eddy-current losses.

loss angle Of a capacitor or dielectric when subjected to alternating electric stress. The angle by which the angle of *lead of the current is less than 90° when the applied voltage is sinusoidal. It is due mainly to dielectric *hysteresis loss.

loss factor 1. The ratio of the average power dissipation to the power dissipation at peak load in a transmission line, circuit, or device.

2. The product of the *power factor and the relative *permittivity of a *dielectric material. It is proportional to the heat generated in a material in a given alternating field.

lossy 1. Denoting an insulator that dissipates more energy than is considered normal for the class of material.

2. Denoting a transmission line designated to have a high degree of attenuation.

loudness The magnitude of the sensation produced when a sound reaches the ear. Although loudness is related to *sound intensity, there is no simple connection between the two. The basis of loudness scales is the *Weber–Fechner law*, which states that the sensation is proportional to the logarithm of the stimulus. In the *decibel scale of intensity level, the sound intensity is logarithmically related to the threshold intensity at the same frequency. This suffers from the disadvantage that the sensitivity of the ear to changes of intensity varies with frequency. The *phon scale of equivalent loudness overcomes this by relating the intensity of a sound to a fixed reference tone of defined intensity and frequency. The phon scale is widely used since it places all sounds in order of their loudness.

loudness level A measure of the strength of a sound as expressed by the *sound pressure level of a pure tone of specified frequency that is judged, by a normal listener, to be equally as loud as the sound.

loudspeaker *Syn.* speaker. A device in which an electrical signal is converted into sound. It is the final unit in any broadcast receiver or sound reproducer. In the most common types of loudspeaker the current is passed through a small coil fixed to the centre of a diaphragm and moving in an annular gap across which is a strong magnetic field. Alternating current in the coil causes the diaphragm to vibrate at the same frequency and emit sound waves. For high efficiency a small diaphragm is used at the mouth of a large exponential *horn. Although the horn gives suitable loading to the diaphragm, it is impractical for most indoor work on account of its size. Instead, a speaker is used having a large conical or elliptical diaphragm with the coil at its apex. The cone is made of stiff paper and is supported round its edge by a metal frame. The magnetic field is produced either by a permanent magnet or an electromag-

net and the coil is held in position in the centre of the gap by a flexible mounting. The cone should be set in a large *baffle to prevent direct passage of sound from front to back and so improve the low-frequency response. In most commercial sound reproducers the cabinet forms the baffle. This type of speaker gives a good response over a moderate range of frequencies but careful design is necessary for good reproduction at high or low frequencies.

low-angle scattering A halo of diffracted radiation immediately surrounding the incident beam, which is dependent only on the size and shape of the scattering particles and is independent of their internal character.

lower sideband *See* sideband.

low frequency (LF) *See* frequency band.

low-pass filter *See* filter.

low voltage In electrical power transmission and distribution. A voltage that does not exceed 250 volts.

LSI Abbreviation for large-scale integration. *See* integrated circuit.

lubrication The reduction of *friction between two solid surfaces sliding over each other by interposing a layer of liquid, or a solid with much lower coefficient of friction.

lumen Symbol: lm. The *SI unit of *luminous flux, defined as the luminous flux emitted by a uniform point source, of intensity one candela (cd), in a cone of solid angle one steradian. Thus 1 lm = $(1/4\pi)$ cd.

luminance Symbol: L_v, L. The brightness, for a specified direction, of a point source of light or a point on a surface that is receiving light. For sources of light it is defined as the luminous intensity, I_v, per unit projected area, i.e.

$$L_v = dI_v/dA \cos\theta,$$

where A is the area and θ is the angle between the surface and the specified direction. For illuminated surfaces it is defined as the illuminance (E_v) per unit solid angle (Ω),

$$L_v = dE_v/d\Omega.$$

The *illuminance is taken over an area perpendicular to the direction of the incident radiation. The general equation of luminous intensity, applying to both a point source and a point receptor, is

$$L_v = d^2\Phi_v/(d\Omega . dA . \cos\theta),$$

where Φ_v is the luminous flux. Luminance is measured in candela per square metre. *Compare* radiance.

luminance signal *See* colour television.

luminescence The emission of electromagnetic radiation from a substance as a result of any nonthermal process. The term is also applied to the radiation itself and is usually used for visible radiation. Luminescence is produced when atoms are excited, as by other radiation, electrons, etc., and then decay to their *ground state. If the luminescence ceases as soon as the source of energy is removed, the phenomenon is *fluorescence*. If it persists, the phenomenon is *phosphorescence*. More precisely, in fluorescence the persistence is less than about 10^{-8} seconds and in phosphorescence it is greater.

If certain solids are subjected to ionizing radiation, electrons may be released within the solid and trapped at *defects. These electrons may be released when the solid is heated and the energy produced is emitted as visible radiation. This is known as *thermoluminescence* (TL). The number of electrons is proportional to the intensity of the incident radiation (i.e. the number of incident photons), hence TL *dosimetry

has been developed to monitor radiation levels wherever radioactive sources are used – hospitals, factories, etc. TL can also be used for archaeological and geological *dating, the radiation sources being, for example, the common radionuclides ^{40}K and the ^{235}U and ^{232}Th decay series.

Luminescence can also be produced by the friction of solids (*triboluminescence*) and chemical reaction (*chemiluminescence*). *Compare* incandescence.

luminosity 1. The attribute of a source of light that gives the visual sensation of brightness. The luminosity depends on the power emitted by the source, i.e. on the *radiant flux, but also on the fact that the sensitivity of the *eye varies for different wavelengths. *Radiant quantities are pure physical quantities based on absolute energy measurements, whereas *luminous quantities depend on some judgement of brightness by an observer and thus on the spectral sensitivity of the eye.

2. Symbol: L. The intrinsic or absolute brightness of a star or other celestial body, equal to the total energy radiated per second from the body. It is related to the body's surface area and *effective temperature*, T_e (i.e. the surface temperature expressed as the temperature of a *black body having the same radius and radiating the same total energy per unit area per second as the body), by a form of the Stefan–Boltzmann law:

$$L = 4\pi R^2 \sigma T_e{}^4,$$

where σ is the Stefan–Boltzmann constant and R the radius of the body. Hence stars with similar T_e but greatly different luminosities must differ in size: they belong to different *luminosity classes*. Luminosity is also related to the absolute *magnitude of a celestial body.

luminous A qualifying adjective denoting physical quantities used in *photometry in which energies of light are evaluated by an observer (*see* luminosity). They are distinguished from their corresponding *radiant quantities by adding a subscript v (for visual) to their symbols.

luminous efficacy 1. Symbol: K. The ratio of the luminous flux, Φ_v, of a radiation to its radiant flux, Φ_e. If monochromatic radiation is considered, the property is called *spectral luminous efficacy*, symbol: $K(\lambda)$, given by the ratio $\Phi_{v,\lambda}/\Phi_{e,\lambda}$.
2. Symbol: η_v, η. The ratio of the luminous flux emitted by a source to the power it consumes. (*Compare* radiant efficiency.)

Luminous efficacy is measured in lumen per watt.

luminous efficiency Symbol: V. A dimensionless quantity defined by the ratio K/K_m, where K is the *luminous efficacy and K_m the maximum spectral luminous efficacy. If monochromatic radiation is considered, the property is called the *spectral luminous efficiency*, symbol: $V(\lambda)$, given by $K(\lambda)/K_m$, where $K(\lambda)$ is the spectral *luminous efficacy.

The term was formerly applied to what is now called *luminous efficacy.

luminous emittance Former name for luminous exitance.

luminous exitance Symbol: M_v, M. At a point on a surface, the *luminous flux leaving the surface per unit area. It is measured in lumen per square metre. *Compare* radiant exitance.

luminous flux Symbol: Φ_v, Φ. The rate of flow of radiant energy as evaluated by the luminous sensation that it produces. The luminous flux is obtained from the *radiant flux of the source corrected according to the effect it has on the observer, i.e. according to the spec-

tral sensitivity of the receptor. Consider, for example, a source of monochromatic radiation of wavelength λ and with radiant flux Φ_e. The luminous flux is then proportional to $\Phi_e V(\lambda)$, where $V(\lambda)$ is the spectral *luminous efficiency. This factor weights the radiant flux according to the sensitivity of a standard observer (in photopic vision) to radiation of wavelength λ. Specifically the luminous flux is given by:
$$\Phi_v = K_m \Phi_e V(\lambda),$$
where K_m is a constant relating the units of luminous flux to those of radiant flux. For *polychromatic radiation the *radiant flux will generally vary with wavelength, and luminous flux can be defined by:
$$\Phi_v = K_m \int (d\Phi_e/d\lambda)V(\lambda)\,d\lambda,$$
where $(d\Phi_e/d\lambda)\,d\lambda$ is the *radiant flux of light with wavelengths in the range $\lambda \rightarrow \lambda + d\lambda$.

The constant K_m has the value of 680 lumens per watt (for photopic vision). It is the maximum spectral *luminous efficacy. Luminous flux is measured in lumens.

luminous intensity Symbol: I_v, I. The *luminous flux emitted per unit solid angle by a point source in a given direction. It is measured in candela. A source may radiate unequally in different directions and the direction has to be specified. If the luminous intensity is averaged over all directions, it is called the *mean spherical intensity*. For extended sources the luminous intensity per unit area, or *luminance, is used. *Compare* radiant intensity.

Lummer–Brodhun photometer *See* photometry.

Lummer–Gehrcke plate An interferometer using an accurately parallel-sided glass or quartz plate of considerable thickness in which multiple reflections occur, giving rise to *interference effects.

It gives a *resolving power of the order of 10^6.

lumped parameter Any parameter of a circuit such as inductance, capacitance, or resistance, that, for the purposes of circuit analysis, can be treated as a single localized parameter throughout the frequency range under consideration.

lux Symbol: lx. The *SI unit of *illumination, defined as the illumination of one lumen uniformly over an area of one metre squared.

Lyman series *See* hydrogen spectrum.

M

Mach angle If a body moves, with a supersonic velocity V, through a fluid from a point X to a point Y in time t, then when the body is at Y the spherical pressure wave originating from X will have a radius ct where c is the local speed of sound in the fluid. Similarly, the pressure waves from the other points between X and Y will have corresponding radii such that all the spherical pressure waves combine to form a right-conical wavefront with its vertex at Y. The semiangle of the cone (β) is called the Mach angle:

$$\beta = \sin^{-1} ct/Vt = \sin^{-1} c/V$$
$$= \sin^{-1} 1/Ma,$$

where Ma is the *Mach number.

machine A device for doing work. In a machine a comparatively small force called the *effort* is used to overcome a larger force (e.g. the weight lifted by a system of pulleys) called the *load*.

The ratio

$$\frac{\text{distance moved by effort}}{\text{distance moved by load}}$$

is the *velocity ratio* of the machine.
The ratio

$$\frac{\text{load}}{\text{effort}}$$

is the *mechanical advantage* or *force ratio* and usually varies with the load.

The principle of work states that: work done by effort = work done on load + work lost in friction.

The fraction,

$$\frac{\text{work done on load}}{\text{work done by effort}}$$

is the *efficiency* and is necessarily less than 1. It is usually multiplied by 100 and expressed as a percentage.

Simple machines include levers, pulleys, gears, gear trains, the inclined plane, and the screw.

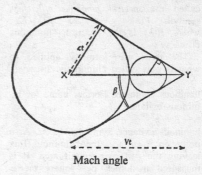

Mach angle

machine code, instructions *See* program.

Mach number Symbol: Ma. A dimensionless number being the ratio of the relative velocity of a body and fluid to the local speed of sound in the fluid. A Mach number in excess of 1 indicates a supersonic velocity; in excess of 5 it is said to be hypersonic. The Mach number appears in all problems of flow in which compressibility is of importance. The resistance to motion of a body moving at high speed in a fluid of small viscosity is, in general, a function of both the Mach and Reynolds numbers. *See also* Mach angle.

macroscopic state The state of matter characterized by the statistical properties of its components. *Kinetic theory is an analysis of the macroscopic state. *Compare* microscopic state.

magic numbers Certain values of the number of protons (Z) in a nucleus or the number of neutrons (N) that produce unusual stability in that nucleus. These values of Z or N are 2, 8, 20, 28, 50, 82, and 126. There is therefore, a tendency for nuclei to prefer magic Z or N: for example, there are six stable nuclides with $Z = 20$, whereas the average number of stable nuclides for a given Z in this part of the periodic table is about two. It is also found that the energy required to remove a nucleon from a nucleus with magic N or Z is higher than for neighbouring nuclei with nonmagic values of N or Z. *See also* shell model.

magnet A body possessing the property of *magnetism. Magnets are either temporary or permanent. *See* electromagnet; permanent magnet.

magnetic amplifier An amplifier in which *transductors are used to produce power amplification of the input signal.

magnetic balance 1. A device used for the direct determination of attraction or repulsion between magnetic poles. One magnet is suspended on a knife-edge system so that it takes up a horizontal position, and the force is applied to one end by bringing a magnet pole near to it, the magnet being restored to horizon-tal by direct addition of weights or the action of a movable rider. The magnets should be long to reduce interference by interaction between their second poles.
2. A type of *fluxmeter in which the force required to prevent movement of a current-carrying coil in a magnetic field is measured.

magnetic bottle Any configuration of magnetic fields used to confine a *plasma, especially a linear device in which the ends are stoppered with *magnetic mirrors. *See* fusion reactor.

magnetic circuit The completely closed path described by a given set of lines of *magnetic flux.

magnetic constant Symbol: μ_0. The *permeability of free space, with the formally defined value:
$$\mu_0 = 4\pi \times 10^{-7} \text{ H m}^{-1}$$
$$= 1.256\,637\,0614 \times 10^{-6} \text{ H m}^{-1}.$$
See rationalization of electric and magnetic quantities.

magnetic crack detection If a magnetizing field is applied to a ferromagnetic body there is often leakage of lines of force, or uneven distribution of magnetization, at points where discontinuities occur at or near the surface. These discontinuities become evident when the surface is painted with a magnetic fluid consisting of very finely divided particles of iron or of magnetic oxide of iron dispersed in oil. The particles concentrate above the discontinuities. *See also* flaw detection.

magnetic declination *See* declination.

magnetic dipole moment *See* magnetic moment.

magnetic disk *See* disk.

magnetic element *See* geomagnetism.

magnetic field The *field of force, surrounding a magnetic pole or a current flowing through a conductor, in which there is a *magnetic flux.

magnetic field strength Symbol: H. A *vector quantity given by the ratio B/μ, where B is the *magnetic flux density and μ the *permeability of the medium. Its magnitude gives the strength of the magnetic field at a point in the direction of the *line of force at that point. The integral of the magnetic field strength along a closed line is equal to the *magnetomotive force. Magnetic field strength is measured in amperes per metre.

magnetic flux Symbol: Φ. The product of a particular area under consideration and the component, normal to the area, of the average magnetic flux density over it. For an element of area dA the flux $d\Phi$ is the scalar product $B \cdot dA$. It is measured in *webers.

magnetic flux density *Syn*. magnetic induction. Symbol: B. The *magnetic flux passing through unit area of a magnetic field in a direction at right angles to the magnetic force. It is a *vector quantity whose magnitude at a point is proportional to the *magnetic field strength and whose direction at that point is that of the magnetic field. It indicates the strength of a magnetic field, often in terms of the effects of the field. For example, the vector product of the magnetic flux density and the current in a conductor gives the force per unit length of conductor. (*See also* Lorentz force.) Magnetic flux density is measured in teslas.

magnetic hysteresis *See* hysteresis.

magnetic induction *See* magnetic flux density.

magnetic intensity Former name for magnetic field strength.

magnetic leakage The loss of *magnetic flux from the core of a transformer or transducer, which reduces the overall efficiency of operation. The leakage is the portion of the total magnetic flux that follows a path such that it is ineffective for the desired purpose.

The *magnetic-leakage coefficient*, symbol: σ, is the ratio of total magnetic flux to effective or useful magnetic flux. It follows that $(\sigma - 1)$ is given by the ratio of leakage flux to useful flux.

magnetic lens *See* electron lens.

magnetic levitation *See* levitation.

magnetic meridians Imaginary lines drawn along the earth's surface in the direction of the horizontal component of the earth's field at all points along their length. They converge on the magnetic poles of the earth. *See* geomagnetism.

magnetic mirror A region of high magnetic field strength that reflects ions from a plasma back into a *magnetic bottle. *See* fusion reactor.

magnetic moment 1. Symbol: m. A property possessed by a *permanent magnet or a current-carrying coil, used as a measure of the magnetic strength. It is the torque experienced when the magnet or coil is set with its axis at right angles to a magnetic field of unit size. The torque can thus be expressed as the *vector product $m \times H$ of magnetic moment and magnetic field strength; m is then often called the *magnetic dipole moment*. It is measured in weber metres. The torque is also expressed as the vector product $m \times B$, where B is the magnetic flux density; m is then often called the *electromagnetic moment*, and is measured in amperes per metre squared.

2. Of a particle. Symbol: μ. A property of a particle arising from its *spin. It is measured in $A\,m^2$ or $J\,T^{-1}$. The electron magnetic moment, symbol: μ_e, is very nearly equal to μ_B, where μ_B is the Bohr *magneton. It has the value

$$9.284\,770 \times 10^{-24}\,J\,T^{-1}.$$

In a system of particles, such as an atom, a particle also has a magnetic moment associated with its orbital motion in the system. The magnetic moment of an orbital electron is equal to $l\mu_B$, where l is the orbital quantum number.

magnetic monopoles Hypothetical magnetic particles (analogous to the electrical particles, the electron and proton) with a magnetic charge of either north or south. They have been postulated on conservation and symmetry principles: an electric particle gives rise to an electric field, and when set into motion gives rise to a magnetic field; a magnetic particle should give rise to a magnetic field and, in motion, produce an electric field. Monopoles would emit and absorb electromagnetic radiation, as do electrons, and could be produced as a pair of monopoles by energetic photons (*see* pair production). Neither *quantum theory nor classical electromagnetic theory bars the existence of the magnetic monopole. *Maxwell's equation would prove completely symmetrical if such particles did exist. They are thought to be more massive than *nucleons. They could be created by the interaction of very high energy particles. In particular, certain *gauge theories predict their existence with Higgs bosons (*see* electroweak theory). Some *grand unified theories also predict monopoles with masses of 10^{16} GeV. Despite these reasons for their existence, and intensive searches for them, no individual monopoles have yet been detected.

magnetic pole strength A measure no longer generally employed for the intensity of a magnet. Magnetic pole strength was originally explained by applying the inverse square law of forces to the poles possessed by a magnet. It can best be understood as the ratio of *magnetic (dipole) moment to magnetic length, and is measured in webers. Modern practice favours the use of magnetic moment in place of this quantity, since the precise location of the poles (regions of concentrated magnetism) is indefinite, and hence the exact magnetic length is unknown.

magnetic potential Former name for magnetomotive force.

magnetic potential difference Symbol: U_m or U. The difference between the magnetic states of two points in a magnetic field. It equals the line integral of the magnetic field strength between the two points. In general, when electric currents are present, this line integral is many-valued and the concept of magnetic potential difference is invalid. It is, however, applicable to regions having boundaries that make it impossible for any closed path to link an electric current.

magnetic quantum number *See* atomic orbital; spin.

magnetic recording In magnetic sound recording, a continuously moving iron-oxide-impregnated plastic tape is longitudinally magnetized so that variations in magnetization represent variations that occur in the audiofrequency currents. If the tape is fed past suitable electromagnets, currents are induced in the coils corresponding to the original magnetizing currents. In practice, microphone currents are amplified electronically and fed to coils surrounding magnetic poles shaped so that a very small length of the recording medium completes the *magnetic circuit. The recording medium is moved at a uniform speed past the recording head. A reproducing head of similar design to the recording head is used to transform the magnetic flux variations into small current variations, which are then amplified and fed to a loudspeaker system. The record is reasonably permanent, but may be easily erased by passing the tape close to an electromagnet carrying a large direct current that magnetizes the material uniformly.

Magnetic recording of information is used extensively in *computer technology. Data can be stored on and retrieved from *magnetic tape and *disk.

magnetic resonance imaging (MRI) A technique that is based on *nuclear magnetic resonance of protons, and is used in diagnostic medicine to produce images (proton-density maps) of the body.

magnetic saturation *See* saturation.

magnetic screening A process whereby an area may be screened from magnetic effects by enclosing it with material of high *permeability.

magnetic shunt A piece of magnetic material mounted near a magnet in an electrical measuring instrument, and having means whereby its position relative to the magnet can be adjusted so that the useful *magnetic flux of the magnet can be varied.

magnetic storms *See* geomagnetism.

magnetic susceptibility *See* susceptibility.

magnetic tape A plastic strip coated on one side with iron oxide, used for *magnetic recording of sound or as a storage medium for information in computing.

In the former case, the tape is ¼″ wide with one, two, or four separate recording tracks.

Tape used in computing is essentially the same as that used in domestic audio and video cassettes. Binary information is stored in the form of rows of magnetized dots, typically 7 or 9 across the tape, and up to tens of thousands per inch along it. The tape is wound on a plastic or metal former, commonly in 2400 ft lengths. Items of data are recorded and retrieved from magnetic tape by read and write heads in a device called a (*magnetic*) *tape unit*, the heads operating in accordance with signals from the computer.

magnetic vector potential *Syn.* vector potential. Symbol: *A*. A quantity defined so that curl $A = B$, where B is the *magnetic flux density.

magnetic viscosity In most ferromagnetic substances, there is a time lag between application of a magnetic field and the resulting magnetization, which is accounted for by the eddy currents induced in the substance. In some materials, however, the persistence and magnitude of the change of magnetization are much too great to be accounted for in this way, and this phenomenon is called magnetic viscosity.

magnetic well A configuration of magnetic fields for containing a *plasma in experimental fusion devices. *See* fusion reactor.

magnetism A property that all materials possess as a result of the motion of their electrons. *Diamagnetism is a weak effect common to all substances and results from the orbital motion of electrons. In certain substances this is masked by a stronger effect due to electron *spin, *paramagnetism. Some paramagnetic materials, such as iron, also display *ferromagnetism (*see also* ferrimagnetism; antiferromagnetism).

A *magnetic field can be produced by an electric current (*see* electromagnet) or by a permanent *magnet. The earth also has a magnetic field (*see* geomagnetism).

magnetite *Syn.* lodestone. A naturally occurring form of ferrous-ferric oxide ($FeO.Fe_2O_3$) that shows magnetic properties.

magnetization Symbol: M. The difference between the ratio of the *magnetic flux density (B) to the *magnetic constant (μ_0) and the *magnetic field strength (H):
$$M = B/\mu_0 - H.$$
It is measured in amperes per metre.

magnetization curves The magnetic properties of ferromagnetic substances are usually studied by drawing curves relating the magnetization of the material to the strength and variations of the magnetizing field. *See* ferromagnetism; hysteresis.

magneto An electrical generator, usually one in which the *magnetic field is provided by a permanent magnet. It can produce periodic high-voltage pulses, and can thus be used to provide the ignition in internal-combustion engines.

magnetobremsstrahlung *See* synchrotron radiation.

magnetocaloric effect *Syn.* thermomagnetic effect. A fall in temperature occurring when a paramagnetic substance suffers *adiabatic demagnetization. It increases as the initial temperature of the substance is lowered so that the effect has been used for the production of temperatures approaching absolute zero. At very low temperatures paramagnetic substances become antiferromagnetic, restricting further cooling.

magnetodamping An increase in internal damping of acoustic vibrations in a metal such as nickel when it is subjected to a strong magnetic field.

magnetohydrodynamics (MHD) The study of the behaviour of a conducting fluid (e.g. an ionized gas, *plasma, or collection of charged particles) under the influence of a *magnetic flux. The motion of the fluid gives rise to an induced electric field that interacts with the applied magnetic field, causing a change in the motion itself.

A *magnetohydrodynamic generator* is a source of electrical power in which a high-speed flame or plasma flows between the poles of a magnet. The free electrons in the flame constitute a current when they flow, under the influence of the magnetic field, between electrodes in the flame. The concentration of free electrons in the flame is increased by adding elements of low *ionization potential, such as sodium or potassium.

magnetometer Any of a variety of instruments for comparing *magnetic field strengths (H) at different places, or for measuring an absolute value of H (*absolute magnetometer*). An early form, still used, consists of a short magnet, freely suspended by a jewelled pivot as in a compass needle, and carrying a long pointer, which moves over a graduated circle. The pivoted needle is deflected from its N–S direction by a magnet placed near to it. The angle of deflection is a function of the field strength. More precise instruments are now available.

magnetomotive force (mmf) Symbol: F_m. The circular integral of the *magnetic field strength, H, round a closed path:
$$F_m = \oint H \, dx.$$
It is measured in amperes.

magneton A fundamental constant, the intrinsic *magnetic moment of an electron. The circulatory current created by the angular momentum p of an electron moving in its orbit produces a magnetic moment $\mu = ep/2m$, where e and m are the charge and mass of the electron. By substituting the quantized relation $p = jh/2\pi$ (h = the Planck constant; j = magnetic quantum number), $\mu = jeh/4\pi m$. When j is taken as unity the quantity $eh/4\pi m$ is called the *Bohr magneton*, symbol: μ_B; its value is

9.274 0780 $\times$ 10^{-24} A m^2.

According to the *wave mechanics of Dirac, the magnetic moment associated with the *spin of the electron would be exactly one Bohr magneton, but *quantum electrodynamics shows that there is a small difference.

The *nuclear magneton*, μ_N, is equal to $(m_e/m_p)\mu_B$, where m_p is the mass of the proton. The value of μ_N is

5.050 8240 $\times$ 10^{-27} A m^2.

The magnetic moment of a proton is, in fact, 2.792 85 nuclear magnetons.

magneto-optical effects Optical phenomena resulting from the presence of a magnetic field. They include the *Zeeman, *Kerr, *Faraday, *Voigt, and *Cotton–Mouton effects.

magnetopause *See* magnetosphere.

magnetoresistance The change in electrical resistance that *ferromagnetic substances undergo when magnetized. It is closely associated with the change in resistivity (*elastoresistance) caused by tension within the elastic limit of the materials, and is also associated with their *magnetostriction. In materials with negative magnetostriction, the effects of tension and longitudinal magnetic field are in opposite directions, resistivity increasing under an applied field.

magnetosphere A region surrounding the earth and most of the other planets in which ionized particles are controlled by the magnetic field of the planet. It is bounded by the *magnetopause* and includes any radiation belts, such as the *Van Allen belts of the earth. The earth's magnetosphere extends to some 60 000 km on the sunward side but is drawn out to many times this distance on the side away from the sun by the *solar wind.

magnetostriction Stresses of compression or extension experienced by a body in a magnetic field and that in ferromagnetic materials are sufficiently great to cause mechanical deformation. Conversely, when such a ferromagnetic material is subjected to mechanical stress, a change in its *permeability occurs. An increase in length of a ferromagnetic rod on application of an axial magnetic field is described as *positive* (or *Joule*) *magnetostriction*; *negative magnetostriction* occurs in materials that decrease in length as the field density increases. (*See also* Guillemin effect; Wiedemann effect.)

Magnetostriction is best demonstrated when a bar of nickel or iron is magnetized by a coil carrying a direct current on which is superimposed an alternating current. If the a.c. frequency coincides with the natural frequency of the rod, the mechanical vibrations have large amplitudes. This is used to generate acoustic waves with frequencies ranging from audible to ultrasonic, depending on the dimensions and mode of vibration of the rod. Such vibrations have many applications, especially at ultrasonic frequencies when they can be used, for example, to hasten certain chemical reactions, break the oxide film on aluminium for soldering, or for cleaning purposes.

Magnetostriction oscillators use this principle (rod plus a.c. coil) to produce frequency-controlled oscillations at frequencies from 25 000 hertz downwards. A *tuned circuit is incorporated so that when its frequency is tuned to correspond to the natural frequency of the ferromagnetic rod, the oscillations set up can be maintained. A magnetostrictive rod can also be used to control the frequency of an oscillator (in a similar manner to a crystal-controlled *piezoelectric oscillator). The frequency of the oscillator is adjusted to be close to the natural frequency of the rod; the induced vibrations in the rod are used to pull the oscillator frequency to the rod's natural frequency and maintain it at a substantially constant value.

There are many other devices whose operation is based on magnetostriction; these include *magnetostriction transducers*, *loudspeakers*, *microphones*, and *filters*.

magnetostriction oscillator *See* magnetostriction.

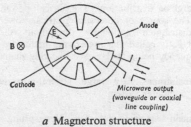

a Magnetron structure

magnetron An *electron tube that produces high-frequency microwave oscillations. The typical structure of a magnetron is shown in Fig. *a*. The cylindrical cathode is surrounded by a co-axial cylindrical anode with *cavity resonators in its inner surface. The whole tube is placed in a uniform magnetic field parallel to the cylindrical axis. Electrons are emitted from the heated cathode and in the absence of the magnetic field would travel radially to the anode under the influence of the

electric field of the anode. The presence of the magnetic field, however, causes them to head into a cycloidal path (*see* Fig. *b*). The maximum distance an electron can travel towards the anode is determined by the magnetic field. The critical field is reached when electrons just fail to reach the anode. A sufficiently large magnetic field turns most of the electrons back towards the cathode, resulting in a sheath of electrons rotating about the cathode.

No field   Weak field   Critical field   Strong field

b Effect of magnetic field on electrons in magnetron

Due to the structure of the anode, the fields associated with this electron cloud induce radio-frequency (r.f.) fields in the cavity resonators, and this field further interacts with the electrons. Depending on the point of interaction with the r.f. field, electrons either travel towards the anode in "spokes" and give up kinetic energy to the field, or they are turned back towards the cathode. The kinetic energy received from electrons by the r.f. field is greater than the power required to turn electrons back to the cathode and the result is a net power gain by the r.f. fields. The closed nature of the circuit provides built-in positive feedback and oscillations can occur. The frequency of the oscillations depends critically on the geometrical structure of the anode and the magnitudes of the electric and magnetic fields. The electrons returned to the cathode cause *back heating* of the cathode reducing the *heater current required when the tube is running, and also stimulate *secondary emission of electrons, which forms a significant portion of the total electron emission.

magnification (optics) Symbol: *M*. When the word is unqualified it refers either to the *magnifying power of an instrument, or to the *lateral magnification* of the image. Lateral magnification is the ratio *y'/y*, where *y* is the height of the object perpendicular to the axis, and *y'* the corresponding height of the image.

magnifying glass *Syn*. magnifier. *See* microscope.

magnifying power (MP) *Syn*. instrument magnification. The ratio of the size of the retinal image of an object seen with an instrument, to the size of the retinal image of the object seen with the unaided eye. This is equivalent to the ratio of the angle subtended at the eye by the image of the object as seen through the instrument, to the angle subtended by the object (*a*) *in situ* for *telescopes, (*b*) when placed 25 cm from the eye for *microscopes; this latter ratio is often called the *angular magnification*.

magnitude A means of expressing the brightness of astronomical bodies, the brighter the body, the lower its magnitude (which may have a negative value). It is based on a logarithmic scale.

The *apparent magnitude*, symbol: *m*, is the magnitude as observed, correcting for atmospheric absorption. Its value depends on the body's *luminosity, its distance, and the amount of light absorption by intervening *interstellar matter. Two bodies of luminous intensities I_1, I_2 will have magnitudes m_1, m_2 related by

$$m_1 - m_2 = 2.5 \log_{10} (I_2/I_1).$$

Thus a difference of five magnitudes means a hundred times the luminous intensity. The reference point of the scale has been fixed as *m* = 0 (in the visible region of the spectrum) when *I* = 2.65 × 10^{-6} lux.

The *absolute magnitude*, *M*, is the apparent magnitude the body would have

at a distance of ten parsecs from the observer. It can be shown that

$$M = (m - 5) + 5 \log_{10} x,$$

where *x* is the distance of the body in parsecs.

Magnox A group of proprietary magnesium alloys used to encase the *fuel elements in certain types of *nuclear reactors. (*See* gas-cooled reactor.) Magnox A consists of magnesium containing 0.8% aluminium with 0.01% beryllium.

Magnus effect If a cylinder or sphere rotates about its axis while at the same time it is in relative motion with a fluid, there is a resultant force on the cylinder or sphere perpendicular to the direction of relative motion. Hence a spinning shell or golf ball is diverted from its direction of propagation.

main (electrical) A conductor or group of conductors used for the transmission and/or distribution of electrical power. The source of domestic electrical power distributed nationally throughout the UK is called the *mains*, the mains frequency being 50 Hz. *See also* ring main.

mainframe Any large general-purpose *computer system.

main-sequence star *See* Hertzsprung–Russell diagram.

main store *See* memory.

majority carrier In a *semiconductor. The type of *carrier constituting more than half of the total charge-carrier concentration.

make-and-break A type of switch that is automatically activated by the circuit in which it is incorporated, and that repetitively makes and breaks the circuit, i.e. rapidly closes and opens the circuit. It is used, for example, in an electric-bell circuit.

Maksutov telescope A telescope consisting of a concave spherical mirror, the spherical *aberration of which is reduced by a convexo-concave *meniscus lens positioned in front of the mirror. The image is either focused onto a curved photographic plate or is formed outside the telescope by an additional optical system such as that used in the Cassegrain telescope (*see* reflecting telescope).

malleability A property of a metal whereby it can be shaped when cold by hammering or rolling. Gold is the most malleable metal.

Malter effect If a layer of semiconductor of high *secondary emission ratio (e.g. caesium oxide) is separated by a thin film of insulator (e.g. aluminium oxide) from a metal plate, it can become strongly positively charged on electron bombardment. The potential may be up to 100 volts with an insulating layer of about 0.1 μm thick.

Malus's law The transmission of *plane-polarized light from a *polarizer through a second identical polarizer varies as $\cos^2\theta$, where θ is the angle between the transmission axes of the two polarizers. The transmission is thus zero when $\theta = 90°$; the polarizers are then said to be *crossed*.

manganin An alloy of 15–25% Mn, 70–86% Cu, and 2–5% Ni, that has a high electrical resistivity (about 38 ohm metre) and low temperature coefficient of resistance. It is used for electrical resistances. Manganin has the advantage over cheaper resistor alloys (*see* constantan) that it gives a very small thermoelectric e.m.f. with copper.

Mangin mirror A diverging *meniscus lens, silvered on the outer convex surface, used with signalling lamps and searchlights to throw out parallel light. A combined refraction and reflection results in a system corrected for spherical aberration and coma.

manometer A device for measuring a fluid pressure or fluid-pressure differences. *See* pressure gauges; liquid-column manometers; micromanometer.

Mariotte's law The name often given in France to *Boyle's law.

mark-space ratio In a pulse waveform, the ratio of the duration of the pulse to the time between pulses. In a perfect *square wave the mark-space ratio is unity.

maser (*m*icrowave *a*mplification by *s*timulated *e*mission of *r*adiation) Any of a class of microwave amplifiers and oscillators that operate by the same principles as the *laser, but with the beam occurring in the *microwave region of the spectrum. The first maser (1951) predated the construction of a laser by several years. Masers generate less *noise than other types of oscillator or amplifier, producing monochromatic *coherent radiation in a narrow beam, which can have very high energy density. A variety of gas masers and solid-state masers exist.

mask A means of shielding selected areas of a semiconductor chip during the manufacture of *semiconductor components and *integrated circuits. The circuit layout is described on a set of photographic masks, which are used during the *photolithography process to define the patterns of openings in the oxide layer through which the various diffusions are made, the windows through which the metal contacts are formed, and the pattern in which the desired metal interconnections are formed.

In the manufacture of *thin-film circuits, metal-foil masks are used to define the pattern of material deposited as a thin film, by vacuum evaporation, onto a substrate.

Mason's hygrometer A *hygrometer of the wet and dry bulb type used as the standard British instrument. The instrument must be exposed to an air draught of 1 to 1.5 metres/second.

mass Newton (1687) defined mass as the quantity of matter of a body, expressed as the product of volume times density. For example, a ball of wool is assumed to have the same amount of matter, whether it is closely compressed or allowed to spread out into a large volume. *Newton's laws of motion assume that mass so defined represents the *inertia* of a body, i.e. its resistance to *acceleration. In principle the masses of two bodies can be compared by comparing their inertias: when subject to equal forces the ratio of the masses is equal to the inverse ratio of their acceleration. Newton's theory of *gravitation assumes that all free bodies have the same acceleration in a gravitational field, hence the *gravitational mass* can be identified with the inertia. This identity is assumed also in the general theory of *relativity.

In practice masses are compared (by weighing or by their inertial properties) indirectly with the international prototype *kilogram.

Einstein (1905) showed by his theory of relativity that the mass of a body m as measured by an observer moving with a speed v with respect to it is given by
$$m = m_0 / \sqrt{(1 - v^2/c^2)},$$
where m_0 is the mass measured by an observer at rest with respect to the body, and c is the speed of light. From

this result, and assuming that energy is rigorously conserved in all processes, he showed that the transfer of energy E entails the transfer of mass m where $E = mc^2$. This leads to the conclusion that mass also is conserved.

Mass must be distinguished from *amount of substance, which is defined in terms of the number of constituent particles in a body. Although different observers with different relative motions determine different values for a mass, they find the same value for the amount of substance, since this in principle is purely a question of counting.

mass absorption coefficient *See* linear absorption coefficient.

mass action, law of The principle that the speed of a chemical reaction between substances is proportional to the product of their concentrations. If a chemical reaction is reversible, it reaches a state of dynamic equilibrium in which the concentrations of all the substances present become constant with time. For a particular reaction
$$aA + bB \rightarrow cC + dD$$
the concentrations [A], [B], [C], and [D] at equilibrium are related by the equation:
$$\frac{[C]^c [D]^d}{[A]^a [B]^b} = K,$$
where K is the *equilibrium constant* for that reaction at a given temperature.

mass defect The difference between the sum of the rest masses of the constituent *nucleons of a particular nucleus and the mass of the nucleus itself. This difference is due to the emission of energy when the nucleus is formed. Energy must therefore be supplied to the nucleus to break it up into its constituents. This energy, the energy equivalent of the mass defect, is the *binding energy of the nucleus.

mass–energy equation Einstein's equation, $E = mc^2$. *See* mass; relativity.

mass excess *See* mass number.

Massieu function Symbol: J. The quantity $-A/T$, where A is the *Helmholtz function and T the thermodynamic temperature.

mass–luminosity law A theoretical law relating the mass m and total outflow of radiation, or luminosity, L of normal stars. The law may be represented by the approximation
$$\log L = 3.3 \log m,$$
where m and L are in *solar units.

mass moments *See* centre of mass.

mass number *Syn.* nucleon number. Symbol: A. The number of *nucleons in the nucleus of a particular atom. It is the number nearest to the atomic mass, m_a, of a nuclide. The difference $(m_a - Am_u)$ is the *mass excess*, where m_u is the unified *atomic mass constant.

mass reactance *See* reactance (acoustic).

mass resistivity The product of the mass and electrical resistance of a conductor, divided by the square of its length. The units are kilogram ohms per metre squared.

mass spectrograph *See* mass spectrum.

mass spectrometer *See* mass spectrum.

mass spectrum The separation of a beam of gaseous ions into components with different values of mass divided by charge. Most measurements are made with singly charged positive ions so the

spectrum is divided simply according to mass.

The apparatus is normally highly evacuated except for the ion source. Ions are separated by various combinations of electric and magnetic fields and are focused onto a detector. A common system is that in which ions are accelerated from rest through a p.d. V, giving kinetic energy $\frac{1}{2}mv^2 = eV$. The ion beam then passes between the poles of a magnet and is deflected into a circular arc of radius R given by:
$$R = mv/Be,$$
where B is the magnetic flux density. Thus ions with different values of m/e are deviated by different amounts.

When the ion beams are detected with a photographic plate the instrument is called a *mass spectrograph*. This is suitable for the precise measurement of the relative masses of the ions, and in particular is used for the determination of the masses of *isotopes.

When the ion beams are detected with an electrometer the instrument is called a *mass spectrometer*. This is suitable for the precise measurement of the relative abundances of the ions. It may be used to find the relative abundances of the isotopes of an element, for chemical analysis, or for the measurement of *appearance potentials.

By using alternating electric fields, it is possible to select ions of different masses according to their times of flight, thus avoiding the need for a heavy magnet and thereby making possible the use of small portable instruments.

mass stopping power *See* stopping power.

master oscillator The frequency of an oscillator depends to a certain extent upon the load on the oscillator. When a high degree of frequency stability is required, it is usual to employ an oscillator of high inherent frequency stability to drive a power amplifier, the latter supplying the power to the load. An oscillator used in this manner is called a master oscillator. When the power amplifier consists of several stages, that which follows the master oscillator is usually designed to operate as a *buffer.

matched load *See* matched termination.

matched termination In a network or transmission line. A termination at which no reflected waves are produced. A load that absorbs all the power incident from a transmission line and forms a matched termination is called a *matched load*. *See also* impedance matching.

matrix A mathematical concept that is similar to a determinant but differs from it in not having a numerical value in the ordinary sense of the term. It obeys the same rules of multiplication, addition, etc. An array of mn numbers set out in m rows and n columns is a matrix of order $m \times n$. The separate numbers are usually called elements. Such arrays of numbers, treated as single entities and manipulated by the rules of matrix algebra, are of use wherever simultaneous equations are found (e.g. changing from one set of Cartesian axes to another set inclined to the first; quantum theory; electrical networks). Matrices are very prominent in the mathematical expression of quantum mechanics.

matrix mechanics A mathematical form of *quantum mechanics that was developed by Born and Heisenberg and originated simultaneously with but independently of *wave mechanics. It is equivalent to wave mechanics, but in it the *wave functions of wave mechanics are replaced by *vectors in a suitable space (Hilbert space) and the observable things of the physical world, e.g. energy,

momenta, coordinates, etc., are represented by *matrices.

The theory involves the idea that a measurement on a system disturbs, to some extent, the system itself. With large systems this is of no consequence, and the system obeys the rules of classical mechanics. On the atomic scale, however, the result depends on the order in which the observations are made. Thus if p denotes an observation of a component of momentum and q an observation of the corresponding coordinate, then $pq \neq qp$. Here p and q are not physical quantities but operators. In matrix mechanics they are matrices and obey the relationship

$$pq - qp = \mathrm{i}h/2\pi,$$

where h is the Planck constant and $\mathrm{i} = \sqrt{-1}$. This leads to the quantum conditions for the system. The matrix elements are connected with the transition probabilities between various states of the system. *See also* uncertainty principle.

matter waves *See* de Broglie waves.

Matthiessen's rule The product of the *resistivity and temperature coefficient of resistance of a metal is the same whether the metal be pure or impure. Normally impurities and alloying elements increase the resistance of a metal markedly, but this effect is accompanied by a corresponding decrease in change of resistance with temperature. The rule is not exact but is often a useful approximation.

maximum and minimum thermometer An alcohol thermometer that records both the highest and the lowest temperatures reached since setting the thermometer. Movement of the mercury in the U-tube, due to the expansion or contraction of the alcohol (or spirit) in the bulb A, causes the mercury to push tiny steel indicators I along the tubes. These

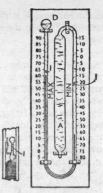

Maximum and minimum thermometer

indicators remain in position if the mercury meniscus recedes, being held against the walls of the tube by tiny springs. The lower ends thus indicate maximum and minimum temperatures. The thermometer is reset by using an external magnet to draw the indicators into contact with the mercury.

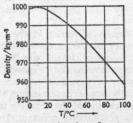

Maximum density of water

maximum density of water Water at 0 °C when heated, contracts until the temperature is 4 °C after which it expands normally (*see* diagram). Owing to the hydrogen bonds between water molecules, ice crystals have a very open three-dimensional tetrahedral structure. When ice melts, this structure collapses and the water molecules become closer packed; small aggregates of molecules can, however, continue to survive up to

4 °C. Thus water has a maximum density at 4 °C.

maximum permissible dose *See* dose.

maxwell Symbol: Mx. The *CGS-electromagnetic unit of magnetic flux, now replaced by the *weber. 1 Mx = 10^{-8} weber.

Maxwell–Boltzmann law *See* distribution of velocities.

Maxwell distribution *See* distribution of velocities.

Maxwell's demon An imaginary creature to whom Maxwell assigned the task of operating a door in a partition dividing a volume containing gas at uniform temperature. The door was opened to enable fast molecules to move (say) from left to right through the partition. In this way, without expenditure of external work, the gas on the right could be made hotter than before and that on the left made cooler.

The concept was presented to illustrate how (in principle) it might be possible to violate the second law of *thermodynamics. According to modern ideas the demon would have to interact with the molecules by means of radiation in order to determine their speeds, and would therefore cause other changes, which would invalidate the argument.

Maxwell's equations A series of classical equations that govern the behaviour of electromagnetic waves in all practical situations. They connect vector quantities applying to any point in a varying electric or magnetic field. The equations are:

curl H	$= \partial D/\partial t + j$
div B	$= 0$
curl E	$= -\partial B/\partial t$
div D	$= \rho$

291

H is the *magnetic field strength, D is the *electric displacement, t is time, j is the *current density, B is the *magnetic flux density, E is the *electric field strength, and ρ is volume density of charge.

From these equations, Maxwell demonstrated that each field vector obeys a wave equation: he showed that where a varying electric field exists, it is accompanied by a varying magnetic field induced at right angles, and vice versa, and the two form an electromagnetic field that could propagate as a transverse wave. He calculated that in a vacuum, the speed of the wave was given by $1/\sqrt{(\varepsilon_0\mu_0)}$, where ε_0 and μ_0 are the *permittivity and *permeability of vacuum. The calculated value for this speed was in remarkable agreement with Fizeau's measured value of the speed of light, and Maxwell concluded that light is propagated as electromagnetic waves. *See also* electromagnetic radiation; magnetic monopole.

Maxwell's formula A formula connecting the relative permittivity ε_r of a medium with its refractive index n. If the medium is not ferromagnetic, the formula is $\varepsilon_r = n^2$.

Maxwell's rule or law Unless otherwise constrained, a movable part of a circuit will be displaced so as to give the maximum possible magnetic flux linkage with the circuit.

Maxwell's thermodynamic relations The equations relating the four thermodynamic variables S, p, T, and V, referring to a given mass of a homogeneous system, namely

$$\left(\frac{\partial T}{\partial V}\right)_S = -\left(\frac{\partial p}{\partial S}\right)_V, \left(\frac{\partial T}{\partial p}\right)_S = \left(\frac{\partial V}{\partial S}\right)_p,$$
$$\left(\frac{\partial V}{\partial T}\right)_p = -\left(\frac{\partial S}{\partial p}\right)_T, \left(\frac{\partial S}{\partial V}\right)_T = \left(\frac{\partial p}{\partial T}\right)_V,$$

where S is the entropy, V is the volume, p is the pressure, T is the thermodynamic temperature.

McLeod gauge A mercury-in-glass vacuum pressure gauge in which a known large volume of gas is compressed into a small volume at which the pressure, now much larger, is measured. Being based on Boyle's law, the gauge cannot be used when condensable vapours are present. It will work down to 10^{-3} pascal and is an absolute instrument.

mean Of n numbers $a_1, a_2, a_3, \ldots a_n$:
  1. Arithmetic mean
     $= (a_1 + a_2 \ldots + a_n)/n$.
(*See* average.)
  2. Geometric mean
     $= (a_1 a_2 a_3 \ldots a_n)^{1/n}$.
  3. Root-mean-square value
$= \sqrt{\{(a_1{}^2 + a_2{}^2 + \ldots + a_n{}^2)/n\}}$.
  4. *See* weighted mean.

mean current density *See* current density.

mean density of matter In the universe. The factor that determines the dynamical behaviour of the universe, i.e. whether it is a continuously expanding open system or a closed system that must eventually stop expanding and contract. It is a function of the *Hubble constant and the gravitational constant. The critical density, which if exceeded will lead to an eventual halt to expansion, is about 5×10^{-27} kg m^{-3} (for a Hubble constant of 55 km s^{-1} Mpc^{-1}).

mean deviation *See* deviation.

mean free path Symbol: λ. **1.** In *kinetic theory, the mean distance that a molecule moves between two successive collisions with other molecules. It is related to the molecular cross section $\pi\sigma^2$ by the relationship:

$$\lambda = 1/\sqrt{2\pi n\sigma^2},$$

where n is the number of molecules per unit volume. The most important means of determining λ is through its connection with viscosity η. According to kinetic theory

$$\lambda = k\eta/\rho u,$$

where ρ = density and u = mean molecular velocity. The value of k lies between $\frac{1}{3}$ and $\frac{1}{2}$ according to the degree of approximation introduced into the theory.

2. In atomic, nuclear, and particle physics, the mean distance that a particle travels in a medium before undergoing a particular type of interaction. For example, there are mean free paths for absorption, elastic scattering, inelastic scattering, fission, etc., for various types of particle and medium. If the number of target particles per unit volume is N and the *cross section for the particular process is σ then $\lambda = 1/N\sigma$.

mean lethal dose *See* median lethal dose.

mean life *Syn.* average life or lifetime. Symbol: τ. 1. The average time for which the unstable nuclei of a radionuclide exist before decaying. It is the reciprocal of the *decay constant and is equal to $T_{1/2}/0.693\ 15$, where $T_{1/2}$ is the *half-life.

2. The average time of survival for an elementary particle, ion, etc., in a given medium or a charge carrier in a *semiconductor.

mean solar time *See* time.

mean spherical intensity *See* luminous intensity.

mean square velocity The average value of the square of all the velocities of a system of particles, given by the relation:

$$C^2 = (n_1 c_1{}^2 + n_2 c_2{}^2 + n_3 c_3{}^2 + \ldots$$
$$n_r c_r{}^2)/n,$$

where n_1 particles have velocity c_1, n_2 particles have velocity c_2, etc., and

$$n = \sum_r^1 n_r$$

is the total number of molecules.

Its value for a gas may be calculated on the *kinetic theory from the expression $p = \frac{1}{3}\rho C^2$, where p and ρ are the pressure and density respectively of the gas. For an ideal gas $C^2 = 3rT$, where r is the gas constant for unit mass of the gas. This expression shows that the mean velocity is dependent only on the temperature of a given gas.

By the Maxwell *distribution of velocities, for the molecules of a gas in a steady state, the mean square velocity has the value $C^2 = 3kT/m$.

Compare mean velocity.

mean sun *See* time.

mean velocity The average value of the velocities of a system of particles, given by the relation

$$\bar{C} = (n_1 c_1 + n_2 c_2 + \ldots n_r c_r)/n,$$

where n_1 particles have velocity c_1, n_2 particles have velocity c_2, etc., and

$$n = \sum_r^1 n_r$$

is the total number of particles.

The Maxwell *distribution of velocities for the molecules of a gas in a steady state yields a value

$$C = 2/\sqrt{(\pi hm)},$$

where $h = 1/(2kT)$.

mechanical advantage *Syn.* force ratio. *See* machine.

mechanical equivalent of heat *Syn.* Joule's equivalent. Symbol: J. Before 1948 the unit generally employed for the measurement of heat and internal energy, in some cases even those involving electrical heating, was the *calorie, in particular the "fifteen-degree calorie" defined over the range from 14.5 °C to 15.5 °C. The usual procedures of calorimetry enabled experimental determinations of such quantities as specific and latent heats to be made in terms of the calorie, but other experiments were required to relate the calorie to the mechanical units of energy, the *joule and the *erg. The mechanical equivalent of heat was defined as the ratio of an amount of work in ergs to the amount of heat in calories to which it is equivalent.

Since 1948 it has been recommended that *all* kinds of energy be measured in "mechanical units" and the mechanical equivalent of heat has been given formally defined values (for the different 1 °C ranges) that represent the experimentally determined values very closely. For the fifteen-degree calorie

$$J = 4.1855 \pm 0.0005 \text{ J cal}_{15}{}^{-1}.$$

The recommendation concerning the unit of heat and internal energy has only been followed generally since the late 1960s. Until this time the calorie employed had been the *international table calorie, whose definition is derived from the formally adopted value for the mechanical equivalent of heat:

$$J = 4.1868 \text{ J cal}_{IT}{}^{-1}.$$

mechanical equivalent of light *Radiant flux expressed in mechanical units, which are equivalent to the unit of *luminous flux, at the wavelength of maximum visibility. It is 0.0015 watts per *lumen at 555 nm. Its reciprocal is also quoted with the same title (660 lumens per watt).

mechanical impedance *See* impedance.

mechanics The branch of science, divided into dynamics, statics, and kinematics, that is concerned with the motion and equilibrium of bodies in a particular frame of reference. *See also* wave mechanics; quantum mechanics; statistical mechanics.

median The central term of a sequence of values arranged in order of magnitude.

median lethal dose (MLD) The absorbed *dose of ionizing radiation that will kill, in a prescribed time, half of a large population of a particular species.

medical physics The application of physics to medicine, as in *radiotherapy, *nuclear medicine, diagnostic physics, dosimetry, and medical electronics. It is an extremely wide field that is constantly expanding with the development of more and more sophisticated medical techniques.

medium frequency (MF) *See* frequency bands.

medium-wave Designating radio waves with wavelengths in the range 0.1–1 km, i.e. with frequencies of 3–0.3 MHz.

mega- Symbol: M. **1.** A prefix meaning 10^6 (i.e. one million); for example, one megahertz (1 MHz) is 10^6 hertz.
2. In situations where the binary number system is used, such as computing, a prefix meaning 2^{20} (i.e. 1 048 576); for example one megabyte is 2^{20} bytes.

megaphone An instrument for amplifying and directing sound. It consists of a conical or rectangular *horn about 30 cm long, the small end of which is held near the mouth of the speaker. The horn increases the efficiency of the voice by providing a suitable loading for it. Provided the solid angle of a conical horn is not large, the wave front of the sound emerging from the open end is almost plane. Owing to the practical limitations on size, the low-frequency components of speech are not radiated efficiently and there is also very little directive effect at these frequencies.

megger A portable insulation tester calibrated directly in megaohms.

Meissner effect *See* superconductivity.

melting *See* fusion.

melting point *See* freezing point.

memory *Syn.* store; storage. A device or medium in which data and *programs can be held for subsequent use by a *computer. It may be either *backing store or *main store* (also called *main memory* or simply *memory*). Main store now consists of *semiconductor memory – either *RAM (random-access memory) or *ROM (read-only memory). Main store is closely associated with the central processor of the computer: program instructions and associated data are stored there temporarily, awaiting use by the processor.

The memory is divided into *storage locations*, each of which can be uniquely identified by its *address*; each location holds the same number of *bits, usually 8, 16, or 32 (*see* word; byte). The processor can retrieve information from a particular location extremely rapidly.

Programs can only be executed when they are in main store. They and their associated data are not held permanently in main store, however, but are kept in larger-capacity backing store (usually magnetic *disks or tapes) until required by the processor.

meniscus A concave or convex upper surface of a liquid column that is due to capillary action.

meniscus lens A convexo-concave or concavo-convex *lens. Such types in spectacle lenses are usually called *deep* meniscus lenses and in general provide a wider field of good definition.

mercury barometer *See* barometer.

mercury-in-glass thermometer A type of thermometer in which mercury acts as the thermometric fluid in a glass bulb attached to a graduated fine capillary tube. During manufacture all air is excluded from the capillary tube, a small bulb being left at the top of the tube as a safeguard against breakage should the thermometer be raised to a temperature beyond the highest value on the graduated scale. The thermometer is calibrated by immersion first in melting ice, then in steam in a *hypsometer, the positions of the mercury meniscus being marked. In the Celsius (centigrade) thermometer, the distance between the marks is divided into 100 equal parts, each part corresponding approximately to a degree Celsius on the mercury in glass scale. The corresponding Fahrenheit scale is often marked as well.

Although it has the advantage of giving a direct reading that is easily read, for accurate work so many corrections have to be applied that this type of thermometer has been replaced by the platinum *resistance thermometer. Since the coefficient of expansion of mercury is not independent of temperature and since the expansion of the glass is not negligible, the thermometer readings can only be corrected to the gas scale by a direct comparison with a *gas thermometer. The chief errors for which correction is necessary are nonuniformity of the bore of the capillary tube; errors in marking the ice and steam points; the effect of external pressure on the bulb; the emergent stem correction. *See also* International Practical Temperature Scale.

mercury switch A switch in which contact is established between two mercury surfaces usually enclosed in a glass tube. Arcing is often suppressed by filling the tube with an inert gas and sometimes a porcelain tube is fused in at the point of contact to eliminate breakage from

295

heat shock. There are many types. Usually they are operated by tilting, and may have delayed make, or break, or both, by the mercury having to flow through a constriction in a side tube. The mercury may flow over a series of contact pools in turn.

mercury-vapour lamp An incandescent arc of mercury vapour between mercury electrodes in an enclosed tube. The light is rich in ultraviolet radiation and in *fluorescent lamps some of this is converted to visible wavelengths by fluorescent powders coated onto the interior of the tube.

mercury-vapour rectifier An electron tube in which a discharge passes in one direction only from a hot-wire cathode to an anode via an atmosphere of ionized mercury vapour. After the initial ionization is achieved, the voltage drop across the tube is only 10 to 15 volts and is almost independent of the current. With a mercury-pool cathode, the voltage drop is 20 to 25 volts.

meridian 1. A *great circle passing through a point on the surface of a body such as the earth (or the *celestial sphere), and through the north and south poles, and crossing the equator at right angles.
2. *See* magnetic meridians.

meridian circle *Syn.* transit circle. *See* telescope.

meridian plane *Syn.* tangential plane. In an optical system. A plane that contains both the optic axis and the *chief ray (i.e. the one passing through the centre of an aperture). It is perpendicular to the *sagittal plane. Rays in a meridian plane, e.g. in oblique *astigmatism, converge to the meridian image point; the focal line at this point is perpendicular to the meridian plane, and is referred to

as the *meridian focal line* or the *tangential focal line*. The corresponding focal surface is known as the tangential surface.

mesh *See* network.

mesh connection A method of connection used in *polyphase a.c. working in which the windings of a transformer, a.c. machine, etc., are all connected in series to form a closed circuit so that a polygon may be used to represent them diagrammatically. A special form is the *delta connection. *Compare* star connection.

meson A collective name given to *elementary particles that can take part in *strong interactions and that have zero or integral spin. By definition, mesons are both *hadrons and *bosons. *Pions* and *kaons* are mesons. Mesons have a substructure composed of a *quark and an antiquark bound together by the exchange of particles known as gluons. *See also* Appendix, Table 7.

mesosphere *See* atmospheric layers.

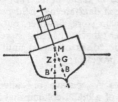

Metacentre

metacentre The point at which a vertical line through the centre of *buoyancy B′ of a tilted ship (or other floating body) intersects the line joining the centre of mass G and the centre of buoyancy B of the upright ship. If G is below M, the force of buoyancy (upwards through B′) together with the weight of the ship (downwards through G) tends to rotate the ship back to the upright position.

metallic crystals Crystals in which a regular arrangement of positive metallic ions is held together by an "atmosphere" of free electrons.

metallizing The covering of an insulating material with a film of metal or other substance to render it electrically conducting. The technique is widely used in solid-state electronics. The conducting film is etched and forms interconnections on *integrated circuits. It is also used in forming *bonding pads for integrated circuits and discrete components.

metal rectifier A *rectifier that depends for its action upon the fact that when a metal is placed in contact with a suitable solid (such as a semiconductor or an oxide of the metal), the resistance offered to the passage of an electric current is very much less when the current flow is in one direction (e.g. from the solid to the metal) than it is in the other direction. Typical materials used are cuprous oxide on copper, and selenium on copper. The potential applied must not exceed a few volts so that for high voltages a series of elements is required. The current passed depends on the area of contact.

metastable state 1. A state of pseudo-equilibrium in which a system, such as a *supersaturated vapour or supercooled liquid (*see* supercooling), has acquired more energy than that normally required for its most stable state, and yet is not unstable. It is often achieved by attaining the state very slowly. A slight disturbance will produce the stable state. Water can be cooled very slowly to a temperature below 0 °C. The addition of a piece of ice will cause the water to freeze rapidly.

2. Symbol: m. A comparatively stable *excited state of a radionuclide that decays into a more stable lower-energy state with the emission of gamma-rays. The nuclide technetium-99m decays into technetium-99, the half-life being 6 hours. It is often an excited state from which all possible transitions to lower states are *forbidden transitions according to the relevant *selection rules.

3. A comparatively stable electronically *excited state of an atom or molecule.

meteor A lump of matter from space that enters the earth's atmosphere and is detected either optically by virtue of its luminosity as a result of friction with air particles, or by radio means by virtue of the trail of ionized gas left in its wake. Isolated meteors are described as *sporadic*. More obvious are the meteor *showers*; between five and a hundred meteors per hour can be observed. Most showers can be associated with former comets, and represent the debris resulting from a break-up. Very little of the estimated 10^6 kg of meteoric material captured by the earth reaches its surface, and only a minute proportion has been recovered. Over 90% of these *meteorites* have a stony constitution; the others are either made up largely of iron and nickel alloys (about 6%) or a mixture of metals and minerals. Stone meteorites are more difficult to find than the so-called irons and stony irons. The larger meteorites can produce craters.

meteorite *See* meteor.

meteorograph A device for recording some or all of the following: temperature, relative humidity, pressure, and wind speed. It is usually carried into the upper air by a balloon or kite.

meteorology The science of the atmosphere, especially with relation to the weather and climate.

meter Any measuring instrument.

method of mixtures A method of calorimetry in which a substance is added to a calorimeter at a different temperature, the mixture being stirred to reach equilibrium at an intermediate temperature. The unknown heat capacity may be calculated by equating the heat lost by one part of the system to the heat gained by the remainder of the system since the law of conservation of energy applies if there is allowance made for heat exchange with the surroundings.

metre Symbol: m. The *SI unit of length, defined (since 1983) as the length of the path travelled by light in vacuum during a time 1/299 792 458 second.

The metre was originally intended to be one ten-millionth of the quadrant from the equator to the north pole through Dunkirk, but difficulties in measurement led to the adoption of the length of a prototype bar instead. Developments in the precision of the measurements of optical wavelengths led to the adoption of a definition in terms of the wavelength of a spectrum line of krypton. The great precision and reliability of atomic *clocks has permitted the current definition in terms of the second.

metre bridge A form of the *Wheatstone bridge in which a uniform resistance wire 1 metre long, which can be tapped at any point along its length, takes the place of two of the four resistors.

metre-kilogram-second electromagnetic system of units (MKS units) A system of absolute units (due to Giorgi (1901), following a suggestion of Maxwell), in which the fundamental units of length, mass, and time are respectively the metre, the kilogram, and the second, and in which the permeability of free space is 10^{-7} henrys per metre. In many electromagnetic equations of practical importance, a factor 4π appears. This factor can be transferred from these equations to others less commonly used by taking the permeability of free space as $4\pi \times 10^{-7}$ henrys per metre and this gives the *rationalized* MKS system of units. *SI units are based on the MKS system and have replaced this system.

metrology The branch of science concerned with the accurate measurements of the three fundamental quantities: mass, length, and time. It is often extended to mean the systematic study of weights and measurements.

MF Abbreviation for medium frequency. *See* frequency bands.

MHD Abbreviation for *magnetohydrodynamics.

mho The reciprocal ohm, formerly used as a unit of conductance. This unit is now replaced by the *siemens.

mica A mineral consisting of complex silicates, characterized by a perfect basal cleavage enabling the crystals to be split into very thin plates. It has a low thermal conductivity and high dielectric strength, being widely used for electrical insulation.

mica capacitor A *capacitor in which *mica is used as the dielectric. Mica capacitors are characterized by low loss and a near-constant capacitance over a wide frequency and temperature range.

Michelson interferometer *See* Michelson–Morley experiment.

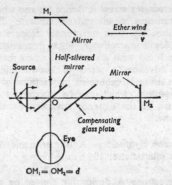

Michelson's interferometer

Michelson–Morley experiment An experiment (1887) that attempted to measure the velocity of the earth through the *ether. Using a Michelson interferometer (*see* diagram) Michelson and Morley attempted to show that there is a difference in the speed of light as measured in the direction of the earth's rotation compared to the speed at right angles to this direction. If there was such a difference the interference fringes observed in the interferometer would be shifted when the instrument was turned through 90°. This shift would correspond to a change of optical pathlength of approximately $2dv^2/c^2$, where v is the velocity of the earth with respect to the ether in the direction OM_2. No shift was observed, indicating the absence of an ether wind.

This fact was of considerable importance and was responsible for the downfall of the ether concept. Attempts to reconcile this concept with the null results of the experiment led to the postulate of the *Lorentz–FitzGerald contraction. The special theory of *relativity rejects the concept of an ether with respect to which there can be determinate motion.

micro- 1. Symbol: μ. A prefix meaning 10^{-6} (i.e. one-millionth); for example, one microsecond (1 μs) is 10^{-6} second.
2. A prefix meaning very small or concerned with very small quantities or objects.

microbalance A *balance capable of weighing very small masses (e.g. down to 10^{-5} mg). It is not practical to use such balances with standard weights. Instead, the beam is made to balance by varying the air pressure in the balance case so altering the upward buoyant force on a bulb fixed on one end of the beam. A manometer is provided for measuring the pressure when the beam is balanced; if the temperature is constant, the density of the gas (and thus the upward buoyant force) is proportional to the pressure.

Microbalances have been used for finding the relative density of a gas by measuring the pressure needed in the case to balance the beam first with the gas and then with oxygen. The density ratio is the inverse ratio of these pressures if the temperature is the same during both measurements.

microcalorimeter A differential calorimeter used for the measurement of very small quantities of heat, such as that evolved by a small quantity of radioactive substance.

microchip A *chip of semiconductor containing complex microcircuits, usually integrated circuits.

microcomputer A compact *computer system in which the central processor is fabricated on a single *chip (or a small number of chips) of semiconductor; this processing unit is called a *microprocessor*. In addition, a microcomputer contains storage and input/output facilities for data and programs on different chips, or possibly on the same chip as

299

the microprocessor. The capability of the system depends not only on the characteristics of the microprocessor but also on the amount of storage provided, the types of *peripheral devices that can be used, the possibility of expanding the system, etc.

microdensitometer A device for automatically measuring and recording small changes in *transmission density across a sample, such as a photographic plate.

microelectronics The branch of *electronics concerned with the design, production, and application of electronic components, circuits, and devices of extremely small dimensions. Increased miniaturization not only reduces size and weight but is also cost-effective, and is extremely desirable particularly in the field of *computers. *Integrated circuits are widely used in microelectronics.

microgravity The condition of near *weightlessness induced by free fall or unpowered space flight.

micromanometer A device for the measurement of very small pressure differences. The U-tube manometer has one arm of the U nearly horizontal so that pressure changes causing only a small difference in vertical height produce easily visible movements of liquid in this arm. In diaphragm gauges the two pressures are applied on either side of the diaphragm and optical methods are used to measure the tiny displacement.

micrometer eyepiece An eyepiece, generally a *Ramsden eyepiece, provided with cross-wires that can be displaced by means of a *micrometer screw. It is used for the measurement of small objects or small separations of objects, lines, etc.

micrometer screw A device for use when measuring small and/or accurate distances, e.g. with micrometer calipers or a depth micrometer. Such instruments are fitted with a drum that when rotated advances a screw of known pitch; the drum is calibrated in fractions of a revolution, which can be interpreted in terms of the distance advanced by the screw.

micron Symbol: μ. A former name for the micrometre; 10^{-6} m.

microphone A device for converting an acoustic signal into an electric signal. It forms the first element of the telephone, the broadcast transmitter, and all forms of electrical sound recorders. The types of microphone most generally used are the *carbon, *crystal, *moving-coil, *capacitor, and *ribbon microphones. Many other types exist for specialized purposes however, such as the magnetostriction, induction-coil, and hot-wire microphones. Most of these use a thin diaphragm that vibrates under the influence of the sound waves. The diaphragm is mechanically coupled to some device, the motion of which changes the properties of a component of an electric circuit or induces an e.m.f. in it. The force exerted by the sound against the diaphragm is usually proportional to the sound pressure, but in the case of the ribbon microphone it is proportional to the particle velocity. For good quality reproduction, resonance in the mechanical system of the microphone should be avoided. This is done by making the resonant frequency of the moving parts either much higher or much lower than the frequency of the sound to be reproduced. Lack of sensitivity is not a great disadvantage since it is usual to amplify the output from the microphone. In all cases a battery or other power supply is needed. Nearly all microphones have directional properties and in many these vary with frequency as a result of diffraction.

microprocessor *See* microcomputer.

microradiography *See* microscope.

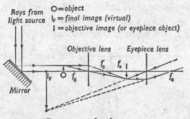

Rays from light source
O=object
I_F = final image (virtual)
I = objective image (or eyepiece object)

Objective lens   Eyepiece lens

f_0 f_e I

I_F O f_0 f_e

Mirror

a Compound microscope

microscope 1. An optical instrument for producing an enlarged image of small objects. The *simple microscope* (or *magnifying glass*) consists of a strong converging lens system, corrected for *chromatic and *spherical aberrations, and used for low-power work. The object is usually placed at the focus of the lens system, producing an image at infinity. The normal unaided eye sees an object most clearly at the near point (25 cm). The *magnifying power is then the ratio $25/f$, where f is the focal length in cm of the lens system.

The *compound microscope* (Fig. *a*) consists essentially of two lens systems and gives a much greater magnification, up to about 1500 diameters. A very short focal length *objective forms a magnified real image of the object, which is further magnified by the *eyepiece acting as a simple microscope. The total magnification is the product of the objective and eyepiece magnification. There is usually a choice of objectives on a compound microscope giving low, medium, and high magnification. Greater magnification can be obtained with oil immersion (*see* resolving power). Several types of eyepiece are in common use, including the *Huygens eyepiece, *Ram-

sden eyepiece (used with measuring microscopes), and *compensating eyepiece (used with *apochromatic objectives). The maximum *resolving power of the optical microscope is between 200 and 300 nm. The *binocular microscope* has two eyepieces. Light from the objective is split into two beams by using prisms. It has the advantage of depth perception and greater eye comfort. There can also be a third eyepiece to which a camera can be attached. The *stereoscopic microscope* has two eyepieces and two objectives, the object being viewed by reflected rather than transmitted light. The magnification is usually of the order of 100 diameters.

Illumination for low-power work with transparent objects is achieved by using a mirror mounted below the object to reflect light from the light source onto the object. For higher magnifications, a substage *condenser, such as the *Abbe condenser or a modified Abbe condenser (variable focus) is necessary. This concentrates the light within a cone of larger angle than that achieved by the mirror, being positioned between the reflecting mirror and the object. Examination of opaque objects requires illumination from above and the objective becomes, in effect, the condenser. The optical microscope can be focused sharply only in one plane so that a two-dimensional image is obtained. If the object is fairly transparent, different depths can be brought into focus, but since material above and below the plane of focus can interact with the light, the image may be blurred. The microscope therefore works best with thin samples viewed by transmitted light, or with flat samples viewed by reflected light. The shape of the object can only be obtained at low magnifications of about 200 diameters.

Details in objects are seen because of varying density regions. With a strongly lit background, a small transparent ob-

301

ject is very difficult to observe. *Dark-field illumination* increases the visibility of small objects. An opaque disc is placed over the centre of the condenser so that the borders of the object are illuminated by marginal rays, which do not enter the objective. Some light is refracted by the specimen, some of which enters the objective. A bright image is thus obtained against a dark background; however, little detail can be seen. (The same effect can be obtained by illuminating the object obliquely so that no direct light enters the objective.)

The refractive index of a transparent specimen varies slightly from point to point. These variations give rise to *diffraction patterns in the focal plane of the objective. The diffracted light passing through the object is one quarter of a wavelength out of *phase with undiffracted light, which has not been transmitted through the object. In *phase-contrast microscopy* a transparent *phase plate* is used to produce a quarter-wavelength shift in the undiffracted light. The phase shift is produced by shaping the surface of the phase plate, which may either have a shallow circular indentation around the centre or a slightly raised disc at the centre. The plate is placed at the focal plane of the objective (Fig. *b*) onto which the image of a substage annular diaphragm falls. The final image has higher contrast due to *interference between the diffracted and undiffracted beams.

In *interference microscopy* a transparent object is placed between two semi-silvered surfaces. Light passing through the object interferes with light that has not passed through, and interference patterns can be observed. It is possible to view opaque objects in a similar way. *Reflecting microscopes* use a reflecting objective rather than the conventional lens system and can focus wavelengths ranging from infrared to ultraviolet at the same point. In *ultraviolet microscopy*

(*see* resolving power) the resolving power of the microscope is increased to 100 nm by using shorter-wavelength ultraviolet radiation. The image is made visible by using a photographic plate to record it. *X-ray microscopy* (or *microradiography*) further increases the resolution, the image being recorded on film or on a fluorescent screen.

2. *See* electron microscope.

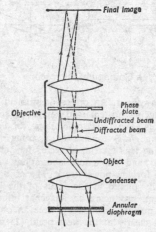

b Phase-contrast microscope

microscopic state The state of matter characterized by the actual properties of each individual elemental component. *Quantum theory is typically an analysis of the microscopic state. *Compare* macroscopic state.

microwave An electromagnetic wave with a wavelength in the range 1 mm to 0.1 m (or sometimes 0.3 m), i.e. with a frequency in the range 300 to 3 GHz (or sometimes 1 GHz). The microwave region of the electromagnetic spectrum thus lies between the infrared and radio regions, overlapping the radio region (*see* frequency band). Microwaves can be generated by devices such as the *maser, *klystron, and *magnetron. They are used in *radar, high-frequency communications, and also for very rapid cooking (in microwave ovens).

microwave background radiation *See* cosmic background radiation.

microwave spectroscopy *See* spectroscopy.

microwave tube An *electron tube that is suitable for use as an amplifier or oscillator at *microwave frequencies. They usually employ *velocity modulation of the electron beam. *Klystrons, *magnetrons, and *travelling-wave tubes are examples.

Mie scattering Scattering of light by spherical particles of diameters comparable with the wavelength; an extension of *Rayleigh scattering, applicable to particles small compared with the wavelength.

migration area The area required for a neutron to slow down from fission energy to thermal energy plus that required to diffuse the energy. The former area is defined formally as one-sixth of the mean square distance between the source and the point where the neutrons reach thermal energy. The diffusion area is formally one-sixth of the mean square distance between the point where the neutron is in thermal equilibrium with the surroundings and the point where it is captured. The *migration length* is the square root of the migration area. *See also* neutron age.

migration of ions The ions of an electrolyte migrate when a current is passed through it, and play a part in the transport of electricity. (*See* electrolytic dissociation.) The cations and anions do not always move at the same velocity, and thus transport different fractions of

the current. Progressive changes take place in the concentration of electrolyte around the electrodes. The fraction of the current carried by either ion is known as its *transport number* (or *transference number*), symbol: *t*.

mil One thousandth of an inch.

Miller effect In an electronic device, the phenomenon whereby the *interelectrode capacitance provides a feedback path between the input and output circuits, which can affect the total input admittance of the device. The total dynamic input capacitance of the device will always be equal to or greater than the sum of the static electrode capacitances because of this effect.

Miller indices *See* rational intercepts, law of.

milli- Symbol: m. A prefix meaning 10^{-3} (i.e. one-thousandth); for example, one millimetre (1 mm) is 10^{-3} metre.

minicomputer Loosely, a medium-sized *computer, usually less capable in terms of performance (and hence cheaper) than a *mainframe computer. There is no clear boundary now between minicomputers and the more sophisticated types of *microcomputers.

minimum deviation *See* deviation.

minimum discernible signal (m.d.s.) The smallest value of input signal to an electronic circuit or device that just produces a discernible change in the output.

Minkowski space-time *See* relativity.

minority carrier In a *semiconductor, the type of *carrier constituting less than half of the total charge-carrier concentration.

minute 1. Symbol: min. A unit of time equal to 60 seconds. Although not an *SI unit, the minute is of practical importance and may be used with the SI units.

2. *Syn.* minute of arc; arc minute. Symbol: '. A unit of angle equal to $1/60$ of a degree, i.e. 0.291 milliradian.

mirror An optical device for producing reflection, generally having a surface that is plane or a portion of a sphere, paraboloid, or ellipsoid. *Concave mirrors* are hollowed out, *convex mirrors* are dome-shaped. The *mirror formulas* generally describe the conjugate focus relations for spherical mirrors. The commonest form is

$$1/v + 1/u = 2/r = 1/f,$$

in which u is the object distance, v the image distance, r the radius of curvature, f the focal length. Objects and images lying in front of the mirror are real and the distances are taken as positive. For a virtual *image, v is negative. For a concave mirror, f is positive; for a convex mirror, it is negative. The magnification, M, is equal to v/u and is positive for inverted images.

A *thick mirror* is a lens with the back surface silvered or possibly a lens in combination with a curved mirror with or without separation.

Although mirrors are free from *chromatic aberration, in general they suffer from *spherical aberration. The paraboloid form (*see* parabolic reflector) focuses parallel rays accurately and is used with reflecting telescopes, searchlight mirrors, etc.; the *ellipsoid mirror* focuses light from one focus to the other focus (both foci are real); the *hyperboloid mirror* reflects light directed to one focus (virtual) to the opposite focus (real). The concave meniscus *Mangin mirror corrects spherical aberration reasonably. The *Schmidt telescope uses a Schmidt corrector to correct spherical aberration of spherical mirrors.

Mirrors are used not only as optical devices but also in the infrared, ultraviolet, and X-ray regions of the spectrum. A mirror may be a finely polished metal surface or a piece of glass with a reflective coating on the front or back surface, often of silver. Silver is an efficient reflector of infrared and ultraviolet. Vacuum-evaporated coatings of aluminium on highly polished substrates are now used in quality mirrors, often with protective coatings of silicon monoxide or magnesium fluoride. The aluminium forms a harder more stable surface than silver and can reflect shorter wavelengths. Mirrors can also be formed of multilayered dielectric films. *See also* magnetic mirror.

mirror nuclides Two nuclides having the same number of nucleons each, but where the number of protons (or neutrons) in one is equal to the number of neutrons (or protons) in the other. The nuclides will have the general form, $_n^m X$ and $_{m-n}^m Y$, and will be the source and product nuclides in a beta-capture or *beta-decay. In the special case of the pair $_n^{2n+1}X$ and $_{n+1}^{2n+1}Y$, they are known as *Wigner nuclides*.

MISFET, MIST *See* field-effect transistor.

mismatch The condition arising when the impedance of a *load is not equal to the output impedance of the source to which it is connected.

missing mass The mass of the universe's hypothetical invisible matter. The existence of this invisible matter is postulated on a number of grounds, including: the unexplained velocity distribution of stars perpendicular to the plane of the Galaxy; the view that the outer regions of galaxies contain matter that contributes mass but not light; studies of galactic dynamics and some cosmological theories suggest that the mean density of the universe exceeds that due to visible matter.

A number of solutions to the missing mass problem have been suggested; these include the existence of many planet-sized bodies and rocks as well as various exotic massive particles, such as massive *neutrinos, *axions, and weakly interacting massive particles (WIMPs).

mixer *Syn.* frequency changer. A device used in conjunction with a beat-frequency oscillator to produce an output having a different frequency from the input. The amplitude of the output bears a fixed relationship to the amplitude of the input (usually approximately linear) and the device is used in a *superheterodyne receiver for changing the frequency of an amplitude-modulated *carrier wave while retaining the modulation characteristics.

mixing ratio *See* humidity.

MKS system *See* metre-kilogram-second electromagnetic system of units.

MLD Abbreviation for *mean lethal dose.

mmf Abbreviation for *magnetomotive force.

mmHg Abbreviation of millimetres of mercury, a former unit of pressure equal to 133.32 pascals. The standard atmosphere was 760 mmHg.

mobility *See* Hall mobility; drift mobility; intrinsic mobility.

mode 1. The value of the abscissa corresponding to the maximum ordinate of a *frequency distribution curve.
2. *Syn.* transmission mode. Any of the several different states of oscillation of

an electromagnetic wave, of given frequency, in a waveguide, cavity resonator, etc. *See* waveguide.

modem Abbreviation of modulator-demodulator. A device that converts the signals from one particular type of equipment into a form suitable for use in another. For example a modem can convert the digital signals from a computer into an analogue form for use over an (analogue) telephone system.

moderator In *nuclear reactors. Material used to slow down the fast neutrons created in a *fission process to the lower velocities appropriate to the type of reactor in use by scattering without appreciable capture.

modulation In general, the alteration or modification of an electronic or acoustic parameter by another. In particular, the process of varying one electronic signal, called the carrier, according to the pattern provided by another signal, as when an audiofrequency signal is impressed onto a higher-frequency *carrier wave for radio transmission. *See* amplitude, frequency, phase, and pulse modulation. *See also* velocity modulation.

modulation factor *See* amplitude modulation; frequency modulation.

modulator 1. Any device that effects the process of *modulation.
2. A device used in radar for generating a succession of short pulses to act as a trigger for the *oscillator.

modulator electrode An electrode used for modulating the flow of current in a device. In a *cathode-ray tube it is the electrode controlling the intensity of the electron beam. In a *field-effect transistor it is the gate electrode(s), which control(s) the conductivity of the channel.

modulus of decay In a system exhibiting *damped oscillations of the form $a = a_0 e^{-\alpha t}$, where a_0 is the initial amplitude and a its value after time t, then the modulus is equal to the time t_1 at which the amplitude has fallen to $1/e$ of its initial value. It is the reciprocal of the damping factor α.

modulus of elasticity The ratio of *stress to *strain for a body obeying *Hooke's law. There are several moduli corresponding to various types of strain:

1. The *Young modulus* (E) =

$$\frac{\text{applied load per unit area of cross section}}{\text{increase in length per unit length}};$$

It applies to tensional stress when the sides of the rod or bar concerned are not constrained. (*See* Poisson ratio.)

2. *Bulk modulus* (or *volume elasticity*) (K) =

$$-\frac{\text{force per unit area}}{\text{change in volume per unit volume}};$$

It applies to compression or dilation, e.g. when a body is subject to changes in hydrostatic pressure. Fluids, as well as solids, have bulk moduli.

3. *Shear* (or *rigidity*) *modulus* (G) =

$$\frac{\text{tangential force per unit area}}{\text{angular deformation}}.$$

Since strain is a ratio and so dimensionless, the moduli have the dimensions of stress, i.e. force/area. The various moduli and the Poisson ratio for an isotropic solid are interrelated. The moduli given in Physical Tables, and most often used, are measured under isothermal conditions; the adiabatic values are always greater.

If stress is not proportional to strain (as in cast metals, marble, concrete, wood), the moduli have to be defined as the ratio of a small change in stress to a small change in strain at a particular value of the stress.

Mohs scale *See* hardness.

moiré pattern The pattern produced by overlying sets of parallel threads or lines, the sets being slightly inclined to one another. The overlaps produce the appearance of dark bands running athwart the individual lines. If the two sets are perfectly regular, these bands are straight, but deviations in either or both give wavy lines as in the characteristic appearance of moiré silk. Transparent diffraction gratings and replicas from them may be compared by superimposing them and examining the resulting moiré pattern.

molar A term now restricted in its meaning to divided by *amount of substance. In practice this means "per mole". Usually a subscript m is added to the symbol of the relevant quantity.

molar conductivity Symbol: j_m. The *conductivity of an *electrolyte solution divided by the concentration of electrolyte present. Note that the word molar in this case means "divided by concentration" and not "divided by amount of substance".

molar gas constant Symbol: R. The constant occurring in the *equation of state for 1 mole of an *ideal gas, namely

$$pV = RT.$$

It may be shown to be a universal constant for all gases. Actual gases obey this equation of state only in the limit as their pressure tends to zero.

The pressure exerted by a gas according to *kinetic theory is shown to be:

$$p = \tfrac{1}{3}mvC^2,$$

where m is the mass of the molecules, v is the number of molecules per cubic metre, and C is the root-mean-square velocity of the molecules. Considering one mole of the gas, L being the number of molecules present (i.e. the Avogadro constant), and V the *molar volume then:

$$pV = \tfrac{1}{3}mLC^2,$$

and V is independent of the nature of the gas at given values of p and T. Hence,

$$RT = \tfrac{2}{3}(\tfrac{1}{2}mLC^2).$$

This expression shows that R is equal to two-thirds of the total translational energy of the molecules in 1 mole of a gas at a temperature of 1 kelvin.

The Boltzmann constant k is given by the ratio R/L and R has the value
8.314 510 J K^{-1} mol^{-1}.

molar heat capacity Symbol: C_m. The *heat capacity of unit *amount of substance of an element, compound, or material. Molar heat capacities are measured in joules per kelvin per mole.

molar latent heat *See* latent heat.

molar volume The volume occupied by 1 mole of a substance. According to *Avogadro's hypothesis all ideal gases have the same molar volume at the same pressure and temperature. The value at STP is
2.241 3837 × 10^{-2} m^3 mol^{-1}.

mole Symbol: mol. The *SI unit of *amount of substance, defined as the amount of substance of a system that contains as many elementary entities as there are atoms in 0.012 kilograms of carbon-12. The elementary entities must be specified, and may be atoms, molecules, ions, electrons, other particles, or specified groups of particles.

molecular beam A collimated beam of atoms or molecules at low pressure, in which all the particles are travelling in the same direction and few collisions occur between them.

Such beams are produced in an apparatus connected to several fast vacuum *pumps. Beams of metal atoms are formed by heating the metal in an oven and allowing the vapour to escape through a small hole. In the case of a permanent gas an unheated enclosure can be used. The vacuum system is usually made of several sections, each connected to a pump and separated by partitions with collimated apertures. Each section is at a lower pressure than the preceeding section. Molecules that do not pass through the apertures are pumped away.

molecular cloud *See* interstellar matter.

molecular flow *Syn.* Knudsen flow. A type of gas flow occurring at low pressures, when the *mean free path of the gas molecules is large compared with the dimensions of the pipe through which the gas is flowing. The rate of gas flow is determined by collisions between the molecules and the wall of the tube, rather than by collisions between molecules. Thus the flow does not depend on the viscosity of the gas.

The ratio of a characteristic dimension of the apparatus through which the gas flows to the mean free path of the gas is known as the *Knudsen number*.

molecular gauges *Syn.* viscosity manometers. Devices used for measuring low gas pressures and whose action is based on the dependence of gas viscosity on pressure at low pressures. In one gauge, a disc is turned rapidly at a uniform speed and a second disc parallel to the first tends to follow its rotation due to the viscous drag of the air. The couple acting on the second disc is a measure of the pressure. The instrument works from about 10^{-1} to 10^{-5} Pa, and is usually calibrated with reference to a *McLeod gauge.

In another form, a flat quartz fibre, fixed at one end, vibrates in the gas and the damping, which is observed, depends on the viscosity. This instrument, called a *decrement gauge* or *quartz-fibre manometer*, is most useful from 1–0.01 Pa.

molecular orbital In an atom the electrons moving around the nucleus have *atomic orbitals that are often represented as a region around the nucleus in which there is a high probability of finding the electron. When molecules are formed, the valence electrons move under the influence of two or more nuclei and their *wave functions are known as molecular orbitals. These can also be represented by regions in space. It is usual to think of molecular orbitals as formed by a combination of atomic orbitals. Two atomic orbitals combine to give two molecular orbitals of different energies and forms. In the lower-energy orbital there is a concentration of charge between the nuclei, which serves to hold them together and thus form the chemical bond; this is called a *bonding orbital*. The orbital of higher energy does not have an internuclear charge concentration and the nuclei tend to repel one another; this is called an *antibonding orbital*. Each molecular orbital can be occupied by two electrons with opposite spins in accordance with the *Pauli exclusion principle.

In hydrogen for example, each hydrogen atom has one electron. When a molecule is formed the pair of electrons occupy the lower-energy bonding orbital forming a stable molecule. The higher-energy antibonding orbital is unoccupied.

molecular polarization When a molecule is subjected to an electric field, there is a small displacement of electrical centres that induces a *dipole in the molecule. If $m = \alpha E$, where m is the electric *dipole moment induced by a field strength E, then the constant α is called the *polarizability* of the molecule.

molecular pump *See* pumps, vacuum.

molecular weight Former name for relative molecular mass.

moment 1. *Syn.* torque. The moment of a force about an axis is the product of the perpendicular distance of the axis from the line of action of the force, and the component of the force in the plane perpendicular to the axis. The moment of a system of coplanar forces about an axis perpendicular to the plane containing them is the algebraic sum of the moments of the separate forces about that axis (anticlockwise moments are taken conventionally to be positive and clockwise ones negative).

2. Moment of momentum about an axis. *See* angular momentum.

3. The moment of a *vector about an axis is similarly defined as in def. **1.** It is a scalar and is given a positive or negative sign as in def. **1.** When dealing with systems in which forces and motions do not all lie in one plane, the concept of moment about a point is needed. The moment of a vector P (e.g. force or momentum) about a point A is a *pseudovector M equal to the *vector product of r and P, where r is any line joining A to any point B on the line of action of P. (The vector product $M = r \times P$ is independent of the position of B.)

The relation between the scalar moment about an axis and the vector moment about a point on that axis is that the scalar is the component of the vector in the direction of the axis.

moment of inertia Of a body about an axis. Symbol: I. The sum of the products of the mass of each particle of the body and the square of its perpendicular distance from the axis. (This addition is replaced by an integration in the case of a continuous body.) For a rigid body moving about a fixed axis, the laws of motion have the same form as those of rectilinear motion, with moment of inertia replacing mass, angular velocity replacing linear velocity, angular momentum replacing linear momentum, etc. Hence the *kinetic energy of a body rotating about a fixed axis with angular velocity ω is $\frac{1}{2}I\omega^2$, which corresponds to $\frac{1}{2}mv^2$ for the kinetic energy of a body of mass m translated with velocity v. *See also* Routh's rule; theorem of parallel axes.

moment of momentum *See* angular moment.

momentum 1. *Syn.* linear momentum. Of a particle. Symbol: *p*. The product of the mass and the velocity of the particle. It is a *vector quantity directed through the particle in the direction of motion. The linear momentum of a body or of a system of particles is the vector sum of the linear momenta of the individual particles. If a body of mass M is translated (*see* translation) with a velocity V, its momentum is MV, which is the momentum of a particle of mass M at the centre of gravity of the body. (*See* Newton's laws of motion, II; conservation of momentum.)
2. *See* angular momentum.

monochord *Syn.* sonometer. A thin metallic wire stretched either horizontally or vertically over two bridges by means of a weight hanging over a pulley or by a spring tensioning device. It is used to study the properties of *stretched strings. A movable bridge provides a convenient means of varying the vibrating length of the wire. In order to increase the volume of the sound as in most stringed instruments, the string and the bridges are mounted on a a hollow box,

the air in which is forced to vibrate to a certain extent by the vibration of the string. The vibration may be excited in any convenient manner, e.g. by plucking, striking, bowing, or by electromagnetic means.

monochromatic radiation Radiation restricted to a very narrow band of wavelengths: ideally one wavelength. Monochromatic light can be isolated with varying degrees of purity by using interference *filters, by isolating a portion of the spectrum by means of a slit, or by the use of *lasers.

monochromator *See* spectrometer.

monoclinic system *See* crystal systems.

monolithic integrated circuit *See* integrated circuit.

monostable A type of circuit having only one stable state, but which can be triggered into a second quasi-stable state by the application of a *trigger pulse. One form consists of a *multivibrator with resistive-capacitive coupling. Monostables are used to provide a fixed duration pulse and can be utilized for pulse stretching, or shortening, or as a delay element.

Morse equation An empirical equation giving the potential energy of two atoms in a molecule as a function of their separation. It has the form:
$$V = D_e\{1 - \exp[-\beta(r - r_0)]\}^2,$$
where V is the potential energy, β is a constant, r is the distance between the atoms, and r_0 is the equilibrium distance, i.e. the bond length. A typical potential energy curve for a diatomic molecule is shown in the diagram. At small separations the energy is very high because of repulsion between the nuclei. At large separations the energy is constant with separation because the mol-

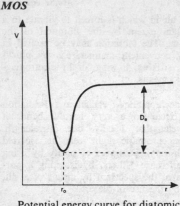

Potential energy curve for diatomic
molecule

ecule has dissociated into two atoms. D_e
is the energy from the minimum to the
dissociation level.

MOS Abbreviation for metal oxide
semiconductor. *See* field-effect transistor; MOS integrated circuit; MOS logic
circuits.

Moseley's law The *X-ray spectrum of a
particular element can be split into several distinct line series: K, L, M, and
N. Moseley's law states that the square
root of the frequency f, of the characteristic *X-rays of one of these series, for
certain elements, is linearly related to
the atomic number Z. A graph of Z
against $\sqrt{f}$ is called a *Moseley diagram*.

The law, which is only approximate,
can be explained in terms of the energies of the electrons in the various inner
shells of the atom.

MOSFET, MOST *See* field-effect transistor.

MOS integrated circuit A type of *integrated circuit based on insulated-gate
*field-effect transistors. MOS circuits
have several advantages and account for
a substantial proportion of all semicon-

ductor devices produced. They usually
have a higher functional *packing density than bipolar *integrated circuits as
MOS transistors are self isolating and
no area-consuming isolation diffusions
are required. MOS transistors may be
used as active load devices thus no
separate process is required to form
resistors, as in bipolar integrated circuits. When used as load devices *pulse
operation of the circuit is easily obtainable using the gate electrodes to activate
the device: power dissipation is greatly
reduced, involving less complicated heating problems. A characteristic of MOS
transistors is their exceptionally high input impedances, allowing the gate electrodes to be used as temporary storage
capacitors (enabling the circuits to be
relatively simple). This is called *dynamic
operation*, and usually these circuits operate above a minimum specified frequency.

The relatively few processing steps required in manufacture, compared to bipolar integrated circuits, enables large
chips to be made thus further increasing
the functional compactness, and reducing the costs.

MOS circuits tend to be slower than
their bipolar counterparts due to their
inherently lower *mutual conductance,
and to the fact that their speed is extremely dependent on the load capacitance.

MOS logic circuits *Logic circuits constructed in *MOS integrated circuits.
They consist of combinations of MOS
*field-effect transistors in series or in
parallel that perform the logic functions
(e.g. act as AND or OR gates; *see* diagram). These are coupled to other MOS
transistors that determine the output
voltages of the circuit. The logic functions are switches, the combination
switch being 'on' when the required input conditions are fulfilled, e.g. both A
and B are high logic levels to produce a

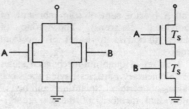

OR function   AND function

high-level output for the AND gate. The high logic level is chosen to be greater than the threshold voltage V_T; the low level is lower. Here the function is represented by a single switch T_S.

Mössbauer effect An effect observed when certain nuclei emit or absorb gamma radiation of low quantum energy. A photon of energy $h\nu$ transfers momentum $h\nu/c$, hence when a stationary nucleus emits or absorbs a quantum, it must recoil to conserve momentum. The kinetic energy of recoil of a free nucleus of mass M is given by:

$$E = (1/2M)(h\nu/c)^2.$$

If a gamma ray is emitted by a nucleus on falling from a given excited state to the ground state, the quantum energy is therefore not sufficient to raise a similar nucleus from the ground to the excited state. Mössbauer showed that for low quantum energies an atom bound in a crystal may remain so bound on emitting or absorbing radiation. In this case the recoil momentum is shared between as many atoms as an acoustic wave will reach during the process of emission or absorption. Hence the mass M in the equation must be replaced by the relatively enormous mass of this large number of atoms, so that E becomes infinitesimal. Such processes are called *recoilless*.

Very many experiments have been done using the nuclide $^{56}_{26}\text{Fe}$. Electron capture in $^{57}_{27}\text{Co}$ produces $^{57}_{26}\text{Fe}$ in an excited state of energy 14.4 keV and life-time 0.1 µs. In iron foils at room temperature about two-thirds of the emissions and absorptions of the 14.4 keV radiation are recoilless. In a *Mössbauer analyser* the source is mounted so that it can be moved at steady speeds (of the order 10^{-5} m s^{-1}) towards or away from the detector. The latter consists of an iron foil, usually highly enriched in the rare isotope ^{57}Fe, behind which is a proportional counter or similar instrument. The movement of the source changes the quantum energy by the Doppler effect. Over a very limited range of speeds corresponding to a line width of the order of 10^{-13} there is a decrease in the count rate as nuclear absorption is added to the normal absorption by the *photoelectric effect. Similar experiments have been done using other gamma emitters, some of which have to be cooled to very low temperatures to give a large enough proportion of recoilless interactions.

The technique can detect fractional changes in energy of the order of one in 10^{16}. It has been used to study: the *Einstein shift in the earth's gravitational field; the quantum states of nuclei; magnetic fields inside ions.

MOS transistor *See* field-effect transistor.

motion *See* Newton's laws of motion.

motor A machine that does work mechanically when it is driven by an electric current. *See* induction motor; synchronous-induction motor; synchronous motor; universal motor.

mouse A computer device that is used as a pointer, i.e. it passes two-dimensional spatial information to a computer. It is moved by hand around a flat surface. These movements are communicated to the computer and lead to corresponding movements of the cursor on

311

a display screen. A mouse has one or more buttons to indicate to the computer that the cursor has reached the desired position.

moving-coil galvanometer *See* galvanometer.

moving-coil instrument *See* ammeter.

moving-coil microphone A type of *microphone in which the diaphragm is connected to a coil and moves it backwards and forwards in a stationary magnetic field, thus inducing an e.m.f.

moving-iron instrument *See* ammeter.

MSI Abbreviation for medium-scale integration. *See* integrated circuit.

multichannel analyser A test instrument that splits an input waveform into a number of channels in respect of a particular parameter of the input. A circuit that sorts a number of pulses into selected ranges of amplitude is known as a *pulse-height analyser*. A circuit that splits an input waveform into its frequency components is known as a *spectrum analyser. In general, a multichannel analyser will have facilities for carrying out both these operations.

multielectrode valve A *thermionic valve that contains two or more sets of electrodes within a single envelope, each set of electrodes having its own independent stream of electrons. The sets of electrodes may have one or more common electrodes (e.g. a common cathode). A typical example is the double-diode-triode, containing the electrode assemblies of two diodes and one triode.

multiplet 1. A group of spectrum states specified by the values of the quantum numbers L (vector sum of orbit and angular momenta of individual electrons)

and S (vector sum of spin momenta of individual electrons). The group of states gives rise to a set of spectrum lines.

2. A set of quantum-mechanical states of *elementary particles having the same value of certain *quantum number(s). Individual members of the set are distinguished by having different values of other quantum numbers. The word multiplet is most commonly used in connection with sets of states that are transformed into each other by operations that form the elements of a *group. *See* isospin; unitary symmetry.

multiplexer *See* multiplex operation.

multiplex operation The use of a single path for the simultaneous transmission of several signals without any loss of identity of an individual signal. The various signals are fed to a *multiplexer*, which allocates the transmission path to the input according to some parameter (e.g. *frequency-division multiplexing or *time-division multiplexing). At the receiving end a *demultiplexer*, operating in sympathy with the multiplexer, reconstructs the original signals at the outputs. The transmission path may be in any of the available media, e.g. wire, waveguide, or radio waves.

multiplication constant or factor (effective) Symbol: k_{eff}. In a *nuclear reactor. The ratio of the number of neutrons liberated in one generation to the number liberated in the previous generation.

multivibrator A form of *oscillator consisting of two linear *inverters coupled so that the input of one is derived from the output of the other. The action of the various types of multivibrator is determined by the coupling used.

Capacitive coupling gives an *astable multivibrator* with two quasistable states; once the oscillations are established the

Multivibrator using resistive coupling

device is free-running, i.e. it can generate a continuous waveform without any *trigger.

Capacitive-resistive coupling gives a *monostable multivibrator.

Resistive coupling (*see* diagram) gives a *bistable multivibrator that has two stable states and can change state on the application of a trigger pulse. *See* flip-flop.

mu-meson Former (and incorrect) name for muon.

muon A negatively charged *lepton similar to the electron except for its mass, which is 206.7683 times greater than that of the electron. It has a mean life of 2.197 09 μs, and decays into an electron, a *neutrino, and an antineutrino:

$$\mu^- \rightarrow e^- + \nu_\mu + \bar{\nu}_e.$$

The decay process of the muon's antiparticle, the *antimuon*, is:

$$\mu^+ \rightarrow e^+ + \bar{\nu}_\mu + \nu_e.$$

It was originally thought to be a meson. *See* elementary particles.

musical scale A series of notes progressing from any given note to its octave by prescribed *intervals chosen for musical effect. Of the very large number of scales used throughout history, only a few survive. The *diatonic scale* consists of eight notes to the octave, some of which are separated by *tones and some by semitones. Later all the tones were divided into semitones, first, for the use in embellishment of the melodic line, and secondly, for harmonic colouring. The extra notes provided are used much less frequently than the others. All notes of the scale bear a relation to the key-note or starting note of the scale.

The *chromatic scale* has thirteen notes to the octave, the notes being the same as those found on the diatonic scale with each tone divided. All the notes of the chromatic scale, however, are used with equal authority, but are still related to the keynote. The *wholetone scale* consists of six notes to the octave, there being no semitones. Only two series of notes are possible. The *pentatonic scale* with five notes to the octave has the spacing tone, 1½ tones, tone, tone, 1½ tones. *See also* temperament.

mutual capacitance A measure of the extent to which two *capacitors can affect each other, expressed in terms of the ratio of the amount of charge transferred to one, to the corresponding potential difference of the other.

mutual conductance *Syn.* transconductance. Symbol: g_m. Of an amplifying device or circuit. The ratio of the incremental change in output current, I_{out}, to the incremental change in input voltage, V_{in}, causing it, the output voltage remaining constant:

$$g_m = \partial I_{out}/\partial V_{in} \qquad (V_{out} \text{ constant}).$$

mutual inductance *See* electromagnetic induction.

myopia (short sight) *See* refraction.

N

NAA Abbreviation for neutron *activation analysis.

nabla *See* del.

nadir *See* celestial sphere.

NAND gate *See* logic circuit.

nano- Symbol: n. A prefix meaning 10^{-9}; for example, one nanometre (1 nm) is 10^{-9} metre.

natural abundance *See* abundance.

natural convection *See* convection (of heat).

natural frequency *See* free vibrations.

natural units A system of units based on *Gaussian or *Heaviside–Lorentz units for electromagnetic quantities. These units are often used in particle physics in place of the generally more widely used *SI units. In natural units, quantities having dimensions of length, mass, and time are given the dimensions of power or energy (usually expressed in *electronvolts), which effectively makes the rationalized *Planck constant and the *speed of light both equal to unity.

n-channel device *See* field-effect transistor.

near infrared or ultraviolet Parts of the infrared or ultraviolet regions of the spectrum of *electromagnetic radiation that are close to the visible region. The regions that are far from the visible region and close to the X-ray and microwave regions, are called the *far ultraviolet* and *far infrared* respectively.

The terms are used rather loosely and it is not possible to give any definite ranges of wavelength. The near infrared is usually the region in which molecules absorb radiation by making transitions between vibrational *energy levels. The far infrared is usually the region in which absorption is due to changes in rotational energy levels.

near point Of the eye. The nearest point for which, with accommodation fully excited, clear vision is obtained. The near point tends to recede with age. It should not be confused with the conventional distance of distinct vision (25 cm).

nebula A cloud of interstellar gas and dust that can be observed either as a luminous patch – a *bright nebula* – or as a dark region against a brighter background – a *dark nebula*.

Various processes can cause a nebula to become visible. Ultraviolet radiation from a nearby source, usually a hot young star, can ionize the interstellar gas atoms, and light (and other radiation) is emitted when the ions interact with free electrons in the nebula. This is called an *emission nebula*, and it has an emission spectrum. With a *reflection nebula*, light from a nearby star or stellar group is scattered by the dust grains. It has essentially the same spectrum as the illuminating star(s). In contrast, dark nebulae have no nearby stars to illuminate them.

Néel temperature *See* antiferromagnetism.

negative 1. A photographically produced image in which the dark and light parts of the subject appear as light and dark, respectively. (*See* photography.) In a colour negative the colours of the object appear as *complementary colours, being converted to the original colours by the colour-printing process. *See* colour photography.
2. Of an electric charge. Having the same polarity as the charge of an electron.
3. Of a body or system. Having a negative charge; having an excess of electrons.

negative bias A potential that is applied to an electrode of an electronic device and is negative with respect to earth potential (or some other fixed reference potential).

negative crystal *See* optically negative crystal.

negative feedback *See* feedback.

negative glow *See* gas-discharge tube.

negative logic *See* logic circuit.

negative magnetostriction *See* magnetostriction.

negative resistance A property of certain electronic devices whereby a portion of the voltage-current characteristic has a negative slope, i.e. the current decreases as the applied voltage increases. Such devices include the *thyristor, the *magnetron, and the *tunnel diode.

negative specific heat capacity A property of certain substances that under stated conditions need heat to be extracted from them if their temperature is to be raised. The most familiar example is a *saturated vapour, for which
$$c = (c_p)_1 + dL/dT - L/T,$$
where $(c_p)_1$ is the specific heat capacity of the liquid at the equilibrium temperature T, L is the specific latent heat of vaporization. When the vapour rises in temperature, it must simultaneously be compressed to keep it saturated since the density of saturated vapour rises with increasing temperature. For steam, the heat of compression is so great that the vapour becomes superheated so that heat must be extracted from it. This occurs through the evaporation of more water, the specific latent heat for this evaporation coming from the supersaturated steam. If the heat of compression is small, the vapour becomes superheat-

ed, condensation of liquid occurring with evolution of heat to the saturated vapour so that the specific heat capacity is positive. If the heat of compression is just sufficient to keep the vapour saturated, the specific heat capacity is zero and the curve of saturated vapour pressure coincides with the curve of adiabatic compression.

nematic structure *See* liquid crystal.

neon tube A *gas-discharge tube containing neon at low pressure, the colour of the glow discharge being red. The striking voltage is between 130 and 170 volts and within a range of current the voltage across the tube remains constant so that for small currents (up to about 100 milliamp) it can be used as a voltage stabilizer. Electrodeless neon tubes will easily glow in the presence of high-frequency currents of high voltage.

neper Symbol: Np. A dimensionless unit used for comparing two currents almost exclusively in telecommunication engineering, the two currents being usually those entering and leaving a transmission line or other transmission network. Two currents I_1 and I_2 are said to differ by N nepers when
$$N = \log_e |I_1/I_2|.$$
If the input and output impedances Z_1 and Z_2 have equal resistances, then the input and output powers, P_1 and P_2 differ by n nepers, where
$$n = \tfrac{1}{2}\log_e(P_1/P_2).$$
One neper equals 8.686 *decibels.

neptunium series *See* radioactive series.

Nernst effect When heat flows through a strip of metal in a magnetic field, the direction of flow being across the lines of force, an e.m.f. is developed perpendicular to both the flow and the lines. The direction of the current the e.m.f. produces depends on the nature of the metal of which the strip is composed. *See* Leduc effect.

Nernst heat theorem Also known as the third law of *thermodynamics. If a chemical change occurs between pure crystalline solids at absolute zero, there is no change in *entropy, i.e. the entropy of the final substance equals that of the initial substances. Planck extended this by stating that the value of the entropy for each condensed phase is zero at absolute zero.

The following deductions may be made from the Nernst heat theorem. (i) The coefficient of expansion of all condensed phases vanishes at absolute zero. (ii) The thermoelectric e.m.f. vanishes at absolute zero, both the Peltier effect and the Thomson effect becoming zero. (iii) The magnetic susceptibility of a paramagnetic crystal vanishes at the absolute zero. (iv) The entropies of the solid and liquid states of helium have the same value at absolute zero.

net radiometer An instrument for measuring the difference in intensity between radiation entering and leaving the earth's surface. The radiation can be direct, diffuse, and reflected solar radiation or infrared radiation from the sky, clouds, and ground. A similar thermopile system is used to that found in *solarimeters, except that both sides of the thermopile are exposed to radiation and the resulting e.m.f. is proportional to the difference in intensity of the incoming and outgoing radiation. *See also* pyrheliometer.

network 1. In electronics. A number of conductors connected together to form a system consisting of a set of interrelated circuits that performs one or more specific functions. The conductors are resistors, capacitors, and inductors, in all forms, i.e. they possess impedance. The behaviour of the network depends on

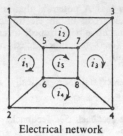

Electrical network

the *network parameters* (or *network constants*), which are the values of the impedances – resistances, capacitances, etc. – of the components of the network, and the manner in which the components are interconnected.

Networks are described as *linear* or *nonlinear*, depending on whether or not there is a linear relation between the voltages and currents. They are *bilateral* or *unilateral* depending on whether they pass currents in both directions or only one direction. A network is described as *passive* if it contains no source or sink of energy (the latter does not include energy dissipated in the resistance elements of the network); otherwise it is described as *active*.

A point in the network at which more than two conductors meet is called a *branch point* or a *node* (e.g. points 1 to 8 in the diagram), and a conducting path between two branch points is called a *branch* (e.g. 1 to 2). A *mesh* is the portion of the network included in any closed conducting loop in the network (e.g. 1-3-7-5-1) and its boundary is the *mesh contour*. The *mesh currents* are the currents that may be considered to circulate round the meshes (Maxwell's cyclic currents). Any branch that is common to two or more meshes is a *common branch* or *mutual branch* (e.g. 5 to 6).

Analysis of linear networks can be achieved by considering the network as a *quadripole, and deriving sets of equations relating the currents, voltages, and impedances at the input and output. It is also possible to apply *Kirchhoff's laws to each mesh in the network in turn.

2. *In computing.* A number of computer systems that are often widely separated but are interconnected in such a way that they can exchange information by following agreed procedures. The computer systems must be able to transmit information onto and receive information from the connected system. The information is sent as an encoded digital signal, and is transmitted along telephone lines, satellite channels, electric cables, etc.

Each computer is not directly linked to every other computer on the network. Direct connections are only made at certain points in the network, called *nodes*. Computing facilities are attached to some or all of the nodes. Nodes may be at a junction of two or more communication lines or at an endpoint of a line. A particular piece of information has to be routed along a set of lines to reach its specified destination, being *switched* at the nodes from one line to another.

Neumann's law *Syn.* Faraday–Neumann law. *See* electromagnetic induction.

neutral 1. Devoid of either positive or negative electric charge.
2. At earth potential.
3. Lacking hue; black, white, or grey.

neutral equilibrium *See* equilibrium.

neutral filter A light filter that absorbs equally all wavelengths: it reduces light intensity without change of relative spectral distribution.

neutralization The provision of negative *feedback in an amplifier to a degree sufficient to neutralize any inherent positive feedback. Positive feedback in an amplifier is usually undesirable since it may give rise to the production of oscillations. Neutralization is commonly employed with radio-frequency amplifiers to counteract the *Miller effect, and also with *push-pull operation to avoid *parasitic oscillations.

neutral temperature For a *thermocouple with one junction maintained at 0 °C, the temperature θ of the hot junction causes the e.m.f. E to vary according to the formula:
$$E = \alpha\theta^2 + \beta\theta,$$
where α, β are constants. The maximum value of E occurs when $\theta = -\beta/2\alpha$; this is called the neutral temperature. It is usual to restrict the use of the thermocouple to the range between 0 °C and the neutral temperature.

neutrino A neutral *elementary particle with *spin ½ that only takes part in *weak interactions. Neutrinos are *fermions and are classified as *leptons. They are generally regarded as stable, but it has been suggested that a neutrino might change into a system of neutrinos and antineutrinos. The rest mass is believed to be zero, so according to the theory of *relativity the particle must move with the speed of light with respect to any observer. A neutrino that transfers mass m also transfers energy mc^2 and momentum mc. Three kinds of neutrino are known, one (ν_e) is associated with the *electron, another (ν_μ) is associated with the *muon, and the third (ν_τ) is associated with the massive *tauon. The antiparticle of the neutrino is the *antineutrino* ($\bar{\nu}$).

The neutrino was first postulated (*Pauli, 1930) to explain the continuous spectrum of *beta rays. It is assumed that there is the same amount of energy available for each *beta decay of a particular nuclide and that this energy is shared according to a statistical law

317

between the electron and a light neutral particle (now classified as the antineutrino, $\bar{\nu}_e$). Later it was shown that the postulated particle would also conserve angular momentum and linear momentum in beta decays.

In addition to beta decay, the electron neutrino is also associated with, for example, *positron decay and *electron capture:

$$^{22}\text{Na} \rightarrow {}^{22}\text{Ne} + e^+ + \nu_e$$
$$^{55}\text{Fe} + e^- \rightarrow {}^{55}\text{Mn} + \nu_e$$

The absorption of antineutrinos in matter by the process

$$^1\text{H} + \bar{\nu}_e \rightarrow n + e^+$$

was first demonstrated by Reines and Cowan. The muon neutrino is generated in such processes as:

$$\pi^+ \rightarrow \mu^+ + \nu_\mu$$
$$\mu^- \rightarrow e + \nu_\mu + \bar{\nu}_e$$

Although the interactions of neutrinos are extremely weak the cross sections increase with energy and reactions can be studied at the enormous energies available with modern *acelerators. In some forms of *grand unified theories, neutrinos are predicted to have a nonzero mass, although no evidence has been found to support this prediction.

neutron An elementary particle with zero charge and a rest mass equal to

$$1.674\,9542 \times 10^{-27}\ \text{kg},$$
i.e. $939.5729\ \text{MeV}/c^2$.

It is a constituent of every atomic *nucleus except that of ordinary hydrogen. Free neutrons decay by *beta decay with a mean life of 914 s. The neutron has *spin ½, *isospin ½, and positive *parity. It is a *fermion and is classified as a *hadron as it has *strong interactions.

Neutrons can be ejected from nuclei by high-energy particles or photons; the energy required is usually about 8 MeV, although sometimes it is less. *Fission is the most productive source. They are detected using all normal detectors of ionizing radiation as a result of the production of secondary particles in *nuclear reactions. The discovery of the neutron (Chadwick, 1932) involved the detection of the tracks of protons ejected by neutrons by elastic collisions in hydrogenous materials.

Unlike other nuclear particles, neutrons are not repelled by the electric charge of a nucleus so they are very effective in causing nuclear reactions. When there is no *threshold energy, the interaction *cross sections become very large at low neutron energies, and the *thermal neutrons produced in great numbers by *nuclear reactors cause nuclear reactions on a large scale. The capture of neutrons by the (n, γ) process produces large quantities of radioactive materials, both useful nuclides such as ^{60}Co for cancer therapy and undesirable by-products.

neutron activation analysis (NAA) *See* activation analysis.

neutron age One-sixth of the mean square displacement of a neutron as it slows down, through a specified energy range, in an infinite homogeneous medium. *See* Fermi age theory.

neutron diffraction A technique for determining the crystal structure of solids by diffracting a beam of neutrons. It is similar in principle to *electron diffraction and can be used in place of *X-ray crystallography. The wavelength of a neutron is related to its velocity by the *de Broglie equation, a neutron with a velocity of about $4 \times 10^3\ \text{m s}^{-1}$ having a wavelength of about 10^{-10} m.

neutron excess *See* isotopic number.

neutron flux *Syn.* neutron flux density. The product of the number of free neutrons per unit volume and their mean speed. The neutron flux in a *power reactor lies in the range 10^{16}–10^{18} per square metre per second.

neutron number Symbol: N. The number of neutrons present in the nucleus of an atom. The neutron number is obtained by subtracting the *atomic number from the *mass number.

neutron star A star that, having exhausted its nuclear sources of energy, has undergone *gravitational collapse. It reaches a stage of electron *degeneracy, but has sufficient mass (> 1.4 solar masses) for further contraction to occur. When the density exceeds 10^7 kg m^{-3} equilibrium between protons, electrons, and neutrons shifts in favour of the neutrons until at densities of 5×10^{10} kg m^{-3} 90% of the protons and electrons have interacted to form neutrons. If the mass of the star is less than 2.0 solar masses, strong repulsive forces between neutrons are set up causing a rapid rise in pressure. Contraction is halted and a stable neutron star is formed. It has a diameter of only 20 to 30 km. *Pulsars are almost certainly rotating neutron stars.

newton Symbol: N. The *SI unit of *force, defined as the force that provides a mass of one kilogram with an acceleration of one metre per second per second.

Newtonian fluid A fluid in which the amount of strain is proportional to the product of the stress times the time. The constant of proportionality is known as the coefficient of viscosity. *See* viscosity; anomalous viscosity.

Newtonian force *Syn.* Coulomb force. A force between points that falls off as the inverse square of the distance between them.

Newtonian frame of reference *See* Newtonian system.

Newtonian mechanics *See* Newton's laws of motion.

Newtonian system *Syn.* Newtonian frame of reference. Any frame of reference relative to which a particle of mass m, subject to a force F, moves in accordance with the equation $F = kma$, where a is the acceleration of the particle, m its mass, and k a universal positive constant equal to unity in *SI units.

Such a frame of reference is one in which the centre of mass of the solar system is fixed, and does not rotate relative to the fixed stars. Any other frame of reference that moves relative to this with a uniform velocity is also a Newtonian system; this is the classical principle of *relativity. For non-Newtonian systems, e.g. a frame of reference fixed to the earth's surface, Newton's equation $F = kma$ can be made to hold if the fictitious inertial *forces are added to F.

Newtonian telescope *See* reflecting telescope.

Newton's formula (for a lens) The distances p and q between two conjugate points and their respective foci are related by $pq = f^2$ (with a suitable sign convention); f is the focal length of the lens. For a mirror, the foci coincide but the relationship is unaltered (except for sign in some conventions).

Newton's law of cooling The rate of loss of heat from a body is proportional to the excess temperature of the body over the temperature of its surroundings. Strictly the law applies only if there is *forced convection, but provided the temperature excess is small, the law is fairly well obeyed, even in the case of free or natural convection.

Newton's law of gravitation *See* gravitation.

Newton's laws of fluid friction 1. The force of resistance, D, opposing the relative motion of a body and a fluid is given by $k_0 A V^2 \rho$, where V is the relative velocity, ρ the density of fluid, A some projected area of the body, and k_0 a constant of proportionality (*compare* drag coefficient). The law was formulated from considerations of the change of momentum in the direction of relative motion.

2. The shearing force between two infinitesimal layers of viscous fluid is proportional to the rate of shear in a direction perpendicular to the direction of motion of the layers. This force is expressed: $F = \eta \partial u / \partial y$, where u is the velocity in the direction of motion and y is perpendicular to the direction of u. η is called the coefficient of *viscosity, and in classical hydrodynamics is a factor peculiar to the molecular nature of the fluid alone.

In the limited range of conditions studied by Newton, law **1** was found to apply to motion in very fluid media such as air and water, while law **2** applied to very viscous liquids. Later research showed that for low values of the *Reynolds number, law **1** applies for all fluids, while in all cases law **2** applies for high values.

Newton's laws of motion In his *Principia* Newton (1687) stated the three fundamental laws of motion, which are the basis of *Newtonian mechanics*.

Law I. Every body perseveres in its state of rest, or uniform motion in a straight line, except in so far as it is compelled to change that state by forces impressed on it. This may be regarded as a definition of force.

Law II. The rate of change of linear momentum is proportional to the force applied, and takes place in the straight line in which that force acts. This definition can be regarded as formulating a suitable way by which forces may be measured, that is, by the acceleration they produce,

$$F = \mathrm{d}(mv)/\mathrm{d}t$$
i.e. $F = ma + v(\mathrm{d}m/\mathrm{d}t)$,
where F = force, m = mass, v = velocity, t = time, and a = acceleration. In the majority of nonrelativistic cases, $\mathrm{d}m/\mathrm{d}t = 0$ (i.e. the mass remains constant), and then

$$F = ma.$$

Law III. Forces are caused by the interactions of pairs of bodies. The force exerted by A upon B and the force exerted by B upon A are: simultaneous; equal in magnitude; opposite in direction; in the same straight line; caused by the same mechanism.

Note: the popular statement of this law in terms of "action and reaction" leads to much misunderstanding. In particular: any two forces that happen to be equal and opposite are supposed to be related by this law, even if they act on the same body; one force, arbitrarily called "reaction", is supposed to be a consequence of the other and to happen subsequently; the two forces are supposed to oppose each other, causing equilibrium; certain forces such as forces exerted by supports or propellants are conventionally called "reactions", causing considerable confusion.

The third law may be illustrated by the following examples. The gravitational force exerted by a body on the earth is equal and opposite to the gravitational force exerted by the earth on the body. The intermolecular repulsive force exerted on the ground by a body resting on it, or hitting it, is equal and opposite to the intermolecular repulsive force exerted on the body by the ground.

A more general system of mechanics has been given by Einstein in his theory of *relativity. This reduces to Newtonian mechanics when all velocities relative to the observer are small compared with those of light. *See also* quantum theory.

Newton's rings Circular interference fringes formed between a lens and a glass plate with which the lens is in contact. There is a central dark spot around which there are concentric dark rings. The radius of the nth ring is given by $\sqrt{(nR\lambda)}$, where R is the radius of curvature of the lens and λ is the wavelength.

Nichrome A heat-resistant alloy with high resistivity that is used in electrical heating elements and resistors. The composition varies but is approximately 62% Ni, 15% Cr, and 23% Fe.

nickel-iron accumulator *See* Edison accumulator.

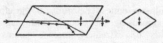

Nicol prism

Nicol prism A prism made of calcite once widely used for polarizing light and analysing *plane-polarized light. The crystal is cut in a special direction, sliced, and recemented together with Canada balsam. Light entering one face undergoes *double refraction; the extraordinary ray passes straight forward through the balsam, whereas the ordinary ray is reflected to the lower face where it is absorbed by a black coating (*see* diagram). The emerging extraordinary ray is plane-polarized with vibrations parallel to the short diagonal of the rhomb-shaped section as viewed from the emergent face.

The Nicol prism has been largely superseded by more effective polarizers. One example is the *Wollaston prism. Two others are the *Glan–Foucault* (or *Glan-air*) *prism* and the *Glan–Thompson*

prism. Both are composed of two calcite halves but the former has a film of air between the sections while the latter is cemented. The beam strikes the surface normally, passes undeviated through the first half and is split at the interface. The Glan–Thompson has the wider angular aperture of the two, but the Glan-air, being uncemented, can handle high-power laser radiation and transmits the broad spectral range of calcite (about 5000 to 230 nm).

NiFe accumulator *See* Edison accumulator.

nile A unit used to measure the departure of a *nuclear reactor from its critical condition. 1 nile corresponds to an *activity of 0.01.

NMR Abbreviation for *nuclear magnetic resonance.

nodal line *See* Eulerian angles.

nodal points *See* centred optical system.

node 1. A point or region in a *standing wave at which some characteristic of the wave motion, such as particle displacement, particle velocity, or pressure amplitude, has a minimum (or zero) value. For standing waves in an air column, displacement nodes are pressure antinodes and conversely. Generally, any kind of wave has two types of disturbance and the nodes for one type are the *antinodes for the other.
  2. In electricity or electronics. A point at which the current or voltage has a minimum value.
  3. *See* network (defs. 1, 2).

noise 1. (general) Sound that is undesired by the recipient, usually a discordant sound such as that produced by a low overflying aircraft or a pneumatic drill. The loudness of a noise is expressed either in decibels or phons and is measured with an audiometer or noise-level meter.

2. Spurious unwanted energy (or the associated voltage) in an electronic or communications system. *Interference often produces noise, but not always (*see* crosstalk; hum). There are two main types of noise, *white noise* and *impulse noise*. White noise has a wide frequency spectrum. It is caused by various sources, the most common being *thermal noise* and *random noise*. Thermal noise is due to the thermodynamic interchange of energy in a material or between a material and its surroundings. Random noise is due to any random transient disturbances. White noise in a communications system gives rise to loudspeaker hiss or television-screen snow. Impulse noise is due to a single momentary disturbance or a number of such disturbances when they are separated from one another in time. In audiofrequency amplifiers this type of noise gives rise to clicks in the loudspeaker. *See also* Schottky noise; Jansky noise; signal-to-noise ratio; noise factor.

noise factor A measure of the *noise introduced into a circuit or device. It is defined as the ratio of the *signal-to-noise power ratio at the source to the corresponding ratio at the output, i.e. the ratio of the actual noise at the output to the noise at the output due only to the source.

noise level A measure of the loudness of a noise, expressed in either *decibels, which indicate the intensity level, or *phons, which measure equivalent loudness. These two units are approximately equal over a large part of the audible range.

no-load Operation of any electrical or electronic circuit, device, machine, etc., under rated operating conditions of volt-

age, speed, etc., but in the absence of a *load.

nondegeneracy The normal state of matter, i.e. matter not cooled to a very low temperature nor subject to excessive stress such that the density is abnormally high. A nondegenerate gas is characterized by Maxwell's law of the *distribution of velocities.

noninductive An electric circuit or winding in which, for the purpose in view, the effect of its inductance is negligible. A circuit completely devoid of inductance is very difficult to obtain.

nonlinear distortion *See* distortion (electrical).

nonlinear network *See* network.

nonlinear optics The study of the effects produced by the electric and magnetic fields of extremely intense beams of light, such as focused *laser beams. The usual classical treatment of the propagation of light – reflection, refraction, superposition, etc. – assumes a linear relationship between the electromagnetic light field and the response of the atomic system constituting the medium. The electric field associated with a focused beam of a high-power laser can be 10^8 volts per metre or more, which is sufficient to generate appreciable nonlinear optical effects.

One effect is the *self-focusing* of laser light. The passage of an intense beam through glass induces local variations in the refractive index of the glass, causing it to act as a converging lens. The beam therefore contracts, its intensity increases, and the contraction process continues. The effect can be sustained until the beam diameter is about 5 μm, when it becomes totally internally reflected.

Frequency mixing is another effect. If two intense laser beams of different frequencies are sent through a suitable dielectric crystal, the nonlinear effects can result in the production of radiation whose frequency is equal to the sum or the difference of the original frequencies.

non-Newtonian fluid *See* anomalous viscosity.

nonreactive An electric circuit or winding in which, for the purpose in view, the *reactance is negligible.

nonreactive load A *load in which the alternating current is in phase with the terminal voltage. *Compare* reactive load.

nonthermal radiation Electromagnetic radiation, such as *synchrotron radiation, that is produced by the acceleration of electrons or other particles but is nonthermal in origin, i.e. it is not *black-body radiation.

nonvortical field *See* curl.

NOR gate *See* logic circuit.

normal The normal to any surface at a point is the line perpendicular to the tangent plane at that point.

normal distribution *Syn.* Gaussian distribution. *See* frequency distribution.

normalization The process of introducing a numerical factor into an equation $y = f(x)$ (in which $y \to 0$ as $x \to \pm \infty$) so that the area under the corresponding graph (if finite) shall be made equal to unity. The process is of importance (*a*) in quantum mechanics (where an extended definition is applicable) and (*b*) in statistics, where the total area under the error equation graph represents the probability of a value of x lying between $+\infty$ and $-\infty$ and must be 1.

(*See* frequency distribution, where $h\pi^{-1/2}$ is the *normalizing factor*.)

normal stress *See* stress.

note 1. A musical sound of specified pitch (frequency) produced by a musical instrument, voice, etc.
2. A sign in musical score representing pitch and time value of a musical sound.

NOT gate *See* logic circuit.

nova A faint *variable star that can undergo a considerable explosion during which the *luminosity increases by up to 100 000 times. The peak luminosity is attained in a few hours or days, then a slow decrease occurs over months or years until the original luminosity is reached.

A nova occurs in a *binary star system where the two components are very close. One component is a *white dwarf. The other star is expanding and is losing mass (hydrogen) to the white dwarf, around which a disc of gas forms. The gas can spiral down to the surface of the white dwarf, and after some 10 000 to 100 000 years enough has accumulated to cause a thermonuclear explosion – the nova outburst. The explosion does not disrupt the binary system, however, and the flow of gas from the other star can continue. *Compare* supernova.

n-p-n transistor *See* transistor.

NTP Abbreviation for normal temperature and pressure (no longer used). *See* STP.

n-type conductivity Conductivity in a *semiconductor caused by a flow of *electrons. *Compare* p-type conductivity.

n-type semiconductor An extrinsic *semiconductor in which the density of conduction *electrons exceeds that of mobile *holes.

nuclear barrier *Syn.* Gamow barrier. A region of high potential energy that a charged particle must pass through in order to enter or leave an atomic nucleus.

nuclear cross section *See* cross section.

nuclear energy Energy released by nuclear reactions, in particular *nuclear fusion and *fission. *See* binding energy; fusion reactor; nuclear reactors.

nuclear energy change *See* Q-value.

nuclear fission *See* fission.

nuclear fusion A *nuclear reaction between light atomic nuclei with the release of energy (*see* binding energy). Such reactions can be caused in the laboratory using an *accelerator to produce beams of deuterons or other light nuclei to bombard suitable targets. In such processes the energy required is incomparably greater than that released. To give a net energy output it is necessary to use *thermonuclear reactions*, i.e. those reactions that occur in a *plasma at very high temperatures.

The greater part of *stellar energy is released by thermonuclear reactions. The processes involve *weak interactions and are only possible because of the great quantity of matter involved. The development of fusion energy on earth requires *strong interactions. This was first achieved in 1952 in the *hydrogen bomb. Considerable efforts are being made to produce a controlled thermonuclear reaction (*see* fusion reactors).

The principal fusion reactions that are likely to be used in any future fusion reactors are:

$$^2H + ^2H \rightarrow ^3He + n \qquad + 3.2 \text{ MeV}$$

$$^2H + ^2H \rightarrow ^3H + ^1H \qquad + 4.0 \text{ MeV}$$

$$^2H + ^3H \rightarrow ^4He + n \qquad + 17.6 \text{ MeV}$$

$$^2H + ^3He \rightarrow ^4He + ^1H \quad + 18.3 \text{ MeV}$$

$$^6Li + ^1H \rightarrow ^3He + ^4He \quad + 4.0 \text{ MeV}$$

$$^6Li + ^3He \rightarrow ^4He + {}$$
$$^4He + ^1H \qquad\qquad + 16.9 \text{ MeV}$$

$$^6Li + ^2H \rightarrow ^7Li + ^1H \qquad + 5.0 \text{ MeV}$$

$$^6Li + ^2H \rightarrow ^3He + {}$$
$$^4He + n \qquad\qquad + 2.6 \text{ MeV}$$

$$^6Li + ^2H \rightarrow 2\,^4He \qquad + 22.4 \text{ MeV}$$

$$^7Li + ^1H \rightarrow 2\,^4He \qquad + 17.5 \text{ MeV}$$

In all of these processes in which neutrons are produced, there will be extra energy released by the interactions of these particles in surrounding substances. Typically 8 MeV is released, usually by means of gamma radiation, on capturing a neutron.

nuclear heat of reaction *See* Q-value.

nuclear isomerism When nuclei exist with the same mass number and atomic number (*see* isobar; isotopes) but show different radioactive properties, they are said to be nuclear isomers. They represent different energy states of the nucleus.

nuclear magnetic resonance (NMR) An effect observed when radio-frequency radiation is absorbed by matter. A nucleus with a *spin has a nuclear *magnetic moment. In the presence of an external magnetic field this magnetic moment precesses (*see* precession) about the field direction. Only certain orientations of the magnetic moment are allowed and each of these has a slightly different energy.

If the nucleus has a spin I there are $2I + 1$ different quantum states due to this quantization; each is characterized by a different value of the magnetic quantum number m, which can have values

$$I, I - 1, \ldots -(I - 1), -I.$$

The difference in energy depends on the strength of the applied magnetic field. The nucleus can make a transition from one state to another with the emission or absorption of electromagnetic radiation according to a selection rule that $\Delta m = \pm 1$.

The technique used is to apply a strong magnetic field ($\sim$2 tesla) to the sample, which is usually a liquid or solid. This field can be varied. A radiofrequency field (1–100 MHz) is imposed at right angles and a small detector coil is wound around the sample. As the magnetic field is varied, the spacing of the energies changes and at a certain value of the magnetic field, this spacing is such that radio-frequency radiation is strongly absorbed. This *resonance produces a signal in the detector coil. A plot of the detected signal against the magnetic field gives an NMR spectrum that can be used for determining nuclear magnetic moments. The energies of the nuclei depend to some extent on the surrounding orbital electrons and this is useful in studying chemical compounds. The technique is widely used in chemistry. It is also now used in medicine to construct images showing the variation of proton density in organs such as the heart, breasts, and lungs.

See also electron-spin resonance.

nuclear magneton *See* magneton.

nuclear medicine The branch of medicine concerned with the application of radioactive nuclei in diagnosis and therapy. The storage of the radioactive materials, instruments for the measurement and visualization of distributions of activity in a patient, interpretation of the results, and the production of suitable compounds for use are interrelated problems calling for cooperation between clinicians and physicists.

A radioisotope will follow the same path inside the body as a nonradioactive, normally ingested isotope of the same element, and will accumulate in the same areas. Measurement of the radioactivity at these areas will indicate any abnormal activity in the body. A high level of radioactivity means that there are overactive cancer cells present.

Equipment commonly used includes *scintillation counters, *Geiger counters, *scanners, and *gamma cameras, often with computer analysis of the outputs of this equipment. *See also* radiotherapy.

nuclear power station A power station using a *nuclear reactor as the source of energy.

nuclear reaction A reaction between an atomic *nucleus and a bombarding particle or photon leading to the creation of a new nucleus and the possible ejection of one or more particles. Nuclear reactions are often represented by enclosing within brackets the symbols for the incoming and outgoing particles or quanta, the initial and final nuclides being shown outside the brackets. For example:

$$^{14}N(\alpha, p)^{17}O$$

represents the reaction:

$$^{14}_{7}N + ^{4}_{2}He = ^{17}_{8}O + ^{1}_{1}H.$$

nuclear reactors *Energy from nuclear fission.* On the whole, the nuclei of atoms of moderate size are more tightly held together than the largest nuclei, so that if the nucleus of a heavy atom can be induced to split into two nuclei of moderate mass, there should be a considerable release of energy (*see* binding energy). The uranium isotope ^{235}U will readily accept a neutron but the nucleus

325

^{236}U so formed is very unstable and one-seventh of the nuclei stabilize by gamma emission while six-sevenths split into two parts (*see* fission). Most of the energy released (about 170 MeV) is in the form of the kinetic energy of these fission fragments. In addition an average of 2.5 neutrons of average energy 2 MeV and some gamma radiation is produced. Further energy is released later by radioactivity of the fission fragments. The total energy released is about 3×10^{-11} joule per atom fissioned, i.e. 6.5×10^{13} joule per kg conserved.

To extract energy in a controlled manner from fissionable nuclei, arrangements must be made for a sufficient proportion of the neutrons released in the fissions to cause further fissions in their turn, so that the process is continuous (*see* chain reaction). At present, the energy released is transferred by heat and is used in the same way as ordinary fuel in order to raise steam, etc.

Types of reactor. A reactor with a large proportion of ^{235}U or plutonium ^{239}Pu in the fuel uses the fast neutrons as they are liberated from the fission; such a reactor is called a *fast reactor*. Natural uranium contains 0.7% of ^{235}U and if the liberated neutrons can be slowed down before they have much chance of meeting the more common ^{238}U atoms, the latter are not likely to absorb them. A high proportion of the neutrons will then travel on until they meet a ^{235}U atom and then cause another fission. To slow the neutrons, a *moderator* is used containing light atoms to which the neutrons will give kinetic energy by collision. As the neutrons eventually acquire energies appropriate to gas molecules at the temperature of the moderator, they are then said to be *thermal neutrons and the reactor is a *thermal reactor*.

Thermal reactors. In a typical thermal reactor, the fuel elements are rods embedded as a regular array in the bulk of the moderator. The typical neutron from a fission process has a good chance of escaping from the relatively thin fuel rod and making many collisions with nuclei in the moderator before again entering a fuel element. Suitable moderators are pure graphite, heavy water (D_2O), and ordinary water (H_2O). Very pure materials are essential as some unwanted nuclei capture neutrons readily. The reactor *core is surrounded by a *reflector* made of suitable material to reduce the escape of neutrons from the surface. Each *fuel element is encased (e.g. in magnesium alloy or stainless steel can) to prevent escape of radioactive fission products. The *coolant, which may be gaseous or liquid, flows along the channels over the canned fuel elements. There is an emission of gamma rays inherent in the fission process and also many of the fission products are intensely radioactive. To protect personnel the assembly is surrounded by a massive *biological shield, of concrete, with an inner iron *thermal shield* to protect the concrete from high temperature caused by absorption of radiation.

To keep the power production steady, *control rods are moved in or out of the assembly. These contain material that captures neutrons readily (e.g. cadmium or boron). The power production can be held steady by allowing the currents in suitably placed *ionization chambers automatically to modify the settings of the rods. Further absorbent rods, the shut-down rods, are driven into the core to stop the reaction, as in an emergency if the control mechanism fails. To attain high thermodynamic efficiency so that a large proportion of the liberated energy can be used, the heat should be extracted from the reactor core at a high temperature. (*See* gas-cooled reactors, boiling-water reactors, pressurized-water reactors, and heavy-water reactors.)

Fast reactors. In fast reactors no moderator is used, the frequency of collisions between neutrons and fissile atoms being increased by enriching the natural uranium fuel with ^{239}Pu or additional ^{235}U atoms that are fissioned by fast neutrons. The fast neutrons thus build up a self-sustaining chain reaction. In these reactors the core is usually surrounded by a blanket of natural uranium into which some of the neutrons are allowed to escape. Under suitable conditions some of these neutrons will be captured by ^{238}U atoms forming ^{239}U atoms, which are converted to ^{239}Pu. As more plutonium can be produced than is required to enrich the fuel in the core, these are called *fast breeder reactors.

nuclear recoil The mechanical recoil suffered by the residual nucleus of an atom on radioactive or other disintegration. It can lead to physical effects such as abnormally high volatilities, or to chemical effects such as initiation of polymerization or molecular rupture. *See also* Mössbauer effect.

nuclear weapons The first nuclear weapon was the *atom bomb* or *A-bomb*, which consisted of two small masses of *fissile material each of which was below the *critical mass. When the bomb was detonated the two subcritical masses were brought rapidly together to form a supercritical mass within which a single *fission set off an uncontrollable *chain reaction. This first bomb consisted of only a few kilograms of uranium-235, but it had an explosive effect of 20 000 tons (20 kilotons) of TNT. Later models of this fission bomb used plutonium to even greater effect, but these weapons are small compared with the *hydrogen* or *fusion* or *H-bomb*. This consists of a fission bomb surrounded by a layer of hydrogenous material, such as lithium deuteride. The fission bomb elevates the temperature of the hydrogen to its *ignition temperature so that a *nuclear fusion reaction takes place with the evolution of enormous quantities of energy – equivalent to tens of megatons of TNT (1 kg of TNT releases about 4 MJ). *See also* fall-out.

nucleon The collective term for a *proton or *neutron, i.e. for a constituent of an atomic nucleus. *See* atom; nucleus; isospin.

nucleonics The practical applications of nuclear science and the techniques associated with these applications.

nucleon number *See* mass number.

nucleosynthesis The creation of the chemical elements by nuclear reactions. Hydrogen and helium were created in the very early universe (*see* big-bang theory). After 100 seconds, protons and neutrons could combine to form deuterium nuclei, which could then combine to form helium nuclei. Most of the helium in the universe today was formed at this time. Once the temperature had dropped sufficiently, electrons and protons could form hydrogen atoms. The heavier elements are synthesized in nuclear reactions occurring largely in stars and in *supernova explosions.

nucleus The most massive part of the atom, having a positive charge given by Ze, where Z is the *atomic number of the element and e the charge on an electron. The radius has been shown to be related to the mass number A of an atom by the formula: $r = cA^{1/3}$, where c is a constant equal to 1.2×10^{-15} m.

Nuclei consist of protons and neutrons collectively called *nucleons. The number of protons in nuclei of the same element is equal to the atomic number Z. The number of neutrons N associated with the Z protons varies within limits, the different numbers of neutrons giving rise to the various isotopes of that element. The total number of nucleons in a given isotopic nucleus is called the mass number A.

A nucleus is completely defined by the value of its atomic number Z and mass number A. This allows an abbreviated form of nomenclature in which a given nucleus is represented by its chemical symbol with Z and A as subscript and superscript respectively, e.g. the common uranium isotopes are $^{235}_{92}U$ and $^{238}_{92}U$.

The nucleons are maintained within a roughly spherical volume by forces inside the nucleus (*see* strong interaction; weak interaction). These attractive binding forces act between pairs of nucleons, being operative over a distance that is less than the nuclear radius. Various theories, such as the *liquid-drop model and the *shell model, have been put forward to explain the structure of the nucleus. In addition the presence of Coulomb repulsive forces between the protons limits the numbers of neutrons that can combine with a given number of protons to give stable nuclei. (*See* magic numbers.)

The mass of a given nucleus is always less than the sum of the rest masses of the constituent nucleons. This is due to the emission of energy on creation of the nucleus. The energy equivalent of the mass lost (*mass defect) is indicative of the degree of cohesion of the nucleons and is known as the *binding energy of the nucleus.

Most naturally occurring atoms have stable nuclei. The naturally occurring radioactive atoms have unstable nuclei giving rise to nuclear transmutations in which the atomic number is altered and the product atom is chemically different. Artificial nuclei are produced by bombarding stable nuclei with high-energy charged particles such as protons, deuterons, etc., or with neutrons. The collision process that occurs is called a *nuclear reaction. *See also* radioactivity; elementary particles; particle physics.

nuclide An atom as characterized by its *atomic number, *mass number, and nuclear energy state. *Compare* isotope.

null method *Syn.* balance method. A method of measurement in which the quantity being measured is balanced by another of a similar kind so that the indicating instrument reading is adjusted to zero (as with a *Wheatstone bridge).

number density Symbol: n. The number of particles, atoms, molecules, etc., per unit volume.

number of poles Of a switch, circuit-breaker, or similar apparatus. The number of different electrical conducting paths that the device closes or opens simultaneously. The device is described as single-pole, double-pole, triple-pole, or multipole if it is suitable for making or breaking an electrical circuit on one pole, two poles, three poles, or more than one pole, respectively.

numerical aperture (NA) A parameter used to express the angle of view of *objectives used in *microscopes and also the light-gathering power of these lenses. It is the product of the refractive index of the medium in which the objective is situated (air, oil, etc.) and the sine of half the angle of view of the objective. As the numerical aperture is increased, the *resolving power of the microscope improves.

Nusselt number Symbol: Nu. The dimensionless group $(hl/K\theta)$, where h is the rate of loss of heat per unit area of a hot body immersed in a fluid, l is a typical dimension of the body, θ is the temperature difference between the body and the fluid, K is the thermal conductivity of the fluid. *See* convection (of heat).

nutation *See* Eulerian angles.

Nyquist noise theorem The law that relates the power P due to thermal *noise in a resistor to the frequency f of the signal. At ordinary temperatures T,

$$dP = kTdf,$$

where k is the *Boltzmann constant.

O

object Extended natural objects consist of points, self-luminous or otherwise, that deliver diverging pencils of light. They are classed as *real objects* when they are delivering rays to some optical system under consideration. Commonly an optical system follows some preceding system that may focus an *image, real or virtual, in front of the second system – this image becomes a real object for the second system. If the second system lies in such a position as to intercept the rays before they have converged to a focus, the real image from the first system becomes the *virtual object* for the second system. Incident rays are convergent when a virtual object is under consideration.

The *object space* is a mathematical conception covering both the region lying in front of the system (real) and behind the system (virtual) in which real or virtual objects may lie, and possessing the same refractive index throughout – that of the preceding region. It completely coexists with the similarly conceived image space.

objective The lens (generally compound), in an optical instrument, that lies nearest to the object viewed. The term is sometimes applied to a mirror used for the same purpose.

oblique astigmatism *See* radial astigmatism.

observable The name used for the measurable things of physical science. In *quantum mechanics they are represented by matrices (matrix form of quantum mechanics) or, alternatively, by operators (wave mechanics).

occlusion The absorption of gases by solids.

occultation The passage of an astronomical body in front of another, especially the moon in front of a star or planet, thus obscuring its light, radio emission, etc. Planets can sometimes occult stars and also their satellites. The precise timing of an occultation provides information about the obscured body and also the obscuring body. *See* eclipse.

octave An *interval having the frequency ratio 2:1.

octode A *thermionic valve having five grids between the cathode and the main anode and an additional anode between the two innermost grids (i.e. a total of eight electrodes).

ocular *See* eyepiece.

odd-even nucleus A nucleus that contains an odd number of *protons and an even number of *neutrons. There are more than fifty stable nuclides with odd-even nuclei.

odd-odd nucleus A nucleus that contains an odd number of both *protons and *neutrons. Most of these nuclei are unstable but $^2_1$H, $^6_3$Li, $^{10}_5$B, and $^{14}_7$N are stable.

odd parity *See* wave function.

oersted Symbol: Oe. The *CGS electromagnetic unit of magnetic field strength. $1 \text{ Oe} = 10^3/4\pi \text{ A m}^{-1}$.

ohm Symbol: Ω. The *SI unit of electric *resistance, defined as the resistance between two points on a conductor through which a current of one ampere flows as a result of a potential difference of one volt applied between the points, the conductor not being the source of electromagnetic force. This unit replaced (1948) the *international ohm* (Ω_{int}), defined as the resistance of a column of mercury of mass 14.4521 grams, length 106.300 cm at 0 °C, and uniform cross section. $1 \Omega_{int} = 1.000 49 \Omega$.

ohmic contact An electrical contact in which the potential difference across it is proportional to the current flowing through it.

ohmic loss Power dissipation in an electrical circuit arising from its resistance, rather than from other causes such as magnetic hysteresis.

Ohm's law The electric current I in a conductor is directly proportional to the potential difference V between its ends, other quantities (especially temperature) remaining constant. As the resistance R is defined by $R = V/I$ the law can be stated as $R =$ constant. (The equation defining R is true even when the resistance is not constant, for example, for a hot filament, so the equation does not express the law.)

Ohm's law is very accurately obeyed by pure metals and alloys over a very large range of currents. It is less reliable for nonmetals.

oil-immersion (microscope). *See* resolving power; immersion objective.

Olbers' paradox A paradox (Olbers, 1826) stating that the generally accepted idea of an infinite number of stars, uniformly distributed in space, would mean that the night sky should glow with uniform brightness. The assumption of a finite number of stars is unnecessary to explain the dark night sky as it is now thought that the recession of galaxies, as indicated by their *redshifts, causes the brightness of distant stars to be greatly diminished.

omega-minus particle Symbol: Ω^-. An *elementary particle classified as a *hyperon. *See* unitary symmetry.

omni-aerial *Syn.* omnidirectional aerial. An aerial of any type that is essentially nondirectional, i.e. is equally effective as an emitter (or collector) of radiation in all directions having the same angle of elevation. *Compare* directive aerial.

ondograph An instrument that produces a graph of an alternating voltage.

opacity The ratio of the *radiant flux incident on an object (e.g. part of an exposed and processed photographic plate) to the flux transmitted. It is the reciprocal of *transmittance. It is a measure of the ability of a solid, liquid, or gaseous body to absorb radiation.

op-amp *See* operational amplifier.

open circuit *See* circuit.

operating point The point on the family of *characteristic curves of an active electronic device, such as a transistor, that represents the magnitudes of voltage and current for the particular operating conditions under consideration.

operational amplifier *Syn.* op-amp. A very high *gain voltage *amplifier, with very high input impedance, usually having a differential input (i.e. its output voltage is proportional to (and very much greater than) the voltage difference between its two inputs). It is invariably used with considerable *feedback, which determines its *transfer characteristics. The feedback circuits make op-amps operate as voltage amplifiers with a gain precisely defined by the values of resistance and/or capacitance used in the feedback circuits, or else enable them to perform mathematical operations such as integration or signal-conditioning functions such as filtering. They are thus used in a very wide range of instrumentation and control applications.

An operational amplifier is usually a multistage device designed for insertion into other equipment. It is supplied as a complete packaged unit, commonly as a single monolithic *integrated circuit.

Oppenheimer–Phillips (O–P) process *See* stripping.

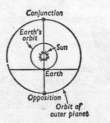

Opposition and conjunction

opposition 1. The positions of any two celestial bodies are in opposition with respect to a third body, such as the earth, when they lie on diametrically opposite sides of the third body (*see* diagram). The sun and moon are in opposition at full moon. Any two celestial bodies are in *conjunction* when a third reference body lies on an extension of the line joining the two bodies. The sun and moon are in conjunction at new moon.

2. Two periodic quantities with the same frequency and waveform are in opposition when they differ in *phase by 180°.

optical activity The ability of certain solutions and crystals to rotate the plane of polarization of *plane-polarized light in proportion to the length of substance traversed and to the concentration in the case of solutions. The angle through which the plane is rotated is the *angle of optical rotation*, symbol: α, and is measured in radians. When looking towards the oncoming light, if the rotation is clockwise the optical activity is called right-handed (*dextrorotatory*); if the rotation is anticlockwise, it is called left-handed (*laevorotatory*).

optical axis *See* optic axis.

optical bench A rigid table, wooden beam, steel girder, etc., that permits the easy slide of mounts for lenses, mirrors, photometer heads, etc., along a straight course and with attached scales to enable the position (or travel) of the optical element to be accurately determined. Designs differ according to the purpose and accuracy required. It is used for focal-length determination, testing of aberrations, photometer slide, interference and diffraction experiments, etc.

optical centre A spot marked on the surface of a lens where the optic axis intersects the surface. For thicker lenses, it is defined as that axial point through which all undeviated rays pass (incident and emergent rays are parallel with or without lateral displacement). Its position depends only on the surface radii and thickness, and is independent of wavelength.

optical density *See* transmission density.

optical disk A light, cheap, and physically compact computer storage device. The most common form uses audio compact-disc technology, and cannot therefore be written to by a computer. Writeable and erasable forms are also available.

optical distance *See* optical path.

optical fibres *See* fibre-optics system.

optical flat A surface that is so flat that there are no irregularities of surface flatness greater than a fraction of the wavelength of light. The surface is tested by observing interference fringes when a known flat is placed in contact (*optical contact*).

optical glass Glass with which special precautions are taken during manufacture to avoid mechanical and optical defects (density, strain, heterogeneity, colour, refractive index, etc.). The lens designer requires a selection of glasses with prescribed refractive indexes and dispersions measured to a high degree of accuracy, in order to provide lenses corrected to various degrees of accuracy for chromatic and spherical aberration. The glass manufacturer therefore provides catalogues of glass varieties such as hard crown, dense barium crown, light barium flint, light flint, etc. These catalogues show the *refractive index for yellow D light (n_0), mean *dispersion ($n_F - n_C$), *Abbe number, etc.

optically negative crystal A crystal in which *double refraction occurs and in which the refractive index (ω) for the ordinary ray is greater than the index (ε) for the extraordinary ray. In an *optically positive crystal*, $\varepsilon > \omega$. This is the case for *uniaxial crystals. In *biaxial crystals, β is nearer to γ than to α (op-

tically negative) or β is nearer to α than to γ (optically positive). Here, α, β, γ are the principal refractive indices, written in ascending order ($\alpha < \beta < \gamma$).

optically positive crystal *See* optically negative crystal.

optical maser Former name for laser.

optical path *Syn.* optical pathlength; optical distance. The distance traversed by light multiplied by the refractive index of the medium (nd). When light passes through different media the total pathlength is
$$n_1 d_1 + n_2 d_2 + \ldots = \Sigma (nd).$$
It is the distance in a vacuum that would contain the same number of waves as occur in the actual path in the medium, so that the optical paths between two wavefronts are all the same.

optical pyrometer A *pyrometer in which the luminous radiation from the hot body is compared with that from a known source. The instrument therefore measures the temperature of a luminous source without thermal contact.

optical rotation *See* optical activity.

optical switch A device whose optical properties (e.g. refractive index and polarizing properties) can be varied by an externally applied field or other influence; electric, magnetic, and surface acoustic wave techniques are all used for this purpose. Light (often a laser beam) can thus be deflected from a detector, so switching the beam.

optical window *See* atmospheric windows.

optic axis 1. The path of rays passing through the centres of the entrance pupil and exit pupil of an optical system (*see* apertures and stops in optical systems). The cardinal points lie on this line.

2. The direction (not a single line) in a doubly refracting crystal in which the ordinary and extraordinary rays apparently do not exhibit *double refraction, while their velocities are equal, i.e. the direction in a crystal along which the polarized components of a ray of light will be transmitted with a single velocity.

optics A branch of physics concerned with the study of light, its production, propagation, measurement, and properties. Since to a first degree of approximation, light travels in straight lines, the ray treatment of light is called *geometric optics as distinct from *physical optics, which attempts to explain the objective phenomena of light. There is an interweaving of optical interest in illuminating engineering, optical engineering, optical working, meteorology, astronomy, etc. *See also* nonlinear optics; optoelectronics.

optoelectronics The convergence of optical and electronic technology in gathering, processing, storing, and displaying information: it is concerned with the generation, processing, and detection of optical signals that represent electrical quantities. Major areas of application include communications and computing. *See also* fibre-optics system; light-emitting diode; semiconductor laser; photodiode; phototransistor.

optoisolator An optoelectronic device whereby two unconnected electric circuits can exchange signals by means of an optical link yet remain electrically isolated.

orbit A curved path, such as that described by a planet or a comet in the field of force of the sun, or by a particle in a field of force. The term is especially used for the locus of an extranuclear electron in an *atom in the Rutherford–Bohr atomic model.

orbital 1. *See* atomic orbital.
2. *See* molecular orbital.

orbital quantum number *Syn.* orbital angular-momentum quantum number; azimuthal quantum number. A *quantum number that governs the orbital angular momentum of a particle, atom, nucleus, etc. The symbol is l for a single entity or L for a whole system. *See* atomic orbital.

orbital velocity The velocity required by a *satellite or spacecraft to enter and maintain a particular orbit around the earth or some other celestial body. The orbital velocity needed for a 24-hour orbit (*see* geostationary orbit) around the earth is approximately 3.2 kilometres per second, at an altitude of about 36 000 km.

order of interference or diffraction A whole number that characterizes a position of an interference fringe according to whether there is interference arising from one, two, three, etc., wavelength difference of path, or according to the direction of the maxima of illumination produced by diffraction.

order of magnitude The value of a number or of a physical quantity given roughly, usually within a power of 10. Thus, 2.3×10^5 and 6.9×10^5 are of the same order of magnitude and 5×10^8 is 3 orders of magnitude greater than either.

ordinary ray *See* double refraction.

OR gate *See* logic circuit.

orthogonal

orthogonal 1. Mutually perpendicular, as in orthogonal axes.
2. Having or involving a set of mutually perpendicular axes as in orthogonal crystals.

orthorhombic (rhombic) system *See* crystal systems.

orthoscopic Free from distortion. Photographic objectives should be orthoscopic as should good-quality magnifiers and eyepieces.

oscillating current (or voltage) A current (or voltage) waveform whose amplitude periodically increases and decreases with time according to some mathematical function. Oscillating waveforms may be sinusoidal, sawtooth, square, etc.

oscillation 1. A vibration.
2. A periodic variation of an electrical quantity, such as current or voltage.
3. A phenomenon that occurs in an electrical circuit if the values of self-inductance and capacitance in the circuit are such that an *oscillating current arises from a disturbance of the electrical equilibrium in the circuit. A circuit in which oscillations can freely take place is called an *oscillatory circuit*. Oscillations that result from the application of a direct voltage input to the circuit and continue until the direct voltage is removed are called *self-sustaining oscillations* (*see* oscillator). *See also* free vibrations; forced oscillations; parasitic oscillations.

oscillator An electric circuit designed specifically to convert direct-current power into alternating-current power, usually at relatively high frequencies. Application of the direct-voltage supply to the circuit is usually sufficient to cause it to oscillate, and for the electrical *oscillations to be maintained until the direct voltage is switched off.

A simple oscillator consists essentially of a frequency-determining device, such as a *resonant circuit, and an active element that supplies power to the resonant circuit and also compensates for damping due to resistive losses. The active element can be considered as supplying a *negative resistance of sufficient value to counterbalance the positive resistance of the resonant circuit. Once started, then, the oscillations will continue.

The effective negative resistance can be provided by the use of positive *feedback to overcome the damping, or by means of an electronic device that exhibits negative resistance on a portion of its characteristic curve. *See also* piezoelectric oscillator; relaxation oscillator.

oscillatory circuit *See* oscillation.

oscillograph An *oscilloscope equipped to make a permanent record (an *oscillogram*) of the parameter being measured.

oscilloscope Short for *cathode-ray oscilloscope.

osmosis A process in which certain kinds of molecules in a liquid are preferentially transmitted by a *semipermeable membrane. There is diffusion of a solvent through the membrane into a more concentrated solution. For example, parchment will allow water molecules to pass but will hinder sugar molecules in a solution. The net effect is that more water passes into the sugar solution than out of it and a hydrostatic pressure builds up, increasing the diffusion of water out of the sugar solution, until ultimately a state of dynamic equilibrium is reached. The hydrostatic pressure then balancing osmosis is called the *osmotic pressure*, symbol: Π.

Van't Hoff showed theoretically that, at great dilution and if the solute molecules do not dissociate, the osmotic pressure of a solution is equal to the pressure that the solute molecules would exert if they were a gas of the same volume. Thus the osmotic pressure of a solution is given by $\Pi V = RT$, where V is the volume of solution containing unit amount of solute.

Electrolytes have a higher osmotic pressure than predicted by this equation. This is because electrolytes in solution dissociate into ions, so increasing the number of entities present. For electrolytes the equation takes the form $\Pi V = iRT$, where i is known as the *van't Hoff factor* and is the ratio of the actual number of entities present in the solution to the number that would be present if no dissociation occurred.

osmotic pressure *See* osmosis.

Ostwald's dilution law A law obtained by applying the law of *mass action to *electrolytic dissociation. If unit amount of acid is dissolved in a volume (V) of water, the acid dissociates into ions according to the equation:

$$HA \rightleftharpoons H^+ + A^-.$$

At equilibrium the concentrations of acid and ions are related by:

$$[H^+][A^-]/[HA] = \alpha^2/(1 - \alpha)V = K,$$

where α is the fraction of acid dissociated and K a constant. K is the equilibrium constant of the reaction and is often called the *dissociation constant*. For weak *electrolytes α is much less than 1 and the law is often stated as $\alpha = \sqrt{KV}$. As the dilution of the electrolyte increases, the degree of dissociation also increases.

Ostwald viscometer An instrument for measuring or comparing the viscosities of liquids, consisting of two bulbs at different heights connected by a capillary tube. The times t_1 and t_2 taken for each liquid to flow out of the upper bulb are measured. The viscosity is then given by $\eta_1/\eta_2 = \rho_1 t_2/\rho_2 t_1$, where ρ is the density.

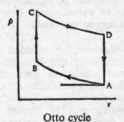

Otto cycle

Otto cycle A four-stroke cycle, two strokes being charging and exhausting processes. After the explosive mixture of air and petrol has been drawn in, it is compressed adiabatically, AB, and fired at B after which the pressure and temperature are increased rapidly at constant volume, BC. The piston then moves out causing adiabatic expansion, CD, after which the exhaust valve opens reducing the pressure to atmospheric, DA. The next inward motion of the piston sweeps out the exhaust gases to complete the cycle.

outgassing 1. In any vacuum system, the removal by heating of some of the air adsorbed on the inside surfaces of the system.

2. The slow deterioration of the vacuum due to the release of adsorbed gases from the interior surfaces of the vacuum system.

out of phase *See* phase.

output 1. The power, voltage, or current delivered by any circuit, device, or plant.

2. The terminals or other place where the signal is delivered.

output impedance The *impedance presented to the *load by a circuit or device.

output transformer A transformer used to couple an output circuit (usually of an *amplifier) to the *load.

overcurrent release *Syn.* overload release. A tripping device that operates when the current exceeds a predetermined value (usually adjustable). A current that causes the release to operate is called an *overcurrent*. The tripping action may be delayed for a definite time. *Compare* undercurrent release.

overdamped *See* damped.

overload (electrical) Any *load that exceeds the rated output of a machine, transformer, or other apparatus. It is expressed numerically as the amount of the excess or may be given as a percentage of the rated output.

overload release *See* overcurrent release.

overshoot *See* pulse.

overtone A constituent of a musical note other than the fundamental or lowest tone. The first overtone is the second *harmonic. Overtone and upper *partial are synonymous terms.

overvoltage *Syn.* excess voltage. Voltage that exceeds the normal value between two conductors or between a conductor and earth.

overvoltage release A tripping device that operates when the voltage exceeds a predetermined value.

Owen bridge A four-arm a.c. bridge used to measure the self-inductance, *L*, of an element in terms of the capacitance, *C*, and resistance, *R*, of other ele-

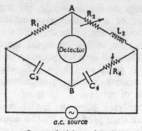

Owen bridge circuit

ments in the arms (*see* diagram). The current in each arm is balanced such that there is no potential difference between points A and B. Then

$$R_1 R_4 C_3 = L_2; \qquad R_1 C_3 = R_2 C_4.$$

These relationships are thus independent of frequency. R_1, C_3, and C_4 are known values. R_2 is varied until a minimal current flows through the detector (a galvanometer, etc.); R_4 is then varied until the current is zero.

oxygen point The temperature of equilibrium between liquid and gaseous oxygen at a pressure of one standard atmosphere, taken as a fixed point in the *International Practical Temperature Scale at 90.188 kelvin.

ozone layer *Syn.* ozonosphere. *See* atmospheric layers.

P

pachimeter An instrument for measuring the elastic shear limit of a solid material.

package A set of computer programs that is directed at some application in general, such as computer graphics, computer-aided design, word processing, statistics, mathematics, or mapping, and that can be tailored to the needs of a particular instance of that applicationn.

packing density 1. The amount of information in a given dimension of a computer storage medium, e.g. the number of *bits per inch of magnetic tape.
2. The number of devices or logic *gates per unit area of an *integrated circuit.

packing fraction The difference between the exact nuclear mass M of a nucleus and its *mass number A, divided by the mass number; $f = (M-A)/A$. The curve of $f \times 10^4$ against A shows a minimum at about $A = 50$, the packing fraction being negative for mass numbers between 16 and 180. Positive values indicate a tendency to instability, and nuclides with these mass numbers ($16 > A > 180$) can be used in *nuclear fusion and *fission processes. *See* binding energy.

pad 1. *See* attenuator.
2. *See* bonding pad.

paired cable *Syn.* twin cable. A cable composed of bundles of *twisted pairs encased in an outer protective sheaf. There may be several thousand pairs in a large cable.

pair production The simultaneous formation of a *positron and an *electron from a *photon. It occurs when a high-energy gamma-ray photon (> 1.02 MeV) passes close to an atomic nucleus. *See also* annihilation.

palaeomagnetism The study of the residual magnetization of certain rocks in order to determine the direction of polarization of the earth's magnetic field at the time of the rock's formation. The age of the rock can be found by radiometric *dating. A graph of polarity versus time shows that the earth's field has

reversed many times during its history (i.e. north and south poles have interchanged) and that there is a variable period of time, a *magnetic interval*, between reversals.

panchromatic film A photographic film that is sensitive to all colours of the visible spectrum.

parabolic reflector *Syn.* paraboloid reflector. A concave paraboloid reflecting surface, i.e. with a shape that results from rotating a parabola about its axis of symmetry. A beam of radiation striking the surface parallel to the axis is reflected to a single point (the focus) no matter how wide the aperture. The surface is thus free of *spherical aberration (but not of *coma). Such surfaces are used in reflecting telescopes and radio and microwave dishes. Conversely, a small source at the focus delivers parallel rays with little divergence. This is used in microscope illuminators, searchlight condensers, directive aerials, etc.

parallax 1. If a remote object is viewed from two points at the end of a baseline, the angle between lines drawn from the object to each end of the base is the parallax. In astronomy, the parallax of remote celestial bodies is measured with respect to various baselines. For example, the *annual parallax* of a star is the maximum angle subtended at the star by the mean radius of the earth's orbit around the sun, the angle earth-sun-star being 90°.
2. The apparent change in the separation between two objects when viewed from different positions. Consider two objects in line with an eye; if the latter moves to the right, the more distant object will appear to have moved to the right of the nearer object. Objects farther away show the same parallactic displacement as the eye displacement.

parallel

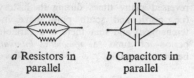

a Resistors in parallel *b* Capacitors in parallel

parallel Involving the simultaneous transfer (or in computing the simultaneous processing) of the individual parts of a whole. Circuit elements connected so that the current divides between them and later reunites, are said to be *in parallel*. For resistors of resistances r_1, r_2, r_3, ... r_n in parallel (Fig. *a*), the total resistance R is given by:

$$1/R = 1/r_1 + 1/r_2 + 1/r_3 + \ldots 1/r_n.$$

The current in any branch is: $i_n = i(R/r_n)$ where i is the total current. For capacitors of capacitances c_1, c_2, c_3, ... c_n in parallel (Fig. *b*), the total capacitance C is given by:

$$C = c_1 + c_2 + c_3 + \ldots c_n.$$

They behave as a large capacitor of the total plate area.

The greatest current a cell can produce (i.e. with zero external resistance) is limited by its internal resistance. Consequently, several cells may be used in parallel to reduce the total internal resistance of the cell circuit so that a large current can be drawn. *Compare* series. *See* shunt.

paramagnetism The property of substances that have a positive magnetic *susceptibility. It is caused by the *spins of electrons, paramagnetic substances having molecules or atoms in which there are unpaired electrons and thus a resulting *magnetic moment. There is also a contribution to the magnetic properties from the orbital motion of the electron. The relative *permeability of a paramagnetic substance is thus greater than that of a vacuum, i.e. it is slightly greater than unity.

A paramagnetic substance is regarded as an assembly of magnetic dipoles that have random orientation. In the presence of a field the magnetization is determined by competition between the effect of the field, in tending to align the magnetic dipoles, and the random thermal agitation. For small fields and high temperatures, the magnetization produced is proportional to the field strength. (At low temperatures or high field strengths, a state of saturation is approached.) As the temperature rises the susceptibility falls according to *Curie's law or the *Curie–Weiss law.

Solids, liquids, and gases can exhibit paramagnetism. Some paramagnetic substances are ferromagnetic below their Curie temperature. (*See* ferromagnetism; diamagnetism; antiferromagnetism; ferrimagnetism.)

Certain metals, such as sodium and potassium, also exhibit a type of paramagnetism resulting from the magnetic moments of free, or nearly free, electrons in their conduction bands. This is characterized by a very small positive susceptibility and a very slight temperature dependence. It is known as *free-electron paramagnetism* or *Pauli paramagnetism*.

parameter A quantity that is constant in a given case but takes a particular value for each case considered.

parametric amplifier A low-noise microwave amplifer in which gain is achieved by periodically varying the *reactance of the device, usually by applying an external voltage to a *varactor. Energy can then be transferred from the external signal so that amplification is achieved.

parasitic capture *Capture of a neutron by an atomic nucleus without any consequent nuclear *fission occurring.

parasitic oscillations Unwanted oscillations that may occur in the circuit of an amplifier or oscillator. Such oscillations usually have a frequency very much higher than the frequencies for which the circuit has been designed, since it is mainly determined by stray inductances and capacitances (e.g. in connecting leads), and interelectrode capacitances.

paraxial rays Rays of light close to the optic axis of a system.

parent *See* daughter product.

parity *Syn.* space-reflection symmetry. Symbol: P. The principle of parity invariance states that no fundamental distinction can be made between left and right; that the laws of physics are the same in a right-handed system of coordinates as they are in left-handed system. This is true for all the phenomena described by classical physics.

A *wave function, $\psi(x, y, z)$, describing a quantum mechanical system is said to have parity $+1$ if it remains unchanged on reflection through the origin, i.e. $\psi(x, y, z) = \psi(-x, -y, -z)$. If $\psi(x, y, z) = -\psi(-x, -y, -z)$ the wave function is said to have parity -1. In general, wave functions do not have a definite parity since $\psi(-x, -y, -z)$ will not usually be proportional to $\psi(x, y, z)$. It is found that the wave functions describing individual elementary particles have a definite symmetry under reflection. This means that an intrinsic parity can be associated with elementary particles.

The parity of the total wave function describing a system of elementary particles is conserved in *strong and *electromagnetic interactions. However, *weak interactions do not exhibit parity invariance and parity is not conserved in these interactions. *Beta decay is an example of a process in which parity is not conserved. Parity invariance requires

that, if in some interaction a particle is produced with left-polarization (i.e. the particle spins in an opposite sense to the direction of motion), then it must also be possible for a right-polarized particle to be produced in a similar interaction and that, on average, equal numbers of each will occur. In beta decay it is found that the electron is always left-polarized.

parsec Symbol: pc. An astronomical unit of length equal to the distance at which a baseline of one *astronomical unit subtends an angle of one second of arc. 1 parsec = $3.085\,677 \times 10^{16}$ metres or about 3.26 light-years.

partial A musical note consists generally of the simultaneous sounding of a group of tones, the frequency of each tone usually being related to the generating tone (namely the fundamental) by the equation $A = nF$, where n is an integer, and A and F are frequencies of a tone of the series and of the fundamental respectively. Tones above the fundamental are known as overtones or upper partials, these two terms being synonymous. Harmonics are partials, but, since partials may be inharmonic, the converse may not be true. The fundamental is known as the first harmonic or first partial; the octave, second note of the harmonic series, as the second harmonic or second partial or first overtone. The quality of a note depends upon the intensity and number of the upper partials in relation to the funadmental.

partial pressure The partial pressure of a gas in a mixture of gases occupying a fixed volume is the pressure that the gas would exert if it alone occupied the total volume. *See also* Dalton's law of partial pressures.

particle physics The study of the structure and properties of *elementary particles and *resonances and their interactions (*see* electromagnetic, strong, and weak interactions; quantum chromodynamics; quantum electrodynamics). The interactions of particles are responsible for their scattering and transformations (decays and reactions). During the first forty years of the twentieth century only the proton and neutron (nucleons) and the electron and positron had been detected; however, the number of particles with a claim to elementarity now approaches 150 (excluding resonances). The task of the particle physicist has been to try to correlate systematically the properties of these particles. *See also* gauge theories; grand unified theories.

particle velocity Symbol: *u*. The alternating component of the velocity of a medium that is transmitting sound. It is thus the total velocity of the medium minus the velocity that is not due to sound propagation. As the velocity is changing regularly with time, the *root-mean-square value is usually taken.

partition function *See* statistical mechanics.

parton A hypothetical pointlike particle postulated to be associated with *quarks in *nucleons. They have been used in *quantum chromodynamics to help in the understanding of high-energy experiments on atomic nuclei.

pascal Symbol: Pa The *SI unit of *pressure, defined as the pressure that results from a source of one newton acting uniformly over an area of one square metre.

Pascal's principle Pressure applied at any point of a fluid at rest is transmitted without loss to all other parts of the fluid. *See* hydraulic press.

Paschen–Back effect An effect similar to the *Zeeman effect, but applicable to magnetic fields so strong that the vectors due to orbital and spin angular momentum of electrons each separately take up their possible orientations relative to the field direction. The split pattern of spectral lines produced is quite different from those of the Zeeman effect, the lines being due to transitions between the quantum states of the electron orbits.

Paschen series *See* hydrogen spectrum.

Paschen's law The *breakdown voltage for a discharge between electrodes in gases is a function of the product of pressure and distance.

passband *See* filter.

passivation Protection of the junctions and surfaces of electronic components, including integrated circuits, from harmful environments. With silicon chips, a protective layer of silicon dioxide is usually formed on the surface.

passive aerial *See* directive aerial.

passive circuit A circuit or *network that contains only *passive components. Such circuits are capable only of attenuation, and are often designed for use as passive *filters.

passive component An electronic component that is not capable of amplifying or control function, e.g. resistors, capacitors, and inductors.

Pauli exclusion principle The principle that no two identical *fermions in any system can be in the same quantum state, that is have the same set of *quantum numbers.

The principle was first proposed (1925) in the form that not more than two electrons in an atom could have the same set of quantum numbers. This hypothesis accounted for the main features of the structure of the *atom and for the *periodic table. An electron in an atom is characterized by four quantum numbers, n, l, m, and s. A particular *atomic orbital, which has fixed values of n, l, and m, can thus contain a maximum of two electrons, since the *spin quantum number s can only be $+\frac{1}{2}$ or $-\frac{1}{2}$. In 1928 Sommerfeld applied the principle to the free electrons in solids and his theory has been greatly developed by later workers (*see* energy levels).

Pauli paramagnetism *See* paramagnetism.

p-channel device *See* field-effect transistor.

p.d. Abbreviation for potential difference.

peak factor The ratio of the *peak value of an alternating or pulsating quantity to its *root-mean-square value. For a *sinusoidal quantity, the peak factor is $\sqrt{2}$.

peak forward voltage The maximum instantaneous voltage applied to a device in the forward direction, i.e. the direction of minimum resistance of the device.

peak inverse voltage The maximum instantaneous voltage applied to a device in the reverse direction, i.e. the direction of maximum resistance of the device. The peak inverse voltage applied to a rectifying device must be less than the *breakdown voltage of the device to prevent avalanche breakdown in a *semiconductor or arc formation in a valve. A rated value is often used to specify the maximum voltage that a device can withstand.

peak value *Syn.* amplitude. **1.** Of an alternating quantity. The maximum positive or negative value. The positive and negative values need not necessarily be equal in magnitude.
2. Of an *impulse voltage (or current). The maximum value of the voltage (or current).

Peltier effect *See* thermoelectric effects.

Peltier element *See* thermoelectric effects.

pencil (of rays) A slender cone or cylinder of rays that traverses an optical system, the pencil being limited by the *aperture stop. The central ray is the *chief ray*.

pendulum A device consisting of a mass, suspended from a fixed point, that oscillates with a known period, T. The various types are:
(1) *Simple pendulum*. A small weight suspended from a point by a light thread. The period of oscillation for small amplitudes of swing is determined by the formula:
$$T = 2\pi\sqrt{(l/g)},$$
where l is the length of the thread and g is the acceleration of *free fall.
(2) *Compound pendulum*. A rigid body of any convenient shape, e.g. a bar, swinging about an axis (usually a knife edge) through any point other than its centre of mass. The period, for small amplitudes of swing, is given by the formula for the simple pendulum in which l is replaced by:
$$\sqrt{(k^2 + h^2)/h},$$
where k is the *radius of gyration about a parallel axis through the centre of mass, and h is the distance of the centre of mass from the axis of swing.
(3) *Horizontal pendulum* A compound pendulum whose axis of rotation is nearly vertical. Used for finding the alteration in the direction of the force of

gravity with time. A massive horizontal pendulum is the basis of a seismograph (known as the *Galitzin pendulum*).

(4) *Conical pendulum* A simple pendulum in which the bob swings in a horizontal circle. The period, for a very small radius of swing only, is the same as for the simple pendulum.

See also compensated pendulum.

Penning ionization The ionization of gas atoms or molecules by collision with *metastable atoms. This process occurs if the *ionization potential of the atom or molecule is less than the energy released when the metastable atom reverts to the ground state. The excess energy is carried away as kinetic energy of the electron. If sufficient energy is available, it is possible to produce ions in excited states.

pentatonic scale *See* musical scale.

pentode A *thermionic valve with five electrodes. It is equivalent to a tetrode with an additional electrode between the screen grid and anode. This electrode, which is of open-mesh design, is at a negative potential with respect to both anode and screen so that it can prevent low-velocity secondary electrons from the anode returning to the screen.

penumbra *See* shadows.

perfect fluid *See* fluid.

perfect gas *See* ideal gas.

perigee The point in the orbit of the moon or an artificial earth satellite that is nearest the earth. The most distant point in an earth orbit is called the *apogee*.

perihelion The point in a solar orbit that is nearest the sun. The orbiting body could be a planet, comet, or artificial satellite. The earth is at perihelion on Jan. 3. The most distant point in a solar orbit is called the *aphelion*. The earth is at aphelion on July 4.

period *Syn.* periodic time. Symbol: T. The time occupied in one complete to and fro movement of a given vibration or oscillation. The period is related to the frequency v, and the angular frequency ω, by:
$$T = 1/v = 2\pi/\omega.$$

periodic table The classification of chemical elements, in tabular form, in the order of their *atomic numbers. The elements show a periodicity of properties, chemically similar elements recurring in a definite order. The sequence of elements is thus broken up into horizontal *periods* and vertical *groups*, the elements in each group showing close chemical analogies, e.g. in valency, chemical properties, etc. One form of periodic table is given in the Appendix, Table 8.

periodic time *See* period.

peripheral devices *Syn.* peripherals. Equipment that can be connected to and controlled by a *computer. Peripherals are external to the central processor of the computer. They may be input devices such as keyboards, output devices such as printers, or backing store such as disk or magnetic-tape units.

periscope An optical instrument to provide a view over or around an obstacle or from a submarine. In its simplest form, it consists of two parallel mirrors at 45° to the direction of view; the top mirror receives light from the object and directs it down to the lower mirror close to the eye of the observer.

permalloy Any of a variety of alloys with a high magnetic *permeability at low magnetic flux density and a low *hysteresis loss.

permanent gas A gas whose *critical constants are such that it remains gaseous under very high pressure at normal temperatures. A gas that cannot be liquefied by pressure alone.

permanent magnet A magnetized mass of steel or other ferromagnetic substances, of high retentivity and stable against reasonable handling. It requires a definite demagnetizing field to destroy the residual magnetism. *See* ferromagnetism.

permanent set The strain remaining in a material after the stresses have been removed.

permeability Symbol: μ. The ratio of the *magnetic flux density in a body or medium to the external *magnetic field strength inducing it, i.e. $\mu = B/H$. It has the unit *henry per metre. The *permeability of free space*, μ_0, is sometimes called the *magnetic constant and has the value $4\pi \times 10^{-7}$ H m^{-1} in the SI system. The *relative permeability*, μ_r, is the ratio μ/μ_0. For most substances μ_r has a constant value. If it is less than unity the material is diamagnetic (*see* diamagnetism); if μ_r exceeds unity it is paramagnetic (*see* paramagnetism). Ferromagnetic substances have high permeabilities, which are not constant but vary with the field strength (*see* ferromagnetism). *See also* permittivity.

permeance Symbol: $\wedge$. The reciprocal of *reluctance. It is measured in henrys.

permittivity Symbol: ε. The ratio of the *electric displacement in a dielectric medium to the applied *electric field strength, i.e. $\varepsilon = D/E$. It indicates the

degree to which the medium can resist the flow of electric charge. It is measured in farads per metre. The *permittivity of free space*, ε_0, is sometimes called the *electric constant. It is equal to $1/(c^2\mu_0)$, where c is the speed of light and μ_0 is the *permeability of free space. It has the value $8.854\,187\,817 \times 10^{-12}$ F m^{-1}.

The *relative permittivity*, ε_r, is the ratio of the permittivity of a medium to the permittivity of free space, i.e. $\varepsilon/\varepsilon_0$. The value varies from unity (for a vacuum) to over 4000 (for *ferroelectrics) but normally does not exceed 10. The quantity ε_r is also called the *dielectric constant* when it is independent of electric field strength and refers to the dielectric medium of a capacitor. Dielectric constant is better defined under these conditions as the ratio of the capacitance of the capacitor to the capacitance it would possess if the dielectric were removed.

persistence *Syn.* afterglow. **1.** The interval of time following excitation during which light is emitted from the screen of a *cathode-ray tube.

2. The faint luminosity, observable in certain gases for a considerable period after the passage of an electric discharge.

personal equation A systematic error of measurement, made by an experienced observer. If the observer tires, or is inexperienced, he ceases to have a personal equation (which can be corrected for), and contributes random errors.

perturbation theory An approximate method of solving a difficult problem if the equations to be solved depart only slightly from those of a problem already solved. For example, the orbit of a single planet round the sun is an ellipse; the perturbing effect of the other planets modifies the orbit slightly in a way cal-

343

culable by this method. The technique finds considerable application in *wave mechanics and in *quantum electrodynamics. Phenomena that are not amenable to solution by perturbation theory are said to be *nonperturbative*.

perveance The space-charge-limited characteristic between electrodes in an electron tube. It is equal to $j/V^{3/2}$, where j is the current density, and V the potential of the collector.

peta- Symbol: P. A prefix denoting 10^{15}; for example, one petajoule (1 PJ) = 10^{15} joules.

Petzval surface *See* curvature of field.

Pfund series *See* hydrogen spectrum.

pH A logarithmic measure of the hydrogen ion (or hydroxonium ion, H_3O^+) concentration of a solution. It equals the logarithm to the base 10 of the reciprocal of the hydrogen (or hydroxonium) ion concentration in moles per dm^3. Thus, if there are 10^{-8} mol dm^{-3} of hydrogen ions present, the solution has a pH of 8. If the pH is greater than 7, the solution is alkaline, and if it is less, the solution is acid.

phase 1. The state of development of a periodic quantity or process, specifically the fraction of the whole *period that has elapsed, measured from some fixed datum. A quantity that varies sinusoidally may be represented by a rotating vector, OB, whose length is proportional to the *peak value of the quantity. OB rotates through 360° about O during the period, T, of the oscillation. The angular velocity, ω, of OB is related to the frequency, ν, of the oscillation by:
$$\nu = 1/T = \omega/2\pi.$$
The phase of the quantity with respect to another such quantity, OA, is given

by the angle, α, between OB and OA. This is called the *phase angle if the two quantities have the same frequency (*see also* phase difference).

Particles in periodic motion due to the passage of a wave are said to be in the same phase of vibration if they are moving in the same direction with the same relative displacement. Particles in a wavefront are in the same phase of vibration and the distance between two wavefronts in which the phases are the same is the wavelength. For the simple harmonic wave,
$$y = a \sin 2\pi(t/T - x/\lambda),$$
the phase difference of the two particles at x_1 and x_2 is $2\pi(x_2-x_1)/\lambda$. When light is reflected at the surface of a denser medium, there is a change of phase of π.

Periodic quantities having the same frequency and *waveform are said to be *in phase* if they reach corresponding values simultaneously; otherwise they are said to be *out of phase*. If the waveforms are not alike, these terms are used in connection with the fundamental components of the waves.

2. One of the separate circuits or windings of a *polyphase system, machine, or other apparatus.

3. *See* phase rule.

phase angle The angle between the two vectors that represent two sinusoidal alternating quantities having the same frequency. (*See* phase.) The term may be used in connection with periodic quantities that are not sinusoidal but that have the same fundamental frequency and it is then the angle between the vectors representing their fundamental components. *See* phase difference.

phase constant *Syn.* phase-change coefficient. *See* propagation coefficient.

phase-contrast microscope *See* microscope.

phased-array radar *See* radar.

phase delay The ratio of the *phase shift undergone by a periodic quantity to its frequency.

phase diagram A graph combining two conditional parameters (e.g. temperature, pressure, entropy, volume) of a substance drawn so that a particular curve represents the boundary between two phases of the substance. For example, on a plot of pressure against volume the *critical temperature isothermal separates the gas phase from the vapour or liquid phase; on a plot of pressure against temperature, the *triple point represents the intersection of the three curves demarcating the solid/liquid, liquid/gas, and solid/gas boundaries.

phase difference 1. Symbol: ϕ. The difference of *phase between two sinusoidal quantities that have the same frequency. It may be expressed as a time or as an angle (the *phase angle).
2. In an instrument transformer. The angle betweeen the reversed secondary vector (current in a current transformer, voltage in a voltage transformer) and the corresponding primary vector. The phase difference is positive or negative according to whether the reversed secondary vector leads or lags the primary vector respectively.

phase discriminator A *detector circuit in which the amplitude of the output wave is a function of the *phase of the input wave.

phase modulation A type of *modulation in which the phase of the *carrier wave is varied about its unmodulated value by an amount proportional to the amplitude of the modulating signal and at a frequency equal to that of the modulating signal, the amplitude of the carrier wave remaining constant.

phase plate *See* microscope (phase-contrast).

phase rule Substances are capable of existing in very different states of aggregation. These states are called *phases*, each phase forming a homogeneous and physically distinct part of the system. Thus, water at ordinary temperatures and pressures can exist in a solid, liquid, or vapour phase. The term *component* is applied to the least number of chemically identifiable substances required to define completely the existing phases. Thus, in the three phases described above, there is only one component, namely water. The phase rule is defined by the equation:

$$F = C - P + 2,$$

where C is the number of components in the system, P the number of phases present, and F the number of *degrees of freedom* of the system (the least number of independent variables defining the state of the system). Thus, if there is only one phase present and one component, namely water, the number of degrees of freedom is two; that is, the system is not completely defined until the pressure and the temperature are fixed. If two phases are present, the number of degrees of freedom is one; that is to say, to each pressure there corresponds a fixed temperature and on the pressure-temperature diagram (*see* phase diagram) there is a definite line separating the liquid from the vapour phases. Similar lines exist separating the solid-liquid phases and the solid-vapour phases. These three curves meet at the *triple point and this represents the only pressure and temperature at which the three phases can exist in equilibrium.

phase shift Of a periodic quantity. Any change that occurs in the *phase of one quantity or in the *phase difference between two or more quantities.

phase space A multidimensional space in which the coordinates represent the variables required to specify the state of the system, in particular a six-dimensional space incorporating three dimensions of position and three of momentum.

phase speed The speed with which wave crests and troughs travel through a medium; in fact the speed with which the phase in a homogeneous train of waves is propagated. It is expressed by λ/T, λ being the wavelength and T the period of vibration. This is equivalent to v/σ, v being the frequency of vibration and σ the *wavenumber. *Compare* group speed.

phase splitter A circuit that has a single input signal and produces two separate outputs with a predetermined *phase difference. An example is the *driver for a *push-pull amplifier.

phase waves *See* de Broglie waves.

phlogiston theory A theory of combustion, refuted by *Lavoisier in the 18th century. All combustible materials were supposed to contain phlogiston, which was released during combustion to leave calx (ash).

phon The unit of equivalent *loudness of a sound, judged subjectively; it is a measure of the intensity level relative to a reference tone of defined intensity and frequency. The accepted reference tone has a *root-mean-square sound pressure of 2×10^{-5} pascal and a frequency of 1000 hertz (this being the threshold intensity at this frequency). A normal observer listens with both ears to the standard tone and the sound being measured alternately. The standard tone is varied in intensity until the observer judges it to be as loud as the sound under test. If then the standard tone is n *decibels above the reference intensity, the equivalent loudness of the sound be-

ing measured is defined as n phons. With the decibel scale of intensity levels, the intensity of a note is referred to its threshold intensity at the same frequency. The decibel and phon scales are not the same since the sensitivity of the ear to changes of intensity varies with frequency. The two scales are nearly the same between 500 and 10 000 hertz, but below this range there is considerable variation.

phonon In a solid the atoms do not vibrate independently but the oscillations are transmitted through the substance as acoustic waves of extremely high frequency f (typically of the order of 10^{12} Hz). The energy transmitted by the waves is quantized; the quantum is called a phonon and has value hf, where h is the Planck constant.

For many purposes the phonons can be treated as if they were gas molecules moving within the space occupied by the solid, the *mean free path being limited by various scattering processes. *Defects and boundaries scatter generally. For large amplitudes the vibrations are *anharmonic and the phonons are then scattered by free electrons and by other phonons. These effects are very important in limiting *thermal conductivity and increasing electrical *resistivity, especially at high temperatures.

See also collective excitations.

phosphor A substance that exhibits *luminescence, emitting light at temperatures below the temperature at which it would exhibit *incandescence. Particular examples are fluorescent substances such as those used on the screens of cathode-ray tubes or in fluorescent lamps.

phosphorescence A type of *luminescence in which light emission may persist for some considerable time after the excitation has ceased.

photocathode A *cathode from which electrons are emitted as a result of the *photoelectric effect.

photocell Any light–electric *transducer. Originally the device was a *photoelectric cell* consisting of a valve diode with a *photocathode and *anode in which a current flows when the photocathode is illuminated. The word is now commonly used to designate a *photoconductive cell*, consisting simply of a slab of *semiconductor with *ohmic contacts fixed at opposite ends. When illuminated with light or other radiation of a suitable wavelength, a marked increase in conductivity occurs due to the generation of charge *carriers (*see* photoconductivity). A photocurrent then flows in an external circuit.

The term photocell is also sometimes · applied to a *photodiode or a *photovoltaic cell.

photochromic substance A substance that changes colour when light falls on it, often reverting to the original colour when the light is removed. The exposure of light generally causes darkening to take place. If the substance is transparent, the fraction of light transmitted is changed.

photoconductive cell *See* photocell.

photoconductivity Enhanced conductivity of certain *semiconductors (in particular grey selenium) as a result of exposure to light or other *electromagnetic radiation. The absorption of a photon in the material increases the energy of an electron in the valence band (*see* energy bands) of the solid. If the photon energy exceeds the *work function of the solid, the electron is liberated by the *photoelectric effect. However, if the photon energy is not sufficient to liberate the electron, it may be enough to increase the electron's energy so that it is excited into the conduction band. This is sometimes called the *internal photoelectric effect*. The presence of extra electrons in the conduction band causes the photoconductivity. *See also* photocell; photodiode.

photodetachment The removal of an electron from a negative ion by a *photon of electromagnetic radiation to give a neutral atom or molecule. The process is the same as for *photoionization of a neutral species and the *ionization potential of the negative ion is equal to the *electron affinity of the atom or molecule.

photodetector Any electronic device, such as a *photocell, *photodiode, or *phototransistor, that detects or responds to light.

photodiode A semiconductor *diode that produces a significant photocurrent when illuminated. There are various types. One form uses a p-n junction that is reverse biased but operated below the *breakdown voltage. When exposed to electromagnetic radiation of suitable frequency, excess charge carriers (electron-hole pairs) are generated as a result of *photoconductivity. The carriers normally recombine rapidly, but those produced in or near the *depletion layer present at the junction can cross the junction and produce a photocurrent. This is superimposed on the normally very small reverse saturation current. A common type of such a device is the *p-i-n photodiode*. This has a layer of intrinsic semiconductor between the p and n regions that wholly contains the depletion layer, allowing devices to be manufactured with a depletion width suitable for optimum sensitivity and frequency response.

photodisintegration *See* photonuclear reaction.

photoelasticity *Syn.* mechanical (or stress) birefringence. The effect whereby a normally transparent isotropic substance can become optically anisotropic under mechanical stress, and thus exhibit *double refraction. Marked effects are thus produced with polarized light. The phenomenon is used by engineers to study the stresses in structures by the examination of models made of transparent plastics.

photoelectric cell *See* photocell.

photoelectric constant The ratio of the *Planck constant to the charge of the electron.

photoelectric effect The liberation of electrons from matter by electromagnetic radiation of certain frequencies. For solids, electrons are only liberated when the wavelength of the radiation is shorter than a certain value (the *photoelectric threshold*). Most solids emit electrons when this value is in the *vacuum-ultraviolet region of the spectrum although some metals (e.g. Na, K, Cs, and Rb) and semiconductors emit for visible and *near-ultraviolet radiation.

Lenard (1902) showed that the maximum speed of the electrons was independent of the intensity and appeared to increase linearly with frequency. Their number was directly proportional to intensity for a given frequency. Einstein explained this behaviour by assuming that the energy of the incident radiation was transferred in discrete amounts (*photons), each of magnitude $h\nu$, where h is the *Planck constant and ν the frequency. Each photon absorbed will eject an electron provided that the photon energy ($h\nu$) exceeds a certain value Φ – the *work function. The maximum kinetic energy of the electrons, E, is then given by $E = h\nu - \Phi$. This is known as the *Einstein photoelectric equation*. The electrons with this maximum kinetic energy are the least strongly bound electrons in the solid. More strongly bound electrons will also be ejected with energies lower than E.

The photoelectric effect does not only apply to solids. Gases and liquids can also emit electrons under the effect of light. For gases each electron is removed from a single atom or molecule and the work function in the Einstein equation is replaced by the *ionization potential. (*See* photoionization.)

Electromagnetic radiation of very short wavelength (X-rays and gamma rays) causes the photoelectric effect in all substances. The Einstein equation applies with Φ replaced by the binding energy of the electron in an inner shell (*see* atomic orbital).

photoelectron spectroscopy A form of *electron spectroscopy for measuring the *ionization potentials of atoms and molecules (by irradiating a sample with monochromatic ultraviolet radiation), or for determining the energies of very strongly bound electrons in the inner *shells of atoms and molecules (by irradiating with monochromatic X-rays). The binding energy of the inner electrons in a particular element depends on the chemical compound of the element. The latter technique is therefore used for qualitative and quantitative analysis.

photoemission Emission of electrons as a result of bombardment by photons, as in *photoionization or the *photoelectric effect.

photofission *See* photonuclear reaction.

photography The production of permanent images by use of sensitized emulsions. Light falling on a photographic emulsion sets up a photoelectric chain of events in which some silver ions belonging to silver salts in the emulsion are converted to neutral silver atoms. In the development of this *latent image*, a chemical reducer liberates further silver atoms clustered round the original atoms giving opaque specks of metallic silver. Subsequent fixing (e.g with hypo, sodium thiosulphate) removes unaffected silver salts so giving a photographic *negative*, which is darkest where the original image was lightest. A *positive* (print) is produced by exposure of a further emulsion, e.g. one coated on a paper backing, to light that has passed through the negative and then chemically processing the paper.

Colour photography is based on a *subtractive process. The film consists of three layers of emulsion, each of which is sensitive to one of three *primary colours: red, green, and blue. Following exposure, the blue components of the subject will be recorded as a latent image in the blue-sensitive emulsion and likewise for the red and green components. The final image in the print or transparency is formed from the correct proportions of cyan, magenta, and yellow, the *complementary colours of red, green, and blue.

See also reciprocity law; exposure; camera.

photoionization Ionization of an atom or molecule by electromagnetic radiation. A *photon of radiation can only remove an electron if its energy ($h\nu$) exceeds the first *ionization potential (I_1) of the atom. The excess energy ($h\nu - I_1$) is taken up by the positive ion and the electron and distributed between their kinetic energies. Since the mass of the ion is always much greater than that of the electron, its extra energy is negligible and thus $E = h\nu - I_1$, where E is the kinetic energy of the electron. This is simply the Einstein photoelectric equation (*see* photoelectric effect) applied to a single atom or molecule. The radiation capable of photoionizing molecules and atoms has energy greater than the

photoelectric threshold, which lies in the ultraviolet region of the spectrum. If the quantum energy of the radiation is high enough, more strongly bound electrons may be removed from the neutral species. For example, electrons may be ejected with energy E_2 ($<E_1$), where $E_2 = h\nu - I_2$ and I_2 is the second *ionization potential. The atom is then left in an electronically *excited state.

photolithography A technique used during the manufacture of *integrated circuits, *semiconductor components, *thin film circuits, and *printed circuits. In this technique a desired pattern is transferred from a photographic *mask onto the substrate material ready for a processing step to be carried out. The clean substrate is first covered with a solution of *photoresist, which is allowed to dry and is then exposed to light or UV through the mask. The depolymerized portions of the photoresist are then washed away with a suitable solvent, the polymerized portion remaining and acting as a barrier to etching substances or as a mask for deposition processes. When the processing step is complete, the photoresist may be stripped off with a suitable solvent.

photolysis The chemical decomposition or dissociation of molecules as a result of the absorption of light or other electromagnetic radiation.

photomagnetism Paramagnetism produced in a substance when it is in a phosphorescent state.

photometer *See* photometry.

photometry Photometry is concerned with measurements of light intensity and amounts of illumination. Two types of measurement are possible. In one, the radiation is evaluated according to its visual effects, i.e. according to judgment

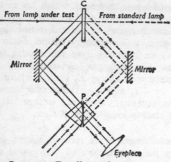

Lummer–Brodhun photometer

by observers (*see* luminosity and spectral luminous efficiency). The physical quantities measured in this way are preceded by the adjective *luminous*. This distinguishes them from physical quantities measured in units of energy, for which the adjective *radiant* is used. *Visual photometry* is the branch of photometry in which the eye is used to make comparisons. In *physical photometry* the measurements are made by physical receptors, such as the *photocell, *thermopile, and *bolometer.

The physical quantities measured in photometry include *luminous intensity, *luminous flux, *illuminance, and *luminance. Practical measurements of luminous intensity are made by comparing the intensity of the lamp under test with that of a standard lamp. The instrument used is called a *photometer*. The general method is to adjust the position of a screen with respect to the two lamps until the intensities of illumination of the screen due to the two lamps are equal. The luminous intensities of the two sources are then proportional to the squares of their respective distances from the screen.

In the (simplified) arrangement of the *Lummer–Brodhun photometer* (see diagram), the light reaching the eyepiece is seen to consist of a central pencil of rays from the left-hand side of the

screen C and an outer bundle from the right-hand side of C. When the intensities of illumination of the two sides of the screen are equal, the field of view seen through the eyepiece will be evenly illuminated. In practice a rather more complicated prism is used at P, and a match is looked for between the two evenly contrasted fields. Other photometers in which a comparison is made between two fields are the *grease-spot photometer and the *shadow photometer.

If the sources of light are of different colours, a *flicker photometer is used. A photometer for measuring the directional characteristics of a source is called a *goniophotometer*. *See also* integrating photometer.

photomicrography The recording of microscope images on photographic media. The recorded image forms a *photomicrograph*.

photomultiplier An *electron multiplier in which the primary electrons causing the cascade are produced by the *photoelectric effect. The cathode of such a tube is a *photocathode, which is illuminated by some means.

photon The *quantum of electromagnetic radiation. It has an energy of $h\nu$ where h is the *Planck constant and ν the frequency of the radiation. For some purposes photons can be considered as *elementary particles travelling at the *speed of light (c) and having a momentum of $h\nu/c$ or $h\lambda$ (where λ is the wavelength). Photons can cause *excitation of atoms and molecules and more energetic ones can cause ionization (*see* photoionization). *See also* Compton effect; photonuclear reaction.

photonuclear reaction *Syn.* photodisintegration. A reaction occurring when a high-energy *photon, such as a gamma ray or X-ray photon, collides with an atomic nucleus. As a result the nucleus disintegrates. In certain cases the photon appears to knock out a neutron (*photoneutron*) or proton (*photoproton*). In other cases the nucleus appears to absorb the photon and then break up as a result of its higher energy – in some cases fission occurs (*photofission*).

photoresist An organic photosensitive material used during *photolithography. Negative photoresists are materials that polymerize due to the action of light; positive photoresists are polymeric materials that are depolymerized by the action of light. The polymerized material acts as a barrier during processing steps.

photosensitivity The property of responding to electromagnetic radiation (especially light), a variety of responses being observed.

phototransistor A bipolar junction *transistor activated by light or UV. The *base electrode is left *floating and the base signal is supplied by excess *carriers generated by illumination of the base. Once equilibrium has been reached, the device behaves like a normal transistor with the base signal being a function of the intensity of the radiation.

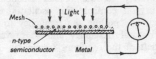

Photovoltaic cell

photovoltaic cell An electronic device that uses the *photovoltaic effect to produce an e.m.f. An example is the *solar cell, the basis of which is an unbiased p-n junction. Other types use a metal-semiconductor junction. These depend for their action on the formation of a potential barrier across the unbiased

junction (*see* Schottky effect); they are often called *rectifier photocells* or *barrier-layer photocells*. A typical structure is shown in the diagram. The contact to the n-type semiconductor is in the form of a mesh to minimize reflection of the incident light.

photovoltaic effect An effect arising when a junction exists between two dissimilar materials and one of the materials is exposed to electromagnetic radiation, usually in the range near-ultraviolet to infrared. The two materials may, for example, be a metal and a semiconductor or two semiconductors of opposite polarity (the combination forming a *Schottky barrier and a *p-n junction respectively). A forward voltage appears across the illuminated junction and power can be delivered to an external circuit (*see* photovoltaic cell).

The incident radiation imparts energy to electrons in the valence band (*see* energy bands), and electron-hole pairs are generated in the depletion region existing around the p-n junction and in the Schottky barrier. As the pairs are produced they are able to cross the junction (due to the inherent field) and produce the forward bias: a migration of electrons into the n-type semiconductor produces a negative bias while an excess of holes migrating into the p-type semiconductor or the metal produces a positive bias.

physical optics *See* optics.

pick-up A *transducer that converts information (usually recorded) into electric signals. The term is particularly applied to the electromechanical devices that reproduce the signals recorded in grooves in gramophone records. Several types of pick-up are in common use.

Crystal pick-ups consist of a *piezoelectric crystal that is stressed by the mechanical vibrations in the grooves of

351

the rotating record, producing a corresponding e.m.f.

Ceramic pick-ups are similar to crystal pick-ups in that their output is also due to the piezoelectric effect. The ceramic materials, e.g. barium titanate, are more reliable and stable under ambient conditions.

Magnetic pick-ups have a small inductance coil in the field of a magnet. Mechanical vibrations in the grooves of the record cause the coil to move and hence the magnetic flux through the coil changes. The induced current in the coil depends on the magnitude of the vibrations and provides the signal for the audio-system.

pico- Symbol: p. A prefix denoting 10^{-12}; for example, 1 picofarad (pF) = 10^{-12} farad.

piezoelectric crystal A crystal that exhibits the *piezoelectric effect. All *ferroelectric crystals are piezoelectric as well as certain nonferroelectric crystals and some ceramics. Examples include quartz crystal, Rochelle salt, and barium titanate (a ceramic).

piezoelectric effect An effect exhibited by *piezoelectric crystals whereby the surfaces become oppositely electrically charged when subject to stress; the sign of the charges changes when a compression of the crystal is changed to a tension. The converse effect, in which the crystal expands along one axis and contracts along the other when subjected to an electrical field, also occurs. The magnitude of the piezoelectric effect depends on the direction of the stress relative to the crystal axes. The maximum effect is obtained when the electrical and mechanical stresses are applied along the X-axis (the electric axis) and Y-axis (the mechanical axis) respectively. The third major axis in a piezoelectric crystal is

the Z-axis (the optic axis). *See* piezoelectric oscillator.

piezoelectric oscillator A highly stable *oscillator in which a *piezoelectric crystal is used to determine the frequency. If an alternating electric field is applied across the crystal, mechanical vibrations result (*see* piezoelectric effect). The crystal is usually cut with its major surface either perpendicular or parallel to the X-axis. It is then mounted between the plates of a capacitor in order to apply the alternating voltage. Normally a metallic film is formed on the crystal faces for this purpose. The crystal is supported by lightweight supports that touch it at mechanical nodes.

Piezoelectric crystals are most conveniently set in vibration by the aid of undamped electric oscillations, which may be generated electronically at any desired frequency and intensity. The connections between the crystal and the oscillator circuit may be made in various ways. In general, the circuits used can be divided into two main types.

In the *crystal oscillator*, the crystal replaces the *tuned circuit in the oscillator, thus providing the resonant frequency of the oscillator. In the *crystal-controlled oscillator*, the crystal is coupled to the oscillator circuit, which is tuned approximately to the crystal frequency. The crystal controls the oscillator frequency by *pulling the frequency to its own natural frequency and so preventing drift of the oscillator frequency.

pi-meson (π-meson). *Syn.* pion. *See* meson.

pinch effect An effect in which an electric current passing through a liquid or gaseous conductor tends to cause its cross section to contract as a result of the electromagnetic forces set up. *See* fusion reactor.

pinch-off *See* field-effect transistor.

p-i-n diode A semiconductor *diode with a region of almost *intrinsic semiconductor between the p-type and n-type regions. *See* IMPATT diode; photodiode.

pin grid array (PGA) A form of package used for complex *integrated circuits, capable of providing up to several hundred connections to one chip. Connections are made by means of an array (i.e. several parallel rows) of output pins on the package periphery, either on two opposite edges or around all four edges. The pins are connected through the casing to the bonding pads of the chip.

pion *See* meson.

Pirani gauge A low-pressure gauge in which electrically heated wire loses heat by conduction through the gas. A constant potential difference is maintained across the wire and its resistance variation with pressure observed. Alternatively, the resistance is kept constant by varying the applied p.d., which may then be measured.

piston gauge *See* free-piston gauge.

pitch 1. A subjective quality of a sound that determines its position in a musical scale. It may be measured as the frequency of the pure tone of specified intensity that is judged by the average normal ear to occupy the same place in the musical scale. Although pitch is measured in terms of frequency, it is also dependent on the loudness and quality of the note. As the intensity is increased, the pitch of a low-frequency note is lowered while that of a high note is raised.
2. Of a screw thread, etc. The distance apart of successive threads (or of successive teeth of a gear wheel).

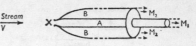

Pitot static tube

Pitot tube An instrument used for measuring the total (static and dynamic) pressure of a fluid stream. It consists essentially of a tube of small bore connected at one end to a *manometer, the other end being open and pointing upstream. Fluid cannot flow through the tube; the pressure registered at the manometer is the stagnation pressure at the nose of the tube, which, by *Bernoulli's theorem, will be $p_0 + \frac{1}{2}\rho V^2$, where p_0 is the static pressure, ρ the density, and V the velocity of the undisturbed stream.

The term Pitot tube is commonly used for the true *Pitot-static* tube (*see* diagram). The tube A represents the Pitot tube connected to the manometer M_1 (which registers the total pressure), the point X being the stagnation point. The tube B is the static tube connected to manometer M_2 registering the static pressure. The difference pressure is the quantity $\frac{1}{2}\rho V^2$. The Pitot-static tube is used to measure the stream velocity and is used on aircraft to measure the relative wind speed.

planar process The most commonly used method of producing junctions in the manufacture of *semiconductor devices. A layer of silicon dioxide is thermally grown on the surface of a silicon substrate of the desired conductivity type. Holes are etched in the oxide layer, using *photolithography, and suitable impurities are diffused into the substrate through these holes to produce a region of opposite polarity. The oxide acts as a barrier to diffusion, except through the holes. The impurities will diffuse into the substrate in directions normal and

353

planck

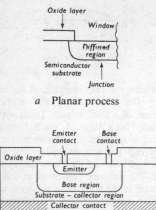

a Planar process

b Planar transistor

parallel to the surface so that the junction meets the surface of the substrate below the oxide (Fig. *a*). Several diffusions can be carried out, one after another. Usually a final layer of oxide is grown over the entire chip (except for the contacts) to provide a stable surface for the silicon and keep surface-leakage currents to a minimum. A planar transistor is shown in Fig. *b*.

planck The unit of *action. It is equal to one joule second.

Planck constant Symbol: *h*. A universal constant having the value $6.626\,076 \times 10^{-34}$ J s. *See* Planck's law.

Planck function Symbol: *Y*. The negative of the *Gibbs function divided by the thermodynamic temperature.

Planck length The length $(Gh/2\pi c^3)^{1/2}$, where *G* is the *gravitational constant, *h* is the *Planck constant, and *c* the speed of light; equal to $1.615\,99 \times 10^{-35}$ m. It arises in theories relating quantum theory to gravitation. *See* Planck mass.

Planck mass The mass $(hc/2\pi G)^{1/2}$, where *h* is the *Planck constant, *c* is the speed of light, and *G* is the *gravitational constant, equal to $2.176\,84 \times 10^{-8}$ kg. *See* Planck length.

Planck's formula *See* black-body radiation; radiation formula.

Planck's law The law that forms the basis of *quantum theory. The energy of *electromagnetic radiation is confined to small indivisible packets or *photons, each of which has an energy *hf*, where *f* is the frequency of the radiation and *h* is the *Planck constant.

Planck time The time $(Gh/2\pi c^5)^{1/2}$ taken for a *photon travelling at the speed of light to travel a distance equal to the *Planck length; equal to $1.708\,63 \times 10^{-43}$ s.

Planck units A system of units used in quantum theories of gravity based on the *Planck length, *Planck mass, and *Planck time. The *gravitational constant, the *speed of light, and the rationalized *Planck constant are all assigned the value of unity, thus all quantities that normally have dimensions involving mass, length, and time become dimensionless in this system.

plane of symmetry The plane across which reflection of each point in a lattice or other system of points will bring the system to self-coincidence.

plane-polarized light *Syn.* linearly polarized light. Light in which the vibrations are rectilinear, parallel to a plane, and transverse to the direction of travel. (In fact all electromagnetic radiation can be plane-polarized.) Light reflected from a surface of polished glass (of refractive index *n*) at an angle of incidence $\tan^{-1} n$ is plane-polarized with vibrations parallel to the surface (*Brewster's* *law*). It is said to be polarized in the plane of incidence, the plane of vibration (electric vector) being perpendicular to the plane of polarization. Plane-polarized light may be produced by reflection or transmission through a pile of plates, by *double refraction in dichroic substances, e.g. tourmaline, Polaroid, by *Nicol prisms, etc. Optically active substances rotate the plane of polarization. *See* optical activity; polarization.

planetary electron An electron orbiting around the nucleus of an *atom.

plane wave A wave in which the *wavefronts form a set of planes, each generally perpendicular to the direction of propagation. It is the simplest example of a three-dimensional wave. *See also* progressive wave.

planoconcave *See* concave.

planoconvex *See* convex.

plan position indicator (PPI) *See* radar.

Planté cell The first primitive accumulator, consisting of rolled lead sheets (Planté plates) dipping into dilute sulphuric acid.

plasma 1. A region of ionized gas in a *gas-discharge tube, containing approximately equal numbers of electrons and positive ions.
2. A highly ionized substance that can be formed at very high temperature (as in stars) or by *photoionization (as in interstellar gas). The atoms present are nearly all fully ionized and the substance consists of electrons and atomic nuclei moving freely. *See* fusion reactors.

plasma oscillations Under certain conditions oscillations of the ions and electrons (independent of the conditions of an external circuit) may be set up in the *plasma of a *gas-discharge tube, and they can cause a scattering of a stream of electrons greater than that explainable by ordinary gas collisions.

plasmon A *collective excitation for quantized *plasma oscillations, in which the free electrons in a metal are treated as a plasma.

plastic deformation A permanent deformation of a solid subjected to a stress. It is sometimes produced in single crystals of metals even by vanishingly small forces.

plastics Materials that, though stable in use at normal temperatures, are plastic at some stage in their manufacture, and can be shaped by the application of heat and pressure. There are two main types of plastic, namely *thermoplastic* and *thermosetting* compositions. The former are mouldable by heat and the shape impressed on them is preserved when they are cooled in the mould. The latter initially soften on heating and then harden on further heating in the mould. Most, if not all, plastic materials are high polymers, i.e. materials composed of giant molecules.

plate An electrode in a *capacitor or *accumulator.

plate tectonics The theory that the earth's surface is composed of a number of large but relatively thin slabs of rigid material that are moving relative to each other. These *plates* extend through the earth's crust into the upper mantle and are supported on the viscous layer known as the asthenosphere. Many of the earth's structures and processes, including oceanic trenches, midocean ridges, major faults, earthquake zones, volcanic belts, mountain building, and continental drift, are explained by the movements and consequent collisions of the plates.

platinum resistance thermometer *See* resistance thermometer.

Plumbicon *See* camera tube.

pneumatics The branch of physics dealing with the dynamic properties of gases.

p-n junction The region at which two *semiconductors of opposite polarity (p-type and n-type) meet. A p-n junction can perform various functions depending on the geometry, the bias conditions, and the doping level in each semiconductor region. Most *diodes, *transistors, etc., utilize the properties of one or more p-n junctions. If the materials are dissimilar, e.g. silicon and germanium, the junction is a *heterojunction*. Normally the same material is used but doped so as to produce two different conductivity types; this is a simple *homojunction*.

Under reverse-bias conditions (i.e. negative bias applied to the p-type semiconductor) very little current flows, until *breakdown occurs (*see* semiconductor; diagram at diode). Under forward-bias conditions, carriers are attracted across the junction into the region of opposite type (where they become *minority carriers) and a current flows in the external circuit. The forward current in a homojunction increases exponentially with the voltage, i.e.

$$I = I_0(e^{eV/kT} - 1),$$

where I_0 = reverse saturation current, e = electronic charge, V = applied voltage, k = Boltzmann constant, and T = thermodynamic temperature. Resistance in the material reduces the rate of rise of current through the device after a few tenths of a volt.

pnpn device A semiconductor device consisting of alternating layers of p-type and n-type semiconductor (almost always silicon), with at least three p-n junctions. Such devices are used for power-switching purposes, examples being the *silicon controlled rectifier and the *silicon controlled switch.

p-n-p transistor *See* transistor.

Pockel effect The *Kerr effect as observed in a *piezoelectric crystal.

Poinsot motion 1. The motion of a rigid body with one fixed point O, acted upon by no forces, or, **2.** the motion of a rigid body relative to the centre of mass O, provided that all the forces acting are equivalent to a single force through the centre of mass. In such motion, the direction and magnitude of the angular momentum vector drawn through O are constant at all times.

point-contact transistor *See* transistor.

point defect *See* defect.

point function A quantity whose value depends on the position of a point in space, e.g. magnetic field, temperature, density.

point group A set of symmetry operations (rotation about an axis, reflection across a plane, or combinations of these), not including translation, that when carried out on a periodic arrangement of points in space brings that system of points to self-coincidence. *See* crystal systems.

poise Symbol: P. The CGS unit of dynamic *viscosity. 1 poise = 0.1 pascal second.

poiseuille Symbol: Pl. A unit of dynamic *viscosity, defined as the viscosity of a liquid that sets up a tangential stress of one newton per square metre across two planes separated by one metre when the velocity of streamlined flow is one metre per second. The unit is identical with the *SI unit of dynamic viscosity (the pascal second), but the name has not received international recognition.

Poiseuille flow The steady laminar flow of a viscous fluid through a pipe of circular cross section such that the velocity distribution has the form of a paraboloid of revolution, the velocity being a maximum at the centre of the pipe, zero at the boundary walls, and the velocity being constant along any line parallel to the axis of the pipe. Assuming *Newton's laws of fluid friction for a viscous fluid, the quantity of fluid flowing per second is given by:

$$Q = \pi(p_1 - p_2)r^4/8\eta l,$$

where p_1, p_2 are the initial and final pressures on a cylinder of fluid, length l, and coefficient of viscosity η, the radius of the pipe being r. This is called the *Poiseuille equation* or the *Hagen–Poiseuille law*.

Poisson distribution A *frequency distribution often applicable in practice to discontinuous variables. It can be applied to a radioactive decay process to predict the probability that a specific event (decay) will occur in a given period. It is the limit of the *binomial distribution. As the number of trials, n, increases, the probability, p, decreases and $np = m$ where m is the average number of times an event occurs in n trials. The probability of r successes in n trials is then: $m^r e^{-m}/r!$.

Poisson equation In SI units:

$$\frac{\partial^2 V}{\partial x^2} + \frac{\partial^2 V}{\partial y^2} + \frac{\partial^2 V}{\partial z^2} = -\frac{\rho}{\varepsilon_0},$$

or $\nabla^2 V = -\rho/\varepsilon$, when V is the electric potential at any point, ρ is the charge density, and ε the permittivity.

Poisson ratio Symbol: μ or ν. The ratio of lateral contracting strain to the elongation strain when a rod is stretched by in-line forces applied to its ends, the sides being free to contract. If the volume does not change under stretching, this ratio = 0.5, but the value is often less in practice.

polar axis A crystal axis of rotation that is not normal to a reflection plane and does not contain a centre of symmetry. Certain crystal properties will be dissimilar at opposite ends of such an axis.

polar coordinates See coordinate.

polar diagram A graphical figure in which a physical quantity, such as the relative field strength radiated in any direction in a given plane by a transmitting aerial, is represented in *polar coordinates. Polar diagrams are usually drawn for planes that, relative to the surface of the earth, are horizontal and vertical. See also lobe.

polarimeter An accurate instrument for measurement of rotation of the plane of polarization (see plane-polarized light) by optically active liquids and solids. See saccharimeter.

polariscope An instrument for studying polarization phenomena. It consists of a *polarizer from which polarized light passes to a transparent substance under investigation and then to a rotatable analyser. The simple (Biot) polariscope, which depends on polarization by reflection, consists of two inclined glass plates, one to polarize the light and the other (rotatable) to analyse. Sometimes the analyser is a *Nicol prism or *Polaroid; sometimes both analyser and polarizer are Nicol prisms.

polarity 1. (general) The condition of a body or system in which there are opposing physical properties at different points.
 2. (magnetic) The distinction between the north- and south-seeking *poles of a magnet.
 3. (electrical) The distinction between the positive and negative parameters (e.g. voltage, charge, current, carrier type) in an electrical circuit or device.

polarization 1. (electrical) In a simple cell consisting of two dissimilar plates in an electrolyte, such as Zn and Cu in dilute H_2SO_4, the current obtained soon falls considerably. This is due to a layer of hydrogen bubbles that collects on the copper plate and not only partially covers the plate and increases the internal resistance of the cell, but also sets up an e.m.f. of opposite direction to that of the cell. This phenomena is known as polarization. To make cells effective for a longer period, some means must be adopted to prevent gas deposition. In the *Leclanché cell a chemical depolarizer is used, which reacts with the hydrogen produced.
 2. (of radiation) Unpolarized (natural or ordinary) *electromagnetic radiation consists of waves in which the vibrations are transverse without any one-sidedness or preferential direction of vibration. Under various circumstances, the direction and characteristics of the vibration are more restricted. In *plane-polarized light or other radiation, the electric field resides entirely in one plane, called the *plane of vibration*. Sometimes the direction of the electric-field vector is not restricted to one plane but rotates with constant angular frequency (as viewed along the direction of propagation), although its magnitude stays constant; this is described as *circu-*

larly polarized radiation. If the electric-field vector not only rotates but also changes its magnitude, then the radiation is *elliptically polarized*. *See also* quarter-wave plate.
 3. *See* dielectric polarization.
 4. *See* molecular polarization.

polarizer A crystal or conglomerate of crystals used to produce *plane-polarized light (pile of plates, *Polaroid, *Nicol prism, etc.).

polarizing angle *Syn.* Brewster angle. When light strikes a glass surface at an angle of incidence given by $\tan^{-1}(n)$, where n is the refractive index, the reflected light is *plane-polarized. At this angle of incidence, the refracted ray makes an angle of 90° with the reflected ray (*Brewster's law*).

Polaroid *Tradename* A thin transparent film containing ultramicroscopic polarizing crystals with their optic axes lined up parallel. One component of polarization is absorbed and the other is transmitted with little loss. There are several different methods of production, the original form using synthetic dichroic crystals; others use stretched polyvinyl alcohol films impregnated with iodine or treated with hydrogen chloride to make the film strongly dichroic (*see* dichroism). Polaroid provides a large area of polarization, used for example in motor car headlights, visors, sunglasses, camera filters. By placing the axis of the Polaroid in an appropriate direction, stray plane-polarized light (produced, for example, by reflection) causing glare, can be removed.

Polaroid (Land) camera *Tradename* A camera that yields finished positive prints or transparencies within about 60 seconds (colour) or 10 seconds (black and white) of the exposure, the developing and processing of the film taking place inside the camera.

polaron An electron coupled to an ion, which arises when an electron is introduced into the conduction band of a perfect ionic crystal inducing lattice polarization around itself.

pole 1. Of electrical apparatus or a circuit. Each of the terminals or lines between which the main circuit voltage exists. (*See* number of poles; quadripole.)
  2. The place toward which lines of magnetic flux converge, or from which they diverge. It usually exists near a surface of magnetic discontinuity, and in the material of higher permeability. A north pole of a magnet is that which is subject to a force towards the north (magnetic) pole of the earth.
  3. The midpoint of a convex or concave mirror. The line joining the centre of *curvature and the pole is the *principal axis* of the mirror.

pole face The end surface of the *core of a magnet through which surface the useful magnetic flux passes. In particular, in an electrical machine it is that surface of the core or pole piece of a *field magnet that directly faces the armature.

pole piece Either of the pieces of ferromagnetic material attached to the ends of an electromagnet or permanent magnet and specially shaped to control the flux in various electrical devices.

polychromatic radiation Electromagnetic radiation with more than one wavelength present.

polyphase system An electrical system or apparatus in which there are two or more alternating supply voltages displaced in *phase relative to each other.

A symmetrical polyphase (*n*-phase) system has *n* sinusoidal voltages of equal magnitude and frequency, with mutual phase differences of $2\pi/n$ radians (or $360/n$ degrees). An exception to this is the two-phase system (quarter-phase) in which the two voltages have a phase difference of 90°. For $n > 2$, the system requires a minimum of *n* line wires.

population inversion *See* laser.

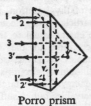

Porro prism

Porro prism A total reflection prism used in the construction of prismatic telescopes and binoculars. The simplest form is a 45°, 90° prism receiving light through the hypotenuse face, and reflecting it back parallel to the original direction after successive internal reflection at the other two faces. This prism inverts in one direction only. The second prism of a binocular completes the inversion at right angles by placing its roof edge at right angles to the first.

port An access point in an electronic circuit, device, network, etc., where signals can be fed in or out or where the variables of the system can be observed or measured.

position vector In mechanics, a line joining a point on the path of a particle to some reference point.

positive crystal *See* optically negative crystal.

positive electron *See* positron.

positive feedback *See* feedback.

positive glow *Syn.* positive column. *See* gas-discharge tube.

positive logic *See* logic circuit.

positive magnetostriction *See* magnetostriction.

positron *Syn.* positive electron. The *antiparticle of the *electron, i.e. an elementary particle with electron mass and positive charge equal to that of the electron. According to the relativistic *wave mechanics of Dirac, space contains a continuum of electrons in states of negative energy. These states are normally unobservable, but if sufficient energy can be given, an electron may be raised into a state of positive energy and become observable. The vacant state of negative energy behaves as a positive particle of positive energy, which is observed as a positron. (*See* pair production; annihilation.)

positronium A short-lived association between a *positron and an *electron, similar to a hydrogen atom. There are two types: *orthopositronium* in which the *spins of the particles are parallel and *parapositronium* in which the spins are antiparallel. The orthopositronium decays with a mean life of about 10^{-7} s to give three photons. The parapositronium has a smaller mean life and produces two photons (*see* annihilation).

post-office box A type of *Wheatstone bridge that consists of a number of resistance coils arranged in a special box and forming three arms of the bridge; the resistance to be measured forms the fourth arm. Each coil is connected between adjacent metal blocks so that it can be shorted out of circuit by inserting a metal plug between the blocks.

Resistances from 0.1 to 10^6 ohms can be measured.

potassium-argon dating *See* dating.

potential 1. Electrostatic, magnetostatic, and gravitational potentials, at a point in the field: the work done in bringing unit positive charge, unit positive pole, or unit mass respectively from infinity (i.e. a place infinitely distance from the causes of the field) to the point. Gravitational potential is always negative but the electrostatic and magnetostatic potentials may be positive or negative. Since these fields are *conservative, the potential is a function only of the position of the point. The difference in potential between two points is the work done in taking the unit object from one point to the other. Potential is a scalar quantity.
2. *See* kinetic potential.
3. *See* thermodynamic potential.

potential barrier The region in a field of force in which the potential is such that a particle, which is subject to the field, encounters opposition to its passage.

potential difference Symbol: V, U. The line integral of the electric field strength between two points. The work done when a charge moves from one to the other of two points (by any path) is equal to the product of the potential difference and the charge. *See* electromotive force.

potential divider A chain of resistors, inductors, or capacitors connected in series and tapped to allow a definite fraction of the voltage across the chain to be obtained across one or more of the individual components. *See* potentiometer.

potential energy Symbol: E_p, V, Φ. The work done in changing a system from some standard configuration to its present state. *See* energy.

potential flow *See* laminar flow.

potential function From the theory of functions of a complex variable, it is shown that a relation of the form:
$$w = \phi + i\psi = f(z),$$
represents a two-dimensional irrotational motion of a fluid in the xy-plane, where z is the complex variable $(x + iy)$, ϕ the *velocity potential, ψ the *stream function, i is $\sqrt{-1}$, and $f(z)$ means a function of z. The complex function $w = (\phi + i\psi)$ is called the potential function; equating the real and imaginary parts of the relation $w = f(z)$ gives the lines of equivelocity potential and the stream lines of the irrotational motion.

potential gradient The rate of change of electric potential, V, at a point with respect to distance x, measured in the direction in which the variation is a maximum. It is measured in volts per metre. The electric field strength, E, is numerically equal to the potential gradient but in the opposite sense:
$$E = -dV/dx.$$

potential scattering *See* scattering.

potential transformer *See* voltage transformer.

potential well A region in a field of force in which the potential decreases abruptly, and on either side of which the potential is greater.

potentiometer 1. A form of *potential divider using a uniform wire as the resistive chain. A moving sliding contact can tap off any potential difference less than that between the ends of the wire. A typical use is in measuring an unknown e.m.f., such as that of a cell, C (*see* diagram). A battery B supplies a steady current along the resistance wire XY. The slider S is moved until the galvanometer G shows no deflection. The distance XS $= l_1$ is noted, the cell is replaced by a standard cell, and a new point of balance l_2 is found. Then, $E_1/E_2 = l_1/l_2$, where E_1 and E_2 are the e.m.f.s of the unknown and standard cell respectively. It will be seen that the true e.m.f. is given, as no current flows through the cell. For precision work, more elaborate forms of potentiometer are available.

2. (pot) Any variable resistor with a third movable contact. The geometry of the device can be arranged so that the output voltage is a particular function (linear, logarithmic, sine or cosine) of the applied voltage.

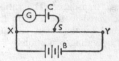

Potentiometer

powder photography A crystal-diffraction method in which the specimen is a randomly orientated crystalline powder (which is usually rotated), in a parallel beam of monochromatic X-radiation (or electrons, neutrons, etc.). *See* X-ray analysis.

power 1. Symbol: P. The rate at which energy is expended or work is done. It is measured in *watts.

The power P developed in a direct-current electric circuit is given by the expression $P = VI$, where V is the potential difference in volts and I is the current in amperes. In an alternating-current circuit, the *active power*, P, is given by $VI \cos \phi$, where V and I are the rms values of the voltage and current and ϕ is the *phase angle between the current and the voltage. The product IV, called the *apparent power*, is measured in volt-amperes. $\cos \phi$ is called the *power factor. $IV \sin \phi$ is called the *re-active power* and is measured in *vars.

2. (optic) The power of a lens or mirror is the reciprocal of the focal length in metres, generally positive if converging and most commonly applied to the *dioptric power* of a lens. For mirrors, the term *catoptric power* is sometimes used. *See also* magnifying power; dispersive power.

power amplification 1. In an *amplifier. The ratio of the power level at the output terminals to that at the input terminals.

2. In a *magnetic amplifier. The product of voltage amplification and current amplification using a specified control circuit.

3. In a *transducer. The ratio of the power delivered to the load to the power absorbed by the input circuit, under specified operating conditions.

power amplifier An *amplifier producing an appreciable current flow into a relatively low impedance or a large increase in output power. It is usually used so that the output is not applied to the input of a further amplifying stage, but to an output *transducer, such as an *aerial or *loudspeaker.

power component The *active current, *active voltage, or *active volt-amperes.

power factor The ratio of the actual *power in watts developed by an a.c. system (as measured by a wattmeter) to the apparent power in volt-amperes (as indicated by voltmeter and ammeter readings). If the voltage and current are *sinusoidal, the power factor is equal to the cosine of the *phase angle between the voltage and current vectors.

power-level difference Symbol: L_P. The quantity

$$\tfrac{1}{2} \log_e (P_1/P_2),$$

where P_1 and P_2 are two powers. The term *sound power level* is used when P_1 and P_2 are *sound powers, P_2 being a reference power. L_P is usually expressed in *nepers or *decibels.

power line *See* transmission line.

power pack A device that converts power from an a.c. or d.c. *power supply, usually the mains, into a form suitable for operating an electronic device.

power station *Syn.* generating station. A complete assemblage of plant, equipment, and the necessary buildings at a place where electric power is generated on a large scale. The main types are: *thermal power station, *nuclear power station, and *hydroelectric power station.

power supply Any source of electrical power in a form suitable for operating electrical or electronic devices. Alternating-current power may be derived from the *mains, either directly or by means of a suitable *transformer. Direct-current power may be supplied from batteries or suitable rectifier/filter circuits. A *bus is frequently used to supply power to several circuits or to several different points in one circuit. Suitable values of voltage are derived from the common supply by some form of *coupling.

power transistor A *transistor designed to operate at relatively high values of power or to produce a relatively high power gain. Power transistors are used for switching or amplification and because of the relatively high power dissipation, which ranges from 1 watt to 100 watt, they usually require some form of temperature control.

Poynting's theorem The rate of energy transfer for *electromagnetic radiation is proportional to the product of the electric and magnetic field strengths, i.e. to the surface integral of the *Poynting vector formed by the components of the field in the plane of the surface.

Poynting vector A *pseudovector giving the direction and magnitude of the rate of energy flow perpendicularly through unit area in an electromagnetic field. It is equal to the vector product of the electric and magnetic field strengths at any point.

PPI Abbreviation for plan position indicator. *See* radar.

Prandtl number The dimensionless group $(C\eta/K\rho)$ occurring in the dimensional analysis of *convection in a fluid due to the presence of a hot body, where C is the heat capacity per unit volume of the fluid, η is the viscosity of the fluid, K is the thermal conductivity of the fluid, ρ is the density of the fluid.

preamplifier An *amplifier used as an earlier stage to the main amplifier. It is frequently placed near the signal source (aerial, pick-up, etc.), being connected by cable to the main amplifier. This improves the *signal-to-noise ratio as amplification of the initial signal occurs before it traverses the path to the main amplifier.

precession If a body is spinning about an axis of symmetry OC (where O is a fixed point) and C is rotating round an axis OZ fixed outside the body, the body is said to be precessing round OZ. OZ is the precession axis. A gyroscope precesses due to an applied torque (called the *precessional torque*). If the moment of inertia of the body about OC is I and its angular velocity is ω, a torque K whose axis is perpendicular to the axis of rotation will produce an angular velocity of precession Ω about an axis

perpendicular to both ω and the torque axis where $\Omega = K/I\omega$. *See* Eulerian angles.

pre-emphasis (and de-emphasis) A technique used to improve the *signal-to-noise ratio in a radiocommunication system employing *frequency modulation. For pre-emphasis, a network is introduced at the transmitter that in effect increases the strength of the higher frequencies of the modulated signal relative to that of the lower frequencies. The original levels are restored by an equivalent process of de-emphasis at the receiver. The technique is also used for tape recordings and gramophone records.

preon A hypothetical entity postulated as the building blocks of *leptons and *quarks. There is no experimental evidence for their existence, nor is there likely to be with the energies of current *accelerators. They do, however, have certain theoretical attractions.

pressure Symbol: p. At a point in a fluid, the force exerted per unit area on an infinitesimal plane situated at the point. In a fluid at rest the pressure at any point is the same in all directions. In a liquid it increases uniformly with depth, h, according to the formula:
$$p = \rho gh,$$
where ρ is the density of the fluid and g is the acceleration of *free fall. In a gas under isothermal conditions it decreases exponentially with height h, according to the formula:
$$p_h = p_0 e^{-(\mu g/RT)h},$$
where μ is the relative molecular mass, R is the *molar gas constant, and T is the thermodynamic temperature.

The SI unit of pressure is the *pascal (i.e. newton per metre squared) but the *bar is also officially recognized. For example, in meteorology, pressure is measured in millibars (mb), where one pascal is equal to 0.01 mb. The atmosphere, millimetre of mercury (mmHg), and torr are now obsolete units of pressure and should only be used for rough comparisons. *See also* standard atmosphere; standard temperature and pressure.

pressure broadening An effect observed in line *spectra whereby high density, and hence pressure, of the emitting or absorbing material causes a broadening of the spectral lines; the higher the density and pressure, the greater the width of the lines. The broadening is caused by collisions with other atoms while the atom is emitting or absorbing radiation. *See also* Doppler broadening; line profile.

pressure coefficient Symbol: β. The quantity relating pressure p and thermodynamic temperature T at constant volume:

$$\beta = (\partial p / \partial T)_V.$$

The *relative pressure coefficient*, symbol: α, is equal to $\beta(1/p)$.

pressure gauges 1. Primary gauges include the liquid-column *manometers and the *free-piston gauge.

2. Secondary gauges include the *Bourdon gauge and *resistance gauge.

3. *See* micromanometers.

4. Vacuum gauges include the *McLeod and *Knudsen gauges, which are primary gauges (although the Knudsen gauge is often made as a secondary gauge), and the *Pirani, *ionization, and *molecular gauges.

pressure head The height of a column of liquid capable of exerting a given pressure, e.g. the head h of liquid, corresponding to a pressure p, is $h\rho g$, where ρ is the density of the liquid and g the acceleration of *free fall.

pressurized-water reactor (PWR) A type of thermal *nuclear reactor in which water under pressure (to prevent boiling) is used as both coolant and moderator. The effectiveness of water as a moderator and the fact that water absorbs neutrons means that the core is compact and that the fuel must be slightly enriched in ^{235}U. The fuel rods of uranium dioxide are clad in a zirconium alloy. Water enters the core at about 280 °C and leaves at 310 °C under a pressure of 16 MPa; a substantial pressure vessel is therefore needed, about 200 mm thick. At this pressure and temperature the coolant does not boil; these conditions are maintained with an electrically heated device called a pressurizer. A strong containment building encloses the pressure vessel and pipework carrying the cooling water; this is designed to contain the coolant should a pipe break. Because depressurized water at these temperatures would rapidly turn to steam, reserve supplies of water are available.

A variant of the PWR is the *boiling-water reactor* (BWR). In this device water is allowed to boil within the core and the resulting steam is separated in the pressure vessel and passed directly to the turbine generators. This avoids the expense of the heat exchangers (steam generators) of the PWR but additional equipment is needed to contain any radioactive gases from the coolant, which might escape from the turbine generator.

Prévost's theory of exchanges (1791) A body emits precisely the same radiant energy as it absorbs when it is in thermal equilibrium with its surroundings (convection and conduction having ceased).

primary cell *See* cell.

primary colours A set of three coloured lights (or pigments) that when mixed in equal proportions produce white light (or black pigment). One group consists of red, green, and blue; cyan (greenish-blue), magenta (reddish-blue), and yellow form another set. Red and cyan, green and magenta, and blue and yellow are pairs of *complementary colours. Three lights of primary colours can be mixed in suitable proportions to produce any other colour, excluding black, by an *additive process. Three pigments, paints, dyes, etc., of primary colours can be mixed to produce any colour, excluding white, by a *subtractive process. *See also* chromaticity.

primary cosmic rays *See* cosmic rays.

primary electrons Electrons incident on a surface distinguished from the secondary electrons that they release. *See* secondary emission.

primary standard A standard used nationally or internationally as the basis for a unit, e.g. the international prototype *kilogram. *Compare* secondary standard.

primary winding The winding on the supply (i.e. input) side of a *transformer, irrespective of whether the transformer is of the step-up or step-down type.

principal axis *See* pole.

principal directions The directions about which there is symmetry of crystal properties, such as refractive indexes, coefficients of thermal expansion, magnetic susceptibility, thermal conductivity, etc.

principal focus *See* focal point.

principal planes (and points) *See* centred optical systems.

principal quantum number *See* atomic orbital.

principal ray *See* apertures and stops in optical systems.

principle A highly general or inclusive law, exemplified in a multitude of cases. Examples include the *Pauli exclusion principle, the Heisenberg *uncertainty principle, *Fermat's principle, and the principle of equivalence (*see* relativity).

printed circuit An electronic circuit together with conducting interconnections fabricated on a thin rigid insulating sheet, usually fibreglass. The supporting sheet plus circuit is called a *printed-circuit board* (PCB). The interconnections are produced by first applying a conducting film, usually copper, to the sheet, coating parts of the film with a protective material using *photolithography, then removing the unprotected metal by etching. This leaves the pattern of interconnections required by the circuit design. The components, usually *integrated circuits, are finally soldered into position between the conducting tracks to complete the circuit.

Doubled-sided PCBs are commonly produced in which both sides of the board have a circuit formed on them, with *feedthroughs to connect the two sides as required. Printed circuits have also been produced with several alternating layers of metal film and thin insulating film mounted on a single board. *See also* edge connector.

printer In computing, a device that converts the coded information from a computer into a readable form printed on paper. There are many types, varying in the method and speed of printing and the quality of the print. Some printers print a single character at a time. Line printers print a complete line of characters at a time. Page printers, such as the laser printer, produce a complete page at a time. The printed characters may have a solid form or may consist of patterns of closely spaced dots.

prism A refracting medium bounded by intersecting plane surfaces that both deviates a beam of light and disperses it into its component colours (*see* dispersion). Prism combinations can be produced to deviate light without dispersion (*see* achromatic prism) or to disperse a beam without producing any mean deviation (*see* direct-vision prism). The main uses of prisms are (*a*) for deflection or small-angle deviation – *narrow-angle* refracting prisms; (*b*) for large-angle deviation and dispersion – in *spectrometers, *refractometers; (*c*) for changing ray direction or inverting or erecting images using total internal reflection (sometimes silvered to permit reflection).

The prismatic effect of *narrow-angle* prisms is given by $P = (n - 1)A$, where n is the refractive index, A the angle of the prism, i.e. the angle made by the refracting faces in a principle section perpendicular to the refracting edge. P is the angle of deviation (usually expressed in degrees) or the power of the prism (expressed in *prism dioptres). Rays are really deflected towards the base direction, while objects viewed through the prism are apparently displaced towards the apex. The *chromatic aberration C of a single prism is ωP, where ω is the dispersive power. For two prisms the chromatic aberration C is $(\omega_1 P_1 + \omega_2 P_2)$ and total power $P = P_1 + P_2$.

For large angle deviation and dispersion, the angle of the prism, A, must be large. Such prisms have a minimum deviation (*see* deviation (angle of)), a property used for the determination of refractive index and of dispersion using spectrometers.

prismatic binoculars Binoculars in which each half consists of a *Kepler telescope, the prisms serving the two-fold purpose of reducing the length of the instrument and also erecting the image. The prisms are two *Porro prisms; one prism inverts in one direction only, the second prism completes the inversion, since its roof edge is crossed at right angles to the roof edge of the former.

prism dioptre A unit of deviating power of a *prism based on a tangent measured in centimetres on a scale placed one metre away or a similar proportion. If θ is the angle of deviation then the power P, in prism dioptres, is equal to $100 \tan \theta$. The unit is used mainly for narrow-angle prisms.

probability The numerical value quantifying the chance that one specified outcome out of several possible outcomes will occur, as a consequence of an unpredictable event.

Independent events. If n is the total number of ways in which an event can occur, and m is the number of ways in which an event can occur in a specified way, then the ratio m/n is the *mathematical* (or *a priori*) *probability*. When a dice is tossed, the mathematical probability of obtaining the number 4 is 1/6. If there are 4 white counters and 5 black counters in a bag, the mathematical probability of drawing out a white is 4/9. In a random sequence of n trials of an event with m favourable outcomes, as n increases indefinitely the ratio m/n has the limit P, where P is the probability.

If in a number of trials an event has occurred n times and failed m times, the probability of success in the next trial is given by $n/(n + m)$. This is the *empirical* (or *a posteriori*) *probability*. The probability of a man not dying at a certain age, based upon past observations (recorded, say, in a mortality table), is an empirical probability.

The probability of a given number of successes in a certain number of trials, when the probability of success in a single trial does not vary, is given by the *binomial distribution of probabilities, P_r*.

Dependent events. If two or more events are so related that the outcome of one affects the outcome of the other or others, the individual probability of each event is calculated in sequence and the product of these gives the *conditional probability*. If there are 4 white counters and 5 black counters in a bag, the probability of picking 3 white counters in the first 3 tries would be:

$$(4 \times 3 \times 2)/(9 \times 8 \times 7).$$

If a variable X can take on a set of discrete random values x_i, each of which has a probability p_i, the relative frequency of occurrence of any one of these values is $x_i p_i$; the cumulative frequency F, up to the value x_n is given by:

$$F(x_n) = \sum_{i=1}^{n} x_i p_i.$$

If X varies continuously, then the cumulative frequency up to a value x_n is given by:

$$F(x)_n = \int_{-\infty}^{x} f(x)\mathrm{d}x,$$

where $f(x)$ represents the relative frequency of any specific value of x. The function $f(x)$ is called the *frequency function* or *probability density function*. The graph of the function $F(x)$ is a normal *frequency distribution.

probability density function *See* probability.

probable error That error such that the chances of the absolute magnitude of the *deviation being greater than or less than it are even. The arithmetic mean of n observations has a probable error that is $n^{-1/2}$ times the probable error of a single observation. (*See* frequency distribution.)

Experimental results are often quoted in scientific literature as $(x \pm \delta)$, δ being a small quantity that may be (*a*) the probable error; (*b*) the standard deviation; (*c*) the error intelligently guessed (*see* errors of measurement).

probe 1. A lead that contains or connects to a measuring or monitoring circuit and is used for testing purposes. The circuit may consist of either active or passive components.

2. A resonant conductor inserted into a *waveguide or *cavity resonator for the purpose of injecting or extracting energy.

processor *See* central processing unit; microprocessor.

product of inertia If the Cartesian axes $OXYZ$ are fixed relative to a rigid body, the product of inertia of the body with respect to the axes OY, OZ is $\sum myz = F$, the summation being carried out for every particle of the body $(x, y,$ and z are the coordinates of a particle of mass m.). There are three such products of inertia: F, $G = \sum mxz$, and $H = \sum mxy$ and also three moments of inertia A, B, C about the axes OX, OY, OZ respectively given by:

$$A = \sum m(y^2 + z^2),$$
$$B = \sum m(z^2 + x^2),$$
$$C = \sum m(x^2 + y^2).$$

The moment of inertia, I, of the body about any axis L through the origin of coordinates is expressible in terms of A, B, C and F, G, H as well as the direction cosines α, β, γ of the axis L with respect to the coordinate axes OX, OY, OZ:

$$I = A\alpha^2 + B\beta^2 + C\gamma^2 - 2F\beta\gamma - 2G\gamma\alpha - 2H\alpha\beta.$$

profile The shape of a wave, pulse, spectral line, etc., found by plotting the amplitude or intensity against time, distance, frequency, or some other temporal or spatial variable.

program A sequence of statements that can be submitted to a *computer system and used to direct the way in which the system behaves. The program must be expressed precisely and unambiguously, and must therefore be written in any of a large variety of artificial *programming languages*. Before a program can be executed by a computer, it must be translated (automatically) into a sequence of *machine instructions*, expressed in the particular *machine code* appropriate to that computer.

programmable ROM *See* PROM.

progressive wave A wave propagated through an infinite homogeneous medium. For any type of wave motion a plane-progressive wave may be represented by the equation of wave motion:
$$\partial^2\xi/\partial t^2 = c^2(\partial^2\xi/\partial x^2),$$
where ξ is the particle displacement at distance x from a fixed point along the direction of propagation, c is the wave velocity, and t is time measured from a fixed instant.

The solution of this equation, for a plane progressive harmonic wave, is given by:
$$\xi = a \sin (2\pi/\lambda)(ct - x),$$
where a is the maximum particle displacement or amplitude and λ is the wavelength. For a given value of x, the displacement ξ varies through a complete cycle when $(2\pi/\lambda)(ct - x)$ changes by 2π radians; the corresponding change in t is T, the *period. Thus, $(2\pi/\lambda)cT = 2\pi$ and $T = \lambda/c$. The frequency f is the reciprocal of T, and, therefore, $1/f = \lambda/c$, or $c = f\lambda$.

The intensity of the wave, I, i.e. the energy passing per unit time per unit area normal to the direction of propagation, is proportional to a^2f^2.

PROM Abbreviation for programmable read-only memory, i.e. programmable ROM. A type of semiconductor memory that is fabricated in a similar way to *ROM. The contents required are, however, added after rather than during manufacture, and cannot be altered from that time. The memory contents are fixed electronically by means of a device called a *PROM programmer. See also* EPROM.

prompt neutron A neutron produced in a reactor by primary fission rather than by decay of a fission product.

propagation coefficient *Syn.* propagation constant. Symbol: P or γ. A measure of the attenuation and phase change of a wave travelling along a *transmission line. It is defined for a uniform line of infinite length supplied at its sending end with a sinusoidal current having a specified frequency. At two points along the line separated by unit distance, let the currents under steady-state conditions be I_1 and I_2, where I_1 is the current at the point nearer the sending end of the line. Then
$$P = \log_e(I_1/I_2)$$
at the specified frequency; note that the vector ratio of the currents is used. P is a complex quantity and may be expressed as
$$P = \alpha + i\beta,$$
where $i = \sqrt{-1}$. The real part, α, is the the *attenuation constant and is measured in *nepers per unit length of line. It measures the transmission losses in the line. The imaginary part, β, is the *phase-change coefficient* or *phase constant* and is measured in radians per unit length of line. It is the *phase difference between I_1 and I_2 introduced by the line.

If at a given instant the displacement at a given point is a maximum and given by p_1, then the instantaneous displacement p_2 at the same instant distance x along the direction of propagation is given by:

$$p_2 = p_1 e^{-Px} = p_1 e^{-(\alpha + i\beta)x}.$$

A line of infinite length is not physically realizable but the conditions in a line of finite length, which is terminated in an impedance equal to its *characteristic impedance, simulate those of the infinite line.

propagation constant *See* propagation coefficient.

propagation loss The energy loss from a beam of *electromagnetic radiation as a result of absorption, scattering, and the spreading out of the beam.

propagation vector *See* wavenumber.

proportional counter *See* proportional region.

proportional region The operation voltage range for a radiation counter in which the *gas multiplication exceeds 1, its value being independent of primary ionization. A counter operating in this region is called a *proportional counter*. The pulse size in the counter is proportional to the number of ions produced as a result of the initial ionizing event. *See* Geiger counter; ionization chamber.

protective relay A *relay that protects electrical apparatus against the damaging effects of abnormal conditions (e.g. overloads and internal faults). When such abnormal conditions occur, the relay causes a *circuit breaker to open so that the faulty apparatus is automatically disconnected from the power supply and from other associated equipment.

proton A positively charged *elementary particle that forms the nucleus of the hydrogen atom and is a constituent particle of all nuclei. It is about 1836 times heavier than the electron. It is a stable

*baryon of mass 938.2796 MeV/c^2 (1.672 623 1 × 10^{-27} kg) having charge $Q = 1$. It has *spin $J = \frac{1}{2}$, *isospin $I = \frac{1}{2}$, and positive *parity. It also has an intrinsic *magnetic moment of 2.793 nuclear *magnetons. Although there are many elementary particles having a lower mass, the proton cannot decay into these as it is the lowest mass particle with baryon number $B = 1$; baryon number is conserved in all interactions.

proton microscope A microscope similar to the *electron microscope but using a beam of protons rather than electrons. This allows a better resolving power and contrast.

proton number *See* atomic number.

proton-proton chain A series of *thermonuclear reactions by which hydrogen nuclei (protons) are converted to helium nuclei with the release of a considerable amount of energy. The reactions require a temperature of 10^7 kelvin. They are almost certainly the major source of energy in the sun and in all stars of lower mass than the sun, and occur in the dense stellar core. There are various sequences that can be followed in the proton-proton chain, but the principal sequence is:

$$^1H + {}^1H \rightarrow {}^2H + \nu + e^+$$
$$^2H + {}^1H \rightarrow {}^3He + \gamma$$
$$^3He + {}^3He \rightarrow {}^4He + 2\ {}^1H,$$

where ν, e^+, and γ are a neutrino, positron, and γ-ray respectively. *Compare* carbon cycle.

proton resonance *Nuclear magnetic resonance of hydrogen nuclei.

proton synchrotron A cyclic *accelerator of very large radius that can accelerate protons to extremely high energies: 70 GeV at Serpukhov, USSR; 450 GeV at CERN, Geneva; and 900 GeV at FNAL, USA. It is basically similar to a *synchrotron. In a synchrotron a fixed orbit is maintained by increasing the magnetic field strength in proportion to the relativistic increase in mass. This leads to a constant angular frequency. This is possible because the electrons are travelling at close to the speed of light at an energy of a few MeV. The equivalent stable orbit at constant angular frequency is not achieved by protons until they have an energy of about 3 GeV. To maintain a fixed orbit up to this energy, the frequency of the accelerating electric field must be varied (it is constant in the synchrotron). The frequency of the field and the particle beam must remain in synchrony and also satisfy the relation $v = \omega r$ where v is the proton velocity, r is the radius of the orbit, and ω is the angular frequency of the protons; $\omega = 2\pi f$, where f is the electric-field frequency. The protons are accelerated by radio-frequency fields between the magnets, and can make several million revolutions in one second. The beam is both focused and maintained in a circular orbit by means of magnets. Strong *focusing is used.

proximity effect An effect that occurs when conductors carrying alternating current are placed close to one another, the distribution of current across the cross section of any one being influenced by the magnetic fields set up by the others. The change in current distribution modifies the *effective resistance of the conductors. The effect is particularly important in coils used at high (e.g. radio) frequencies.

pseudoscalar A quantity defined by a single magnitude but distinguished from a *scalar as it is an odd function, i.e. its sign is reversed on reversing the signs of a set of coordinate axes. Examples are pressure and volume.

pseudovector *Syn* axial vector. A quantity defined in terms of a magnitude and the direction of an axis. Pseudovectors obey the same rules of algebra and calculus as polar *vectors but are even functions, i.e. signs of the components are unchanged on reversing the directions of the coordinate axes. Examples include area, torque, and magnetic flux density. The vector product of two polar vectors is a pseudovector.

psi/J particle *See* J/psi particle.

psychrometer A *hygrometer in which a strong draught is obtained past the wet and dry bulbs either by whirling on a sling or by a fan. The best-known type is the *Assmann psychrometer* in which the necessary ventilation is provided by a clockwork-driven fan.

p-type conductivity Conductivity in a *semiconductor caused by the effective movement of mobile *holes.

p-type semiconductor An *extrinsic semiconductor in which the density of mobile *holes exceeds that of conduction *electrons. *See also* semiconductor.

pulley Any of a number of simple *machines in which forces are transmitted by ropes, belts, etc., used in conjunction with pulley wheels and axles. The wheels usually have a grooved rim in which the rope, etc., can run. In a frictionless (i.e. theoretical) system, the force (or pull) in any part of a continuous rope has a constant value, F. Consider a system of vertically aligned equal-sized wheels, connected by a continuous rope, the topmost wheel fixed to a support. A weight w, attached to the lowest wheel, is to be raised. If the number of supporting ropes is n, then $nF = w$. The mechanical advantage, i.e. the ratio of load to effort, is then equal to n.

pulling An effect occurring when an electronic oscillator is coupled to a circuit in which there is another independent oscillation: the oscillator frequency tends to change towards that of the independent oscillation. The tendency is particularly strong if the two frequencies differ by only a small percentage. Complete synchronization can sometimes be achieved. Pulling is used to control the frequency of an oscillator, as in crystal-controlled oscillators.

pulsar A member of a class of astronomical objects that all possess extreme characteristics, including (*a*) their energy output varies with exceptional regularity at a fast rate; (*b*) they have small dimensions (20–30 km across) and are the densest observable form of matter; (*c*) their energy per unit area is very large. Pulsars are generally believed to be examples of *neutron stars.

Pulsars are rapidly spinning bodies emitting energy either in the radio or the X-ray regions of the spectrum. (Some radio pulsars have also been detected at optical and gamma-ray wavelengths.) The period of the pulsed output represents the rotation rate. In the case of radio pulsars, the radio waves (which are polarized) originate from the pulsar's extremely strong magnetic field. As the pulsar rotates, the radio beam may be observed as a pulsed signal as it sweeps past the earth. The period of most radio pulsars lies between 0.1 and 4 seconds, indicating that these pulsars are rotating in the range 10–0.25 times per second. All, however, are gradually slowing down as they lose rotational energy; the rate of slowdown varies considerably. A small group of radio pulsars rotate even more rapidly. Their pulse rate is a few milliseconds and they are thus called *millisecond pulsars*.

The millisecond pulsars and a few other radio pulsars are in close orbit with another star, i.e. they are binary stars. In contrast, all X-ray pulsars are in binary systems. The X-rays are generated when gas, transferred from the companion star, falls onto the pulsar. The gas flow affects the pulsar's spin, and as a result all X-ray pulsars are gradually speeding up. The periods of some X-ray pulsars are a few seconds but many have longer periods, amounting to several minutes.

It is believed that some but not all pulsars originate in *supernova explosions.

pulsating current An electric current that varies in magnitude in a regularly recurring manner. The term implies that the current is unidirectional.

pulsating star *See* variable star.

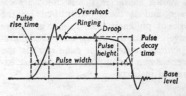

Practical rectangular pulse

pulse A single transient disturbance manifest as an isolated wave, or one of a series of transient disturbances recurring at regular intervals, or a short train of high-frequency waves, as used in *echo-sounding and *radar. A single pulse consists of a voltage or a current that increases from zero to a maximum value and then decreases to zero in a comparatively short time. A pulse is described as rectangular, square, triangular, etc., according to the geometrical shape of the pulse when its instantaneous value is plotted as a function of time.

In practice a perfect geometrical shape is never achieved and a practical rectangular pulse is shown in the diagram. The magnitude of the pulse normally has a constant value, ignoring any spikes or ripples, and is called the *pulse height*. A practical pulse has a finite *rise time*, usually occurring between 10% and 90% of the pulse height, and a finite *decay time*, occurring between the same limits. The *pulse width* is the time between the rise and decay time. A practical pulse frequently rises to a value above the pulse height and falls to the pulse height with damped oscillations. This is called *overshoot* and *ringing*. A similar phenomenon occurs as the pulse decays to the base level. *Droop* can occur in a rectangular pulse when the pulse height falls slightly below the nominal value. This is particularly associated with inductively coupled circuits (*see* inductive coupling).

A group of regularly recurring pulses of similar characteristics is called a *pulse train*, and is usually identified by the type of pulses in the train, e.g. square wave, sawtooth wave, etc. The *pulse-repetition frequency* (or *pulse rate*) is the average number of pulses per second, expressed in hertz.

pulse-code modulation (PCM) *See* pulse modulation.

pulse generator An electronic circuit or device that produces current or voltage pulses of a desired waveform.

pulse height *See* pulse.

pulse-height analyser *See* multichannel analyser.

pulse-height discriminator An electronic circuit that selects and passes *pulses whose amplitude lies between specified limits.

pulse modulation A form of *modulation in which (most commonly) a *pulse train is used as the carrier. Information

is conveyed by modulating some parameter of the pulses with a set of discrete instantaneous samples of the message signal. For instance, the pulse height can be modulated in accordance with the amplitude of the modulating signal (*pulse-amplitude modulation*), or the time of occurrence of the leading or trailing edge can be varied from its unmodulated position (*pulse-width modulation*).

In *pulse-code modulation*, the amplitude (usually) of the modulating signal is sampled and a digital code is used to represent the sampled value: sample amplitudes falling within specified ranges of values are assigned different discrete values, each of which is represented by a specific pattern of pulses. The signal is thus transmitted in the form of a digital code (i.e. as a stream of *bits), which is converted back to the analogue signal at the receiving point.

pulse operation Any means of operation of electronic circuits or devices in which the energy is transferred in the form of pulses.

pulse regeneration In most forms of *pulse operation the pulses can get distorted by circuits or circuit elements. Pulse regeneration is the process of restoring the original form, timing, and magnitude to a *pulse or pulse train.

pulse-repetition frequency *Syn.* pulse rate. *See* pulse.

pulse shaper Any circuit or device that is used to change any of the characteristics of a *pulse.

pulse train *See* pulse.

pulse width *See* pulse.

pumps, vacuum Modern kinetic vacuum systems usually use two pumps in tandem. The backing pump or forepump, which works directly to the atmosphere, is one of the many forms of rotary oil pump. These pumps reduce the pressure to between 100 and 0.1 Pa according to the type. The second pump is usually a *diffusion* or *condensation pump*, which is rather similar in principle to the *filter pump but uses a jet of oil vapour instead of water. There may be several such jets acting in tandem.

Molecular pumps work on the principle that a gas molecule striking a rapidly moving surface may be given a high velocity in the direction of the exit pipe. These pumps need a backing-pump. *See also* air-pumps; ion pump; sorption pump.

punch through A type of *breakdown that can occur in both bipolar junction *transistors and *field-effect transistors. If the collector-base voltage applied to a bipolar transistor is increased, the *depletion layer associated with the collector-base junction spreads into the base region. At a sufficiently high voltage, the *punch-through voltage*, the depletion layer reaches the emitter region and a direct conducting path is formed from emitter to collector; carriers from the emitter "punch through" to the collector and breakdown occurs.

In an FET, a similar process occurs. When the drain voltage reaches a sufficiently high value, the depletion layer associated with the drain spreads across the substrate and meets the source junction; carriers can then punch through the substrate.

push-pull operation The use in a circuit of two matched devices in such a way that they operate with a 180° *phase difference. The output circuits combine the separate outputs in phase. One means of achieving the desired phase shift in the inputs is a transformer-coupled input circuit. Transformer coupling of the outputs of the matched

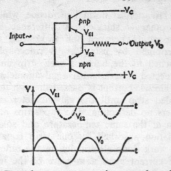

Complementary transistor push–pull operation

devices combines them so that the resultant a.c. output is in phase. *Complementary transistors may also be used (*see* diagram), in which case no phase shift is required in the inputs.

Push-pull circuits are frequently used for *class A and *class B amplification; these are then called *push-pull amplifiers* or *balanced amplifiers*.

PWR Abbreviation for pressurized-water reactor.

pyknometer A form of *relative-density bottle consisting of a bulb B joined to two capillary tubes, T_1 and T_2, on one of which (T_2) is a reference mark. Liquid is drawn in through T_1 and fills B before reaching the reference mark on T_2. The density of the liquid can be measured by determining the mass of the accurately defined volume.

pyranometer *See* solarimeter.

pyrheliometer An instrument for measuring the intensity of direct solar radiation at normal incidence, diffuse radiation being excluded. It can also measure radiation from a selected part of the sky.

In the *Ångstrom pyrheliometer* two identical strips of blackened platinum are mounted so that one is exposed to

373

the radiation at normal incidence, while the other is shielded. A difference in temperature between the two strips is determined by using a *thermocouple attached to the back of each strip and connected in series with a galvanometer. An electric current is passed through the shaded strip until the galvanometer registers no deflection. The two strips are then at the same temperature and solar radiation absorbed by one strip equals the work done electrically on the other. The current gives a measure of the intensity. Each strip is exposed in turn to the radiation and a mean value is found for the required current. It is a standard instrument requiring no external calibration.

In the *Linke-Fuessner actinometer* (a pyrheliometer), the temperature difference between a blackened surface, exposed to a narrow angle of solar radiation, and a reference point is determined. The surface consists of a *thermopile of similar design to that used in a *solarimeter. The surface increases in temperature until its rate of loss of heat by all causes is equal to the rate of gain of heat from radiation. This increase in temperature should depend only on the intensity of the radiation, and must be independent of external conditions, such as ambient temperature, and wind speed.

pyroelectricity The development of opposite electric charges at the ends of polar axes in certain crystals when there is a change of temperature. Such crystals (e.g. tourmaline, lithium sulphate) do not possess a centre of symmetry.

pyrometer An instrument for measuring high temperatures. There are several types:

(1) The *optical pyrometer, which depends for its action on Planck's formula (*see* radiation formula). Instruments of this type are used for the measurement of high temperatures on the *international temperature scale.

(2) The *total radiation pyrometer*, which depends for its action on the *Stefan–Boltzmann law $E = \sigma T^4$. The most convenient form is the *Féry total radiation pyrometer.

(3) *See* resistance pyrometer.

Q

QCD Abbreviation for *quantum chromodynamics.

QED Abbreviation for *quantum electrodynamics.

Q-factor *Syn.* quality factor. Symbol: Q. A measure of the quality of performance of a resonant system, especially a *resonant circuit. It indicates the ability of the system to produce a large output at the resonant frequency.

For a resonant circuit comprising resistance R, capacitance C, and inductance L,
$$Q = 1/R\sqrt{(L/C)}.$$
In the case of a simple series resonant circuit at the resonant frequency, $\omega_0 L = 1/\omega_0 C$, where ω_0 is 2π times the resonant frequency. Thus,
$$Q = \omega_0 L/R \text{ or } Q = 1/\omega_0 CR.$$
This is effectively the ratio of the total inductive or total capacitive *reactance to the total series resistance at resonance. In the case of a simple parallel resonant circuit,
$$\omega_0 = \sqrt{(1/LC)}\sqrt{(1 - 1/Q^2)}.$$
If Q is large, this reduces to the value for a series resonant circuit.

The *selectivity of a resonant circuit is given by $1/Q$.

A single reactive component may be capable of resonance (i.e. if the self-inductance of a coil or self-capacitance of a capacitor is large enough). For a single component, Q is the ratio of reactance to effective series resistance, which for an inductance Q is equal to $\omega_0 L/R$ and for a capacitor to $1/\omega_0 CR$.

QFD Abbreviation for quantum flavourdynamics (*see* electroweak theory).

Quadrant electrometer

quadrant electrometer A type of *electrometer in which a light foil-covered vane, supported by a quartz fibre, moves within hollow quadrantal segments of a cylindrical metal box. Opposite quadrants are connected together, but insulated from the case of the instrument. A mirror is carried to reflect a spot of light and measure deflections of the vane. If the vane hangs symmetrically within the quadrants when the needle and both pairs of quadrants are at zero potential (*see* diagram), then the deflection θ of the needle is given by:
$$\theta = k_1(V_A - V_B),$$
where V_A and V_B are the potentials of the pairs of quadrants; if one pair of quadrants is earthed, then
$$\theta = k_2 V_A,$$
where k_1 and k_2 are constants and characteristic of the instrument. It is assumed that the voltage applied to the vane is large compared with V_A and V_B.

quadrature Two periodic quantities having the same *frequency and *waveform are said to be in quadrature when they differ in phase by 90°, i.e. one wave reaches its maximum value when the other passes through zero.

quadrature component *See* reactive current, reactive voltage.

quadripole An electrical *network that has only four terminals – a pair of input terminals and a pair of output terminals. Its behaviour is usually described by the impedances presented at its terminals at specified frequencies. A common arrangement, used especially for *filters and *attenuators, consists of a number of impedances arranged in series and in parallel in a ladder-shaped network. This arrangement can be broken down for analysis into identical T-shaped and π-shaped sections, each with the same characteristic impedance.

quadrupole A distribution of charge or magnetization equivalent to two equal electric or magnetic *dipoles arranged very close together and set in opposite directions. The potential falls off as the inverse cube of the distance. For an arbitrary distribution of charges, the potential V_r at a distance r is given by
$$V_r = \frac{e}{4\pi\varepsilon_0 r} + \frac{p}{4\pi\varepsilon_0 r^2} + \frac{q}{4\pi\varepsilon_0 r^3} + \cdots$$
The first term gives the coulomb potential, e being the net charge. The second gives the dipole potential, p being the net dipole moment. Similarly, the third gives the quadrupole potential, when q is the *quadrupole moment*. ε_0 is the electric constant.

quality (in sound) 1. The fidelity of reproduction of a sound.
2. *See* timbre.

quality factor 1. *See* Q-factor.
2. *See* dose.

quantity of electricity Symbol: Q. The time integral of electric current, i.e. $\int I dt$, equivalent to the electric charge.

quantity of heat Symbol: Q. *See* heat.

quantity of light Symbol: Q. The time integral of the *luminous flux, i.e. $\int \Phi dt$.

quantization The process, used in electronics and computing, of constructing a set of discrete values that represents a quantity that varies continuously. One example is the measurement of the amplitude of a signal at discrete intervals of time, when the signal itself varies continuously with time. Another is the measurement of the brightness of small picture elements (*pixels) making up a picture, which may be regarded as space-continuous.

quantized *See* quantum theory; quantum number.

quantum The smallest amount of energy that a system can gain or lose. The change in energy corresponding to a quantum is very small and only noticeable on an atomic scale. *See* quantum theory.

quantum chromodynamics (QCD) A *quantum field theory that is a *gauge theory of the *strong interactions based on the exchange of massless gluons between quarks and antiquarks. QCD is similar to *quantum electrodynamics (QED), the quantum field theory of electromagnetic interactions, but with the gluon as the analogue of the photon and with a quantum number known as *colour* replacing that of electric charge. Each quark type (or flavour) comes in three colours (red, blue, and green, say), where colour is simply a convenient label and has no connection with ordinary colour. Unlike the photon in QED, which is electrically neutral, gluons in QCD carry colour and can therefore interact with themselves. Particles that carry colour are believed not to be able to exist as free particles. Instead, quarks and gluons are permanently confined in-

side *hadrons (strongly interacting particles, such as the proton and the neutron).

The gluon self-interaction leads to the property known as *asymptotic freedom*, in which the interaction strength for the strong interactions decreases as the momentum transfer involved in an interaction increases. This allows perturbation theory to be used and quantitative comparisons to be made with experiment, similar to (but less precise than) those possible in QED. QCD has been tested successfully in high energy muon-nucleon scattering experiments and in proton–antiproton and electron–positron collisions at high energies. Strong evidence for the existence of colour comes from measurements of the interaction rates for $e^+e^- \rightarrow$ hadrons and $e^+e^- \rightarrow \mu^+\mu^-$. The relative rate for these two processes is found to be a factor of three larger than would be expected without colour; this factor measures directly the number of colours (i.e. three) for each quark flavour. Energetic quarks and gluons produced in very high energy particle collisions undergo a process known as *hadronization* or *fragmentation*, in which the quark or gluon becomes a collimated jet of hadrons (mostly pions) aligned with the original quark or gluon direction, and it is these jets of particles that are observed experimentally. In e^+e^- collisions, interactions are observed in which three separated jets or hadrons are produced. These are interpreted as being due to the underlying process $e^+e^- \rightarrow q\bar{q}g$, with the subsequent fragmentation of the quark, antiquark, and gluon into the observed jets of hadrons; such interactions provide direct evidence for the existence of the gluon.

quantum discontinuity The discontinuous emission or absorption of energy accompanying a quantum jump.

quantum efficiency *Syn.* quantum yield. A measure of the efficiency of a device in the use of light or other electromagnetic radiation. It is the proportion of photons of a specified frequency, incident on the device, that can induce a reaction of a particular type. It can, for example, be the proportion of incident photons that generate photoelectrons in a photocell, or that are involved in producing a photographic image.

a Feynman diagrams

quantum electrodynamics (QED) A relativistic quantum-mechanical theory of *electromagnetic interactions. QED's descriptions of the photon-mediated electromagnetic interactions have been verified over a great range of distances and have led to highly accurate predictions. QED is a *gauge theory. In QED, the electromagnetic force can be derived by requiring that the equations describing the motion of a charged particle remain unchanged in the course of local symmetry operations. Specifically, if the phase of the *wave function by which a charged particle is described is altered independently at every point in space, QED requires that the electromagnetic interaction and its mediating photon exist in order to maintain symmetry.

In the *Feynman propagator* approach, the scattering of electrons and photons is described by a matrix (the scattering matrix), which is written as an infinite sum of terms corresponding to all the possible ways the particles can interact by the exchange of virtual electrons and photons (*see* virtual particle). Each term may be represented by a diagram (called a *Feynman diagram*). These diagrams are built up from vertices representing the emission of a (virtual) photon by an

electron, and propagators that represent the exchange of virtual photons or electrons, as in Fig. *a*. Fig. *b* shows the first few diagrams for electron-electron scattering.

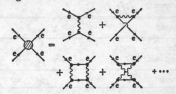

b Feynman diagrams for electron-electron scattering

In these diagrams all lines joining two vertices are propagators. The lines having a vertex at only one of their ends and the other end free, represent the physical particles before and after the interaction. A set of simple rules enables the contribution to the scattering matrix to be calculated from each of these diagrams.

quantum field theory A quantum mechanical theory in which particles are represented by fields whose normal modes of oscillation are quantized. *Elementary particle interactions are described by relativistically invariant theories of quantized fields (i.e. by *relativistic quantum field theories*). In *quantum electrodynamics, for example, charged particles can emit or absorb a *photon, the quantum of the electromagnetic field. Quantum field theories naturally predict the existence of *antiparticles and both particles and antiparticles can be created or destroyed; a photon, for example, can be converted into an electron plus its antiparticle, the positron. Quantum field theories provide a proof of the connection between spin and statistics underlying the *Pauli exclusion principle. *See also* electroweak theory; gauge theory; quantum chromodynamics.

Table of Conserved Quantum Numbers

Quantum Number — Interaction	Angular Momentum J, J_3	Charge Q	Baryon Number B	Isospin I	Isospin Q.N. I_3	Strangeness S	Parity P	C-Parity C	G-Parity G	Lepton Numbers l_e, l_μ, l_τ
Strong	√	√	√	√	√	√	√	√	√	√
Electromagnetic	√	√	√	×	√	√	√	√	×	√
Weak	√	√	√	×	×	×	×	×	×	√

quantum flavourdynamics (QFD) *See* electroweak theory.

quantum Hall effect *See* Hall effect.

quantum jump Transmission from one *stationary state of an atom or molecule to another, accompanying the emission or absorption of energy.

quantum mechanics A mathematical physical theory that grew out of Planck's *quantum theory and deals with the mechanics of atomic and related systems in terms of quantities that can be measured. The subject developed in several mathematical forms, including *wave mechanics (Schrödinger) and *matrix mechanics (Born and Heisenberg), all of which are equivalent.

quantum number In *quantum mechanics, it is often found that the properties of a physical system, such as its angular momentum and energy, can only take certain discrete values. Where this occurs the property is said to be *quantized* and its various possible values are labelled by a set of numbers called quantum numbers. For example, according to Bohr's theory of the *atom an electron moving in a circular orbit could not occupy any orbit at any distance from the nucleus but only an orbit for which its *angular momentum (*mvr*) was equal to

$nh/2\pi$, where n is an integer (0, 1, 2, 3, etc.) and h is the Planck constant. Thus the property of angular momentum is quantized and n is a quantum number that gives its possible values. The Bohr theory has now been superseded by a more sophisticated theory in which the idea of orbits is replaced by that of regions in which the electron may move, characterized by quantum numbers n, l, and m (*see* atomic orbital).

Properties of *elementary particles are also described by quantum numbers. For example, an electron has the property known as *spin, and can exist in two possible energy states depending on whether this spin is set parallel or antiparallel to a certain direction. The two states are conveniently characterized by quantum numbers $+\frac{1}{2}$ and $-\frac{1}{2}$. Similarly properties such as *charge, *isospin, *strangeness, *parity, and *hypercharge are characterized by quantum numbers. In interactions between particles, a particular quantum number may be conserved, i.e. the sum of the quantum numbers of the particles before and after the interaction remains the same. It is the type of interaction – *strong, *electromagnetic, *weak – that determines whether the quantum number is conserved (*see* table). *See* energy level.

quantum of action *See* quantum theory.

quantum state *See* stationary state.

quantum statistics Statistics concerned with the equilibrium distribution of elementary particles of a particular type among the various quantized energy states. It is assumed that these particles are indistinguishable.

In *Fermi–Dirac statistics*, the *Pauli exclusion principle is obeyed so that no two identical *fermions can be in the same quantum mechanical state. The exchange of two identical fermions (e.g. two electrons) does not affect the probability of distribution but it does involve a change in the sign of the *wave function.

The *Fermi–Dirac distribution law* gives f_E, the average number of identical fermions in a state of energy E:

$$f_E = 1/[e^{\alpha + E/kT} + 1],$$

where k is the *Boltzmann constant, T is the *thermodynamic temperature, and α is a quantity depending on temperature and the concentration of particles. For the valence electrons in a solid, α takes the form $-E_1/kT$, where E_1 is the *Fermi level.

In *Bose–Einstein statistics*, the Pauli exclusion principle is not obeyed so that any number of identical *bosons can be in the same state. The exchange of two bosons of the same type affects neither the probability of distribution nor the sign of the wave function.

The *Bose–Einstein distribution law* gives f_E, the average number of identical *bosons in a state of energy E:

$$f_E = 1/[e^{\alpha + E/kT} - 1].$$

The formula can be applied to photons, considered as quasi-particles, provided that the quantity α (which conserves the number of particles) is zero. Planck's formula for the energy distribution of *black-body radiation was derived from this law by Bose.

At high temperatures and low concentrations both the quantum distribution laws tend to the classical distribution:

$$f_E = Ae^{-E/kT}.$$

quantum theory A departure from the classical mechanics of Newton involving the principle that certain physical quantities can only assume discrete values. In quantum theory, introduced by Planck (1900), certain conditions are imposed on these quantities to restrict their values; the quantities are then said to be *quantized*.

Up to 1900, physics was based on Newtonian mechanics. Large-scale systems are usually adequately described. Several problems, however, could not be solved, in particular, the explanation of the curves of energy against wavelength for *black-body radiation, with their characteristic maxima. These attempts were based on the idea that the enclosure producing the radiation contained a number of *standing waves and that the energy of an oscillator is kT, where k is the *Boltzmann constant and T the thermodynamic temperature. It is a consequence of classical theory that the energy does not depend on the frequency of the oscillator. This inability to explain the phenomenon has been called the *ultraviolet catastrophe*.

Planck tackled the problem by discarding the idea that an oscillator can gain or lose energy continuously. He suggested that it could only change by some discrete amount, which he called a *quantum*. This unit of energy is given by $h\nu$, where ν is the frequency and h is the *Planck constant. h has dimensions of energy × time, or *action, and was called the *quantum of action*. According to Planck an oscillator could only change its energy by an integral number of quanta, i.e. by $h\nu$, $2h\nu$, $3h\nu$, etc. This meant that the radiation in an enclosure had certain discrete energies and by considering the statistical distribution of oscillators with respect to their energies, he was able to derive the Planck *radiation formula.

The idea of quanta of energy was applied to other problems in physics. In

Flavour	Mass (GeV/c^2)	Charge Q	Isospin I_3	Strangeness S	Charm C	Bottomness B	Topness T
d	≈ 0.3	$-\frac{1}{3}$	$-\frac{1}{2}$	0	0	0	0
u	≈ 0.3	$+\frac{2}{3}$	$+\frac{2}{3}$	0	0	0	0
s	≈ 0.5	$-\frac{1}{3}$	0	-1	0	0	0
c	≈ 1.5	$+\frac{2}{3}$	0	0	$+1$	0	0
b	≈ 5.0	$-\frac{1}{3}$	0	0	0	-1	0
t	> 90	$+\frac{2}{3}$	0	0	0	0	$+1$

1905 Einstein explained features of the *photoelectric effect by assuming that light was absorbed in quanta (*photons). A further advance was made by Bohr (1913) in his theory of atomic spectra (*see* atom; hydrogen spectrum) in which he assumed that the atom can only exist in certain energy states and that light is emitted or absorbed as a result of a change from one state to another. He used the idea that the angular momentum of an orbiting electron could only assume discrete values (i.e. was quantized). A refinement of Bohr's theory was introduced by Sommerfeld in an attempt to account for fine structure in spectra. Other successes of quantum theory were its explanations of the *Compton effect and *Stark effect. Later developments involved the formulation of *quantum mechanics.

quantum yield *See* quantum efficiency.

quark A fundamental constituent of *hadrons, i.e. of particles that take part in *strong interactions. Quarks are never seen as free particles (*see* quark confinement) but their existence has been demonstrated in high-energy scattering experiments and by symmetries in the properties of observed hadrons. They are regarded as elementary *fermions, with *spin $\frac{1}{2}$, *baryon number $\frac{1}{3}$, strangeness 0 or -1 and *charm 0 or $+1$. They are classified in six *flavours*

[up (u), charm (c), and top (t), each with charge $\frac{2}{3}$ the proton charge; down (d), strange (s), and bottom (b), each with $-\frac{1}{3}$ the proton charge]. Each type has an antiquark with reversed signs of charge, baryon number, strangeness, and charm. The top quark has not been observed experimentally but there are strong theoretical arguments for its existence. The top quark mass is known to be greater than about 90 GeV/c^2, too heavy to be detected in current high-energy particle accelerators.

The fractional charges of quarks are never observed in hadrons, since the quarks form combinations in which the sum of their charges is zero or integral. Hadrons can be either *baryons or *mesons. Essentially baryons are composed of three quarks while mesons are composed of a quark-antiquark pair. These components are bound together within the hadron by the exchange of particles known as *gluons*. Gluons are neutral massless gauge bosons. (*See* quantum chromodynamics.)

The quarks and antiquarks with zero strangeness and zero charm are the u, d, ū, and d̄. They form the combinations:

proton (uud), antiproton (ūūd̄);

neutron (uud), antineutron (ūd̄d̄);

pions: π^+ (ud̄), π^- (ūd), π^0 (dd̄, uū).

The charge and spin of these particles is the sum of the charge and spin of the component quarks and/or antiquarks.

In the strange baryons (e.g. the Λ and Σ particles), one or more of the quarks are the s flavour. In the strange mesons (e.g. the K mesons), either the quark or antiquark is strange. Similarly, the presence of one or more c quarks leads to the charmed baryons and a c or c̄ to the charmed mesons.

It has been found useful to introduce a further subdivision of quarks, each flavour coming in three *colours* (red, green, blue). Colour as used here serves simply as a convenient label and is unconnected with ordinary colour. A baryon comprises a red, a green, and a blue quark and a meson comprises a red and antired, a blue and antiblue, or a green and antigreen quark and antiquark. In analogy with combinations of the three *primary colours of light, hadrons carry no net colour, i.e. they are 'colourless' or 'white'. Only colourless objects can exist as free particles. The characteristics of the six quark flavours are shown in the table.

quark confinement The theory that *quarks can never exist in the free state, which is substantiated by lack of experimental evidence for isolated quarks. The explanation given for this phenomenon in the *gauge theory known as *quantum chromodynamics (QCD), by which quarks are described, is that quark interactions become weaker as they come closer together and fall to zero when the distance between them is zero. The converse of this proposition is that the attractive forces between quarks become stronger as they move apart; as this process has no limit, quarks can never separate from each other.

In some theories, it is postulated that at very high temperatures, as might have prevailed in the early universe, quarks can separate; the temperature at which this occurs is called the *deconfinement temperature*.

quarter-phase *See* two-phase; polyphase system.

quarter-wavelength line A *transmission line one quarter of a wavelength long, used particularly as an *impedance-matching device in systems operating at the higher radio frequencies.

quarter-wave plate A thin double-refracting optical element, often of quartz or mica, that can be used to change the *polarization of an incident wave. It is cut parallel to the optic axis and its thickness is such that it introduces a quarter-wavelength path difference, i.e. a relative phase shift of 90°, between the ordinary and extraordinary waves. When plane-polarized light, at 45° to either principal axis, is incident on the plate, it is converted to circularly polarized light (and vice versa).

quartz The most abundant mineral, consisting of crystalline silicon dioxide (silica, SiO_2) and having diverse physical properties and uses. It is a *piezoelectric crystal, much used as a *piezoelectric oscillator. It also produces *double refraction, being an optically positive uniaxial crystal (*see* optically negative crystal). It rotates the plane of polarization to the left or right according to the variety, and to different extents for different colours. In addition, it transmits wavelengths between 180 nm (in the UV) and 4000 nm (in the IR). A variety of quartz prisms, lenses, and other optical elements make use of these properties. Quartz fibres – extremely fine filaments of quartz – are used as torsion threads in delicate instruments (*see also* molecular gauges).

quartz-crystal clock *See* clocks.

quartz-crystal oscillator *See* piezoelectric oscillator.

quartz-iodine lamp *Syn.* quartz-halogen lamp. A compact high-intensity electric light consisting of a fused-quartz bulb filled with an inert gas containing iodine (or sometimes bromine) vapour, and a tungsten filament operated at a higher temperature than that of conventional lamps. Quartz is used to withstand the high temperature attained by the bulb.

quasars *Syn.* quasi-stellar objects (QSOs). Members of a class of astronomical bodies that lie well beyond our Galaxy and emit an immense amount of energy from a compact region of space. They were first detected (1963) as the bright optical counterparts to some powerful radio sources, but in fact only about 1% of quasars are radio sources. The greater part of the energy output is in the infrared region of the spectrum. They are also strong X-ray sources.

Quasars all have large *redshifts. The first value determined was $z = 0.158$; some have recently been detected with exceptionally large redshifts, $z > 4$. The quasar redshifts are now generally interpreted as being due to the *Doppler effect arising from the expansion of the universe. This explanation makes quasars extremely distant: those with the largest redshifts are the most distant objects, and hence the youngest objects, observed in the universe. Why there are so few quasars at shorter distances is not yet known.

Quasar spectra are dominated by bright emission lines superimposed on a continuum. The quasar redshift is determined from the emission spectra. Many quasar spectra also show absorption lines, which sometimes have a wide range of redshifts up to the value of the emission redshift. Those close to the emission redshift possibly arise from matter close to the quasar.

To be visible at such great distances, quasars must be exceedingly luminous: many have absolute *magnitudes exceeding −27. Their light-producing region, however, has been found to be extremely small (in some cases less than a light-day). Quasars are now thought to be the energetic cores of galaxies. The only process known to be efficient enough to generate the quasar energy output is some form of accretion of matter onto a supermassive *black hole at the heart of the quasar.

quench A capacitor, resistor, or combination of the two, placed across a contact to an inductive circuit to inhibit sparking when the current ceases. Typically, a quench is employed across the make-and-break contacts of an induction coil.

quiescent current In any circuit, the current flowing in the circuit under conditions of zero applied signal.

Quincke's tubes *See* Herschel–Quincke tube.

Q-value *Syn.* nuclear energy change; nuclear heat of reaction. The amount of energy produced in a *nuclear reaction, often expressed in megaelectronvolts (MeV).

R

rad A former unit of absorbed *dose of radiation, equal to 0.01 joule per kilogram of material. It has been replaced by the *gray, equal to 100 rad.

radar (RAdio Direction And Ranging) A system for locating distant objects by means of reflected radio waves, usually of microwave frequencies. Modern systems are highly sophisticated and can produce precise and detailed information about stationary and moving objects. Radar is used for navigation and guidance of aircraft, ships, etc., and is also used in meteorology and astronomy and for military purposes.

A radar system consists of a source of microwave power, a transmitting aerial emitting a narrow beam together with a receiving aerial, a receiver to detect the echo, and a *cathode-ray tube – the *radar indicator* – to display the output in suitable form.

Pulse radar systems transmit short bursts of microwaves, and the reflected pulse is received in the interval between the transmitted pulses. The same aerial is generally used for transmission and reception. *Continuous wave* (CW) systems transmit energy continuously, a small proportion being reflected by the target and returned to the transmitter. This uses less bandwidth than the conventional pulse systems.

Doppler radar employs the *Doppler effect to distinguish between stationary and moving targets. The change in frequency between transmitted and received waves is measured to give the velocity. *V-beam radar* can determine range, bearing, and height of an object. It uses two transmitters simultaneously, the fan-shaped beams rotating continuously. One beam is vertical, the other inclined to it at ground level.

In any of the above systems the direction and distance of the target is given by the direction of the receiving aerial, and the time between transmission of the signal and reception of the echo. The transmitting and receiving aerials are usually dish aerials, steered so as to scan an area. A common procedure is to rotate the aerials in a horizontal plane, and produce a synchronous circular scan on the radar indicator. A target is displayed as a luminous spot on a radial line. Such a presentation is called a *plan position indicator* (PPI).

Phased-array radar is a highly sophisticated system. It typically consists of a flat rectangular arrangement of small identical radiating elements. Each is fed a microwave signal of equal amplitude but the relative phase of the signals can be altered electronically across the face of the array. This allows the direction of the radar beam to be altered rapidly: there is no mechanical movement of the aerials, the beam being steered through the principle of wave interference. The same set of delays that steered the transmitted beam brings all the constituent signals of the echo back into phase at the array, ready for processing. Computer control enables several hundred targets to be tracked simultaneously.

radial astigmatism *Astigmatism due to oblique incidence on a lens system.

radian An SI unit of angle. One radian encloses an arc equal to the radius of a concentric circle. 2π radians $= 360°$; 1 radian $= 57.296°$.

radiance Symbol: L_e, L. **1.** For a point of radiant energy, the *radiant intensity, in a specified direction, per unit projected area:
$$L_e = dI_e/dA \cdot \cos\theta,$$
where A is the area and θ is the angle between the specified direction and the surface.
2. For a point on a surface that is receiving radiant energy, the *irradiance (E_e) per unit solid angle (Ω):
$$L_e = dE_e/d\Omega.$$
The irradiance is taken over an area perpendicular to the direction of incident radiation.

Radiance is measured in watts per steradian per square meter. *Compare* luminance.

radiant A qualifying adjective denoting pure physical quantities used in *photometry in which *electromagnetic radiation is evaluated in energy units. Radiant quantities are distinguished from their corresponding *luminous quantities by adding a subscript e (for energy) to their symbols.

radiant efficiency Symbol: η_e, η. The ratio of the *radiant flux emitted by a source of radiation to the power consumed.

radiant energy Energy in the form of *radiation. The total power emitted or received by a body in the form of radiation is the *radiant energy flux*, or simply *radiant flux.

radiant exitance Symbol: M_e, M. The *radiant flux leaving a surface per unit area. It was formerly called the *radiant emittance*. It is measured in watts per square metre. *Compare* luminous exitance.

radiant exposure Symbol: H_e, H. **1.** The surface density of the total radiant energy received by a material.

2. A measure of the total energy of the radiation incident on a surface per unit area, expressed by the product of the *irradiance, E_e and the irradiation time: $H_e = E_e \int dt$. It is measured in joules per square metre. *Compare* light exposure.

radiant flux *Syn.* radiant power. Symbol: Φ_e, Φ. The total power emitted or received by a body in the form of *radiation. The term is usually applied to the transfer of energy in the form of *electromagnetic radiation as opposed to particles, but it is not usually applied to radio waves. It is measured in watts. *Compare* luminous flux.

radiant flux density Symbol: ϕ. Either the *irradiance or *radiant exitance of a surface. For *radiant flux Φ, it is given by $\Phi = \int \phi dS$ for area S.

radiant intensity Symbol: I_e, I. The *radiant flux (Φ_e) emitted per unit solid angle (Ω) by a point source in a given direction: $I_e = d\Phi_e/d\Omega$. It is measured in watts per steradian. *Compare* luminous intensity.

radiant power *See* radiant flux.

radiation Anything propagated as rays, waves, or a stream of particles but especially light and other electromagnetic waves, sound waves, and the emissions from radioactive substances.

radiation belts Regions within a planet's *magnetosphere in which energetic particles – mainly electrons and protons – are trapped by the planet's magnetic field. *See* Van Allen belts.

radiation diagram *See* radiation pattern.

radiation formula The formula, devised by *Planck, to express the distribution of energy in the normal spectrum of black-body radiation. Its usual form is:

$$8\pi chd\lambda/\lambda^5(\exp[ch/k\lambda T] - 1),$$

which represents the amount of energy per unit volume in the range of wavelengths between λ and $\lambda + d\lambda$; c = speed of light, h = the *Planck constant, k = the *Boltzmann constant, T = thermodynamic temperature.

radiation impedance *See* radiation resistance.

radiation pattern *Syn.* radiation diagram. A graphical representation of the distribution in space of radiation from any source, especially an *aerial; for a particular aerial, the pattern for transmission is identical to that for reception. The graph is normally in polar coordinates (*see* polar diagram).

radiation physics The study of radiation, particularly *ionizing radiation, and the physical effects it can have on matter.

radiation pressure 1. The very small pressure exerted upon a surface exposed to electromagnetic radiation. The transfer of energy E is associated with the transfer of momentum p, where $p = E/c$, c being the speed of light; hence any body that absorbs, reflects, refracts, or scatters electromagnetic radiation experiences a force. If a parallel beam of intensity I (power/area) is totally absorbed by a surface normal to the beam, the radiation pressure is I/c. If the radiation is perfectly reflected straight back, the pressure is double this. For uniformly diffused radiation the pressure is $\rho/3$ where ρ is the energy density of the radiation.

2. The steady pressure exerted on a surface by sound waves. This pressure is to be distinguished from the oscillatory change of pressure observed at the displacement node of the *standing waves. Radiation pressure, P, can be expressed as:
$$P = L(1 + \gamma)/c,$$
where L is the *sound intensity, c the speed of sound, and γ the ratio of the specific heat capacities. This assumes an adiabatic process takes place.

radiation pyrometer A *pyrometer that depends for its action upon the effect of thermal radiation from a hot body. In one form, heat rays from the hot body are focused upon a sensitive *thermocouple. The e.m.f. produced in the latter is a function of the temperature of the hot body and is either measured by means of a potentiometer or is utilized to produce a deflection of a galvanometer or a millivoltmeter.

radiation resistance 1. At a surface vibrating in a medium. The portion of the total *resistance (unit-area, acoustical, or mechanical) due to the radiation of sound energy into the medium.

The mean power radiated per unit area of a plane wavefront is:
$$\tfrac{1}{2}\rho c \xi^2_{max} = \tfrac{1}{2}\rho c f^2 a^2,$$
where ρ is the density of the medium, c the speed of sound in it, ξ_{max} is the maximum sound particle speed, which is equal to fa (f being the frequency and a the amplitude of the vibration). In the analogous electrical case, this equation represents the power of dissipation in a circuit of resistance ρc. The quantity ρc is usually known as the radiation resistance or impedance of a medium that is transmitting plane waves.

The characteristic impedance for a spherical wave of radius r is expressed as:
$$z = \rho c(X' + iY'),$$
where
$$X' = k^2 r^2/(k^2 r^2 + 1)$$
$$Y' = kr/(k^2 r^2 + 1)$$
$$k = 2\pi/\lambda = 2\pi f/c$$
(λ being the wavelength).

The first term $\rho c X'$ is a resistance term and the second term is a reactance. If r is very large, z will reduce to ρc as for plane waves. For a place at a small distance (compared with λ) from the source, kr is small and the radiation impedance or resistance is approximately
$$\rho c k^2 r^2 = 4\pi^2 f^2 \rho r^2/c.$$
This impedance may be included whenever there is a change from plane to spherical waves.

2. A fictitious resistance in which the power radiated away from or collected by an *aerial is considered to be dissipated. When the *ohmic losses in the elements themselves are zero, the radiation resistance is the resistive part of the impedance presented at the feed point.

radiative capture *See* capture.

radiative collision A collision that takes place between two charged particles and from which *electromagnetic radiation is

385

emitted due to the conversion of part of the kinetic energy. *See* bremsstrahlung.

radio 1. The use of *electromagnetic radiation to transmit or receive electrical impulses or signals without connecting wires. Also the process of transmitting or receiving such signals. The term is usually confined to the communications system transmitting audio information (wireless).

2. A *radio receiver.

3. Denoting electromagnetic radiation in the frequency range 3 kHz to 300 GHz. (*See* radio frequency.)

4. A prefix denoting *radioactivity.

radioactive series Most of the natural radionuclides have *atomic numbers (Z) in the range $Z = 81$ to $Z = 92$. These substances can be grouped into three radioactive series: the *uranium series*, *thorium series*, and *actinium series*. The *mass numbers of these radionuclides can be represented by a set of numbers: $4n$ (thorium series), $4n + 2$ (uranium series), and $4n + 3$ (actinium series), where n is an integer between 51 and 59. The *parent nuclides at the head of the series are long-lived with *half-lives in the region $10^9 – 10^{10}$ years. They are uranium-238 (uranium series), thorium-232 (thorium series), and uranium-235 (actinium series). The other members of the three series are formed mainly by *alpha decay or *beta decay of the preceding nuclide. The final substances are all stable isotopes of lead.

There is a fourth radioactive series, the *neptunium series* ($4n + 1$) in which the half-lives of the three nuclides at the top of the series, plutonium-241, americium-241, and neptunium-237 are much shorter than the parents of the other three series. These substances have therefore either disappeared from the earth's surface or are present in negligible quantities. They can be synthesized.

radioactive tracer A definite quantity of *radioisotope introduced into a biological or mechanical system so that its path through the system and its concentration in particular areas can be determined by measuring the radioactivity with a *Geiger counter, *gamma camera, or a similar device. A compound containing a radioisotope and used as a tracer is said to be *labelled*.

radioactive waste Solid, liquid, and gaseous waste products from nuclear reactors, uranium processing plants, hospitals, etc., that are radioactive. Because the radioactivity of some materials persists for thousands of years, their disposal must be undertaken with great care. High-level waste (spent nuclear fuel, etc.) needs artificial cooling and is therefore stored by its producers for several decades before disposal. Intermediate-level waste (reactor components, filters, sludges, etc., from processing plants) is solidified and stored mixed with concrete in steel drums in power stations prior to burial in deep mines or beneath the seabed in concrete chambers, where it cannot contaminate ground water. Low-level waste (solids and liquids contaminated by traces of radioactivity) presents few problems; in the UK, since 1988 it has been disposed of in steel drums in concrete-lined trenches at Driggs in Cumbria. This work has been undertaken by Nirex Ltd, a company set up jointly by the government and the nuclear industry. Other countries make similar arrangements. Until 1983, when it was suspended by international agreement, low- and intermediate-level wastes were disposed of in the deep Atlantic in steel drums cast in concrete. In addition some very dilute low-level gaseous and liquid wastes have been discharged into the air and the sea.

radioactivity The spontaneous disintegration of the nuclei of some nuclides (called radionuclides) with emission of *alpha particles or *beta particles, sometimes accompanied by a *gamma ray. The processes involved in *alpha decay and *beta decay alter the chemical nature of the atom involved, because of the change in *atomic number, and usually result in a more stable nucleus. Specific energy changes take place in the nucleus during a disintegration and any excess energy possessed by the nucleus after the expulsion of an α- or β-particle is emitted by gamma radiation or *internal conversion. Another particle that can be emitted by the nucleus is the *positron (an antielectron), the disintegration process being analogous to β-decay. It is possible for a γ-ray alone to be emitted, when a *metastable state of a radionuclide decays to a lower energy state of the same nuclide. A radionuclide can disintegrate into two different energy states of the same nucleus, forming a pair of nuclear *isomers. This occurs in the uranium series (*see* radioactive series). Electron *capture is another disintegration process.

Natural radioactivity is the disintegration of naturally occurring radionuclides. It was discovered in 1896 by Becquerel. Apart from the members of the *radioactive series, some other naturally occurring elements are known to contain radioisotopes. These include carbon, lutecium, neodymium, potassium, rhenium, rubidium, samarium, and scandium.

Artificial radioactivity was first demonstrated in 1934 by the Joliot-Curies. They showed that the bombardment of nuclei (boron and aluminium) by α-particles produced *artificial radionuclides*. These substances decay by the same processes as natural radionuclides. It has been possible to create many nuclides having an atomic number greater than that of uranium (92). These are the *transuranic elements*, produced by bombarding heavy stable atoms with high-energy protons, neutrons, deuterons, carbon atoms, etc. They are all radioactive.

The *activity of both natural and artificial radionuclides decreases exponentially with time. The time take for half a given number of atoms of a particular nuclide to be transformed is called the *half-life. It can vary from 1.5×10^{-8} seconds up to 10^{17} years. The fraction of atoms decaying in a certain time is not truly constant. Radioactive decay is a statistical phenomenon and the half-life is an average value of very many disintegrations. The activity of a given specimen is measured by determining the ionizing ability of its characteristic radiation, using one of a number of *counters or *detectors.

radio astronomy The study of astronomy through the radio signals emitted by some celestial bodies. The signals originate from sources of nonthermal radiation and are associated with bodies both within and beyond our Galaxy. A *radio telescope is used for observing. Radio sources within our Galaxy include *pulsars, *supernova remnants, and HI and HII regions – regions of neutral predominantly atomic hydrogen and predominantly ionized hydrogen respectively – in interstellar space. Extragalactic radio sources include *quasars and the highly active radio galaxies.

radiobalance An instrument in which the heating due to the absorption of radiation is neutralized by the cooling due to the *Peltier effect at one junction of a thermocouple, thus enabling the amount of incident radiation to be measured absolutely.

radiocarbon dating *See* dating.

radio frequency (r.f.) Any frequency electromagnetic radiation in the *frequency band 3 kilohertz to 300 gigahertz, or of alternating currents in this frequency range.

radio-frequency heating *Dielectric heating or *induction heating, when the alternating field has a frequency greater than about 25 kHz.

radiogenic Resulting from radioactive *decay.

radiogoniometer An apparatus by means of which the bearing of radio waves incident upon a fixed (i.e. nonrotating) aerial system to which it is connected may be determined. It consists fundamentally of two fixed coils mounted with their axes at right angles and a third coil that can be rotated inside the other two.

radiography The production of shadow photographs (*radiographs*) of the internal structure of bodies opaque to visible light by the radiation from X-rays, or by gamma-rays from radioactive substances.

radio interferometer *See* radio telescope.

radioisotope An *isotope of an element that undergoes *disintegration.

radiology The study and application of X-rays, gamma rays, and other penetrating *ionizing radiation.

radiolucent *See* radiopaque.

radioluminescence The emission of visible electromagnetic radiation from a radioactive substance.

radiolysis The chemical decomposition of materials into ions, excited atoms and molecules, etc., by *ionizing radiation.

radiometer An instrument for measuring the total energy or power received from a body in the form of radiation. It is used in physical *photometry, astronomy, and meteorology. The term is used in particular for instruments that detect and measure radiation in the infrared, visible, and near ultraviolet. *See* thermopile; bolometer; net radiometer; pyrheliometer.

radiometric age The age of a geological or archaeological specimen, determined by radiometric *dating.

radionuclide Any radioactive *nuclide.

radiopaque Opaque to radiation, especially X- and gamma rays. Radiopaque substances, such as bones, are visible on a radiograph (*see* radiography). *Radiotransparent* substances, such as skin, are transparent to radiation and are not visible on a radiograph. If a medium is almost entirely transparent to radiation, it is *radiolucent*.

radio receiver *Syn.* radio; wireless. A device that converts radio signals into audible signals. A simple receiver consists of a receiving aerial, a tuner that can be adjusted to the desired *carrier frequency, preamplifier, detector, audiofrequency amplifier, and a loudspeaker. A common refinement of a simple receiver is the *superheterodyne receiver. Radios can detect *frequency-modulated (FM) signals and/or *amplitude-modulated (AM) signals. High-fidelity devices usually contain extra circuits associated with the audiofrequency amplifier to restore the bass and treble response of the output to that of the original audible source.

radiosonde system A compact apparatus comprising a *meteorograph and radio transmitter that is carried into the earth's atmosphere by a balloon and transmits radio signals indicative of the temperature, pressure, and humidity.

radio source An extraterrestrial source detected with a *radio telescope. Many sources are extended. *See* radio astronomy.

radiospectroscope An apparatus for displaying (usually on a cathode-ray tube) an analysis of the radio-frequency energy arriving at an aerial. The wavelengths in actual use for transmission at any particular time are shown, and by the height and spread of the trace on the tube face, some indication of the field strength and modulation is given.

radio telescope A type of telescope used in *radio astronomy to record and measure the radio-frequency emissions from celestial radio sources. All radio telescopes consist of an *aerial or system of aerials, connected by *feeders to one or more *receivers. The aerials may be in the form of large metal *dishes or simple linear *dipoles. The receiver outputs may be displayed, or recorded for computer analysis.

The steerable dish consists of a wire mesh approximately shaped and mounted so that it can be directed to any part of the sky without serious distortion. A pencil beam of radiation is received from a small but not highly defined area of sky. A well-known example is the telescope at Jodrell Bank near Manchester. It has a diameter of 76 metres, an altazimuth mounting (*see* telescope), and is efficient at wavelengths above 0.1 metres.

The *radio interferometer* is a form of radio telescope that consists of two or more fixed or steerable radio aerials separated by a known distance and connected to the same radio receiver. *Interference occurs between waves from a radio source that are received by the aerials. The position of the source can thus be determined. If large numbers of aerials are used, they are generally arranged in parallel rows or in two rows at right angles to each other. Alternatively the method of *aperture synthesis may be employed. The interferometer is more sensitive than a single dish as it can detect radiation from sources of small angular diameter. As resolution increases, however, large-scale structure is lost.

radiotherapy The use of beams of *ionizing radiation, such as X-rays, energetic electrons, and the stream of gamma rays from the radioisotope cobalt-60, in the treatment of cancer. A sufficient amount of radiation must be given to kill the cancer cells without harming intervening tissue. This is achieved by irradiating from several different directions with a narrow beam so that the tumour receives the maximum dose.

radiotransparent *See* radiopaque.

radio waves *Electromagnetic radiation of *radio frequency, used in *radio and *television broadcasting and other communications systems. *See* aerial; modulation; ionosphere; radio astronomy.

radio window *See* atmospheric windows.

radius of curvature *See* curvature.

radius of gyration The root mean square distance of the mass distribution of a rigid body from its axis of rotation. For a rigid body of *moment of inertia I about a given axis, $I = mk^2$, where m is the mass and k is the radius of gyration with respect to this axis.

radius vector A line joining a point on a curve to a reference point (such as the origin of polar coordinates). *See* coordinate.

rainbow The continuous spectrum of sunlight seen as one or more circular arcs in the sky when the light falls on raindrops. The observer's back must be towards the sun. The primary bow, which is usually the only one seen, makes a mean angle of about 41° with the line from the sun through the head of the observer. If the background is very dark, the secondary bow may be seen at about 52° from this line.

The primary bow is spread over about 2° with the red on the outside. The bow is formed at minimum deviation by light, which undergoes one partial internal reflection in a raindrop and is deviated and dispersed on entering and leaving. The secondary bow is spread over more than 3° with the violet on the outside. The light undergoes two partial reflections in a drop so it is far less intense than the primary bow.

RAM Abbreviation for random-access memory. A type of *semiconductor memory used in computers for which both recording and retrieval of data by the user is possible – i.e. the user can both write to and read from RAM (*compare* ROM). The basic storage elements – often called *cells* – are microscopic devices fabricated as an *integrated circuit. A single cell can store one *bit – either a binary 0 or a binary 1. Very large capacity memories can be produced. The cells are formed in a rectangular array so that each one can be uniquely identified by its row and column. Any cell can thus be accessed directly in any order (and extremely rapidly), i.e. there is *random acess to the cells. To preserve the cell contents, RAM requires its power supply to be maintained.

RAM devices can be classified as *static RAM* or *dynamic RAM* (DRAM). Static RAM is fabricated from either bipolar or MOS components (*see* integrated circuit); each cell is formed by an electronic *latch whose contents remain fixed until written to. Dynamic RAM is fabricated from MOS components, the cells utilizing the charge stored on a capacitor as a temporary store; leakage currents require the cell contents to be "refreshed" at regular intervals (typically every millisecond). Compared with static RAMs, DRAMs have larger cell densities, lower power consumption, but slower access times. There may be many thousands of cells on one RAM chip: a 64 K RAM chip stores a total of 64 kilobits (i.e. 65 536 bits) of data.

Raman effect *Syn*. Raman scattering. A *scattering effect in which the light scattered from molecules differs in wavelength from the incident light (generally longer but may be shorter). The effect is distinguished from *fluorescence in that it is not a resonance effect – the incident light differs in wavelength from that of the absorption band of the substance. In addition, the scattered light has a much smaller intensity than most fluorescent light.

The changes in wavelength of the weak scattered light are characteristic of the scattering material; they correspond to specific energy-level differences in the substance associated with molecular rotation and vibration. The Raman effect is therefore used as a powerful analytical tool in the process known as *Raman spectroscopy*. Laser light is used as the incident light and the Raman spectrum of a sample is recorded and analysed.

Raman scattering *See* Raman effect.

Raman spectroscopy *See* Raman effect.

ramjet A propulsion engine in which a fuel burns in air that has been compressed by the forward motion of the engine only. It consists of a suitably shaped duct into which fuel is fed at a controlled rate, the combustion products being expanded in a nozzle.

Ramsden eyepiece A type of *eyepiece that, in its most elementary form, con-

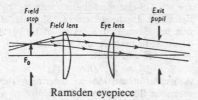

Ramsden eyepiece

sists of two planoconvex lenses of equal focal length, each equal to the separation of the curved surfaces facing one another. Commonly the separation is reduced to ⅔ of the lens focal length. It is better than the *Huygens eyepiece for spherical aberration, distortion, and longitudinal chromatic aberration, but suffers from lateral chromatism. The Ramsden eyepiece is used in measuring instruments, such as microscopes and spectrometers, with cross-wires or a *graticule in the plane of the field stop. The achromatized Ramsden is called the *Kellner eyepiece. *See* achromatic lens.

random access A method of retrieval or storage of data in which the individual storage locations can be accessed (read or written to) directly, in any order. There is random access with *disk storage. *See also* RAM; ROM.

random noise *See* noise.

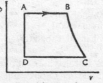

Rankine cycle

Rankine cycle An ideal steam-engine cycle since it is theoretically reversible. It differs from the *Carnot cycle in that a separate boiler and condenser are used. Beginning at A with the working substance as water in the boiler, AB represents isothermal expansion on boiling at constant pressure (*see* diagram); BC the adiabatic expansion of the steam in the cylinder or turbine during which it cools to the temperature of the condenser; and CD the isothermal compression on condensing at constant pressure. DA represents the transfer of the cold water to the boiler and at A the water is heated to the temperature of the boiler. The first three stages are identical with those of the Carnot cycle.

In later forms of this cycle the steam from the boiler enters a *superheater* in which the temperature is raised at constant pressure before entering the turbine. This increases the *efficiency and reduces the harmful effects of steam condensing in the turbine.

Rankine temperature Symbol: °R. An obsolete thermodynamic temperature scale linked to the Fahrenheit degree. Absolute zero on this scale is -459.67 °F, therefore °R = °F + 459.67. The ice point is 491.7 °R, often taken as 492 °R. 1K = 1.8 °R.

rarefaction The converse of *compression.

raster scan A method of producing pictorial images, the picture being built up line by line. It is used in *television and in most computer-graphics displays.

ratio Of a *transformer. For a single-phase power transformer, the ratio of the e.m.f.s induced in the primary and secondary windings. The following definitions are used in practice since they make allowance for special methods of connecting the windings (e.g. in *polyphase working) and for special applications:

1. *Voltage ratio of a power transformer.* The ratio of the voltage between terminals on the higher-voltage side to the voltage between terminals on the lower-voltage side at *no-load.

2. *Turns ratio of a transformer* (general). The ratio of the number of turns in the phase winding associated with the higher-voltage side to the number of turns in the corresponding phase winding associated with the lower-voltage side. For a single-phase transformer, the voltage ratio is substantially equal to the turns ratio but this is not generally the case for a polyphase transformer. It is possible for the latter type to have a turns ratio that is less than unity while the voltage ratio is greater than unity.

3. *Of an *instrument transformer.* The ratio of the primary terminal voltage (or primary current) to the secondary terminal voltage (or secondary current) under specified load conditions.

rational intercepts, law of If the edges formed by the intersections of three faces on a crystal are chosen as axes of reference OX, OY, OZ and a fourth face intersects these axes in A, B and C, then any other face on the crystal will intercept the axes in A', B', C' such that $OA/OA' = h$, $OB/OB' = k$, $OC/OC' = l$ where h, k, and l are rational whole numbers rarely exceeding 6, and where (h, k, l) are said to be the *Miller indices* of the face, relative to the axes OA, OB, OC.

rationalization of electric and magnetic quantities A technique that has been used to modify electrical and magnetic equations to provide a more rational "common sense" approach. The technique is best explained by reference to three examples. The magnitude of the *magnetic flux density B at a point distance r from an infinitely long straight wire carrying a current I and situated in a vacuum can be derived from Ampere's theorem as:

$$B = 2\mu_k I/r,$$

where μ_k is the *permeability of free space. This constant has unit value in the *CGS-electromagnetic system by definition. In the system of *SI units, it can be shown that μ_k is equal to 10^{-7} SI units. In practice this is written as $\mu_k = 10^{-7} K$ and hence:

$$B = 2\mu_k I/Kr \qquad (1)$$

A similar discussion to determine the magnetic flux density at the centre of a flat circular coil of radius r and N turns yields:

$$B = 2\pi\mu_k NI/Kr \qquad (2)$$

A completely different kind of analysis gives the capacitance C of an isolated evacuated sphere of radius r as:

$$C = K\varepsilon_k r \qquad (3)$$

where ε_k is the *permittivity of free space.

On further consideration, each of these three formulae has an irrational appearance. The magnetic field around a point on a straight wire is well known to be circular, and yet the quantity 2π that characterizes the concept of circularity does not occur in equation (1). The magnetic field at the centre of a flat circular coil is quite uniform and parallel, and yet equation (2) contains the 2π that was expected in equation (1). The electric field emanating from a sphere is, of course, three-dimensionally symmetrical yet no 4π appears in equation (3). Each of the equations can be thrown into a rational form by putting $K = 4\pi$. Thus

for a straight wire $B = \mu_k I/2\pi r$
for a circular coil $B = \mu_k NI/2r$
for a spherical capacitor
         $C = 4\pi\varepsilon_k r$

0π implies linearity, 2π implies circular symmetry, and 4π implies spherical symmetry. The permeability and permittivity of free space have magnitudes changed by a factor of 4π, and are termed the *magnetic and *electric constants.

Many formulae are unaffected by the process of rationalization, particularly those (such as Ohm's law) that are purely concerned with electric currents. *See also* Heaviside–Lorentz units.

ray A mathematical concept to give a first-order representation of the rectilinear propagation of light and basic to geometric optics theory. In an isotropic medium, the rays are normal to the wavefront; in *double refraction, the ordinary rays arc normal to the wavefront, while it is only in special cases that the extraordinary rays are so. In general, rays are the shortest optical paths between wavefronts. *See also* pencil (of rays).

Rayleigh criterion *See* resolving power.

Rayleigh disc A device used for the comparison of *sound intensities. It is based upon the principle that a light disc tends to set itself at right angles to the direction of an air stream whether the stream is alternating or direct. A small disc is suspended by a torsion thread so as to lie at an angle to the opening of a cylindrical resonator when it is unexcited. When the resonator is excited, the alternating air flow round the disc causes it to rotate. For small angles of deflection, the rotation is proportional to the sound intensity in the resonator tube and consequently to the intensity in the undisturbed field. To increase sensitivity, the disc is suspended in the connecting neck of a double resonator (*see* Helmholtz resonator). The disc diameter should be small compared with the wavelength of the incident sound.

Rayleigh–Jeans formula *See* black-body radiation.

Rayleigh limit To prevent detectable deterioration in the quality of an image, the optical path differences should not exceed $\lambda/4$.

Rayleigh refractometer *See* refractive index.

Rayleigh scattering The scattering of light by particles of dimensions small compared with the wavelength of light. For plane-polarized incident light of wavelength λ, the scattered intensity bears to the incident intensity the ratio

$$\frac{I}{I_0} = \frac{\pi^2 \sin^2 \theta}{r^2} (\varepsilon_r - 1)^2 \frac{V^2}{\lambda^4},$$

where θ is the angle between the electric vector of the incident beam and the direction of viewing, r is the distance from the particle to the point at which observations are made, and the particle has volume V and is of material of dielectric constant ε_r relative to the surroundings. For unpolarized light, $\sin^2\theta$ is replaced in the formula by

$$\tfrac{1}{2}(1 + \cos^2\phi),$$

where ϕ is the angle between the incident light and the direction of observation.

The fourth power dependence on wavelength means that blue light is much more strongly scattered than red light from a medium containing very fine particles. This is called the *Tyndall effect*. It accounts for the bluish appearance of smoke and of clear sky when the observation is not along the direction of illumination. The setting sun, seen through a considerable thickness of atmosphere, appears reddish because the light has been robbed of much of the blue end of the spectrum.

Rayleigh's law In magnetic materials subjected to magnetic fields low in comparison with the maximum coercive force, the *hysteresis loss in a cycle varies directly as the cube of the magnetic flux density. The law ceases to be

valid at the field value at which the
*Barkhausen effect takes place.

reactance 1. (electrical) Symbol: X. If an
alternating e.m.f. is applied to a circuit
the total opposition to the flow of an
alternating current is called the *impedance. The part of the impedance that is
not due to pure *resistance is called the
reactance and is caused by the presence
of *capacitance or *inductance. If the
alternating e.m.f. is given by:
$$E = E_0 \cos 2\pi ft = E_0 \cos \omega t$$
(ω = angular frequency), then the peak
value of the current in a series circuit
with resistance R and inductance L is
I_0, where
$$I_0 = E_0 / \sqrt{[R^2 + (\omega L)^2]}.$$
$\sqrt{[R^2 + (\omega L)^2]}$ is the impedance and
ωL is the reactance, in this case the *inductive reactance*. Similarly for a series
circuit with resistance R and capacitance
C,
$$I_0 = E_0 / \sqrt{[R^2 + (1/\omega^2 C^2)]}.$$
Here $1/\omega C$ is the *capacitive reactance*.
The reactance is the imaginary part of
the complex impedance Z, i.e.
$$Z = R + iX.$$
It is measured in ohms.
 2. (acoustic) Symbol: X_a. The magnitude of the imaginary part of the acoustic *impedance Z_a. If the reactance is
caused solely by inertia, it is called the
acoustic mass reactance. If it is due to
stiffness, it is called the *acoustic stiffness
reactance*. The product of acoustic mass
reactance and angular frequency is called the *acoustic mass* (Symbol: m_a). The
product of acoustic stiffness reactance
and angular frequency is called the
acoustic stiffness (Symbol: S_a). For an
enclosure of volume V with dimensions
that are small compared with the wavelength of sound, the acoustic stiffness is
given by $\rho c^2 / V$, where ρ is the density
and c the speed of sound in the me-
dium.
 3. (mechanical) Symbol: X_m. The
magnitude of the imaginary part of the

mechanical *impedance Z_m. If the reactance is caused by inertia, it is termed
the *mechanical mass reactance*. If it is
due to stiffness, it is called the *mechanical stiffness reactance*.

reactance coil *See* inductor.

reactance drop *See* voltage drop.

reactance transformer A device consisting of pure reactances arranged in a
suitable circuit. It is commonly employed for matching impedances at radio frequencies.

reaction 1. An interaction between
atoms or molecules (chemical reaction)
or between nuclides (*nuclear reaction).
 2. *See* Newton's laws of motion, III.

reactive current The component of an
alternating current that is in quadrature
with the voltage, the current and voltage
being regarded as vector quantities.
Compare active current.

reactive load The *load in which the
current and the voltage at the terminals
are out of phase with each other. *Compare* nonreactive load.

reactive power *See* power.

reactive voltage The component of an
alternating voltage that is in quadrature
with the current, the voltage and current
being regarded as vector quantities.
Compare active voltage.

reactive volt-amperes The product of the
current and the *reactive voltage, or the
product of the voltage and the *reactive
current.

reactivity An indication of the departure
of a *nuclear reactor from the condition
in which the reaction can just take place
(the *critical* condition). The reactivity is

defined by the expression $(1 - 1/k)$, where k is the ratio of the number of neutrons produced in a generation to the total number absorbed or lost; it is called the effective *multiplication constant. $k > 1$ implies a supercritical condition for the reactor. $k < 1$ implies a subcritical condition.

reactor 1. An electrical device possessing *reactance and selected for use because of that property.
2. *See* nuclear reactor.

read-only memory *See* ROM.

real image *See* image.

Réaumur scale A temperature scale in which the ice point is taken as 0° R and the steam point as 80° R.

receiver The part of a communication system that converts the transmitted waves into perceptible signals of the desired form. *See* radio receiver; television; superheterodyne receiver.

reciprocal lattice A theoretical lattice associated with a crystal lattice. If *a*, *b*, and *c* are the sides of a *unit cell of the real lattice then *a′*, *b′*, and *c′* define the reciprocal lattice, where

$$a' = \frac{b \times c}{a \cdot (b \times c)}, \quad b' = \frac{c \times a}{a \cdot (b \times c)},$$

and

$$c' = \frac{a \times b}{a \cdot (b \times c)}.$$

Reciprocal lattices are used in crystallography and in theories of the solid state.

reciprocal theorem (Maxwell) If a force F applied to one point in an elastic system produces a deflection d at another point, the same force F applied at the second point in the direction of the original deflection produces a deflection d at the first point in the direction in which F was first applied.

reciprocity failure *See* reciprocity law.

reciprocity law The density of processed photographic material is a function of exposure (= *illuminance $\times$ time) only, for a standard procedure of processing. When the illuminance (or exposure time) varies greatly from that to which a given emulsion is most sensitive, the reciprocity law no longer holds (*reciprocity failure*); the image density is then less than expected.

reciprocity relations (Onsager) If two flows (of heat, electricity, matter, etc.) J_1, J_2, produced by gradients or forces X_1, X_2, so interact that

$$J_1 = L_{11}X_1 + L_{12}X_2$$
$$J_2 = L_{21}X_1 + L_{22}X_2,$$

then, subject to a condition restricting magnitudes, $L_{12} = L_{21}$; and similarly for three or more interacting flows.

These relations can be used, for example, to establish relations between thermoelectric coefficients or justify the use of a single mutual inductance between two circuits.

recombination rate The rate at which electrons and holes in a *semiconductor recombine, tending to restore the system to thermal equilibrium. An electron in the conduction band may recombine with a hole in the valence band. Alternatively, there may be electron capture or hole capture by a suitable acceptor or donor impurity in the semiconductor.

rectifier An electrical device that permits current to flow in only one direction and can thus make alternating into direct current. It operates either by suppressing or attenuating alternate half-cycles of the current waveform or by reversing them. The most common rectifiers are semiconductor *diodes.

rectifier instrument A d.c. instrument that can be made suitable for a.c.

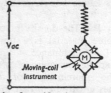

Circuit of rectifier instrument for use as an a.c. voltmeter

measurements by using a *rectifier to convert the alternating current to a unidirectional current. A common arrangement consists of four *diodes and a *moving-coil instrument, M, connected to form a bridge circuit (*see* diagram). Instruments of this type are usually calibrated to read *root-mean-square values on a.c. supplies of sinusoidal waveform.

rectifier photocell *See* photovoltaic cell.

rectilinear propagation The progress of light in straight lines in an isotropic medium; the *ray is the geometrical representation of it. On account of the wave character of light, it is only approximately true (*see* diffraction). According to the general theory of *relativity, light rays travelling through free space are deflected towards any massive bodies they may pass.

red giant A type of cool *giant star emitting light in the red region of the spectrum. A normal star expands to a red giant as it exhausts its nuclear fuel. *See* stellar spectra; Hertzsprung–Russell diagram.

redshift A shift in the spectral lines of an astronomical body towards longer wavelength values relative to the wavelengths of these lines in the terrestrial spectrum; optical lines are shifted towards the red end of the visible spectrum. The *redshift parameter*, z, is given by the ratio $(\lambda' - \lambda)/\lambda'$, where λ and λ' are the wavelengths of a spectral line from a terrestrial and astronomical source respectively.

Redshifts of objects within our Galaxy are due to the Doppler effect, arising from the movement of a star or other source away from the observer. The value of z is then v/c, where v is the velocity of recession and c the speed of light. Redshifts of extragalactic objects (e.g. other galaxies and *quasars), are also interpreted in terms of the Doppler effect, which for these objects arises from the expansion of the universe. Since the recessional velocities for these objects can be very great, the relativistic expression for redshift must be used:
$$z = [(c + v)/(c - v)]^{1/2} - 1.$$
See also Hubble constant.

reduced distance A distance in a medium divided by the refractive index of the medium. It may be regarded as an air-equivalent distance, and conjugate focus relations of refraction from one medium to another, using reduced distances, are the same as for a thin lens in air.

reduced equation of state An *equation of state in which the variables, p, V, and T are expressed as fractions of the *critical pressure, *critical volume, and *critical temperature respectively, these fractions being known as the *reduced pressure* (α), *volume* (β), and *temperature* (γ). The reduced form of the *van der Waals equation is:
$$(\alpha + (3/\beta^2))(3\beta - 1) = 8\gamma.$$

reduced mass Let a small particle of mass m be attracted by, and describe a closed orbit about, a heavier one of mass M. Then the equations describing the motion, assuming M to be fixed, may be transformed into those holding if M is not fixed (in which case both particles revolve around their common centre of gravity) by replacing m by μ, the reduced mass, where

$$1/\mu = 1/m + 1/M.$$

reduced pressure, temperature, volume *See* reduced equation of state.

redundancy 1. In a transmission system. The existence of information in excess of that required for the essential information to be transmitted. This is often deliberately included to allow for loss in the transmission system.

2. In electronic circuits or systems. The inclusion of extra components or circuits to increase the reliability of the system, i.e. if a fault should develop, the function of that part of the circuit may be taken over by the redundant circuits or components provided.

reed A thin bar of metal or cane clamped at one end and set into transverse vibration generally by a flow of air. The frequency of the note emitted depends on the material of the reed and its dimensions. The reed may vibrate against an air slot completely stopping the flow of air at certain parts of its vibration, or it may vibrate freely through a slot of dimensions slightly larger than the reed tongue. The reed is the generating source in several musical instruments. In brass instruments of the orchestra the mouthpiece consists of a cup or cone-shaped orifice that fits over the player's lips, the latter forming a twin reed. For high frequencies the player must compress his lips more tightly and increase the blowing pressure. The vocal chords of the larynx are often considered to be a pair of free reeds, the pitch produced being altered by tension and wind pressure. However, the range of tension and thickness available seem too small for the vibrations covering two octaves over which a normal voice may operate. It is suggested that the sound is rather produced by a sort of *jet tone in which the sides of the vocal chords also take part. This may be the explanation of sound production by the players' lips in brass instruments.

reed relay *See* relay.

reference tone An accepted standard pure tone of known intensity and frequency. Some such tone is necessary as the basis of any scale of sound intensity or loudness.

reflectance Symbol: ρ. The ratio of the *radiant or *luminous flux reflected by a body to the incident flux. The term may be qualified by the adjectives *specular, diffuse,* and *total* according to the nature of the reflecting surface.

reflecting microscopes *See* microscope.

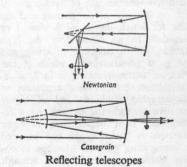

Newtonian

Cassegrain

Reflecting telescopes

reflecting telescope *Syn.* reflector. An optical (or infrared) *telescope with a large-aperture concave mirror, usually paraboloid, for gathering and focusing light from astronomical bodies. There is no chromatic *aberration and very little spherical aberration and coma, all of which occur in the *refracting telescope.

The mirror is supported on its back surface, and has either an equatorial mounting, with one axis parallel to the earth's axis, or an altazimuth mounting, with one axis vertical (*see* telescope).

There are several types of reflecting telescopes that differ from each other by the additional optical system used to bring the image to a convenient point, where it can be viewed through an eyepiece or recorded photographically or electronically. In the *Newtonian telescope* a small prism or flat mirror deviates the light from the concave mirror to the side of the telescope where it is viewed or recorded. In the *Cassegrain telescope* a small hyperboloid convex mirror has one focus coincident with that of the coaxial concave mirror. The other focus lies at the pole of the concave mirror. Light passing through a hole in the mirror at this point is viewed or recorded. Both types have high-quality definition at the centre of the field and are very stable.

The intensity of the image depends mainly on the amount of light collected (i.e. on the area of the mirror) and the time for which the image is viewed or recorded. In an observatory the high stability of the telescope structure, and the great precision with which it can track an astronomical object in its diurnal motion, permits long exposures of photographic emulsions and of highly sensitive photoelectric devices. As a result, very faint objects can now be detected. A *spectrograph can be used in conjunction with the telescope to obtain *stellar spectra. *See also* Schmidt telescope; Maksutov telescope.

reflection 1. (Of light) The process occurring when light strikes a surface of separation of two different media such that some is thrown back into the original medium. If the surface is smooth, reflection is *regular*, otherwise it is *diffuse* and the light is scattered. The two

laws of reflection, namely that incident ray, normal, and reflected ray lie in the same plane, and the angle of incidence (with the normal) is equal to the *angle of reflection* (with the normal), suffice to determine the position and attributes of the image, whether at plane, curved, or multiple mirrors, etc. Reflection can be regarded mathematically as a special case of *refraction ($n' = -n$). For normal incidence, the fraction of light reflected is equal to:

$$(n - 1)^2/(n + 1)^2,$$

where n is the relative *refractive index. For other angles of incidence, the fraction depends on the plane of *polarization. *Selective reflection* is said to occur when certain wavelengths are reflected more strongly than others. When reflection occurs at a denser medium, there is a change of phase of π; when it occurs at a less dense medium, there is no change of phase. *See also* total internal reflection.

Other forms of electromagnetic radiation, such as infrared and radio waves, also undergo reflection. In the case of X-rays, reflection can only occur at *grazing incidence* i.e. for very small *glancing angles, up to a critical angle.

2. (Of sound) A similar process occurring for a sound wave, the geometrical laws of reflection of sound waves being the same as those of light waves. The apparent differences between light and sound in reflection are merely questions of scale. The typical wavelength of sound being 100 000 times the wavelength of light it requires the dimensions of the reflecting surface to be 100 000 times the corresponding one in light to produce diffuse reflection or scattering. A mirror or lens to produce concentration of sound must be enormous compared with mirrors and lenses used in optical work. The same remark applies to gratings in sound diffraction (*see* acoustic grating). Plane waves when reflected may produce *standing waves.

If the wavelength of the sound is small compared with the dimensions of the reflector, the ordinary geometrical laws of optics are applicable. The reflection of sound waves on a large scale produces *echoes (see also acoustics). The surface of water forms a good reflector for sound waves in air, since the specific acoustic *impedances or resistances are widely different from each other.

reflection coefficient 1. A uniform *transmission line is said to be correctly terminated (or matched) when the terminating impedance is equal to the *characteristic impedance of the line (symbol Z_0). If the terminating impedance (Z_R) differs from Z_0, the actual current in the line at the termination under steady-state conditions may be regarded as the *vector sum of two currents, one being the current that would flow if Z_R is made equal to Z_0 (called the incident current), the other being a current that is reflected from Z_R (called the return current or reflected current). The vector ratio of the return current to the incident current is called the reflection coefficient. In terms of the impedances Z_0 and Z_R, the reflection coefficient is:

$$(Z_0 - Z_R)/(Z_0 + Z_R).$$

2. *See* acoustic absorption coefficient.

reflection density Symbol: D. The logarithm to base ten of the reciprocal of the *reflectance, i.e. $D = -\log_{10}\rho$.

reflection loss In communications. The loss in power (expressed in decibels) that occurs when a load is not correctly matched to the source, measured against the maximum power obtained with correct matching.

reflector 1. *See* reflecting telescope.

2. A layer of material surrounding the *core of a *nuclear reactor, whose purpose is to scatter some of the escaping *neutrons back into the core.

3. *See* directive aerial.

refracting angle (and edge) The angle formed by the two refracting surfaces of a prism in the principal section, i.e. a section perpendicular to the *refracting edge*, which is the edge formed by the intersection of the two refracting surfaces. This region is also called the apex of the prism.

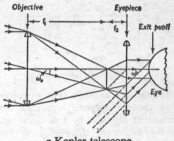

a Kepler telescope

refracting telescope *Syn.* refractor. An optical telescope consisting essentially of two lens systems. The *objective is a convex lens of long focal length, f_1; the *eyepiece is a lens of short focal length, f_2.

In the *Kepler telescope* (Fig. *a*), the eyepiece is convex and in normal adjustment the lenses are separated by the sum of the focal lengths producing a real but inverted image at infinity. The telescope is thus suitable for astronomical use or for making physical measurements. The magnification is the ratio (f_1/f_2) of the focal length of the lenses. This is equivalent to the ratios of the angles ω/ω_0. Normally a *Huygens, *Ramsden, or *Kellner eyepiece is used. An additional lenticular erecting system as in the *Fraunhofer eyepiece, makes the instrument suitable for terrestrial purposes. (*See also* prism binoculars.)

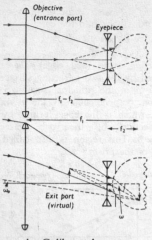

Objective
(entrance port)

Eyepiece

$f_1 - f_2$

f_1

f_2

Exit port
(virtual)

ω

b Galilean telescope

The *Galilean telescope* (Fig. *b*) consists of a concave eyepiece separated from the convex objective by a distance equal to the difference $(f_1 - f_2)$ in focal lengths. This produces an erect image, but the *exit port is virtual and lies inside the instrument. As the eye cannot be placed here, its best position is as near the eyepiece as possible thus restricting the field of view. The eye pupil acts as the *exit pupil. Under faint illumination, the pupil of the eye expands and the resulting increased exit pupil renders the telescope more efficient at night than the Kepler telescope. The magnification is the ratio (f_1/f_2) of the focal lengths and is rarely greater than six.

The brightness of the image depends on how much light can be collected by the objective, i.e. on its area. The observation of faint objects is therefore considerably improved by using a large-diameter objective and focusing the real image onto a photographic plate, exposing the plate for a long period. Large objective lenses are very difficult to grind and to mount. In addition, chro-

matic and spherical *aberration, coma, etc., have to be reduced to a minimum. Most of these problems are removed in *reflecting telescopes. *See also* telescope.

refraction 1. (Of light) The change of direction that a ray undergoes when it enters another transparent medium. The *laws of refraction* are (1) the incident ray, normal, and refracted ray all lie in the same plane, and (2) *Snell's law, $\sin i/\sin i' = n$ (a constant; *see* refractive index). According to the wave theory, the direction of the wavefront is altered because of the change of velocity. The action of prisms, lenses, etc., is explained by refraction. (*See also* dispersion.)

The refractive defects of the eye include *myopia* (short sight), in which light is focused in front of the retina, and *hypermetropia* (long sight), in which light is focused behind the retina. Myopia is corrected by using suitable concave lenses and hypermetropia by using convex lenses.

2. (Of sound) The change in direction in sound waves on reaching a boundary or point at which the velocity changes. Suppose the velocities of sound in two different media are C_1 and C_2; and the angles of incidence and refraction of a plane sound wave θ_1 and θ_2. The geometrical law of refraction will follow directly from the fact that the velocity in each medium is independent of the direction of the wavefront. Thus,

$$C_1/C_2 = \sin\theta_1/\sin\theta_2.$$

Refraction of sound occurs not only by a complete change of medium, but also by the gradual change of the properties of the same medium, for example by wind or temperature gradients. Such temperature or wind refraction in the atmosphere is analogous to the optical phenomena of mirage and has a very important influence on the range of transmission of sound in the atmosphere.

3. (Of lines of force) Electrical lines of force are refracted on passing at an angle from one dielectric medium to another. The tangents of the angles of incidence and refraction are in the ratio of the relative *permittivities.

refractive index Symbol: n. The ratio of the sine of the angle of incidence to the sine of the angle of refraction. If the first medium is a vacuum, the value is the *absolute refractive index*. The absolute refractive index is thus the ratio of the speed of light in a vacuum to the *phase speed of light in the medium, i.e. $n = c_0/c$. The value of the ratio for two media is the *relative refractive index*. If n_{12} is the relative index from medium 1 to medium 2, and n_{23} from medium 2 to medium 3, etc., then

$$n_{12} \times n_{21} = 1; \text{ i.e. } n_{12} = 1/n_{21}$$
$$n_{12} \times n_{23} \times n_{31} = 1;$$
$$\text{i.e. } n_{23} = n_{13}/n_{12}.$$

In general,

$$n_{12} \times n_{23} \times n_{34} \times \ldots n_{k1} = 1.$$

The relative refractive index is the ratio of the phase speeds of light in the two media, i.e. $n_{12} = c_1/c_2$. Commonly, the term refractive index refers to the value for sodium yellow ($\lambda = 589.3$ nm) relative to air whose absolute index is 1.000 29. *Dispersion of light arises on account of differences of refractive index for different colours, i.e. for different wavelengths.

Refractive index can be measured by finding the angle of minimum *deviation of a solid in the form of a prism. (A hollow glass prism may be used for a liquid.) Alternatively the *critical angle of incidence on an interface between two media can be found.

Interference methods can also be used. Light of wavelength λ_0 in a vacuum will have wavelength $\lambda = \lambda_0/n$ when in a medium of refractive index n. A length l in this medium contains $l/\lambda = ln/\lambda_0$ wavelengths. If therefore light from a given source is divided into two

channels of equal length, one in a vacuum and the other in the given medium, there is an effective path difference of $(n - 1)l/\lambda_0$ and on reuniting the two beams, interference results. The *Jamin refractometer* divides the light by using reflections at the front and back surfaces of oblique thick glass blocks and uses two parallel tubes, one evacuated and the other slowly filled with gas while fringes are counted passing across the eyepiece.

The *Rayleigh refractometer* uses a pair of parallel slits across a collimating lens to give two beams through the gas tubes. The interference fringes are compared with a fixed fringe system and a compensating device of inclined glass plates enables the effective path difference to be compensated and the achromatic (uncoloured) fringes of the two systems to be brought to coincidence.

refractivity An optical quantity equal to $(n - 1)$ where n is the refractive index.

refractometer An instrument for more or less direct measurement of *refractive index.

refractor *See* refracting telescope.

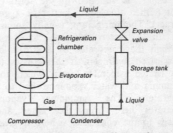

a Vapour-compression cycle

refrigerator A form of *heat pump used to maintain a chamber at a lower temperature than those of its surroundings. Most commercial refrigerators use either a *vapour-compression cycle* or a *vapour-absorption cycle* (which has no moving parts). In the vapour-compression cycle (Fig. *a*), a volatile refrigerant, e.g. ammonia, sulphur dioxide, or a chlorofluorocarbon (CFC), boils in an evaporator extracting heat from the chamber to be refrigerated as it does so. The vapour from the evaporator is compressed before passing to a condenser, in which it gives off its heat to the surroundings. The liquid from the condenser passes to a storage tank before being expanded through a valve to a low-pressure liquid, which again enters the evaporator for the cycle to be repeated.

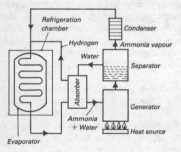

b Vapour-absorption cycle

In the vapour-absorption cycle (Fig. *b*), energy is supplied in the form of heat (either an electric heater or a gas burner). The refrigerant, usually ammonia in a water solution, is drawn through the evaporator by a stream of pressurized hydrogen. It then passes to the heated generator, from which the ammonia and water vapour are led to a separator. The ammonia vapour separates here from the water and passes to the condenser, where it gives off its heat to the surroundings, becoming a liquid. The liquid ammonia is then mixed with the hydrogen gas, which carries it to the evaporator again. An absorber, between the evaporator and the heated generator, uses the water from

the separator to dissolve the ammonia vapour before it enters the generator.

regelation If two blocks of melting ice are pressed firmly together the increase in pressure lowers the melting point and so the ice at the contact faces melts, taking its latent heat from the neighbouring ice whose temperature thus falls below 0 °C. When the pressure is removed the film of water previously formed freezes, giving its latent heat to the neighbouring ice and thus joining the two blocks into a single block of ice. This process is known as regelation.

regenerative braking *See* electric braking.

regenerative cooling A process used, e.g. in the *Linde process for liquefying air, in which compressed gas is cooled by expansion through a nozzle and the cool expanded gas is used to cool the oncoming compressed gas in a heat exchanger before it is cooled by expansion.

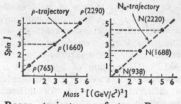

Regge trajectory of two Regge poles, mass being given in brackets

Regge pole model A theoretical model used to describe the scattering of *elementary particles at high energies. In general, it is found that the *strong interactions involved in such processes cannot be described in terms of the exchange of a single elementary particle. Although the contribution from the exchange of low-mass particles is usually the most important contribution to the *scattering amplitude, the contributions from the exchange of the higher-mass *resonances cannot be neglected. Mathematically it is possible to describe the collective effect of exchanging all these particles in terms of the exchange of a few objects called Regge poles whose *spins increase with their effective masses. The path traced out by the spin of a Regge pole as its "mass" varies is called a *Regge trajectory*. On a graph of spin against the square of the mass (*see* diagram), Regge trajectories are found to be approximately straight lines. The individual particles represented by a Regge pole have all quantum numbers the same except for their spins, which differ by $\Delta J = 2n$, where n is an integer.

register A semiconductor device that acts as a storage location in the processing unit of a computer. A register usually stores a single *word or sometimes a *byte or *bit, the information being held only temporarily before it can be operated on. Storage and retrieval of the information must be extremely rapid. A register thus normally consists of a group of *flip-flops, each storing one bit.

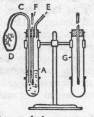

Regnault hygrometer

Regnault hygrometer A *hygrometer of the dew-point type consisting of two silver vessels A and G (*see* diagram), mounted side by side. Air may be blown from D through a tube C dipping into ethoxyethane (ether) contained in A. This causes the ether to evaporate

through E thus cooling the tube A until eventually, at the dew point, moisture condenses on the outside of A giving it a dull appearance compared with the surface of G. This temperature, and that at which the dullness disappears on allowing the apparatus to stand, are noted on the thermometer F, the mean giving the dew point; this, in conjunction with the room temperature, enables the relative humidity of the air to be calculated.

regulation Of electrical generators, transformers, and power transmission lines. The changes that take place in the available voltage due to internal resistance (for direct curent) or to internal impedance (for alternating current) when the load is changed under specified conditions.

rejector A parallel *resonant circuit. The *impedance of a circuit comprising inductance and capacitance in parallel has a maximum value at one particular frequency. The maximum impedance is called the *dynamic impedance. Compare* acceptor.

relative aperture *See* f-number.

relative atomic mass Symbol: A_r. The average mass per atom of a given specimen of an element, expressed in unified *atomic mass units. The value depends on the isotopes present in the specimen. The natural isotopic composition is assumed unless otherwise stated. Formerly called *atomic weight*.

relative density *See* density.

relative-density bottle A small flask with a perforated glass stopper that may be completely filled with a liquid. In order to determine the relative *density of the liquid, the bottle is weighed empty (m_1), full of liquid (m_2), and finally, full of water (m_3). The relative density of the liquid is then

$$(m_2 - m_1)/(m_3 - m_1).$$

Ingenious modifications of the procedure enable the relative density of powders, and of quantities of liquid insufficient to fill the bottle, to be found. *Compare* pyknometer.

relative humidity *See* humidity.

relative molecular mass Symbol: M_r. The average mass of a molecule or other molecular entity, expressed in unified *atomic mass units. It is equal to the sum of the *relative atomic masses of the constituent atoms. Formerly called *molecular weight*.

relative permeability *See* permeability.

relative permittivity *See* permittivity.

relative pressure coefficient *See* pressure coefficient.

relative velocity The velocity of A relative to B is the velocity that B, supposing himself at rest, assigns to A. If A and B are moving in the same direction the relative velocity of A to B is $v_A - v_B$; if moving in opposite directions, it is $v_A + v_B$. This only applies when v_A and v_B are very small compared to the speed of light. *See* relativity.

relativistic particle A particle the speed of which with respect to a particular observer is not small compared with the speed of light, such that the observer must use the theory of *relativity instead of classical physics. For example, a particle has a mass m_0 when measured by any observer at rest with respect to it, but an observer with relative speed v obtains the value m given by:

$$m = m_0 (1 - v^2/c^2)^{-1/2},$$

where c is the speed of light. The particle is not itself changed. It is the large

relative motion of the particle and observer that requires the more exact theory.

relativistic quantum field theories *See* quantum field theory.

relativity Two theories pioneered by Einstein, the *special theory of relativity* (1905) and the *general theory of relativity* (1915).

The special theory gives a unified account of the laws of mechanics and of electromagnetism (including optics). Before 1905 the purely relative nature of uniform motion had in part been recognized in mechanics, although Newton had considered time to be absolute and also postulated absolute space. In electromagnetism the *ether was supposed to provide an absolute basis with respect to which motion could be determined. (*See also* Galilean transformation equations; Michelson–Morley experiment; Newton's laws of motion.) Einstein rejected the concepts of absolute space and time, and made two postulates. (1) The laws of nature are the same for all observers in uniform relative motion. (2) The speed of light is the same for all such observers, independently of the relative motions of sources and detectors. He showed that these postulates were equivalent to the requirement that the coordinates of space and time used by different observers should be related by the *Lorentz transformation equations. The theory has several important consequences.

The transformation of time implies that two events that are simultaneous according to one observer will not necessarily be so according to another in uniform relative motion. This does not affect in any way the sequence of related events so does not violate any concepts of causation. It will appear to two observers in uniform relative motion that each other's clock runs slowly. This is the phenomenon of *time dilation*; for example, an observer moving with respect to a radioactive source finds a longer decay time than that found by an observer at rest with respect to it, according to:

$$T_v = T_0/(1 - v^2/c^2)^{1/2},$$

where T_v is the mean life measured by an observer at relative speed v, T_0 is the mean life measured by an observer relatively at rest, and c is the speed of light.

Among the results of relativity optics is the deduction of the exact form of the *Doppler effect (see also* redshift).

In relativity mechanics, mass, momentum, and energy are all conserved. An observer with speed v with respect to a particle determines its mass to be m while an observer at rest with respect to the particle measures the *rest mass* m_0, such that:

$$m = m_0/(1 - v^2/c^2)^{1/2}.$$

This formula has been verified in innumerable experiments. One consequence is that no body can be accelerated from a speed below c with respect to any observer to one above c, since this would require infinite energy. Einstein deduced that the transfer of energy δE by any process entailed the transfer of mass δm, where $\delta E = \delta m c^2$, hence he concluded that the total energy E of any system of mass m would be given by:

$$E = mc^2$$

(*see* conservation of mass and energy). The kinetic energy of a particle as determined by an observer with relative speed v is thus $(m - m_0)c^2$, which tends to the classical value $\frac{1}{2}mv^2$ if $v \ll c$.

Attempts to express *quantum theory in terms consistent with the requirements of relativity were begun by Sommerfeld (1915). Eventually Dirac (1928) gave a relativistic formulation of the *wave mechanics of conserved particles (*fermions). This explained the concepts of *spin and the associated magnetic moment, which had been postulated to

account for certain details of spectra. The theory led to results of extremely great importance for the theory of elementary particles (*see* annihilation, antiparticle, pair production), the theory of *beta decay, and for *quantum statistics. The *Klein–Gordon equation is the relativistic wave equation for *bosons.

A mathematical formulation of the special theory of relativity was given by Minkowski. It is based on the idea that an event is specified by four coordinates: three spatial coordinates and one time coordinate. These coordinates define a four-dimensional space and the motion of a particle can be described by a curve in this space, which is called *Minkowski space-time. See* four-dimensional continuum.

The special theory of relativity is concerned with relative motion between nonaccelerated frames of reference. The general theory deals with general relative motion between accelerated frames of reference. In accelerated systems of reference, certain fictitious forces are observed, such as the centrifugal and Coriolis forces found in rotating systems. These are known as fictitious forces because they disappear when the observer transforms to a nonaccelerated system. For example, to an observer in a car rounding a bend at constant velocity, objects in the car appear to suffer a force acting outwards. To an observer outside the car, this is simply their tendency to continue moving in a straight line. The inertia of the objects is seen to cause a fictitious force and the observer can distinguish between noninertial (accelerated) and inertial (nonaccelerated) frames of reference.

A further point is that, to the observer in the car, all the objects are given the same acceleration irrespective of their mass. This implies a connection between the fictitious forces arising from accelerated systems and forces due to gravity, where the acceleration produced is independent of the mass (*see* free fall). For example, a person in a sealed container could not easily determine whether he was being driven towards the floor by gravity or if the container was in space and being accelerated upwards by a rocket. Observations extended in space and time could distinguish between these alternatives, but otherwise they are indistinguishable. This leads to the *principle of equivalence* from which it follows that the inertial mass is the same as the gravitational mass.

A further principle used in the general theory is that the laws of mechanics are the same in inertial and noninertial frames of reference.

The equivalence between a gravitational field and the fictitious forces in noninertial systems can be expressed by using *Riemannian space-time*, which differs from the Minkowski space-time of the special theory. In special relativity the motion of a particle that is not acted on by any forces is represented by a straight line in Minkowski space-time. In general relativity, using Riemannian space-time, the motion is represented by a line that is no longer straight (in the Euclidean sense) but is the line giving the shortest distance. Such a line is called a *geodesic*. Thus space-time is said to be curved. The fact that gravitational effects occur near masses is introduced by the postulate that the presence of matter produces this curvature of space-time. This curvature of space-time controls the natural motions of bodies.

The predictions of general relativity only differ from Newton's theory by small amounts and most tests of the theory have been carried out through observations in astronomy. For example, it explains the shift in the perihelion of Mercury, the bending of light or other electromagnetic radiation in the presence of large bodies, and the *Einstein shift.

Very close agreement between the predictions of general relativity and their accurately measured values have now been obtained. *See also* gravitational waves.

relaxation oscillations 1. Oscillations characterized by a sawtooth type of waveform, the vibrating system apparently relaxing at each peak and returning quickly to the zero position from which the build-up recommences. Such oscillations can only be maintained by the existence of an effectively steady applied unidirectional force.

2. Oscillations of a system to which an impulse is applied intermittently. In this sense, the waveform of a system in relaxation starts a succession of short trains of damped oscillations, for which the amplitude is renewed from time to time.

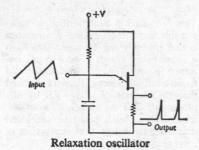

Relaxation oscillator

relaxation oscillator An oscillator in which one or more voltages or currents change suddenly at least once during each cycle. The circuit is arranged so that during each cycle energy is stored in and then discharged from a reactive element (e.g. a capacitor or inductor), the two processes occupying very different time intervals. An oscillator of this type has an asymmetrical output waveform that is far from being sinusoidal, a *sawtooth waveform being commonly generated. Square or triangular waveforms can also be produced by means of a suitable circuit. The output waveform is very rich in harmonics and for some purposes this is particularly useful. Common types of relaxation oscillator include the *multivibrator and *unijunction transistor (*see* diagram).

relaxation time Symbol: τ. **1.** The time required for the *electric polarization of any point of a suitably charged dielectric to fall from its original value to $1/e$ of that value, due to the electric conductivity of the dielectric.
2. Generally, the time required for an exponential variable to decrease to $1/e$ of its initial value.
3. The time required for a gas, in which the Maxwellian distribution of speeds has been temporarily disturbed, to recover that state.

relay An electrical device in which one electrical phenomenon (current, voltage, etc.) controls the switching on or off of an independent electrical phenomenon. There are many types of relay, most of which are either electromagnetic or solid-state relays.

The *armature relay* is one form of electromagnetic relay in which a coil wound on a soft-iron core attracts a pivoted armature that operates contacts or tilts a mercury switch; there are several variants. Another electromagnetic device is the *reed relay*, which has a coil wound around a glass envelope containing fixed contacts and a centrally placed reed contact in the form of a thin flat metal strip. When the coil is energized the reed is deflected, either making or breaking contact.

In a solid-state relay isolation between input and output terminals is provided using a *light-emitting diode (LED) in conjunction with a *photodetector. The switching is achieved using a *silicon controlled rectifier (SCR) or more commonly two SCRs (a triac). This type of relay is compatible with *digital circuit-

ry and has a wide variety of uses with such circuits. Isolation may also be achieved by transformer coupling on the input.

reluctance Symbol: R. The ratio of the *magnetomotive force, F_m, applied to a magnetic circuit or component, to the magnetic flux Φ, in that circuit, i.e. $R = F_m/\Phi$. It has the units henry^{-1}. Its reciprocal is the *permeance.

reluctivity The reciprocal of magnetic *permeability.

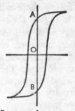

Hysteresis curve

remanence *Syn.* retentivity. The residual *magnetic flux density in a substance when the magnetizing field strength is returned to zero. It is represented by OA and OB in the hysteresis curve shown. *See* hysteresis.

renewable energy sources Sources of energy that do not use up the earth's finite mineral resources. They therefore exclude all fossil fuels and fission fuels. Because the combustion of fossil fuels emits carbon dioxide into the atmosphere and thus increases the *greenhouse effect and because fission fuels are thought by some to be hazardous and expensive and to cause problems of *radioactive waste disposal, renewable energy sources are regarded as environmentally desirable. The renewable forms of energy currently being explored and developed are solar energy, wind energy, tidal energy, wave energy, geothermal energy, and biomass energy. Hydroelectric energy is already widely exploited and has limited scope for further expansion; fusion energy is a virtually renewable source that offers enormous reserves of energy once the problems of *fusion reactors have been solved.

Solar energy is already being used, especially in hot countries. With some 10^{17} joules per second of solar power falling on the earth there remains great scope for widening its use. Two methods are used. The commonest is to heat water flowing through special panels on a building's roof. The second method is to use *solar cells. If these are to have a commercial future there will have to be a considerable fall in their cost.

Wind energy is a pollution-free cheap source. It does, however, require a great deal of land and a method of storing electricity for use when the wind drops. About 12 wind turbines are already feeding electricity to the UK national grid, mostly making use of Atlantic winds on the west coast. If all the usable sites in the UK could be harnessed for this purpose, some 20% of the country's energy could probably be wind-generated.

Tidal energy makes use of the tides to collect water behind a barrage, which is later released to turn a turbogenerator. Tidal power stations are in use in the USSR and France (on the River Rance). One is proposed for the Severn estuary, which, with its tidal rise of 8.8 m, could generate 7% of the UK electricity.

Wave energy uses the energy of the waves to make a string of floats (called Salter ducks) bob up and down. This bobbing motion is harnessed to turn a generator. Off the coast of the UK there are probably sufficient suitable sites to generate over 100 GW of electricity, once technical problems have been solved.

In geothermal energy, geysers and hot springs are tapped. Iceland, Italy, New Zealand, and the USA all make use of this form of energy. However, in the UK the very deep drilling required to reach usable sources involves considerable technical problems.

Biomass energy relies on the combustion of biofuels, such as methane generated by sewage or by farm, industrial, or household organic waste. Biofuels can also include specially cultivated organisms or crops (e.g. cane sugar) grown for their energy potential.

It is estimated that by 2025 some 20% of the UK's energy could be met by renewable sources.

repeater A device, used especially in telegraphic and telephonic circuits, that receives signals in one circuit and automatically delivers corresponding signals to one or more other circuits. A repeater usually amplifies the signal or may perform pulse regeneration on transmitted pulses.

residual *See* deviation.

resilience The amount of potential energy stored in an elastic substance by means of elastic deformation. It is usually defined as the work required to deform an elastic body to the elastic limit divided by the volume of the body.

resistance 1. (electrical) Symbol: R. The ratio of the potential difference across a conductor to the current flowing through it. (*See* Ohm's law.) If the current is alternating the resistance is the real part of the electrical *impedance Z, i.e. $Z = R + iX$, where X is the *reactance. Resistance characterizes a dissipation of energy as opposed to its storage. It is measured in ohms. *See also* resistor; temperature coefficient of resistance.

2. (acoustic) Symbol: R_a. The real part of the acoustic *impedance Z_a.

3. (mechanical) Symbol: R_m. The real part of the mechanical *impedance Z_m.

4. (thermal) Symbol: *R*. (i) The reciprocal of the *thermal conductance. The units are $m^2 K W^{-1}$. (ii) The ratio of the temperature difference between two points and the mean rate of flow of entropy between them. The units are $K^2 W^{-1}$.

resistance drop *See* voltage drop.

resistance gauge A gauge used for measuring high fluid pressures by means of the change in electrical resistance produced in manganin or mercury by those pressures. These gauges are calibrated with reference to the *free-piston gauge.

resistance strain gauge An instrument for measuring structural strains by the increase in electrical resistance of a wire or grid of wires attached to, or supported by, the structure.

resistance thermometers Thermometers in which the change in electrical resistance of a wire is used as the thermometric property. A small coil of wire, usually platinum but of other metals or carbon for use at low temperatures, is wound on a mica former and enclosed in a silica or porcelain sheath. The change in resistance is determined by placing the coil in one arm of a *Wheatstone bridge. As the measuring galvanometer is often at a great distance, duplicate leads are added to the other balancing arm to compensate for temperature variations in the leads. The platinum resistance thermometer is useful over a very wide range from −200 °C to over 1200 °C.

resistivity 1. Symbol: *ρ*. The electrical quantity defined by the equation $\rho = RA/l$, where *R* is the resistance of a wire of length *l* and cross-sectional area *A*. It is measured in ohm metres. The reciprocal of resistivity is *conductivity.

The product of the resistivity and the density is sometimes called the *mass resistivity*.

2. Symbol: φ. The reciprocal of the *thermal conductivity. It is measured in $m K W^{-1}$.

resistor A piece of electrical apparatus possessing *resistance and selected for use because of that property. *Carbon resistors* are widely used in electronic circuits. They consist of finely ground carbon particles mixed with a ceramic material, encapsulated into insulated tubes. The casing has a set of coloured stripes denoting the value of the resistance. For more precise values of resistance, *wire-wound* and *film resistors* are used. In the former, a *constantan or *maganin wire of uniform cross section is wound into a suitable shape; in the latter, a thin uniform layer of resistive material is deposited in a continuous pattern on an insulating core.

resistor-transistor logic A family of integrated *logic circuits in which the input is via a resistor into the *base of an inverting *transistor. They tend to be slow and susceptible to noise and are now little used.

resolution 1. The separation of a vector into its components.

2. The amount of information or detail revealed in an image produced by a telescope, microscope, computer, etc. It can be expressed in numerical terms. *See also* resolving power.

resolving power A measure of the ability of a *telescope or *microscope to produce detectably separate images of objects that are close together, or for a spectroscopic instrument, to separate two wavelengths that are very nearly equal.

The resolving power of an optical telescope is measured by the angular

a Two patterns overlapping completely

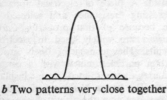

b Two patterns very close together

c Two patterns sufficiently far apart
to be separately distinguished

separation of two point sources that are just detectably separated by the instrument. The smaller this angle, the greater is the resolving power. The image of a point source formed by the primary mirror or lens will consist of a *diffraction pattern that has a central bright spot surrounded by alternate dark and light rings (*Airy rings). Two point sources close together will give rise to two overlapping diffraction patterns. A cross-section of the variation in intensity of the light in such a diffraction pattern is illustrated in Fig. *a*, while Figs. *b* and *c* show the intensity produced by two patterns overlapping to different extents.

Rayleigh proposed that a reasonable criterion for resolution of two point sources was that the inner dark ring of one diffraction pattern should coincide with the centre of the second diffraction pattern, a condition that is known as the *Rayleigh criterion*. This leads to the condition that two point sources are resolved by a telescope mirror or lens of aperture D, provided their angular separation in radians is not less than $1.22\lambda/D$ where λ is the wavelength. The Rayleigh criterion, which is quite arbitrary, corresponds in the combined diffraction pattern to an intensity ratio saddle to peak (AB/CD) of 0.81, for sources of equal intensity.

The *Abbe criterion* for resolution is that the angular separation should not be less than λ/D. The *Dawes rule*, derived experimentally, is that the separation should be at least $1100/D$ arc seconds. Both are less stringent than the Rayleigh criterion.

The Rayleigh criterion can be applied not only to optical but also to infrared and radio telescopes, and indicates the need for a large aperture to give the necessary resolving power. Objects will only be resolved, however, if there is sufficient magnification to produce the necessary separation of the images.

The resolving power of a microscope is measured by the actual distance between two object points that can be detectably separated by the instrument. The greater the resolving power the smaller will this distance be. Application of the Rayleigh criterion shows that the least separation for resolution is:

$$0.61 \; \lambda/n \sin i,$$

where λ is the wavelength of the light used, n the refractive index of the medium between object and objective, and i the semi-angle subtended at the object by the edges of the objective. The quantity ($n \sin i$) is the *numerical aperture of the objective. Abbe concluded that a separation of $0.5\,\lambda/n \sin i$ gave a more practical figure.

For good resolution the expression shows that the need is for higher numerical aperture and shorter wavelength. The numerical aperture may be increased by filling the space between object and objective with a medium of higher refractive index than air. This process is called *oil-immersion*, the medium usually chosen being cedar-wood oil. A wide-angle objective can also be used as long as no objectionable *aberrations are introduced in the images. A good high-power microscope objective will have a numerical aperture of perhaps 1.6, corresponding to about 200 nm as the least separation for resolution.

The second way of increasing the resolving power is to decrease the wavelength of the radiation used, as occurs in *ultraviolet microscopy*. Again in the *electron microscope use is made of the fact that the wavelength associated with a swiftly moving electron is much smaller than the wavelength of light. By using suitably focused electron beams, photographic images are obtained that bring down the limit of resolution to as little as 0.2 nm.

In a spectroscopic instrument, the requirement is for a detectable separation of wavelengths that are very nearly equal. In such cases, the *chromatic resolving power* is measured as the ratio of the wavelength studied to the difference in wavelengths that can just be separated.

If a prism *spectrometer is used then, assuming the Rayleigh criterion, the resolving power is given by the expression ($t \, dn/d\lambda$), where t is the maximum thickness of prism traversed by the beam and $dn/d\lambda$ is the ratio of change in refractive index to change in wavelength for the material of the prism. In a simple laboratory spectrometer of this type a resolving power of about 10^3 is possible.

If a *diffraction grating is used, the resolving power is the product of the total number of lines illuminated and the order number of the spectrum being used. For a 7.6 cm concave grating used in the second order a resolving power of about 10^5 is theoretically possible.

resonance 1. A condition in which a vibrating system responds with maximum amplitude to an alternating driving force. The condition exists when the frequency of the driving force coincides with the natural undamped oscillatory frequency of the system. (*See* forced vibrations.)

2. A condition existing when an oscillatory electric circuit responds with maximum amplitude to an external signal of angular frequency ω. The *impedance of an a.c. circuit with inductance L and capacitance C in series is given by:

$$Z = \sqrt{\{R^2 + [(\omega L) - (1/\omega C)]^2\}}.$$

When $\omega L = 1/\omega C$ a condition is achieved in which the impedance depends on the resistance alone, and as the resistance may be quite low for high values of L and C the current flowing will be high. This is the condition for resonance. Similar resonance can also be obtained when the capacitance and inductance are in parallel. (*See* resonant circuit; tuned circuit.)

3. *See* resonances.

resonance cross section *See* cross section.

resonances Extremely short-lived *elementary particles that decay by *strong interaction in about 10^{-24} second. They are thus hadrons. Resonances may be regarded as *excited states of the more stable particles. If the energy of colliding particles is slowly increased, a sudden peak in particle production is observed at the resonance energy; production drops off as the energy is further increased. This indicates that a particle exists with an effective mass equal to the combined relativistic masses of the colliding particles. Resonances cannot be observed directly but appear as spreads in the masses of the more stable hadrons.

Over one hundred *baryon and *meson resonances are known. Baryon resonances are usually denoted by the symbol of a similar baryon of lower mass, followed in brackets by the approximate resonance mass in MeV; examples include

N(1450), N(1520), N(1535),

Λ(1405), Σ(1385), Ξ(1530).

N is the symbol for the *nucleon. Most meson resonances have individual symbols, but in the following examples the mass is again shown in brackets:

ρ(775), ω(784), K*(892),

η′(958), φ(1019), f(1260).

η′ is spin 0 only; f meson is spin 1.

resonance scattering *See* scattering.

resonant cavity *See* cavity resonator.

resonant circuit A circuit that contains both inductance and capacitance so arranged that the circuit is capable of *resonance. The frequency at which resonance occurs – the *resonant frequency* – depends on the values of the circuit elements and their arrangement.

A *series resonant circuit* contains the inductance and capacitance in series. Resonance occurs at the minimum *impedance of the circuit, and a very large current is produced at the resonant frequency. The circuit is said to accept that frequency. Although large voltages are developed across the individual elements, these are out of *phase so that the total voltage developed is relatively low.

A *parallel resonant circuit* contains the circuit elements arranged in parallel. Resonance occurs at or near the maximum impedance of the circuit. The currents in each branch can be very large but are out of phase. This results in a minimal overall current at the resonant frequency, a voltage maximum being produced. The circuit is said to reject that frequency. Parallel resonant circuits are thus also called *antiresonant circuits*. *See also* tuned circuit.

resonant frequency *See* resonant circuit; forced vibrations.

rest energy The energy equivalent of the *rest mass of a body or particle, usually given in electronvolts.

restitution *See* coefficient of restitution.

rest mass The *mass of a body as determined by an observer who is at rest with respect to it. *See* relativistic particle; relativity.

resultant The sum of a number of *vectors (e.g. forces, velocities) found according to the polygon of vectors. The resultant of a system of forces applied to a body is the single force that has the same translational effect as the system itself.

resultant tone *See* combination tones.

retentivity *See* remanence.

reticle (or reticule) *See* graticule.

retina The inner coat of the eye, consisting of nerve fibres and endings sensitive to light (rods and cones).

reverberation The persistence of audible sound after the source has been cut off. If the time difference between reception of direct sound and its echo is less than about 1/15th of a second, true reverberation occurs and a number of echoes gives the sensation of a continuous sound of diminishing intensity. In the *acoustics of large halls the *reverberation time* is of considerable importance. It is defined as the time for the energy density to fall to the threshold of audibility from a value 10^6 times as great, i.e. a fall of 60 *decibels. Optimum periods generally lie between 1 and 2.5 seconds. The value should be low for speech and light music but high for orchestral music. In general, the optimum period is proportional to the linear dimensions of the room.

reverberation chamber *Syn.* echo chamber. A room that has a very long *reverberation time and is carefully designed to allow a uniform energy distribution of sound to be produced. For a room to have a long reverberation time there must be very little absorption by the exposed surfaces. Thus the walls and ceiling are usually plastered and then painted to ensure uniformity of surface throughout the room. However, these highly reflecting surfaces tend to produce *standing waves, which disturb the measurement of the energy distribution. Standing waves may be avoided by using a room in which no two walls are parallel or by having a large steel reflector rotating silently in the room. It is also customary to use a revolving source that produces a *warble tone and to take measurements at various points in the room. All sound from outside the chamber must be excluded and elaborate soundproofing is necessary.

reverberation time *See* reverberation.

reversal 1. (photographic) The transformation of a photographic negative into a positive (*see* photography).

2. (spectroscopic) Bright emission lines in the spectrum of a discharge tube or flame may be *reversed*, i.e. apparently transformed into dark absorption lines, when intense white light traverses the source and enters the spectroscope, due to selective absorption by the gas or vapour at the same frequencies as it emits.

reverse bias *See* reverse direction.

reverse direction *Syn.* inverse direction. The direction in which an electrical or electronic device has the larger resistance. Voltage applied in the reverse direction is the *reverse voltage* or *reverse bias* and the current flowing is the *reverse current*.

reversible change A change that is carried out so that the system is in *equilibrium at any instant and so that an infinitesimal change in the factor effecting the change causes every feature of the forward process to be completely reversed. This often means that the change must be carried out infinitely slowly since any kinetic energy could not change sign, and there must be no friction since this would always result in the evolution of heat. Such a process is never realizable in practice but close approximations to it can be attained. All practical processes involve *irreversible changes*, i.e. changes in which the system is not in equilibrium at all instants during the change.

reversible engine A *heat engine that operates reversibly, in the sense described under reversible change. *See* Carnot cycle; Carnot's theorem.

revolution 1. Motion of a body about an axis or around an external point. In astronomy the word is restricted to orbital motion of a body about its centre of mass, the term rotation being used for axial motion.

2. One complete cycle of such motion.

Reynolds' law The pressure head h required to maintain a liquid flow at constant speed v through a pipe of length l and radius r is given by the equation:
$$h = klv^p/r^q.$$
The constants k, p, and q are known as the *Unwin coefficients*. $p \approx 1$; $q \approx 2$.

Reynolds number Symbol: *Re*. A dimensionless quantity equal to $\rho vl/\eta$, where ρ is the density of a fluid of viscosity η, in motion with speed v relative to some

413

solid characterized by the linear dimension, *l*. For steady flow through a system with a given geometry, the flowlines take the same form at a given value of the Reynolds number. Thus flow of air through an orifice will be geometrically similar to that of water through a similar orifice if the dimensions and speed are chosen to give identical values of the Reynolds number.

rheology The study of the deformation and flow of matter.

rheometer An instrument designed to measure the flow properties of solid materials by investigating the relationship between stress, strain, and time.

rheostat A variable *resistor connected into a circuit, in series, to vary the current flowing in the circuit. It often consists of a linear or circular wire-wound resistor with a sliding contact, or it may have a number of small resistors that can each be brought into the circuit by a rotary switch. The word is usually applied to physically large devices. Small rheostats are usually called *potentiometers.

rheostat braking *See* electric braking.

rhombic system *See* crystal systems.

rhombohedral system *See* crystal systems.

ribbon microphone A type of *microphone that makes use of the simple dynamo principle that when a conductor moves perpendicular to a magnetic field, an e.m.f. is induced in it. The conductor in this case is a very thin strip of aluminium alloy a few millimetres wide, loosely fixed in a strong magnetic field parallel to the plane of the strip. The resulting force on the ribbon due to a sound wave is proportional to the differ-

ence in pressure between the front and back of the ribbon. When the acoustic path difference to the two sides is much smaller than a quarter of a wavelength, the resultant pressure on the ribbon is proportional to the product of the particle velocity and the frequency. If the resonant frequency of the ribbon is lower than the frequency of the sound, the resulting e.m.f. is then independent of frequency. Sound waves originating in the plane of the ribbon arrive at its front and back faces in phase so that no resultant force is produced. Thus the microphone has strong directional characteristics that can be used to reduce the pick-up of unwanted *noise.

Richardson–Dushman equation *Syn.* Richardson's equation. The basic equation of *thermionic emission relating the temperature of a body to the number of electrons it emits. It has the form:
$$j = AT^2 e^{-b/T},$$
where *j* is the emitted current density, *A* and *b* are constants, and *T* the thermodynamic temperature. *A* depends on the nature of the metal surface and *b* can be put equal to ϕ/k, where ϕ is the *work function and *k* the *Boltzmann constant.

Riemannian space-time *See* relativity.

Righi effect *See* Leduc effect.

right ascension *See* celestial sphere.

right-hand rule For dynamo. *See* Fleming's rules.

rigid body A body in which the distance between every pair of particles remains constant under the action of any forces. An abstract but useful concept in mechanics.

rigidity modulus *See* modulus of elasticity.

ring current A strong electric current flowing westwards in the earth's upper atmosphere. It arises from the net flow of electrons (eastwards) and protons (westwards) trapped in the *Van Allen belts, and causes variations in the normal pattern of the earth's magnetic field.

ringing *See* pulse.

ring main 1. An electric main that is closed upon itself to form a ring. If the ring is supplied by a power station at one point only, then between that point and any other point in the ring to which a consumer may be connected there are two independent electrical paths. Hence, in the event of a fault in the ring, the latter can be broken and the faulty section disconnected without interrupting the supply to the consumer. 2. A domestic wiring system in which individual outlets have their own fuses, a number of such outlets being connected in parallel to a ring circuit.

ring winding *Syn.* toroidal winding; Gramme winding. A winding in an electrical machine in which the coils are wound on an annular magnetic core, one side of each turn being threaded through the magnetic ring.

ripple (electrical) An a.c. component superimposed on a d.c. component, resulting in variations in the instantaneous value of a unidirectional current or voltage. The term is used particularly in connection with the output of a *rectifier. The ratio of the *root-mean-square value of the ripple (i.e. a.c. component) to the mean value (i.e. d.c. component) is called the *ripple factor*. A circuit (usually a type of *filter) designed to reduce the magnitude of a ripple is called a *smoothing circuit*.

ripples On a fluid surface. Waves of small amplitude on the surface of a fluid for which the wavelength is small and the effects of the *surface tension of the fluid are important. The velocity of ripples of length λ on a fluid of density ρ and surface tension σ is given by:

$$\sqrt{(\lambda g/2\pi + 2\pi\sigma/\lambda\rho)},$$

g being the acceleration of *free fall.

ripple tank The similarity between a plane section of a three-dimensional sound wave and *ripples on a water surface has been used to study the motion of sound waves. Ripples are produced in a rectangular tank by rods that can be dipped just under the water surface. The action of a continuous note is represented by attaching the rod to an electrically maintained tuning fork. For demonstration purposes the ripple tank is usually fitted with a glass bottom through which light is projected from beneath and a 45° mirror above the tank to reflect the light horizontally onto a ground-glass screen. If the light is interrupted at the same frequency as that of the tuning fork, the ripples appear stationary. Reflection may be studied by placing suitable objects in the tank, reflection from the sides of the tank being damped out by the use of shelving beaches. The ripples travel more slowly in shallow water and so give the effect of a denser medium. Refraction can be studied by placing objects beneath the water surface, the effect of an acoustic lens being obtained by placing an optical lens just below the surface. Diffraction is shown by using suitable obstacles and apertures. If two dipping rods are used, the *superposition principle may be demonstrated.

rise time *See* pulse.

rms value *See* root-mean-square value.

Rochon prism A double-image prism consisting of two quartz prisms, the first to receive the light, cut parallel to the axis, the second with optic axis at right angles; their deviations are in opposition. The ordinary ray passes through undeviated and is achromatic; the extraordinary ray is deflected (doubling is produced). The prism can be used to produce plane polarization by intercepting the extraordinary ray.

rocket A missile or space vehicle powered by ejecting gas, that carries both its own fuel and oxidant (if required). Rockets are therefore independent of the earth's atmosphere and are the power systems used in space flights. Most space flights have been powered by chemical rockets in which the thrust is obtained by the expansion occurring when a solid or liquid fuel (e.g. alcohol) reacts chemically with the oxidant (e.g. liquid oxygen). Most rockets are multistage devices, the first, or booster stage, being jettisoned in the less dense region of the upper atmosphere. This makes the vehicle lighter and therefore easier to accelerate to its *escape speed, and it also reduces friction heating as the escape is not reached until the air has very low density.

rods and cones *See* colour vision.

roentgen (or **röntgen**) Symbol: R. A unit of exposure *dose of X- or gamma rays such that one roentgen produces in air a charge of 2.58×10^{-4} C on all the ions of one sign, when all the electrons released in a volume of air of mass 1 kg are completely stopped. $1 R = 2.58 \times 10^{-4}$ C kg^{-1}. The roentgen is rarely now used as exposure in SI units is measured in coulombs per kilogram.

rolling friction *See* friction.

ROM Abbreviation for read-only memory. A type of *semiconductor memory used, especially in computers, for the storage of information that does not require modification. ROM is fabricated in a similar way to *RAM, with a rectangular array of storage elements, but the contents of the storage elements are fixed during manufacture. They can only be read. Programmable ROM does exist (*see* PROM; EPROM), but the contents cannot be modified within a computer. As with RAM, any storage element in ROM can can be uniquely identified and directly accessed in any order, i.e. there is *random access to the storage elements.

roof prism *Syn*. Amici prism. *See* direct-vision prism.

root-mean-square (rms) value *Syn*. effective value; virtual value. The square root of the mean value of the squares of the instantaneous values of a current, voltage, or other periodic quantity during one complete cycle; it is the effective value of current or voltage in an alternating current provided the resistance is constant. The rms value of a sine wave is the peak value divided by $\sqrt{2}$.

root-mean-square (rms) velocity The square root of the *mean square velocity.

rotameter An instrument that measures the rate of flow of fluids. It consists of a small float that, due to the motion of the fluid, moves vertically in a transparent calibrated tube. The height of the float gives a measure of the speed of the fluid.

rotating sector A device used in conjunction with a *pyrometer when measuring very high temperatures in order to cut by a known fraction the amount of radiation incident on the pyrometer.

rotation 1. The turning or spinning motion of a body, geometrical configuration, etc., about an axis that passes through it. *See also* irrotational motion.
2. One complete cycle of such motion.
3. (rot) *See* curl.

rotational field *See* curl.

rotational quantum number *See* energy level.

rotation of plane of polarization *See* optical activity.

rotation photography A crystal-diffraction method in which a single crystal is allowed to rotate about an axis normal to an incident beam of monochromatic X-rays (or electrons, neutrons, etc.), the photographic film, plane or cylindrical, being stationary.

rotation spectrum *See* spectrum.

rotatory dispersion A form of *dispersion arising when the amount of rotation of the plane of polarization differs for different wavelengths. *See* optical activity.

rotor The rotating part of an electrical machine. The term is usually applied only to an a.c. (as distinct from a d.c.) machine. *Compare* stator. *See also* induction motor.

Routh's rule The moment of inertia of a uniform solid body about an axis of symmetry is given by the product of the mass and the sum of squares of the other semiaxes, divided by 3, 4, or 5 according to whether the body is rectangular, elliptical, or ellipsoidal.

The circle is a special case of the ellipse. The rule works for a circular or elliptical cylinder about the central axis only, but for circular or elliptical discs it works for all three axes of symmetry.

For example, for a circular disc of radius a and mass M, the moment of inertia about an axis through the centre of the disc and lying (a) perpendicular to the disc, (b) in the plane of the disc, is:

(a) $\frac{1}{4}M(a^2 + a^2) = \frac{1}{2}Ma^2$
(b) $\frac{1}{4}Ma^2$.

See Lees' rule.

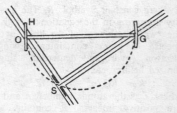

Rowland mounting

Rowland mounting A mounting for a concave *diffraction grating that has the plate-holder H and the concave grating G mounted on carriages running on two rails fixed at right angles, with the slit S mounted at their intersection. O is the centre of curvature of the grating (*see* diagram). It can be shown that all rays diffracted at a given angle from such a grating will be brought to a focus at a point on the broken circle (the *Rowland circle*), of diameter equal to the radius of curvature of the grating. As carrier G moves towards S, successively higher orders are brought into focus. The plate at H records only a small part of the spectrum at one time and it should be bent to lie along the circle. In practice a number of plates may be arranged along an arc of this circle.

Since the grating acts by reflection and does not require any lenses, it is suitable for use with ultraviolet radiation; by working at grazing incidence it can be used with soft X-rays.

rubidium-strontium dating *See* dating.

Russell–Saunders coupling *See* coupling.

rutherford Symbol: rd. A unit of *activity. It is the quantity of a nuclide required to produce 10^6 disintegrations per second. 1 rutherford = 10^6 *becquerel.

rydberg *See* atomic unit of energy.

Rydberg constant Symbol: R. The quantity appearing in the equation that gives the *wavenumbers k of the lines in the spectra of atoms containing a single electron (hydrogen, deuterium, singly ionized helium, etc.):
$$k = R(1/n^2 - 1/m^2)Z^2,$$
where Z is the atomic number, n and m are positive integers. R contains the *reduced mass of the electron as a factor so its value is slightly different for each type of atom. Assuming the actual rest mass of the electron (i.e. the mass of the nucleus is regarded as infinite), the value is
$$R = 1.097\,373\,153\,4 \times 10^7 \text{ m}^{-1}.$$
See hydrogen spectrum; Rydberg energy.

Rydberg energy The quantity
$$Rhc = me^4/8\varepsilon_0^2 h^2,$$
where R is the *Rydberg constant, h is the Planck constant, c the speed of light, ε_0 the electric constant, and m the mass of the electron. Subject to certain small corrections, this represents the binding energy of the electron in a hydrogen atom in the ground state, the nuclear mass being regarded as infinite. Value = 13.6058 eV.

Rydberg spectrum An absorption *spectrum of a gas taken with ultraviolet radiation and used for determining *ionization potential. The spectrum contains a large number of lines, each corresponding to excitation of an electron from its normal orbit (in the *ground state) to an allowed orbit further from

the nucleus (an *excited state). The lines form a series (*see* hydrogen spectrum) and become closer together as the energy increases. At one particular energy they merge into a continuum. This is the energy required to ionize the atom or molecule.

S

saccharimeter An instrument for measuring the rotation of the plane of polarization of optically active solutions, especially sugars.

sagittal coma *See* coma.

sagittal plane In an optical system. A plane that contains the *chief ray and is perpendicular to the *meridian plane, which contains the chief ray and the optic axis. The sagittal plane generally changes slope as the chief ray is deviated by the optical elements of the system; hence there are several sagittal planes for various regions within the system. All skew rays from an object point and lying in a sagittal plane are called *sagittal rays*.

Saha equation An equation that predicts the degree of thermal ionization in a gas at constant pressure. The equilibrium constant of the ionization reaction is shown to be proportional to both the thermodynamic temperature and the *ionization potential of the species being ionized.

sampling A technique of measuring only some portions of a signal, the resultant set of discrete values being taken as representative of the whole. For the information contained in a signal to be contained in the output sample value without significant loss, the rate of sampling of a periodic quantity must be at least twice the frequency.

satellite A natural or artificial body orbiting another body so large that the centre of mass of the system is well within the larger body. Six of the major planets possess natural satellites.

Many artificial satellites have been launched, either to orbit the earth itself, or to orbit other bodies of the solar system. They fall into two classes:

(i) *Information satellites* are designed to provide information concerning the earth, other celestial objects, or space itself, and to relay it back by radio. They carry a variety of measuring instruments and cameras, plus support equipment to control and power them and store data prior to transmission. The information supplied may, for example, be for weather forecasting or navigation, or it may concern the resources, atmosphere, and physical features of the earth or a great variety of topics in astronomy.

(ii) *Communications satellites* are designed to provide high-capacity communications links between widely separated locations on the earth's surface. Telephone services and live TV broadcasting are achieved by transmission of radio signals from one point on earth to a satellite, where it is amplified before being relayed back (at a different frequency) to another or other locations on earth. The orbits of communications satellites lie above the earth's *atmosphere. High-frequency radio waves (microwaves), which can penetrate the *ionosphere, must therefore be used. Most satellites are in a *geostationary orbit in the plane of the earth's equator. An orbiting geostationary satellite appears stationary to observers on the ground and it can cover an extensive area of the earth's surface.

The orbits of satellites are approximately ellipses with the centre of the primary at one focus. (*See* Kepler's

laws.) Departures from the ideal laws are caused by attractions by other bodies, imperfect spherical symmetry of the primary, or friction in the upper atmosphere. For the limiting case of a circular orbit of radius r around a body of mass m the period T is given by:

$$T^2 = 4\pi^2 r^3 / Gm,$$

where G is the gravitational constant.

saturable reactor *See* transductor.

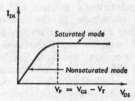
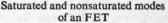

Saturated and nonsaturated modes of an FET

saturated mode The operation of a *field-effect transistor in the portion of the characteristic beyond pinch-off, i.e. $V_{DS} \geqslant V_p$, where V_p is the pinch-off voltage (*see* diagram). The drain current I_{DS} is independent of the drain voltage V_{DS} in this region. *Nonsaturated mode* is operation of the device in the portion of its characteristic below pinch-off.

saturated vapour A vapour in equilibrium with the liquid or solid phase. The pressure exerted by a saturated vapour is dependent upon temperature (*see* Clausius–Clapeyron equation; triple point) and upon the curvature of a liquid surface. Values of this *equilibrium vapour pressure*, or *saturated* or *saturation vapour pressure* (SVP), are given for flat liquid surfaces. If the actual vapour pressure is greater than that for equilibrium, the vapour is *supersaturated. If the pressure of the vapour is less than the equilibrium value, the liquid or solid will evaporate.

saturated vapour pressure (SVP) The pressure exerted by a *saturated vapour.

saturation 1. (magnetic) The degree of magnetization of a substance that cannot be exceeded however strong the applied magnetizing field. In this state all the domains (*see* ferromagnetism) are assumed to be fully orientated along the lines of force of the magnetizing field.
 2. (electronic) A condition in which the output current of an electronic device is substantially constant and independent of voltage. In the case of a *field-effect transistor, saturation is an inherent function and produces the maximum current inherent to the device. With a bipolar junction *transistor, saturation occurs because the output from the collector electrode is limited by the elements of the external circuit; changing these alters the magnitude of the *saturation current* drawn from the transistor.
 3. (light) *See* colour.

saturation vapour pressure (SVP) *See* saturated vapour.

saturation voltage The residual voltage between the collector and emitter of a bipolar junction *transistor, under specified conditions of base current, when the collector current is limited by an external circuit. *See* saturation.

SAW devices *See* surface acoustic wave devices.

sawtooth waveform A periodic waveform whose amplitude varies approximately linearly between two values, the time taken in one direction being very much longer than the time taken in the other (*see* diagram). Sawtooth waveforms are usually produced by *relaxation oscillators and are frequently used to provide a *time base.

scalar A quantity defined by a single magnitude (as distinct from a *vector, which has magnitude and direction and thus needs three numbers to define it). Examples are mass, time, and wavelength.

scalar product *See* vector.

scaler A device that produces an output pulse when a specified number of input pulses have been received. It is frequently used for counting purposes, particularly in conjunction with *Geiger counters and *scintillation counters.

scanning The process of exploring an area or volume in a methodical manner, in order to produce a variable electrical output whose instantaneous value depends on the information contained in the small area examined at each instant. The information can then be reproduced by a suitable receiver. The technique is most often used in *television, *facsimile transmission, and *radar.

scanning electron microscope (SEM) *See* electron microscope.

scanning-transmission electron microscope (STEM) *See* electron microscope.

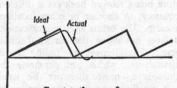

Sawtooth waveform

scanning tunnelling microscope (STM) A type of electron microscope based on the *tunnel effect and used primarily for studies of surfaces. If two electrical conductors are brought very close together, it is possible for electrons to tunnel from one to the other. The sample to be analysed in the microscope forms one conductor and a very fine metal tip forms the other; if the sample is non-conducting it must be coated with a thin layer of conducting material. An electric potential is applied between the conductors and electrons tunnel across, producing a small current. The tip is scanned over the surface of the sample, and in the process is moved up and down so that the current remains constant. The tip therefore remains at a constant distance from the sample surface and its vertical movements are processed by computer to provide a topographical map of the surface. Horizontal and vertical resolution is approximately 0.2 nm and 0.01 nm respectively.

scattering 1. The deflection of light by fine particles of solid, liquid, or gaseous matter from the main direction of a beam. If the particles are relatively large, reflection and refraction as well as diffraction play a part; if the particles are small (smaller than a wavelength or as small as a molecule), the effect is diffractive. In *Rayleigh scattering the scattered intensity for small particles varies as $(1/\lambda^4)$ so that white light scattered by very small particles is bluish (Tyndall effect). Chalk dust particles are whitish – they are larger and reflection effects are more noticeable. The blue of the sky is due to scattering by air molecules, and the red sun is due to the removal of the blue by scattering from the direct beam.

2. The deflection of sound waves by a reflecting surface whose dimensions are greater than the wavelength of the wave. The effectiveness of a small reflector thus improves with smaller wavelengths. Rayleigh showed that, as with light, the intensity of the scattered wave varies inversely as $(1/\lambda^4)$, where λ is the wavelength of the incident sound. The *harmonics of an incident sound will be scattered in increasing proportion towards the shorter wavelengths, i.e. the higher frequencies, the *octave of a note scattered from a small object being sixteen times more intense than in the incident sound.

3. The deflection of radiation resulting from the interaction of individual particles or photons with the nuclei or electrons in the material through which the radiation is passing or with the photons of another radiation field. *Inelastic scattering* occurs as a result of an inelastic *collision in which net changes occur in the internal energies of the participating systems and in the sum of their kinetic energies before and after the collision; there is no such energy change in *elastic scattering*, which occurs as a result of an elastic collision.

Thomson scattering is the scattering of electromagnetic radiation by free (or loosely bound) electrons, which can be explained by classical physics or nonrelativistic quantum theory in terms of the forced vibrations of the electrons of an atom that is absorbing radiation. These oscillating electric charges become the source of electromagnetic radiation of lower energy than the incident radiation and it is emitted in all directions. If I_0 is the intensity per unit area of the incident radiation, the total intensity, I, of the scattered radiation is given by:

$$(8\pi/3)(e^2/mc^2)^2 I_0,$$

where e and m are the electronic charge and mass and c is the speed of light. The quotient I/I_0 has the dimensions of an area and is the scattering *cross section of the electron.

The elastic scattering of photons by electrons, known as the *Compton effect, produces a reduction of energy of the photons. The *Raman effect involves scattering of photons by molecules. The Coulomb field of the nucleus produces *Coulomb scattering of particles due to electrostatic repulsion. (For scattering of high-energy particles, *see* scattering amplitude, strong interactions.)

Resonance scattering occurs at high energies when the incident wave can penetrate the nucleus and interact with its interior. If the wave is reflected at the nuclear surface, *potential scattering* takes place. *Shadow scattering* results from the interference of the incident wave and scattered waves.

scattering amplitude A mathematical function specifying the *wave functions of elementary particles scattered in a collision. One of the most important ways in which the interactions between particles can be investigated experimentally is by allowing a beam of high-velocity particles to collide with other particles. As a result of the collisions that occur, some of the particles are deflected and in some cases the final particles may have different *quantum numbers from the initial ones (*see* strong interactions). Details of the scattering, such as the angular distribution of the final particles, will depend on the nature of the forces that act during the collision. Scattering amplitudes describe these scattering processes theoretically. From these amplitudes, the angular distribution of the final particles can be calculated. Scattering amplitudes are usually written in terms of relativistically invariant quantities.

scattering cross section Symbol: σ_s. A measure of the probability that a specified particle will be scattered by a specified nucleus or other entity through an angle greater than or equal to a specified angle, θ. (*See* cross section; scattering.) The *differential scattering cross section* is a measure of the probability of scattering through an angle lying between θ and $\theta + d\theta$.

schlieren method A method of exhibiting inhomogeneities in transparent media (e.g. flaws in glass, convection currents, shock waves, pulses of sound). It depends on the use of special illumination to make changes in refractive index apparent as the density of the medium changes.

Schmidt number Symbol: *Sc*. A dimensionless parameter equal to the ratio v/D, where v is the *kinematic viscosity and D the diffusion coefficient.

Schmidt telescope (or camera) A wide-field astronomical telescope that uses a thin figured transparent plate – a *Schmidt corrector* – at the centre of curvature of a short-focus spherical primary mirror to remove *spherical aberration; *coma and oblique *astigmatism are also negligible. The corrector plate is shaped like a convex lens near the centre and like a concave lens near the periphery. A large field of view can be sharply focused on a curved surface, to which a photographic plate can be sprung.

Schmitt trigger A type of *bistable circuit that gives a constant high-voltage output when the input signal exceeds a specific voltage, and falls to a constant low-voltage output below this value. A constant output is obtained irrespective of the input waveform, which may be sinusoidal, sawtooth, etc. This type of circuit is frequently used as a *trigger circuit and in binary *logic circuits to maintain the logical 1 and logical 0 levels.

Schottky barrier *See* Schottky effect; Schottky diode.

Schottky defect *See* defect.

Schottky diode A *diode consisting of a metal-semiconductor junction (*Schottky barrier*), which has rectifying characteristics similar to a *p-n junction. When a forward bias is applied, *majority carriers with sufficient energy can cross the barrier and a current flows by the process of thermionic emission (*see* Schottky effect). A Schottky diode differs from a p-n junction diode in that the *diode forward voltage is different (lower for commonly used materials) and there is no charge stored when the diode is forward biased. The device can therefore be turned off very rapidly by the application of reverse bias, as the *storage time is negligible.

Schottky effect A reduction of the *work function of a solid due to the application of an external electric field, leading to a consequent increase in its *thermionic emission. The presence of an accelerating field lowers the potential energy of electrons outside the solid. This leads to a distortion of the potential barrier and consequent lowering of the work function. In the case of a metal there is also a contribution from *image potentials. The lowering of the work function increases the electron current due to thermionic emission. In the Schottky effect the electrons leaving the solid pass over the potential barrier, as opposed to tunnelling through it. (*See* tunnel effect; field emission.)

If the vacuum is replaced by a semiconductor the metal-semiconductor junction is called a *Schottky barrier*. Similar effects occur although generally the decrease in work function is less.

Schottky noise *Syn.* shot noise. Variations in the current output from an electronic device that arise due to the random manner in which electrons or holes are emitted from an electrode (e.g. the collector of a *transistor or the source of an FET).

Schrödinger wave equation (1926) The basic nonrelativistic equation of *wave mechanics expressing the behaviour of a particle in a field of force. The time-dependent equation describing *progressive waves, applicable to the motion of free particles, is

$$\nabla^2 \psi - \frac{4\pi m}{ih}\left(\frac{\partial \psi}{\partial t}\right) - \frac{8\pi^2 mU}{h^2}\,\psi = \phi$$

where ∇^2 is the *Laplace operator, m is the mass of the particle, U is the potential energy, h is the Planck constant, and $i = \sqrt{-1}$. The *wave function ψ is a function of the coordinates and time.

A particle bound in a system, such as an electron in an atom, is analysed by the time-independent form of this equation describing *standing waves,

$$\nabla^2 \psi + \frac{8\pi^2 m}{h^2}(E - U)\psi = \phi$$

where ψ is now a function of the coordinates only.

The quantity ψ is usually a complex function. The most commonly accepted meaning is that the value of $|\psi|^2 \mathrm{d}V$ at a point represents the probability of finding the particle in the element of volume $\mathrm{d}V$ at this point.

Schrödinger's time-independent equation can be solved analytically for a number of simple systems but for most problems it is necessary to use *perturbation theory or other approximate methods. The time-dependent equation is of the first order in time but of the second order with respect to the coordinates, hence it is not consistent with *relativity. The solutions for bound systems give three *quantum numbers, corresponding to three coordinates, and an approximate relativistic correction is possible by including a fourth *spin quantum number.

Schwarzschild radius The critical radius to which matter in space must be compressed in order to form a *black hole. It is given by $2GM/c^2$, where M is the mass, G the gravitational constant, and c the speed of light. It is true for a nonrotating black hole. The surface having such a radius is the *event horizon* of the black hole and defines the boundary from inside which neither mass nor radiation can escape. The enormous gravitational tidal forces inside the black hole draw matter towards the centre where it is destroyed in a region of infinite curvature, a space–time *singularity*, where the known laws of physics break down.

scintillation 1. The production of small flashes of light, near ultraviolet and near infrared, from certain materials (*scintillators*) as a result of the impact of radiation. Each incident particle or interacting quantum produces one flash.

2. Rapid irregular variations in the intensity of light or other radiation as it passes through an irregular medium. Starlight is deflected by mobile irregularities in the refractive index of the earth's atmosphere so that a star's image wanders rapidly about its mean position, i.e. the star *twinkles*.

scintillation counter A detector of ionizing radiation based on *scintillation. The scintillator may be an inorganic crystal (commonly sodium iodide with a small quantity of thallium iodide), an organic substance, a plastic, or a liquid. On the passage of an ionizing particle or the interaction of a high-energy photon there is a scintillation that is detected by one or more *photomultipliers. The magnitude of the observed pulse is proportional to the energy given up to the scintillator by the particle or quantum, hence the spectrum of the radiation can be studied using a *multi-channel analyser. Alternatively, the device can be used as a simple counter of highly ionizing particles, such as alpha particles, the counter circuit being designed so that weakly ionizing events are not recorded.

scope Short for oscilloscope.

scotophor A material, such as potassium chloride, that darkens under electron bombardment and recovers on heating. It is used on screens of cathode-ray tubes (in place of the usual phosphor) when long persistence is required.

SCR Abbreviation for *silicon controlled rectifier.

screen 1. The front surface of a TV, VDU, or other *cathode-ray tube, suitably coated, on which the visible pattern is displayed.
2. *See* shielding.

screen grid *See* thermionic valve.

screening *See* shielding.

screening constant A number than when subtracted from the *atomic number, Z, gives an effective atomic number. It allows for the screening effect on the nuclear charge, Ze, of the inner electron shells.

screw dislocation *See* defect.

SCS Abbreviation for *silicon controlled switch.

S-drop *See* strange matter.

search coil *See* exploring coil.

second 1. Symbol: s. The *SI unit of time defined as the duration of 9 192 631 770 periods of the radiation corresponding to the transition between two hyperfine levels of the *ground state of the caesium-133 atom.
2. *Syn.* arc second; second of arc. Symbol: ". A unit of angle equal to 1/3600 degree.

secondary *See* secondary winding.

secondary cell *See* cell.

secondary cosmic rays *See* cosmic rays.

secondary electron An electron emitted from a material as a result of *secondary emission.

secondary emission The process occurring when an electron moving at sufficiently high velocity strikes a metal surface, the impact causing other electrons to escape from the surface. The total energy of one incident electron is often sufficient to eject several secondary electrons. The principle is used in the *electron multiplier and in *storage tubes. Secondary emission is also produced by the impact of positive ions on surfaces.

secondary spectrum *See* achromatic lens.

secondary standard 1. A copy of a *primary standard for which the difference between the copy and the primary standard is known.

2. A quantity accurately known in terms of the primary standard and used as a unit.

secondary waves *See* Huygen's principle.

secondary winding The winding on the load (i.e. output) side of a *transformer irrespective of whether the transformer is of the step-up or step-down type.

sedimentation The free fall of particles in suspension in a liquid. If the particles are spherical, they attain terminal velocities dependent on the relative density of solid and liquid, the viscosity of the latter, and the size of the particles (*see* Stokes' law). This result is used to separate particles of different sizes. *See also* centrifuge; ultracentrifuge.

Seebeck effect *See* thermoelectric effects.

seed crystal A small crystal on which crystallization can begin, e.g. in a supersaturated solution or a supercooled liquid.

Segrè chart A chart of the nuclides in which the number of protons in nuclides is plotted against the number of neutrons. Stable examples will be found on, or in the vicinity of, a line with a gradient of 1: the gradient diminishes somewhat for nuclides of high atomic number.

Seidel aberrations Five aberrations due to sphericity of surface, namely *spherical aberration, *coma, *astigmatism, *curvature of field, and *distortion. The approximation $\sin \theta \simeq \theta$ used in the simple paraxial theory of geometric optics becomes unsatisfactory for rays away from the optic axis. Using the first two terms in the sine expansion, i.e. $\sin \theta \simeq \theta - \theta^3/3!$, Seidel deduced correcting terms to the paraxial theory. The Seidel aberrations are associated with these terms.

seismograph An instrument used to register the movement of the ground due to distant earthquakes, underground nuclear explosions, etc., consisting in principle of a massive pendulum set into motion by the force developed at its point of suspension.

seismology The study of the structure of the earth by means of the waves (*seismic waves*) produced by earthquakes, explosions, etc. *Seismographs at various points on the earth's surface record the arrival of the different kinds of waves that travel through the earth and along its surface. *P-waves* (longitudinal waves) and the slower *S-waves* (shear waves) travel deep into the earth, where they are refracted at the boundaries between layers of different density. The source of the earthquake is called its *focus* and the

425

nearest point on the earth's surface to the focus is the *epicentre*.

selection rules Rules derived by *quantum mechanics specifying the transitions that may occur between different *quantum states of a system. For example, in a change between two quantum states of vibration of a molecule the selection rule is that the vibrational quantum number can only change by one unit, i.e. $\Delta v = \pm 1$. Transitions that follow the selection rules are *allowed transitions*. *Forbidden transitions* are ones in which the rules are not followed and they are very unlikely, but sometimes not impossible.

selective absorption The absorption of radiation of certain wavelengths in preference to others, as in coloured glasses and pigments. The colour is determined by the remaining light transmitted (glasses and filters) or reflected (pigments) after absorption. All substances show strong absorption at some spectral region, which, for many transparent colourless optical media, occurs in the infrared and ultraviolet regions.

selective fading *See* fading.

selective radiation Radiation from a body with a relative spectral energy distribution differing from that of a *black body, or of a *grey body in which the radiation for all wavelengths is a constant fraction of that given by a black body. Although all substances show selective temperature radiation over the visible portion, many give an approximately grey radiation. With such, it is possible to specify a temperature for a black body that will give roughly a similar distribution of spectral energy as the luminous selective radiator (*see* colour temperature). Incandescent gases and vapours show more or less extreme selectivity.

selective reflection The strong reflection of certain wavelengths in preference to others, which occurs for all substances, although the wavelengths may be outside the visible spectrum; the wavelengths reflected generally correspond with those of the absorption bands. Metallic reflection is responsible for surface colour as distinct from pigment or body colour, which arises from absorption of light penetrating the substance prior to its reflection.

selectivity The ability of a radio receiver to discriminate against signals having *carrier frequencies different from that to which the receiver is tuned.

selenium cell A *photovoltaic cell consisting of a thin layer of light-sensitive grey selenium coated onto a metal disc and covered with a film of gold or platinum that is sufficiently thin to allow light to pass through. It is used in *exposure meters and can be built into cameras.

self-capacitance The inherent capacitance between individual turns of an inductance coil or a resistor. To a first approximation it may be represented as a single capacitance connected in parallel with the coil or resistor.

self-conjugate particle *See* antiparticle.

self-excited 1. Of an electrical machine (generator). Having the *field magnets wholly or substantially excited by a magnetizing current generated in the machine itself.

2. Of an *oscillator. Being in a state in which the oscillations build up to a steady output value, following application of power to the circuit, without any separate input of the required output frequency.

self-inductance *See* electromagnetic induction.

Sellmeier equation A mathematically deduced formula to give the variation of refractive index, n, with wavelength, λ, in the neighbourhood of an absorption band (λ_0):

$$n^2 = 1 + A_0\lambda^2/(\lambda^2 - \lambda_0^2),$$

where A_0 is a constant for a given material. It fails within the region of the absorption band, agreeing better in the regions of wavelength for which the substance is transparent.

SEM Abbreviation for *scanning electron microscope.

semiconductor A material having a *resistivity between that of conductors and insulators and usually having a negative *temperature coefficient of resistance. *Intrinsic semiconductors* are perfect crystals (no defects or impurities) in which the energy gap between the conduction band and the valence band is from a few tenths of an electronvolt up to 2 eV (*see* energy bands). At a given temperature a specific number of electrons will be thermally excited into the conduction band. They leave behind an equal number of vacant states in the valence band. The action of an applied field causes conduction both in the conduction band and in the valence band. The conduction electrons are accelerated by the applied field and carry charge through the semiconductor. The electrons in the valence band move to occupy the adjacent vacancy and the net effect is that the vacancy moves through the material as if it were a positive charge. The vacancies are known as *holes and are treated as positive charge carriers.

Extrinsic semiconductors are those materials in which the presence of impurities (and also defects) in the crystal lattice determine the properties of the semiconductor. The presence of impurities affects the conductivity significantly. The semiconductor properties depend on the type of impurity.

Donor impurities are atoms that have more valence electrons than are required to complete the bonds with neighbouring atoms. The presence of these atoms affects the distribution of quantum states in the immediate vicinity, and states are formed in the forbidden band, close to the conduction band. At absolute zero each of these *donor states* will contain an electron. Within a few tens of degrees from room temperature only about half will be occupied. The electrons from the donor states fill most of the holes in the valence band and enter states in the conduction band. The number of electrons in the conduction band, n_e, exceeds the number of holes in the valence band, n_p, and the semiconductor is said to be *n-type*. The product $n_e n_p$ equals n_i^2, where n_i is the number of free electrons (or holes) in the intrinsic material. No conduction can occur by electrons moving between donor states, but the electrons in the conduction band and holes in the valence band can move. Thus as the temperature rises the conductivity increases until nearly all the donor states are empty. It then falls as there are more collisions between electrons and *phonons. At higher temperatures the conductivity rises again as electrons are thermally excited across the energy gap.

Acceptor impurities are atoms that have fewer valence electrons than are required to complete the bonds with neighbouring atoms and they therefore accept electrons from any available source to complete the bonds. These extra electrons are almost as tightly bound to the atom as the valence electrons and the presence of acceptor impurities results in states just above the valence band. The electrons in the valence band only need a small increment of energy to occupy the *acceptor states* and provide the source of electrons for the acceptor atom. Mobile holes are left in the valence band, but the electrons are bound to the acceptor atom, which is ionized by the electron capture. Mobile holes thus predominate and the semiconductor is known as *p-type*.

The effect of the impurity is to move the *Fermi level near to the conduction band in n-type, and near to the valence band in p-type as the distribution of available states has changed.

The conductivity of an extrinsic semiconductor will depend on the type and amount of impurity present and this may be controlled by adding impurities. The process is known as *doping* and the amount of impurity is the *doping level*. This is usually carried out by *diffusion or *ion implantation of the impurity into the crystal. At conditions of thermal equilibrium a dynamic equilibrium exists in the semiconductor. Mobile charges move around the crystal in a random manner due to scattering by the nuclei in the crystal lattice. The crystal retains overall charge neutrality and the number of charge carriers remains essentially constant, but a continuous process of regeneration and recombination occurs as thermally excited electrons fall back into the valence band and combine with holes.

The carriers that predominate in a particular semiconductor are called the *majority carriers* (e.g. electrons in n-type) and the others are *minority carriers* (e.g. holes in n-type). In an extrinsic semiconductor the number of majority carriers is approximately equal to the number of impurity atoms. If extra charge carriers are generated by, for example, light energy, these will have a limited lifetime. The *bulk lifetime* of excess carriers is the average time interval between generation and recombination of minority carriers in the bulk of a homogeneous semiconductor. Recombination also takes place near the surface

of a semiconductor; the mechanism is rather different from the direct recombination occurring in the bulk of the material.

If an electric field is applied to the semiconductor, charge carriers move under the influence of the field, but still undergo scattering processes. The result of the field is to impose a drift in one direction onto the random motion of the carrier. The *drift mobility* is the average drift velocity of carriers per unit electric field, and the average distance travelled by a carrier during its lifetime is the *diffusion length*. The mobility of electrons is about three times that for holes.

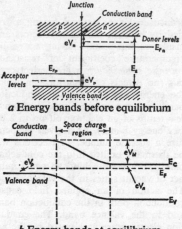

a Energy bands before equilibrium

b Energy bands at equilibrium

Junctions between semiconductors of different types form a fundamental part of modern electronic components and circuits. If two semiconductors are brought into contact with no external applied field, thermal equilibrium will be established between the two, and the condition of zero net current requires the Fermi level to be constant throughout the sample. Fig. *a* shows the situation before equilibrium is established between samples of p- and n-type semiconductor. Electrons and holes will cross the junction until an equilibrium is set up by the establishment of two *space-charge regions preventing further flow of carriers. The space-charge regions are due to ionized impurities on either side of the junction, and a voltage drop is formed across the junction, while maintaining the overall neutrality of the sample. The *diffusion potential* or *built-in potential* V_{bi} is given by:

$$eV_{bi} = E_g - e(V_n + V_p)$$
$$= E_{Fn} - E_{Fp}$$

(Figs. *a* and *b*), where *e* is the charge of an electron.

The space-charge region around the junction forms a *depletion layer, the width of which depends on the electric field in the interface and the numbers of acceptor and donor ions on either side of the junction.

If a voltage V is applied across the junction, the electrostatic potential is given by $(V_{bi} + V)$ for reverse bias (positive voltage on the n-region) and $(V_{bi} - V)$ for forward bias (positive voltage on the p-region). The depletion width increases with reverse bias as the holes in the p-region are attracted to the negative electrode and vice versa in the n-region, and very little current flows across the junction. The reverse-bias current flowing is due to minority holes in the n-region and electrons in the p-region crossing the junction.

Under conditions of forward bias the depletion width decreases as the built-in potential is reduced by the applied field, and the current through the sample increases exponentially with voltage for a few tenths of a volt (*see* diode).

Commonly used semiconductor materials are elements falling into group 4 of the periodic table, such as silicon or germanium. The donor and acceptor impurities are group 5 and group 3 elements, differing in valency by only one electron. Certain compounds, such as gallium arsenide (which has a total of 8

valence electrons), also make excellent semiconductors. These materials are classified as 3–5 or 2–6 depending on their position in the periodic table. Suitable impurities in these cases would be 2, 4, and 6 or 3 and 5 respectively.

semiconductor counter A *photodiode used as a radiation *counter.

semiconductor device Any electronic circuit or device whose essential characteristics are due to the flow of charge *carriers within a *semiconductor.

semiconductor diode *See* diode.

semiconductor laser *Syn.* diode laser; injection laser. A small robust cheap and flexible type of *laser fabricated from *semiconductor material. It consists essentially of a *p-n junction diode, being a refinement of the *light-emitting diode.

Laser action is nearly impossible to produce in silicon, and semiconductor materials such as gallium arsenide must be used. Under a forward bias, electrons flow across the junction from the n-side to the p-side, where they form an excess minority-carrier concentration; the process is called electron *injection. These electrons can recombine with *holes in the p-side, emitting a photon by spontaneous emission; the photon energy is approximately equal to the energy gap between conduction and valence bands, E_g. At sufficiently large values of applied voltage, great numbers of electrons can cross the junction. Above a threshold current, stimulated emission (*see* laser) can occur: an electron excited into the conduction band is induced to emit a photon by a photon from a previous recombination event. As the injection current increases, the stimulated emission increases. The photon produced by stimulated emission matches the incident photon in both energy and phase; a narrow range of wavelengths is produced by the device as a whole.

The diode is usually constructed to have two flat parallel ends of the crystal, which are perpendicular to a flat p-n junction. These ends act as partially reflecting mirrors and the light can be reflected back into the p-n region, causing further amplification. The diode acts as a resonant cavity, the light and its reflection being in phase. An intense laser beam emerges from the mirror ends of the crystal. A very high current density is required, and to prevent overheating at room temperature the beam has to be pulsed.

A continuous laser beam can be achieved by using a modified crystal. A region of pure GaAs is made adjacent to a region of aluminium gallium arsenide in which some of the gallium atoms in the GaAs crystals have been replaced by aluminium atoms. The junction between these two regions of similar crystal structure (a *heterojunction*) can be used to reduce the threshold current required to achieve laser action, and a continuous laser beam is possible. The output power is tens of milliwatts at wavelengths between 900–700 nm and the efficiency can be as high as 10%.

A range of different materials and modifications to the basic structure have now been produced to provide lasers of different wavelengths and to optimize the operation. Semiconductor lasers have an extensive range of applications.

semiconductor memory *Syn.* solid-state memory. Any of various types of cheap compact computer *memory composed of one or more *integrated circuits fabricated in *semiconductor material (usually silicon). The integrated circuit comprises a rectangular array of possibly thousands of microscopic electronic devices, each of which can store one bit of data – either a binary 1 or a binary 0. Data can be accessed extremely rapidly. The different types include *RAM, *ROM, *PROM, and *EPROM There is *random access to all these types.

semipermeable membrane A membrane used in dialysis and *osmosis that allows certain molecules in a fluid to pass through while stopping others. In general, the larger molecules are stopped.

semitone The smallest pitch interval between successive tones of the present Western *musical scales. In the scale of equal *temperament this frequency interval is ideally $2^{1/12}$.

semitransparent cathode A type of *photocathode in which electrons are emitted from the opposite surface to that on which the radiation falls.

sensation level A measure of the intensity I of a sound with reference to the minimum audible intensity I_0. If the sensation level L is measured in decibels,

$$L = 10 \log_{10}(I/I_0).$$

I_0 is usually taken as 2.5 picowatts per square metre.

sensitivity 1. Generally, the response of a physical device to a unit change in the input.

2. Of a measuring instrument. The magnitude of the change of deflection or indicated value that is produced by a given change in the measured quantity.

3. Of a radio receiver. A measure of the ability of the receiver to respond to weak input signals. Quantitatively, it is the smallest input at the receiver that will produce a certain output under specified conditions, particularly a designated *signal-to-noise ratio.

sensor *See* transducer.

separately excited Of an electrical machine (generator). Having the *field magnets wholly, or substantially, excited by a magnetizing current that is obtained from a source other than the machine itself. Thus the excitation current is entirely independent of the load current supplied by the machine and reasonably satisfactory operation can be obtained over a range of voltage from zero to the maximum that the machine can generate.

separation energy The energy needed to remove one proton or neutron from the nucleus of a particular nuclide.

serial 1. Involving the sequential transfer or processing of the individual parts of a whole. For example, in *serial transmission*, the individual *bits making up a unit of data are transmitted one after the other along the same path.

2. *Syn.* sequential. Involving the occurrence of two or more computing events or activities such that one must finish before the next begins. For example, in *serial* (or *sequential*) *access*, data are read from the storage medium – usually magnetic tape – in the order in which they occur until the required item or storage location is found.

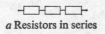

a Resistors in series

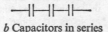

b Capacitors in series

series Pieces of electrical apparatus are *in series* when they are connected so that one current flows in turn through each of them. For conductors of resistances r_1, r_2, r_3, ... r_n in series, the total resistance, R, is given by:

$$R = r_1 + r_2 + r_3 + \ldots + r_n.$$

For capacitances c_1, c_2, c_3, ... c_n in series:

$$1/C = 1/c_1 + 1/c_2 + 1/c_3 + \ldots + 1/c_n.$$

Cells in series add their e.m.f.s. *Compare* parallel.

series-characteristic motor *Syn.* inverse-speed motor. Any electric motor, the speed of which decreases substantially with increase of load, as with a *series-wound or heavily *compound-wound motor. *Compare* shunt-characteristic motor.

series-parallel connection 1. An arrangement in which electrical machines or electronic devices or circuits may be connected in series or in parallel as alternatives.

2. An arrangement in which electrical machines or electronic devices or circuits are connected so that some are in series and some are in parallel with one another.

series resonant circuit *See* resonant circuit.

series-wound machine A d.c. machine in which the *field magnets are wholly, or substantially, excited by a winding that is connected in *series with the armature winding, or alternatively is connected so as to carry a current proportional to that in the armature winding. *Compare* compound-wound machine; shunt-wound machine.

servomechanism In general, a closed-sequence control system that is automatic and power-amplifying. For example, in servo-assisted brakes in motor vehicles a small foot pressure in the brake pedal activates a servomechanism that applies a much greater pressure to the brake shoes or pads.

sextant An instrument for measuring angles (up to 120°) between two objects, and particularly the angle between an

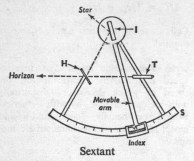

Sextant

astronomical body and the horizon (i.e. its altitude). The horizon is observed through the upper clear half of a fixed horizon glass H by applying the eye to the eyepiece of the telescope T. The index glass I is rotated until the image of a star is reflected from I into the lower silvered half of H, and thence to T. The required angle is read from the graduated scale S at the point indicated by an arm attached to I.

Seyfert galaxy Any of a group of *galaxies with an exceptionally bright central region. Most are otherwise normal nearby spiral galaxies. Although the central region is an optical source, most of the emission is in the infrared, with strong UV and X-ray emission. This emission is substantially nonthermal. It is thought that the central activity could be powered by a *black hole, and that Seyferts are a less powerful example of the *quasar phenomenon.

shadow mask A perforated metal sheet placed between the *electron guns and the screen of some *colour-television picture tubes to allow the correct selection of colours on the screen.

shadow photometer A simple photometer that uses a rod in front of a screen; the two light sources to be compared are adjusted until the shadows thrown just touch and match in intensity. The *lu-minous intensities are then in the ratios of the squares of the distances of the sources from the shadows thrown by each.

shadows The shape and relative density of shadows can be explained for most practical purposes in terms of geometric optics. A point source delivers rays to an obstacle, the extremities of which limit light that may pass; the shadow throughout is complete (*umbra*). With an extended source the shadow shows variation of density – umbra and *penumbra*. When the edges of geometric shadows are examined more carefully, *diffraction phenomena are evident and these are more marked as the size of the obstacle is decreased.

shadow scattering *See* scattering.

shear *See* strain; stress.

shear modulus *See* modulus of elasticity.

shear wave A *transverse wave that travels without compression of the medium. The S waves propagated through the solid body of the earth in earthquakes are an example (*see* seismology).

shell *See* atomic orbital; electron shell; shell model.

shell model Any of the models of the nucleus in which the interactions between *nucleons are approximated by assuming the nucleons move in a single central potential (much as the electrons in the *atom move in the electrostatic field of the nucleus). By solving the *Schrödinger equation corresponding to this potential, a set of possible *quantum states are obtained. The sets of states corresponding to the same energy are called *shells*. Being *fermions, the nucleons must obey the *Pauli exclusion principle. Therefore no two nucleons can occupy the same quantum state. The nucleons will try to reach the lowest energy possible by filling the lowest energy states first. The more nucleons there are, the greater the number of shells that will be filled. This is exactly analogous to the filling of atomic shells by orbital electrons in heavy elements. The shell model is useful in explaining why nuclei with certain proton or neutron numbers (called *magic numbers) are more stable than others. These are thought to correspond to nuclei in which the number of nucleons is just sufficient to fill a given number of shells.

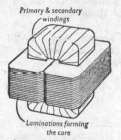

Primary & secondary windings

Laminations forming the core

Single-phase shell-type transformer

shell-type transformer A *transformer in which the *core encloses the greater part of the windings. The core is made of *laminations that are usually built up around the windings, the latter having been assembled beforehand. *Compare* core-type transformer.

Shenstone effect A considerable increase in photoelectric emission produced in some metals, such as bismuth, after an electric current has passed through.

SHF Abbreviation for superhigh frequency. *See* frequency bands.

shield 1. A mass of material (such as concrete, etc.) surrounding a *nuclear reactor core, or other source of radiation, to absorb neutrons or other dangerous radiations. A *biological shield* is specially designed to protect laboratory workers and plant operators from harmful radiations.

2. *See* shielding.

shielding *Syn.* screening. Removal of the influence of an external energy field by surrounding a region with a *shield* or *screen* of suitable material. In the case of an electric field, earthed metal walls are required. Stray magnetic fields are removed by using a shield of high *permeability.

shift register A digital circuit, a *register, used to displace a set of information, stored in the form of pulses, either to the right or left. If the information is a pattern of *bits representing a binary number, a shift in position to the left (or right) is equivalent to multiplying (or dividing) by a power of two. Shift registers are extensively used in *computers.

SHM Abbreviation for *simple harmonic motion.

shock waves Waves of compression and rarefaction that originate in the neighbourhood of sharp points or roughness on obstacles exposed to the flow of a compressible fluid at high speeds. The local compressions and rarefactions set up in this way are not instantly reversible and are propagated out into the fluid as, in effect, sound waves. They transport momentum from the vicinity of the obstacle to a distance or convert kinetic energy into internal energy. *See also* sonic boom.

short circuit An electrical connection of relatively very low resistance made intentionally or otherwise between two points in a circuit.

433

short-range force *See* force.

short-wave Designating radio waves with wavelengths in the range 10–100 metres, i.e. in the high-frequency band.

shot noise *See* Schottky noise.

shower A group of *elementary particles and *photons arising from the impact of a single particle of high energy. *See* cosmic rays.

shunt 1. (general) If two electrical devices or circuits are connected in *parallel, either one is said to be in shunt with the other.
2. *Syn.* instrument shunt. A four-terminal resistor, of low value, that is connected in parallel with an instrument such as an ammeter so that only a fraction of the main circuit current flows through the instrument, thereby increasing the range of the latter.
3. *See* magnetic shunt.

shunt-wound machine A d.c. machine in which the *field magnets are wholly, or substantially, excited by a winding that is connected in *shunt with the armature winding. *Compare* compound-wound machine; series-wound machine.

sideband In *modulation. A band of frequencies embracing either all the upper or all the lower *side frequencies, called the *upper* and *lower sideband* respectively.

side frequency Any frequency produced as a result of *modulation. For example, in *amplitude modulation if a *carrier wave of frequency f_c is modulated by a sinusoidal signal of frequency f_s ($f_s \ll f_c$), the resulting wave has three components, the frequencies of which are f_c, $f_c + f_s$ (the upper side frequency), and $f_c - f_s$ (the lower side frequency).

sidereal day, year, and time *See* time.

siemens Symbol: S. The *SI unit of electrical *conductance, defined as the conductance of an element that possesses a resistance of one ohm. The unit used to be called the mho or reciprocal ohm.

Siemens' electrodynamometer An *electrodynamic instrument in which the torque produced by the electromagnetic forces is balanced against the torque of a spiral spring by adjustment of a calibrated torsion head attached to the spring. It may be an ammeter, voltmeter, or wattmeter and can be used with direct or alternating current.

sievert Symbol: Sv. The *SI unit of *dose equivalent, used for protection purposes in the case of ionizing radiation. It has replaced the rem: 1 Sv = 100 rem. In terms of other SI units, 1 Sv = 1 J kg^{-1}.

sigma particle Symbol: Σ. An *elementary particle, charged or neutral, classified as a *hyperon.

sigma pile A neutron source plus *moderator, but without any fissile material, used to analyse the properties of the moderator.

signal The variable parameter, such as current or voltage, by means of which information is conveyed through an electronic circuit or system.

signal generator Any circuit or device producing an electrical parameter that is adjustable and controllable. Most commonly it supplies a specified voltage with variable amplitude, frequency, and waveform. The term is reserved for continuous-wave generators, especially sine-wave generators. *See also* pulse generators.

signal level The magnitude of a signal at any point in a transmission system, usually with reference to some arbitrarily chosen value.

signal-to-noise ratio (S/N ratio) The ratio of one parameter of a wanted signal to the same parameter of the *noise at any point in an electronic circuit, device, or transmission system. It is often measured in *decibels.

signature A collection of symbols that can be used for identification purposes or to select a particular interpretation. One example is a group of spectral lines, usually in an emission spectrum, that identifies an atom or molecule, and possibly its ionization state and isotopic form.

silent discharge An inaudible electrical discharge that takes place at high voltage and involves a relatively high dissipation of energy. Such a discharge readily takes place from a conductor that has a sharp point.

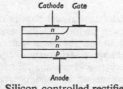

a Silicon controlled rectifier

silicon controlled rectifier (SCR) A *semiconductor *rectifier whose forward anode-cathode current is controlled by means of a signal applied to a third electrode, called the *gate*. The device is constructed as a four-layer *chip of semiconductor material forming three *p-n junctions, with *ohmic contacts to three of the layers (Fig. *a*).

A voltage is applied across the device between anode and cathode. With reverse bias (positive voltage applied to cathode), the device is effectively "off".

With forward bias, no current will flow until a signal is applied to the gate (the anode voltage being sufficiently high). Current then flows through the device, and continues even after the cessation of applied signal to the gate. The current may only be cut off by reducing the anode voltage to near zero, or by reducing the current through the device to a low value. The minimum current for continuation of conduction is the *holding current*. The SCR is the solid-state equivalent of the *thyratron valve and was originally called a *thyristor*. The current-voltage characteristic (Fig. *b*) is similar to that of the thyratron.

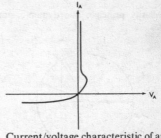

b   Current/voltage characteristic of an SCR

The most important applications of SCRs are in a.c. control systems and solid-state *relays. If an a.c. signal is applied to the anode, the device may be switched on at any desired portion of the positive half-cycle using a *trigger pulse, or by illuminating the gate region. The device will automatically switch off at the end of the half-cycle when the anode voltage drops below the turn-off level. If the gate turn-on current is supplied at the beginning of the positive half-cycle, a single SCR acts as a half-wave rectifier. Full-wave rectification can be achieved by using two SCRs in antiparallel connection. A version of the SCR for switching a.c. is known as the *triac*. It is almost equivalent to two antiparallel SCRs made in a single chip, but is able to operate with a single gate connection.

silicon controlled switch (SCS) A switch fabricated in a silicon chip. Like the *silicon controlled rectifier, it is a *pnpn device but has connections to all four semiconductor layers of the device. Under conditions of forward bias, it can be turned on by a voltage pulse to the control gate electrode (connected to the inner p-type layer – see Fig. *a* at silicon controlled rectifier) and can be turned off by a voltage pulse to a second gate electrode that is connected to the inner n-type layer.

similarity principle *See* dynamic similarity.

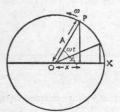

Simple harmonic motion

simple harmonic motion (SHM) The motion of a body subjected to a restoring force directly proportional to the displacement from a fixed point in the line of motion. The force equation is

$$m\ddot{x} = -kx,$$

where m is the inertia of the body, and k is the restoring force per unit displacement. Putting k/m equal to ω^2, the solution of the equation is

$$x = A \cos(\omega t + \alpha),$$

where A is the maximum displacement, ω is the *angular frequency ($\omega = 2\pi f$), and α is the angle determining the displacement at $t = 0$ and is called the *epoch*. The period, T, is given by $T = 2\pi/\omega$.

SHM may be represented graphically as the projection onto a straight line of the path of a particle travelling in a circle with uniform angular velocity (*see* diagram). The amplitude A is the radius of the circle and the angle α is the angular displacement from the fixed line OX at $t = 0$. At time t the displacement, x, is given by:

$$x = A \cos(\omega t + \alpha).$$

simplex operation The operation of a communications channel between two points in which signals or data can be carried in one direction only.

simulator A device, especially one under the control of a computer, that performs a *simulation*, i.e. that mimics the behaviour of an actual system but is made from components that are easier, cheaper, or more convenient to manufacture. Simulators are frequently used for solving complex problems such as weather forecasting, and as a design aid and training tool. Simulation is the major application of *analogue computers.

simultaneity In classical physics different observers in uniform relative motion are assumed to use the same scale of time (*see* Galilean transformation equations) hence any two events that are found to be simultaneous by one observer will also be found to be so by any other observer in uniform relative motion. According to the theory of *relativity this is not generally true. Two events occuring in different places may be found to be simultaneous by one observer, but will not be simultaneous according to an observer moving uniformly relative to the first one. This does not affect the temporal sequence of related events nor imply any failure of determination.

sine condition If n and n' are refractive indexes of media in front of and beyond

Sine condition

a surface, y and y' the object and image sizes, α and α' the angles made between the conjugate rays and the axis passing through the axial feet of the object and image, then

$$ny \sin \alpha = n'y' \sin \alpha'$$

is the sine condition. For *paraxial rays (α, α' both small and equal to α_p, α_p'), the sine condition becomes

$$ny\alpha_p = n'y'\alpha_p'.$$

For an optical system to be free of *coma, it is necessary that

$$\sin \alpha / \sin \alpha' = \alpha_p / \alpha_p' = \text{const.}$$

This is also sometimes referred to as the sine condition. If there is no *spherical aberration, compliancy with the above condition will be both necessary and sufficient for zero coma.

sine galvanometer An instrument similar to the *tangent galvanometer except that the coil and scale are rotated together, while current is flowing, to return the needle to zero. The current is proportional to the sine of the angle of rotation.

sine wave *See* sinusoidal; equivalent sine wave.

singing Unwanted self-sustained oscillations in a communications system. The term is used particularly in connection with telephone lines in which *repeaters are incorporated.

singing arc An arc that emits a musical note as a result of impressed oscillations that cause a variation in the heating effect. The oscillations are set up by shunting the arc with an inductance and capacitance in series.

single crystal Any crystal in which the atomic planes of the same kind are sufficiently parallel to diffract a collimated beam of incident radiation cooperatively and thus to give single spots in the diffraction pattern.

single-phase Of an electrical system or apparatus. Having only one alternating voltage. *Compare* polyphase system.

single-shot multivibrator *See* monostable.

single-sideband transmission (SST) The transmission of one only of the two *sidebands produced by the *amplitude modulation of a *carrier wave. The carrier and the other sideband are usually suppressed at the transmitter. At the receiver, it is necessary to reintroduce the carrier artificially by combining the sideband with a locally generated oscillation. The frequency of the latter should be as nearly as possible equal to that of the original carrier, but this requirement is not so stringent as in the case of *double-sideband transmission*. The main advantages of SST over the method in which the carrier and the two sidebands are transmitted are the reduction in transmitter power (since the carrier and one sideband are not transmitted) and the reduction of the *bandwidth required for the transmission of signals within a specified *frequency band.

singularity *See* Schwarzschild radius.

sinusoidal Of a periodic quantity. Having a *waveform that is the same as that of a sine function. It is only the shape of the graph that is significant. For example, two e.m.f.s represented by

$$e_1 = E_1 \sin \omega t \text{ and } e_2 = E_2 \cos \omega t$$

would both be described as sinusoidal e.m.f.s, and are both *sine waves*.

siphon An inverted U-tube with one limb longer than the other. The shorter

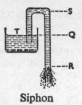

Siphon

limb dips into a liquid and, when the siphon is completely full of the liquid, it emerges from the lower end. This is a convenient method for removing water from a receptacle that cannot be emptied conveniently in any other way. The simple explanation is that the pressures at P and Q are equal, and that the excess pressure due to the column of liquid QR causes the liquid flow. Since the liquid has to rise a distance PS, this must be less than the barometric height of the liquid. Some liquids, however, will siphon in a vacuum, provided that the liquid is free from dissolved gas. This suggests that the cohesive forces between molecules also contribute to the action.

SI units (Système International d'Unités). The internationally agreed system of *coherent units that is now in use for all scientific and most technological purposes in many countries (including the UK). SI units are based on the *MKS system and replace the units used in the *CGS system and the f.p.s. system (Imperial units). SI units are now of two kinds: *base units* and *derived units*. There are seven base units for the seven dimensionally independent physical quantities shown in the Appendix, Table 2. The base units are arbitrarily defined in terms of reproducible physical phenomena or prototypes. Table 4 gives the derived units with special names and symbols. Two units, the radian and steradian, were originally given the status of *supplementary units*, but are now regarded as dimensionless derived

units. All these units are defined in this dictionary in their alphabetical place. The SI unit of any other quantity is derived by multiplication and/or division of the base units without introducing any numerical factors.

SI units are used with a set of prefixes to form decimal multiples and submultiples of the units. The Appendix, Table 3 lists these prefixes and their symbols.

A number of conventions are observed in the printing or writing of SI units. Symbols for a prefix are written next to the symbol for the unit without a space, e.g. cm. A space is left between symbols for units in derived units, e.g. N m for newton metre. The letter *s* is never added to a symbol to indicate a plural: 10 ms indicates ten milliseconds, not ten metres. Compound prefixes are never used, e.g. nm for 10^{-9} metre is correct, mμm is incorrect. A symbol for a unit with a prefix attached is regarded as a single symbol, which can be raised to a power without using brackets, e.g. cm^2 means $(0.01 \text{ m})^2$ and not 0.01 m^2.

The word gram has a special place in SI units; although it is not an SI unit itself, prefixes are attached to the symbol g and not to kg, e.g. 10^3 kg is written Mg and not kkg.

When writing numbers with SI units, the digits are arranged in groups of three but a space rather than a comma is placed between each group; e.g. 10^5 is written 100 000 not 100,000 and 10^{-5} is written 0.000 01. If a number consists of only 4 digits it may be written without spaces, e.g. 1000 or 0.0001.

SI units make quite clear the relationship between a physical quantity (say length *l*) and the units in which it is expressed, i.e.

physical quantity = numerical value × unit

or, in the case of length, $l = n$ m, where *n* is a numerical value and m is the agreed symbol for the SI unit of length, the metre. This equation should be treated algebraically, e.g. the axis of a graph or a column of a table giving numerical values in metres should be labelled l/m. If the axis or column gives the values of $1/l$, it should be labelled m/l.

Certain units outside the SI system have been retained because of their practical importance (e.g. day, degree (of arc), tonne) or their use in specialized fields (e.g. bar, parsec, electronvolt). SI prefixes can be attached to these units, and in some cases compound units can be formed from these non-SI units and SI units.

skiatron A cathode-ray tube in which the usual coating of fluorescent substances is replaced by a screen of alkali halide crystals, which become darkened under electron bombardment. It can be arranged so that the trace remains on the screen until erased.

skin effect A nonuniform distribution of electric current over the cross section of a conductor when carrying an alternating current. The current density is greater at the surface of the conductor than at its centre. It is due to electromagnetic (inductive) effects and becomes more pronounced as the frequency of the current is increased. It results in a greater *I^2R loss in the conductor than that occurring when the current is uniformly distributed. In consequence, the *effective resistance of a conductor when carrying alternating current is greater than the true resistance, and the high-frequency resistance is greater than its d.c. or low-frequency resistance. With very high frequencies hollow conductors may be used, or stranded conductors formed of many fine filaments (litzendraht wire) may be employed.

skin friction The resistance or drag experienced by a body in relative motion with a fluid, due to the *laminar motion of the fluid and the large rate of shear of the fluid close to the body boundaries.

sky wave *See* ionosphere.

slip Of an *induction motor. The ratio of the difference between the *synchronous speed and actual speed of the motor to the synchronous speed. It is usually expressed as a percentage.

slip ring *Syn.* collector ring. A ring, usually made of copper, that is connected to and rotates with a winding (as, for example, in certain types of electrical machines), so that the winding may be connected to an external circuit by means of a *brush or brushes resting on the surface of the ring.

slip-ring rotor *See* induction motor.

slope resistance *Syn.* electrode a.c. resistance; electrode differential resistance. Of a specified electrode in an electronic device. The ratio of a very small change in the voltage applied to the particular electrode to the corresponding change in the current in that electrode, the voltages of all other electrodes being maintained constant at known values. For example, the *collector slope resistance is given by:
$$r_c = \partial V_c / \partial I_c,$$
where V_c is the collector voltage and I_c is the collector current; the base and emitter voltages are held constant.

slow-down density Symbol: q. A measure of the rate at which *neutrons lose energy by collisions in a *nuclear reactor, expressed by the number of neutrons per unit volume falling below a certain energy per unit time. *See* Fermi age theory.

slow neutron A neutron with a kinetic energy that does not exceed a few electronvolts. The term is sometimes loosely applied to a *thermal neutron.

slow vibration direction The direction of the electric vector of the ray of light that travels with least velocity in a crystal and therefore corresponds to the largest refractive index.

small-signal parameters If the behaviour of an electronic device is represented by instantaneous values of current and voltage appearing at the terminals of a four-terminal *network, the small-signal parameters are the coefficients of the equations representing this behaviour for small values of input. *See also* transistor parameters.

smart fluid *See* electrorheological fluid.

smoothing circuit *See* ripple.

Snell's law The law of *refraction:
$$\sin i / \sin i' = C,$$
where i is the angle of incidence, i' is the angle of refraction, and C is a constant, now known to be the ratio, n/n', of the *refractive indices of the initial medium and the refracting medium.

snow 1. Crystals of ice, forming loose structures usually of a hexagonal pattern, produced in the atmosphere by direct condensation of water vapour to the solid phase. The flakes form on suitable minute nuclei, probably clay particles. *See also* hail.
2. An unwanted pattern resembling falling snow that appears on a TV or radar screen, usually when the received signal is absent or weak. (With a colour TV the pattern is coloured.) It is caused by electrical *noise in the receiver.

Soddy and Fajans' rule *Syn.* displacement rule. The emission of an α-particle from a radioactive *nuclide causes a decrease of two in atomic number, and the emission of a β-particle increases the atomic number by unity, with consequent changes in the position of the substance in the periodic table.

soft radiation *Ionizing radiation with a low degree of penetration, especially X-rays of relatively long wavelength. *Compare* hard radiation.

soft-vacuum tube A vacuum tube in which the degree of the vacuum is such that ionization of the residual gas influences the electrical characteristics of the tube. *Compare* hard-vacuum tube; gas-filled tube.

software The *programs associated with a *computer system. There are two basic forms. *Systems software* is an essential accompaniment to the *hardware of the computer system and includes programs (such as the *operating system*) that are essential to the effective use of the system. *Applications software* relates to the role that the computer system plays within a given organization.

solar cell Any device that uses solar radiation to drive an electric current. One type, used to power equipment in spacecraft and artificial *satellites, is essentially a *semiconductor in which a voltage is set up across the p-n junction when photons from the sun fall on the surface (*see* photovoltaic cell). Other types of solar cells, used in desert regions, water-heating systems, etc., consist of complex *thermopiles, one set of junctions being illuminated by solar radiation. A *solar battery* consists of several solar cells. A *solar panel* is a large flat array of solar cells attached to the outside of a spacecraft or satellite and orientated to receive the maximum solar radiation.

solar constant The rate of reception of solar energy by unit area at a specified distance from the sun, the radiation striking the surface normally. A recent measurement of the solar constant at the earth's mean distance from the sun gives a value of

$$1.353 \ (\pm \ 1.5\%) \ \text{kW m}^{-2}.$$

Ground measurements (using *pyrheliometers) must be corrected for atmospheric absorption. Ideally, measurements should be made from a satellite.

solar energy *See* renewable energy sources; solar cell.

solarimeter *Syn.* pyranometer. An instrument for measuring the total solar radiation intensity (sun plus sky) received on a horizontal surface. In the *Moll–Gorczynski solarimeter* a *thermopile is used to measure the intensity. Incoming radiation raises the temperature of one set of thin junctions above the ambient temperature at which the other set is maintained. This temperature difference results in an e.m.f., which is a measure of the radiation intensity and is independent of ambient temperature, wind velocity, etc. In the *Eppley pyranometer*, a central white disc is surrounded by a concentric black ring. Exposure to solar radiation causes the black ring to rise in temperature, the difference between the black and white surfaces being measured by thermocouples. The resulting signal is proportional to radiation intensity. *See also* net radiometer; pyrheliometer.

solar mass *See* solar units.

solar panel *See* solar cell.

solar time *See* time.

solar units A set of units, including *solar mass* (symbol: $M_\odot$), *solar radius* (symbol: $R_\odot$), and *solar luminosity* (symbol: $L_\odot$), in which certain properties of the sun such as the mass, radius, and luminosity are taken as unity; the same properties of other stars can then be compared to those of the sun.

solar wind A stream of ionized particles, mainly protons and electrons, that flow from the sun in all directions. The average particle energy is much lower than that of *cosmic rays, the velocity being between $250-900$ km s^{-1} at the distance of the earth's orbit. The number of particles and their velocity increase following solar activity such as sunspots and solar flares. The solar wind causes the shape of the earth's magnetic field to be unsymmetrical (*see* magnetosphere). Changes in the intensity of the wind produce fluctuations in the magnetic field, causing magnetic storms and affecting radio communications. *See also* aurora; Van Allen belts.

solenoid A coil of wire with a length that is large compared with its diameter. At a point inside the solenoid and on its axis, where the ends subtend semiangles θ_1, θ_2, the magnetic flux density **B** is given by:

$$\boldsymbol{B} = \tfrac{1}{2}\mu_0 nI(\cos \ \theta_1 \ + \ \cos \ \theta_2),$$

where n is the number of turns per unit length, I the current, and μ_0 the *magnetic constant. (The formula ignores end effects.)

solenoidal A *vector field that has no divergence in a region of space is solenoidal in that region. In such a region, the lines of force (or of flow) either form closed curves (e.g. the magnetic field of a current) or terminate at infinity or on bounding surfaces (e.g. the electric field between capacitor plates).

solid angle Symbol: Ω or ω. An area is said to subtend in three dimensions a solid angle at an outside point. The solid angle is measured by the area subtended (by projection) on a sphere of

441

unit radius or by the ratio of the area (A) intercepted on a sphere of radius r to the square of the radius (A/r^2). The unit of solid angle is the *steradian. The solid angle completely surrounding a point is 4π steradians. If a small area (dA) is at a distance R from a point and its normal makes an angle θ with a line drawn to the point, the solid angle formed by the area and point is ($dA \cos \theta)/R^2$.

solidification curve *See* ice line.

solid solution A homogeneous mixture of two solids whose composition may vary within certain limits. Solid solutions are found in certain *alloys. Atoms or ions of one component (the solute) are absorbed into the crystal lattice of the other component (the solvent), usually by substitution of solute atoms or ions and either on a random or regular basis. At a certain composition each component can form regular individual lattices, known as *superlattices*.

solid-state device An electronic device consisting chiefly or exclusively of *semiconducting materials or components.

solid-state memory *See* semiconductor memory.

solid-state physics The branch of physics concerned with the structure and properties of solids and the phenomena associated with solids. These phenomena include electrical conductivity, especially in *semiconductors, *superconductivity, *photoconductivity, *photoelectric effect, and *field emission. The properties of solids and the associated phenomena are often dependent on the structure of the solid.

solid-state relay *See* relay.

soliton A solitary wave state with the properties of a stable particle-like entity that can occur in both classical and quantum areas of physics, such as fluid mechanics, optics, solid-state devices, and particle physics. In particle physics a magnetic monopole may be regarded as a soliton.

solstice 1. One of the two points at which the ecliptic is furthest north or south of the celestial equator. (*See* celestial sphere.)
2. One of the two days of the year when the sun is at these points (approx. June 21 or Dec. 22), marking the day with either the greatest or the least number of daylight hours. *Compare* equinox.

solute A substance dissolved in a pure liquid; the pure liquid is called the *solvent*, and the intimate mixture produced is called the *solution*.

solution *See* solute.

solvent *See* solute.

sonar A contraction of *so*und *na*vigation *r*anging. A method of locating underwater objects by transmitting an ultrasonic pulse and detecting the reflected pulse. The time taken for the pulse to travel to the object and return gives an indication of the depth of the object.

sonde A small *telemetry system in a balloon, rocket, or satellite, used in meteorology, astronomy, etc. *See also* radiosonde system.

sone A unit of loudness. The loudness L in sones is defined by the formula:
$$10 \log_{10} L = (P - 40)\log_{10} 2,$$
where P is the equivalent loudness measured in *phons. The scale has been chosen so that a sound of x sones seems to the listener to be k times as loud as

a sound of x/k sones. Experiment has shown this to be justified for loudness between ¼ and 250 sones.

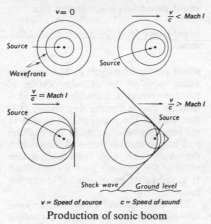

v = Speed of source   c = Speed of sound

Production of sonic boom

sonic boom The noise originating from the backward projected *shock waves set up by an aircraft travelling at greater than the speed of sound (Mach 1). A stationary source of sound emits a series of concentric *wavefronts, the radius of which will increase with time. For a moving sound source, the wavefronts crowd together in the direction of motion until, at Mach 1, they become tangential to a line perpendicular to the direction of motion. At speeds above Mach 1, two conical lines, tangential to the wavefronts, delineate the shock waves set up by atmospheric-pressure discontinuities. In level flight, the intersection of the shock-wave cone with the ground forms a hyperbola at all points on which the sonic boom is heard simultaneously. A double report is heard as both the nose and tail of the aircraft pass through the sound barrier.

sonometer *See* monochord.

sorption pump A type of vacuum *pump that relies for its action on the *adsorp-tion of gas by a material, usually charcoal or a *molecular sieve. The adsorbent is contained in a bulb connected to the system and cooled by liquid air. Pumps of this type are used at fairly high pressures in place of rotary pumps and can also be used at low pressures ($< 10^{-6}$ Pa) to remove small quantities of gas.

sound Periodic mechanical vibrations of a medium by means of which sound energy is carried through that medium. (Sound cannot travel through a vacuum.) The word sound is also applied to the sensation felt when such vibrations fall on the ear. As the human ear has a restricted range of perception of *pitch, sound is strictly limited to those vibrations whose frequencies lie between about 20 and 20 000 hertz (audiofrequencies). There is, however, no essential objective peculiarity about those vibrations whose frequency lies below or above this region; these infrasonic and ultrasonic vibrations have the same basic properties as sound waves and can be included in the same study. (*See also* acoustic wave; ultrasonics.)

The source of sound executes vibrations. In order to do so, it must possess *elasticity, in the sense that when it is displaced from its position of rest and is released, a force comes into play directed towards that position. It must also possess *inertia, in the sense that in returning, it overshoots its equilibrium position and oscillates to and fro.

The propagation of the sound into the surrounding medium involves portions of the medium in vibration (*see* compression), but because of a progressive lag in the taking up of these vibrations, the disturbance is propagated with a finite speed known as the *speed of sound. Sound travels through a fluid medium as a *longitudinal wave motion. In solids longitudinal, *transverse, and *torsional waves all occur. Finally the waves are picked up by receivers, which may be of a biological or physical nature.

The study of sound finds a number of applications: in meteorology and sound ranging (as atmospheric acoustics), in depth sounding and the detection of submarine objects, and in prospecting by means of echoes; in architecture, as acoustics of buildings; in the design of musical instruments, and as the scientific basis of music and voice production (phonetics).

sound absorption coefficient *See* acoustic absorption coefficient.

sound-energy flux *Syn.* sound power, or, more generally, acoustic power. Symbol P or P_a. The rate of flow of sound energy. For a plane or spherical progressive wave with a speed c in a medium of density ρ it can be shown that at any point:
$$P = p^2 A/\rho c,$$
where p is the root-mean-square *sound pressure at that point and A is the area through which the flux passes. Sound-energy flux is measured in *watts.

sound-energy reflection coefficient *See* acoustic absorption coefficient.

sounding balloons *Syn.* balloon sondes. Small balloons used for carrying recording instruments into the earth's atmosphere.

sound insulation *See* reverberation chamber; dead room.

sound intensity Symbol: I or L. The rate of flow of sound energy through unit area normal to the direction of flow. It is related to the root-mean-square *sound pressure p and the density ρ of the medium by:
$$I = p^2/\rho c,$$

where c is the speed of the sound. The SI unit is watt per square metre.

sound power *See* sound-energy flux.

sound-power level *See* power-level difference.

sound pressure Symbol: p. The instantaneous value of the periodic portion of the pressure at a particular point in a medium that is transmitting sound. It is that part of the pressure that is due to the propagation of sound in the medium and has an average value of zero over a period of time. The *root-mean-square value of the pressure level is often used. It is measured in pascals or sometimes bars.

The more general term *acoustic pressure* is used when referring to the whole range of *acoustic waves.

sound-pressure level Symbol: L_p. A dimensionless quantity given by the natural logarithm of the ratio of the *sound pressure, p, to a reference sound pressure, p_0, i.e. by:
$$\log_e(p/p_0)$$
$$= \log_e 10 \times \log_{10}(p/p_0),$$
where p_0 is either stated or taken to be 2×10^{-5} Pa in air and 0.1 Pa in water. Sound pressure levels are measured in *decibels (dB) or *nepers (Np):
$$L_p = 1 \text{ dB when}$$
$$20 \log_{10}(p/p_0) = 1$$
$$1 \text{ dB} = (\log_e 10)/20 \text{ Np}.$$

soundtrack A track at one side of a *videotape or cine film on which sound signals can be recorded in order to provide simultaneous reproduction of the sound associated with the vision projection. The soundtrack on film consists of variations in width or density of a silver image, ideally related in linear fashion in frequency and amplitude to the original sound. In reproduction, a beam of light projected through the sound track

444

is amplitude-modulated by these variations. The modulation is converted into an audiofrequency signal by a photocell. The signals are amplified sufficiently to operate a loudspeaker system. In practice, the volume range of the recorded signal is compressed to a range of about 40 to 50 decibels, particularly to raise the lower sound levels above the inherent noise level of the system. The frequency range normally recorded is from 50 to 12 000 hertz.

sound wave *See* sound; acoustic wave.

source 1. The point at which lines of flux originate in a *vector field; e.g. a point at which there is a positive electric charge in an electrostatic field. A *sink* is a point at which lines of flux terminate.

2. In classical hydrodynamics, a point at which fluid is continually emitted and from which the flow is radial and uniform. A negative source is called a *sink*. The strength of a source (m) is the quantity of fluid emitted in unit time. The quantity emitted is sometimes expressed as $4\pi m'$; the symbol m' then defines the strength.

3. The electrode in a *field-effect transistor that supplies charge *carriers (electrons or holes) to the interelectrode space.

4. Any energy-producing device, e.g. current source, power source.

5. An object that emits light, such as the sun, an electric light, a gas-discharge tube, an arc, or a spark.

source impedance The impedance of any energy source presented to the input terminals of a circuit or device. An ideal voltage source will have zero source impedance, whereas an ideal current source will have infinite source impedance.

space-charge region In any device, a region in which the net *charge density is

significantly different from zero. For example, in a *semiconductor or *thermionic valve, space-charge regions can exist in equilibrium under zero applied-bias condition, forming potential barriers; these barriers must be overcome by the applied bias before current flows.

space group A set of operations (rotation about an axis, reflection across a plane, translation, or combinations of these) that when carried out on a periodic arrangement of points in space brings the system of points to self-coincidence.

space lattice *See* Bravais lattice.

space quantization *See* spin.

space-reflection symmetry *See* parity.

space-time continuum *See* four-dimensional continuum.

spallation A particularly vigorous nuclear reaction caused by a bombardment of a target by high-energy particles, resulting in the emission of a number of nucleons.

spark A visible disruptive discharge of electricity between two places at opposite high potential. It is preceded by ionization of the path. There is a rapid heating effect of the air through which the spark passes, which creates a sharp crackling noise. The distance a spark will travel is determined by the shape of the electrodes and the p.d. between them.

spark chamber A device in which the tracks of charged particles are made visible and their location in space accurately recorded. It was developed from the *spark counter. It consists essentially of a stack of narrowly spaced thin metal plates or grids in a gaseous atmosphere, partially surrounded by one or more auxiliary particle detectors. Any charged particle detected in one of the auxiliary devices triggers the rapid application of a high-voltage pulse to the stack of plates. The passage of the particle through the plates is marked by a series of spark discharges along its path. The track is recorded electronically or photographically. Subsequent events, such as collisions and disintegrations, can also be recorded. Use of the auxiliary devices leads to a select triggering of the spark chamber for a particular type of energy or radiation.

spark counter A type of particle detector used in the detection and measurement of heavily ionizing particles, especially α-particles. The counter consists of a pair of electrodes in the form of a wire or mesh anode in close proximity to a metal-plate cathode. A high potential difference is applied across the electrodes, its value being just less than that required to cause a discharge across the air gap. If a charged particle approaches the anode, the field between the electrodes is increased sufficiently to cause a spark discharge. At the moment of discharge the anode potential drops significantly. Particles may be detected by the noise as sparks occur. The number of particles may be measured by photography or counting circuits designed to respond to the change in voltage across the anode load resistor.

spark discharge *See* conduction in gases.

spark gap An arrangement of electrodes specially designed so that a disruptive discharge takes place between them when the applied voltage exceeds a predetermined value.

sparkover *See* flashover.

spark photography Any form of photography in which a spark provides the illumination. In spark photography the camera lens is left open and the duration of the spark is controlled to give the correct exposure.

spatial filtering A technique whereby an optical image can be improved by filtering (i.e. removing) certain spatial frequencies from it (*see* spatial period). The image may, for example, be a micrograph or one transmitted from a planetary spaceprobe. It is modified so that the information of interest is made more accessible. A mask is used to filter the required components from the diffraction pattern of the image. For example, removal of low spatial frequencies (with a high-pass filter) will enhance the sharp edges in an image at the expense of regions in which the intensity is uniform or changing only slowly. Other filters can enhance contrast or remove extraneous lines from a composite picture or patterns of dots from halftone pictures.

spatial frequency *See* spatial period.

spatial period The distance over which a regular pattern repeats itself. The *spatial frequency* of the pattern is the reciprocal of the spatial period. The pattern may, for example, be a series of lines of equal width and spacing as found on a diffraction grating or it may be a diffraction pattern.

speaker Short for loudspeaker.

special relativity *See* relativity.

specific The use of the adjective *specific* to qualify the name of an extensive physical property is now restricted to the meaning "per unit mass", as in the term specific heat capacity. When the physical quantity is denoted by a capital letter (e.g. *L* for the latent heat), the specific quantity is denoted by the corresponding lower-case letter (*l* for the specific latent heat).

Formerly the word had other meanings, which are now deprecated. Many terms using these meanings have been renamed. For example, specific gravity is now called relative humidity, specific resistance is now called resistivity.

specific activity Symbol: *a*. The *activity per unit mass of a radionuclide.

specific charge The ratio of the charge on an *elementary particle to the mass of that particle; the charge per unit mass of a particle. *See also* e/m.

specific conductance The former name for *conductivity (electrical).

specific gravity The former name for relative *density.

specific heat capacity Symbol: *c*. The *heat capacity per unit mass; the quantity of heat required to raise the temperature of one kilogram of a substance by one kelvin. It is measured in J kg⁻¹ K⁻¹. A theory of the variation of the specific heat capacities of solids with temperature has been given by *Debye using the quantum theory. (*See also* Dulong and Petit's law.)

For solids and liquids the value of the specific heat capacity is determined at constant pressure. For a gas, there are two principal specific heat capacities depending on the way in which the temperature is increased. If the pressure is kept constant the specific heat capacity at constant pressure, c_p, is obtained; if the volume is kept constant, the specific heat capacity at constant volume, c_V, is obtained. The specific heat capacity at constant pressure always exceeds that at constant volume by the work done in expansion. For a solid:

$$c_p - c_V = A c_p^2 T,$$

where *A* is a constant. For an ideal gas in which the internal energy is independent of the volume:

$$c_p - c_V = nR,$$

where *n* is the number of moles per kilogram and *R* is the *molar gas constant. These equations are derived from the general thermodynamic equation

$$c_p - c_V = T \left\{ \frac{\partial p}{\partial T} \right\}_V \left\{ \frac{\partial v}{\partial T} \right\}_p,$$

while

$$\left(\frac{\partial c_V}{\partial v} \right)_T = T \left(\frac{\partial^2 p}{\partial T^2} \right)_V$$

and

$$\left(\frac{\partial c_p}{\partial p} \right)_T = -T \left(\frac{\partial^2 p}{\partial T^2} \right)_p$$

give the variation of specific heat capacities with volume and pressure. On the kinetic theory the *molar heat capacity at constant volume of a gas is given by $F \times R/2$ assuming the equipartition of energy, where *F* is the number of *degrees of freedom. Although this result is by no means true for real gases, the specific heat capacity does increase greatly with the atomicity of the molecule. The ratio c_p/c_V of a gas always exceeds unity and is a constant denoted by the symbol γ. *See also* gamma; negative specific heat capacities.

specific inductive capacity A former name for relative *permittivity.

specific latent heat of fusion *Syn.* enthalpy of melting. Symbol: l_f or ΔH_s^l. The quantity of heat required to change the state of unit mass of a substance from the solid to the liquid state at the melting point. It is measured in joules per kilogram. *See also* Clausius–Clapeyron equation.

specific latent heat of sublimation *Syn.* enthalpy of sublimation. Symbol: l_s or

ΔH_s^g. The quantity of heat required to change unit mass of a substance from the solid to the vapour state without change of temperature.

specific latent heat of vaporization *Syn.* enthalpy of evaporation. Symbol l_v or ΔH_s^f. The quantity of heat required to change unit mass of a substance from the liquid to the vapour state at the boiling point. It is measured in joules per kilogram, and is given by the *Clausius–Clapeyron equation.

The value of l_v decreases as the temperature increases, eventually becoming zero at the *critical temperature. For all substances the variation of specific latent heat with temperature is given by the thermodynamic formula:

$$l_v = T \frac{dp}{dT} (v_2 - v_1),$$

where c_1 and c_2 are the *specific heat capacities of the liquid and vapour respectively and v_1 and v_2 are their specific volumes.

specific optical rotary power Symbol: α_D. A measure of the *optical activity of a substance in solution. It is given by the equation:
$$\alpha_D = \alpha V / ml,$$
where α is the angle of rotation of the plane of polarization when light traverses a pathlength l of a solution containing a mass of substance m in a volume V. The units are $m^2 \, kg^{-1}$.

specific reluctance The former name for *reluctivity.

specific resistance The former name for *resistivity.

specific volume Symbol: v. The volume of unit mass of a substance; the reciprocal of density.

spectral 1. Of or relating to a spectrum.
   2. Of or relating to a particular frequency or wavelength.

spectral class *See* stellar spectra.

spectral lines *See* spectrum; hyperfine structure of spectral lines.

spectral luminous efficacy *See* luminous efficacy.

spectral type *See* stellar spectra.

spectrograph *See* spectrometer.

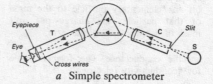

a Simple spectrometer

spectrometer 1. An instrument for producing, examining, or recording a *spectrum. When an emission spectrum is investigated, the radiation from the source is passed through a *collimator, which produces a parallel beam of radiation. This is deviated and dispersed by a prism or *diffraction grating and the angular deviation depends on the wavelengths present. The refracted or diffracted radiation is then observed or recorded in some way so that the angular deviation can be measured.

In a simple spectrometer (Fig. *a*) light from the source S is collimated by C and dispersed by the prism. The telescope T is used to observe this light and it can be rotated around the prism table and the angular deviation measured. By a suitable calibration, wavelengths can be measured. The angles of prisms and refractive indexes can also be determined.

Many modern spectrometers use diffraction gratings. In Fig. *b*, radiation enters at the slit S_1, and is reflected by a concave grating G through S_2 onto a detector D, such as a *photomultiplier. In this instrument both slits are kept constant and the grating is rotated through a range of angles. A collimator is not essential because the grating is ruled on a concave mirror and focuses the radiation. At each particular angle of the grating, radiation of a particular wavelength is focused onto the exit slit. A graph of the angle of the grating (as abscissa) against the response of the photomultiplier can, with suitable calibration, give a curve of wavelength against intensity of radiation.

When an absorption spectrum is investigated, a source of radiation with a continuous spectrum is usually used. The radiation may be passed through the sample and then into the spectrometer so that the distribution of intensity with wavelength is found. Alternatively, the radiation can be passed into the spectrometer so that one particular wavelength of radiation is selected and passed through the sample onto the detector. The wavelength can be varied by changing the position of the grating or prism. In this way the spectrometer is used to isolate one wavelength of radiation, and is called a *monochromator*.

Any spectrometer that records a spectrum is called a *spectrograph*; the record is called a *spectrogram*. The design of spectrographs depends on their application and the wavelength of radiation for which they are used. In the visible region glass prisms and transmission gratings can be used. Gratings are usually used in other wavelength regions.

The instruments used in detection of the radiation consist either of an electronic imaging device or a photographic plate, sensitive to the wavelengths under investigation. Highly sensitive electronic devices, such as *CCD detectors, can detect light, ultraviolet, and often X-rays; the information obtained is fed to a computer for analysis. Other electronic detectors include *bolometers,

and devices based on *photoconductivity or the *photovoltaic effect.

Spectrometers are often called *spectroscopes*. A *spectrophotometer* is an instrument for measuring the intensity of each wavelength in a spectrum, and the term is often used synonymously with spectrometer. (*See also* spectroscopy.)

2. A similar instrument for determining the distribution of energies in a beam of particles, such as electrons (*see* electron spectroscopy) or ions (*see* mass spectrometer).

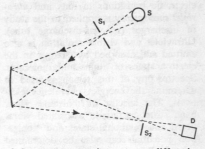

b Spectrometer using concave diffraction grating

spectrophotometer *See* spectrometer.

spectroscope *See* spectrometer.

spectroscopy The technique of producing *spectra, analysing their constituent wavelengths, and using them for chemical analysis or the determination of energy levels and molecular structure. A spectrum is formed by the emission or absorption of electromagnetic radiation accompanying changes between the quantum states of atoms and molecules. The frequency of the radiation depends on the type of states involved and spectroscopic techniques are used over a very wide range of frequencies and yield a wide variety of information.

Gamma-ray spectroscopy involves the measurement of the distribution of energies in *gamma rays emitted from nuclei. The energy levels of nuclei are also found from a spectrum of absorbed gamma rays in the *Mössbauer effect. *X-ray emission occurs when orbital electrons make transitions from outer orbits to vacant inner orbits. It gives information on electronic states in atoms and molecules, and is often used for studying *energy bands in solids. The absorption and emission of ultraviolet radiation are also used for the determination of electronic states. In the far *ultraviolet region it is associated with electronic transitions in ions and *ultraviolet spectroscopy* is applied to the study of discharges (*see* gas-discharge tube). Ultraviolet and visible radiation is also emitted and absorbed by changes in the electron states of atoms and molecules. Spectroscopy of this region is used to determine the energies of valence electrons. The spectra of molecules show band structure due to associated vibrational and rotational states and *ionization potentials can also be determined (*see* Rydberg spectrum). Visible and ultraviolet spectroscopy is a widely used analytical technique. Elements and compounds present in a sample can be detected by the presence of characteristic lines in their emission or absorption spectrum. A quantitative determination can often be made from the intensities of the lines.

At longer wavelengths, in the near *infrared region, the photon energies correspond to changes between vibrational states of molecules. *Infrared spectroscopy* gives information on the vibration at frequencies and force constants of chemical bonds and on the potential energy curves of molecules (*see* Morse equation).

Certain groups of atoms in molecules tend to absorb at characteristic frequencies and infrared spectroscopy is widely used by chemists for identifying compounds. Far infrared and *microwave radiation is absorbed during changes in the rotational states of molecules. Spectroscopy in this region can give moments of inertia of molecules and their bond lengths and shapes. *Microwave spectroscopy* is particularly useful in the study of large molecules.

At still lower frequencies (and lower photon energies) the states studied are those due to the *spin of electrons and nuclei in applied magnetic fields. *Electron-spin resonance spectroscopy is used to study paramagnetic materials and *nuclear-magnetic resonance spectroscopy gives nuclear magnetic moments. *See also* spectrometer; Raman spectroscopy; electron spectroscopy; hydrogen spectrum.

spectrum 1. Any particular distribution of *electromagnetic radiation, such as the display of colours (violet, blue, green, yellow, orange, red) produced when white light is dispersed by a prism or *diffraction grating. The term is also applied to a plot of the intensity of electromagnetic radiation against wavelength, frequency, or quantum energy, or to a photographic or electronically produced record of dispersed electromagnetic radiation.

Spectra can be obtained with a *spectrometer or spectrophotometer. The spectrum is characteristic of the radiation itself in that it specifies the wavelengths that are present and their intensities. It is also characteristic of the substance that is emitting or absorbing the radiation (*see* emission spectrum; absorption spectrum). Absorption spectra are usually simpler than emission spectra.

A spectrum in which there is a continuous region of radiation emitted or absorbed is a *continuous spectrum*. An example is the spectrum of visible and infrared radiation emitted by a *black ultraviolet body. The spectra of gaseous atoms often contain a number of sharp bright lines of emitted radiation or a number of dark lines on a continuous background due to absorption. Such spectra are called *line spectra* and result because the atoms make transitions between states of definite energy. *Band spectra* contain a series of *bands*, i.e. a series of regularly spaced lines that are very close together and may not be resolved by the apparatus used to disperse the radiation. They are characteristic of emission or absorption by molecules.

Spectra are also classified according to whether the radiation is in the X-ray, ultraviolet, visible, infrared, or microwave region of the spectrum and according to whether the process producing or absorbing the radiation involves a change in electronic, vibrational, or rotational states. Thus in the visible and ultraviolet regions, changes occur between electronic states and an *electronic spectrum* is obtained. These are line spectra in the case of atoms and band spectra in the case of molecules. The bands are formed because in making the transition from one electronic state to another, the molecule may end up in any of a number of possible vibrational states, each differing in energy. In the *near-infrared region spectra result from changes from a rotational state in one vibrational state to a rotational state in another vibrational state. A spectrum of this type is called a *vibration-rotation spectrum*. Changes between rotational states in the same vibrational state lead to absorption or emission of radiation in the far-infrared and *microwave regions and the formation of a *rotation spectrum*. At the other end of the scale, ultraviolet spectra are electronic spectra of ions. X-ray spectra involve changes of electronic states of inner tightly bound electrons.

2. Any distribution of energies, momenta, velocities, etc., in a system of particles, as in a mass spectrum* (*see* mass spectrometer) or an electron spectrum (*see* electron spectroscopy).

spectrum analyser An instrument for measuring the energy distribution with frequency for any waveform. It may be used to determine the frequency response and distortion of any transmission system by comparison of input and output waveforms. *Multichannel analysers are commonly used as spectrum analysers.

speculum A copper-tin alloy (Cu 67%, Sn 33%) used for metal mirrors and reflection diffraction gratings. It takes a high polish and does not tarnish readily.

speed 1. *See* velocity.
2. A value specifying the sensitivity of a photographic material to light.
3. A measure of the light-transmitting power of a lens, commonly indicated by its *f-number: as the f-number decreases, the speed increases.

speed of light in vacuum *Syn.* speed of electromagnetic radiation (in vacuum). (Use of the word velocity rather than speed is discouraged.) Symbol: *c*. A fundamental constant now defined as
$2.997\ 924\ 58 \times 10^8\ \mathrm{m\,s^{-1}}$ (exactly).
This value has been recommended since 1975 for universal use. It is the speed at which not only light but all *electromagnetic radiation travels in a vacuum. The speed decreases when the radiation enters a material medium.

Measurements of the finite speed of light began with an astronomical method (Römer, 1676), and have continued up to the present with diverse terrestrial methods of ever-increasing accuracy and precision. Radio and radar methods agree with the more recent optical determinations.

See also relativity (special theory); metre.

speed of sound Symbol: *c*. The speed of sound in dry air at STP is $331.4\ \mathrm{m\,s^{-1}}$. In sea water it is $1540\ \mathrm{m\,s^{-1}}$ and in fresh water 1410 m s⁻¹. The transmission of sound may involve either longitudinal, transverse, or torsional wave motion, according to the medium. The speed with which the waves travel is dependent on the fundamental physical quantities, elasticity and density. The speed of sound of small amplitude elastic waves in any extended medium is given by the equation:

$$c = \sqrt{(K/\rho)},$$

where K is the appropriate *elastic constant and ρ is the normal density of the medium. For gases or liquids K is the bulk *modulus of elasticity while for solids the Young, axial, or shear modulus must be introduced.

In general, there are two elasticities for any fluid medium, the adiabatic and the isothermal, and the ratio betwen them is equal to the ratio between the two specific heat capacities (γ). For gases, the compressions and rarefactions in sound waves take place so rapidly that the changes are adiabatic and therefore the speed of sound $c = \sqrt{(\gamma p/\rho)}$, where p is the gas pressure.

From the above equation, the speed of sound in a gas depends upon the temperature in accordance with the relation:

$$c_\theta = c_0 \sqrt{(1 + \alpha\theta)},$$

where c_0 and c_θ are the speeds of sound at 0 °C and θ °C and α is the coefficient of expansion of the gas. Also, since p/ρ is constant at constant temperature for any gas, the speed is independent of the pressure, except at very high pressures. In addition, the speed is independent of the frequency, except at high frequencies (*see* dispersion of sound). The speed depends upon the nature of the gas since γ and p are involved in the equation. Hence, moisture and impurities affect the speed.

sphere gap A spark gap having spherical electrodes. A sphere gap can be used to measure extra high voltages with great reliability.

spherical aberration An *aberration that can occur in optical systems: when rays

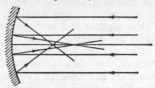

a Spherical aberration (mirror)

are traced after reflection or refraction at large-aperture surfaces (commonly spherical), they do not unite accurately at a focus (Fig. *a*, *b*). When the outer zones focus within the focal point for *paraxial rays, the spherical aberration is said to be positive (*see* caustic curve). The image of a point appears as a circular disc. The best focus is between the two extreme foci, at which point the image is the *circle of least confusion*. For a particular zone of a lens or mirror, the *longitudinal spherical aberration* is the axial distance between the paraxial focus and the intersection of a zonal ray with the axis.

First-order theory of spherical aberration is based on the employment of the first two terms in the expansion of the sine (*see* Seidel aberrations). In practice, it may be eliminated from mirrors by grinding them to the shape of a paraboloid (for use with distant, e.g. celestial, objects) or of an ellipsoid (for finite object distances). Alternatively, a Schmidt corrector can be used, as in the *Schmidt telescope. With a lens, the aberration will be small if the angles of incidence on the lens surfaces are small. The necessary deviation of the ray should thus be shared equally between the two lens faces. A telescope objective is therefore roughly planoconvex with the convex face towards the object. In a microscope ojective, a number of lens components are used so that the deviation of the rays is shared between the refracting surfaces.

The term spherical aberration is also used to embrace all the aberrations due to the sphericity (form) of surfaces, e.g. spherical aberration, coma, radial astigmatism, curvature of field, and distortion (Seidel aberrations).

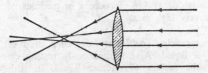

b Spherical aberration (lens)

spherical lens A lens with one or both surfaces that are portions of spheres. Likewise a *spherical mirror* has a surface that is a portion of a sphere. Such surfaces are mechanically easier to produce than others, such as ellipsoid, paraboloid, etc., and are therefore most commonly used, although the employment of aspherical or deformed spherical surfaces is increasing. Most of the problems of lens design are concerned with correcting the *aberrations produced.

spherocylindrical lens *See* cylindrical lens.

spherometer An instrument for measuring the curvature of the surfaces of lenses and mirrors.

spike A current or voltage transient of extremely short duration.

spin The intrinsic angular momentum of an *elementary particle or group of particles. Bohr's theory of the *atom predicts that lines in the spectra of the alkali metals should be single. In fact, they consist of closely spaced *doublets. To explain this Pauli, in 1925, suggested that each electron could exist in two

states with the same orbital motion. Uhlenbeck and Goudsmit interpreted these states as due to the spin of the electron about an axis. The electron is assumed to have an intrinsic angular momentum in addition to any angular momentum due to its orbital motion. This intrinsic angular momentum is called spin. It is quantized in values of

$$\sqrt{s(s + 1)}h/2\pi,$$

where s is the *spin quantum number* and h the *Planck constant. For an electron the component of spin in a given direction can have values of $+\frac{1}{2}$ and $-\frac{1}{2}$, leading to the two possible states. An electron with spin behaves like a small magnet with an intrinsic *magnetic moment (*see also* magneton). The two states of different energy result from interaction between the magnetic field due to the electron's spin and that caused by its orbital motion. There are two closely spaced states resulting from the two possible spin directions and these lead to the two lines in the doublet.

In an external magnetic field the angular momentum vector of the electron precesses (*see* precession) around the field direction. Not all orientations of the vector to the field direction are allowed; there is quantization so that the component of the angular momentum along the direction is restricted to certain values of $h/2\pi$. The angular momentum vector has allowed directions such that the component is $m_s(h/2\pi)$, where m_s is the *magnetic spin quantum number*. For a given value of s, m_s has the values $s, (s - 1), \ldots -s$. For example, when $s = 1$, m_s is 1, 0, and -1. The electron has a spin of $\frac{1}{2}$ and thus m_s is $+\frac{1}{2}$ and $-\frac{1}{2}$. Thus the components of its spin angular momentum along the field direction are $\pm\frac{1}{2}(h/2\pi)$. This phenomenon is called *space quantization*.

The resultant spin of a number of particles is the vector sum of the spins (*s*) of the individual particles and is given the symbol *S*. For example, in an atom two electrons with spins of ½ could combine to give a resultant spin of $S = ½ + ½ = 1$ or a resultant of $S = ½ - ½ = 0$.

Alternative symbols used for spin are *J* (for elementary particles) and *I* (for nuclei). Most elementary particles have a nonzero spin, which may either be integral or half integral. (*See* boson; fermion.) The spin of a nucleus is the resultant of the spins of its constituent nucleons.

spin glass A type of alloy containing a small proportion (0.1–10%) of a magnetic metal, such as iron or magnesium, and a nonmagnetic metal, such as gold or copper, in which the magnetic-metal atoms are distributed randomly in the crystal lattice. Examples include AuFe and CuMn. These alloys are 'glasses' in the sense that they have a random distribution of magnetic atoms; the 'spin' refers to the magnetic property of the atoms. Spin glasses have complicated magnetic behaviour, which is difficult to treat by conventional theory because the random distribution of magnetic atoms means that there is a lack of regular order in the lattice.

spin-paired electrons Two electrons with opposing *spins in an *atomic orbital. *See also* Pauli exclusion principle.

spin quantum number Symbol: *s*, m_s, *J*, *I*. *See* spin.

spin-statistics theorem A theorem in relativistic *quantum field theory stating that half-integer *spins can only be quantized consistently if they obey Fermi–Dirac statistics and integer spins can only be quantized consistently if they obey Bose–Einstein statistics (*see* quantum statistics). This theorem, which has been rigorously proved in many

ways, provides the basis for the *Pauli exclusion principle.

spontaneous fission (S.F.) Nuclear *fission that takes place independently of external circumstances and is not initiated by the impact of a neutron, an energetic particle, or a photon. It is a form of *radioactivity and obeys the exponential decay law. It occurs in nuclides of very high mass.

spreading coefficient Imagine a drop of a liquid, A, instantaneously resting on the surface of another liquid, B, with which it is immiscible. The three forces acting on unit length are numerically equal to the surface tensions T_A and T_B of the liquid and their interfacial surface tension T_{AB}. The condition for equilibrium is that these three forces shall balance. If the drop does not spread, $T_B < T_A + T_{AB}$. The quantity $(T_B - T_A - T_{AB})$ is called the spreading coefficient and if this is positive, the drop will spread.

spreading resistance Of a *semiconductor device. The component of resistance due to the bulk of the semiconductor away from the junctions and contacts.

spring balance A device with which a force is measured by the extension produced in a helical spring. It is used in weighing. The extension produced is directly proportional to the force (weight).

spurious response An unwanted output from an electronic circuit, device, or transducer in the absense of an input signal or as a result of the presence of an unwanted input signal.

sputtering The evaporation of particles of the cathode in a *gas-discharge tube during the discharge of electricity, due to bombardment by positive ions. It can be used to coat a nonconductor close to the cathode with a thin adhesive metallic film. Typically the gas pressure is between 150 and 1.5 Pa and the voltage between cathode and anode is between 1 and 20 kV.

square wave A *pulse train consisting of rectangular pulses with *mark-space ratio equal to unity.

squegging oscillator An *oscillator in which the oscillations build up in amplitude to a peak value then fall to zero. *See also* blocking oscillator.

squid (*s*uperconducting *q*uantum *i*nterference *d*evice) Any of a family of superconducting devices that are capable of measuring extremely small magnetic fields, voltages, and currents. Their action is based on the d.c. Josephson current flowing across *Josephson junctions in certain configurations, the current in such a device being highly sensitive to an external magnetic field.

squirrel-cage rotor *See* induction motor.

stable circuit A circuit that does not produce any unwanted oscillations under any operating conditions.

stable equilibrium *See* equilibrium.

standard atmosphere An internationally established reference for pressure, defined as 101 325 pascals. Although this was formerly used as a unit of pressure – the *atmosphere*, symbol: atm – the standard atmosphere should not now be regarded as a unit. Atmospheric pressure fluctuates about the standard value.

standard cell An electric cell used as a voltage-reference standard. *See* Clark cell; Weston standard cell.

standard deviation *See* deviation.

standard illuminant An illuminating source set up for standard colorimetry, i.e. to enable measurements of the colours of non-self-luminous samples to be determined. The samples are illuminated at 45° to the normal and viewed normally. Three standards are prescribed having *colour temperatures of 2848 K, 4800 K, and 6600 K.

standardization 1. The process of relating a physical magnitude (e.g. a weight), or the indication of a meter (e.g. one supposedly reading current) to the standard unit of that quantity.
2. Establishment of an international, national, or industrial agreement concerning the specification or production of electronic, electrical, mechanical, and other types of components, or of equipment in general. Among other advantages, this greatly increases the ability to interchange components and devices.

standard model *See* electroweak theory.

standard temperature and pressure (STP or s.t.p.). A standard condition for the reduction of gas temperatures and pressures. Standard temperature is now 298.15 K; it was formelry 273.15 K, i.e. 0 °C. The standard pressure for gases was formerly 101 325 pascals. It is now recommended that the standard pressure for reporting thermodynamic data be 10^5 Pa (1 bar); normal boiling points however may still be reported on the basis of a pressure of 101 325 Pa.

standing wave *Syn.* stationary wave. A wave incident normally on a boundary of the transmitting medium is reflected either wholly or partially according to the boundary conditions, the reflected wave being superimposed on the incident wave and thereby creating an interference pattern of *nodes and antinodes. If the incident wave is represented by:

$$\xi = a \sin(2\pi/\lambda)(ct - x),$$

then for total reflection from a rigid boundary the reflected wave would be:

$$\zeta = -a \sin(2\pi/\lambda)(ct + x)$$

and the combination would be

$$\xi = a \sin(2\pi/\lambda)(ct - x) - a \sin(2\pi/\lambda)(ct + x),$$

i.e.

$$\xi = -2a \sin(2\pi/\lambda)x \cos(2\pi/\lambda)ct.$$

This is a standing wave, i.e. a wave that remains stationary, the displacement being always zero at $x = 0$, $\lambda/2$, λ, $3\lambda/2$, etc. (nodes), and vibrating with amplitude $2a$ at $x = \lambda/4$, $3\lambda/4$, $5\lambda/4$, etc. (antinodes).

In contrast, at a free boundary the phase of the reflected wave is the same as that of the incident wave, thus creating an antinode at the boundary. The standing-wave pattern has, however, the same spacings between nodes and antinodes as before.

In practice the reflection may be only partial, thus creating minimum deflections at the nodes instead of the zero deflections.

Each kind of wave has two types of disturbance. Electromagnetic waves have electric and magnetic fields; sound has pressure variations and particle displacements; surface waves in liquids have longitudinal and transverse displacements; stretched strings have particle displacements and variations of tension. Generally the nodes for one type of disturbance are antinodes for the other and the energy density (averaged over a cycle) is uniform in a system of standing waves, energy being interchanged between the two types of antinode during a cycle.

Stanhope lens A thick biconvex lens magnifier, with front surface of radius two-thirds the thickness and a concentric back surface with radius one-third the thickness (glass assumed). The ob-

ject to be viewed is put in contact with the front surface.

Stanton number Symbol: St. A dimensionless parameter defined by the equation:

$$St = h/\rho c_p v,$$

where h is the *heat transfer coefficient, ρ is the density of the fluid, c_p is the specific heat capacity, and v is the speed of flow. See convection (of heat).

star A self-luminous celestial body. Its energy is generated by *nuclear-fusion reactions in its core, and this energy is transported to the surface where it is radiated away into space. The inwardly directed gravitational force in a star balances the outwardly directed gas and radiation pressure, maintaining it in a state of hydrostatic equilibrium. As the star ages its interior structure and chemical composition change. See also magnitude; stellar energy; stellar spectra; Hertzsprung–Russell diagram; white dwarf; neutron star; black hole; celestial sphere.

star connection A method of connection used in *polyphase a.c. working in which the windings of a transformer, a.c. machine, etc., each have one end connected to a common junction the latter being called the *star point*. In three-phase working, the windings may be represented by the symbol Y or T and hence it is also known as the Y or T connection. *Compare* mesh connection.

Stark effect The wavelength of light emitted by atoms is altered by the application of a strong transverse electric field to the source, the spectrum lines being split up into a number of sharply defined components. The displacements are symmetrical about the position of the undisplaced line, and are proportional to the field strength up to about 100 000 volts per cm.

Stark–Einstein equation The formula for the energy per mole E absorbed in a photochemical reaction. If f is the frequency of the absorbed radiation $E = hLf$ where h is the *Planck constant and L the *Avogadro constant.

stat- A prefix that, when attached to the name of a practical electrical unit, denotes the corresponding unit in the CGS-electrostatic system. This system of units is no longer employed.

static 1. Electrical disturbance in a radio or TV system, such as a crackling or hissing sound at a loudspeaker. It is produced by electrostatic induction arising from atmospheric conditions, such as lightning flashes.
2. Electric sparks or crackling produced by friction.
3. Not changing or incapable of being changed over a period of time; undisturbed or causing no disturbance or movement.

static friction *See* friction.

statics The branch of *mechanics dealing with bodies at rest relative to some given frame of reference, with the forces between them, and with the equilibrium of the system. *Hydrostatics* is a branch of statics dealing with the equilibrium of fluids and with their stationary interactions (e.g. pressure, flotation) with solid bodies.

static tube An instrument for measuring the static, or undisturbed, pressure in a fluid flow. *See* Pitot tube.

stationary orbit *See* geostationary orbit.

stationary state In the *quantum theory, or *quantum mechanics. The state of an

atom or other system that is fixed, or determined, by a given set of quantum numbers. It is one of the various quantum states that can be assumed by an atom.

stationary-time principle *See* Fermat's principle.

stationary wave *See* standing wave.

statistical error A fluctuation from an average value that arises from the randomness of the associated event. The event may, for example, be radioactive decay or electron emission. If the average value is n then the statistical error is of the order of $\sqrt{n}$.

statistical mechanics The theory in which the properties of macroscopic systems are predicted by the statistical behaviour of their constituent particles. For example, if a large collection of molecules is considered, its total energy is the sum of all the individual energies of the molecules. These in turn are energies of vibration, rotation, and translation, and electronic energy. According to *quantum theory a molecule can only have certain allowed energies; it can be thought of as occupying any of a number of *energy levels. Consequently the system as a whole can also have any number of possible energy levels, E_1, E_2, E_3, ... If a large collection of systems is considered, each containing the same amount of substance, there will be a distribution of systems over energy levels: N_1 will have energy E_1, N_2 energy E_2, etc. According to the Maxwell–Boltzmann distribution law:

$$\frac{N_i}{N} = \frac{g_i e^{-E_i/kT}}{\sum_i g_i e^{-E_i/kT}},$$

where N_i systems have energy E_i, N is the total number of systems, and g_i is the *statistical weight of this energy level. The expression:

$$\sum_i g_i e^{-E_i/kT}$$

is called the *partition function* and has the symbol Z. A collection of systems of this type is called a *canonical assembly* or *ensemble*. The average energy of a system, E, is $\Sigma N_i E_i / \Sigma N_i$ and consequently:

$$E = kT^2 (\partial \log_e Z / \partial T).$$

In statistical mechanics it is assumed that this average instantaneous value of a property over a large number of systems is the same as the average value of this property for one system over a period of time. Thus the expression gives the *internal energy of the system. The partition function of the canonical assembly is related to the energy levels of the individual molecules by the equations:

$$Z = z^L \text{ and } z = \Sigma_j g_j \exp(-\varepsilon_j /kT)$$

Here z is the partition function of the assembly of molecules with energy ε_1, ε_2, ε_3, etc. Usually the partition function of 1 mole is considered and L is the *Avogadro constant. In principle, statistical mechanics can be used to obtain thermodynamic properties of a system from a knowledge of the energy levels of its components. However, in many cases it is difficult to evaluate the partition functions because of interactions between the particles. *See also* quantum statistics.

statistical weight *Syn.* degeneracy. Symbol: g. If a system has a number of possible quantized states and more than one distinct state has the same energy level, this energy level is said to be *degenerate* and its statistical weight is the number of states having that energy level. *See also* statistical mechanics.

stator The portion of an electrical machine that includes the nonrotating magnetic parts and the windings associated with them. The term is normally used only in connection with a.c. machines. *Compare* rotor.

steady state The state reached in a system under steady conditions. The steady state is obtained after all *transients, produced by one or more recent changes in existing conditions, have died away.

steam calorimeter A calorimeter in which the amount of heat supplied is calculated from the mass of steam condensed on the body under test. *See* Joly's steam calorimeter.

steam engine A machine that takes heat from a steam boiler, performs external work, and rejects a smaller amount of heat to a condenser. *See* Rankine cycle.

steam line The curve showing the variation of the boiling point of water with pressure, i.e. the variation of the saturation pressure of water vapour with pressure.

steam point The temperature of equilibrium between the liquid and vapour phases of water at standard pressure. Its former importance was as the upper fixed point on the Celsius scale of temperature. Thermodynamic temperature is now based on the triple point of water, and its value (273.16 K) has been chosen to make the steam point equal to 100 °C within the limits of experimental measurement. *Compare* ice point.

steam turbine *See* turbine.

steel *See* alloy.

steerable aerial A *directive aerial whose direction of maximum radiation or sensitivity can be altered. This can be achieved mechanically, as when a dish aerial is tilted and/or rotated, or electronically, as in phased-array *radar.

Stefan–Boltzmann constant (Formerly Stefan's constant) *See* Stefan–Boltzmann law.

Stefan–Boltzmann law A formula relating the radiant flux per unit area emitted by a *black body (*radiant exitance) to the temperature. It has the form $M_e = \sigma T^4$, where M_e is the radiant exitance and σ is the *Stefan–Boltzmann constant*:

$$\sigma = 2\pi^5 k^4 / 15 h^3 c^2,$$

where k is the Boltzmann constant, c the speed of light in vacuum, and h is the *Planck constant. σ has the value 5.670 51 $\times$ 10^{-8} W m^{-2} K^{-4}.

The law was originally deduced by Stefan from the results of experiments by Dulong and Petit. Boltzmann later gave a proof based on thermodynamics.

Stefan's constant Former name for Stefan–Boltzmann constant.

Stefan's law Former name for Stefan–Boltzmann law.

stellar energy Stars emit radiation of great intensity for enormous periods of time, for example the sun radiates with a power about 3.6 $\times$ 10^{26} W, equivalent to 4 $\times$ 10^9 kg s^{-1}, and is believed to have emitted for at least 5 $\times$ 10^9 years. The major part of stellar energy is generated by *nuclear-fusion processes, but gravitational contraction is needed to generate the high temperatures (of order 10^7 K) at which fusion can occur, and for certain later stages of evolution.

The primary process of thermonuclear fusion in the sun and all main-sequence stars cooler than the sun is the *proton-proton chain reaction. The *carbon cycle predominates in hot stars. In both processes hydrogen is converted to helium with considerable energy release. At higher temperatures other fusion reactions can occur. For example, at about 10^8 K helium nuclei will fuse to form carbon-12 nuclei.

stellar evolution *See* Hertzsprung–Russell diagram.

stellar spectra Stars emit radiation over a wide range of wavelengths, the maximum amount of energy being emitted at a particular wavelength. This wavelength will occur at the red end of the visible spectrum if the energy emitted is not very high (a cool star). An energetic star (a hot star) emits at the blue end of the spectrum.

Various groups of emission and absorption lines appear in a star's spectrum depending on temperature. Stars are usually classified according to their spectra and can be grouped into *spectral types*; the colours and typical temperatures are as follows:

O	hottest blue, 40 kK
B	hot blue, 20 kK
A	blue, blue-white, 9 kK
F	white, 7 kK
G	yellow, 6 kK
K	orange-red, 4.5 kK
M	coolest red, 3 kK

Ionized-helium lines, neutral-helium lines, and hydrogen and ionized-metal lines are dominant in O, B, and A stars respectively. In F and G stars metallic lines strengthen. In K and M stars metallic lines are strong and molecular bands are present. Stellar spectra therefore indicate not only temperature but chemical composition.

St. Elmo's fire The brush discharge from the pointed parts of ships or aircraft when in a strong atmospheric electrical field.

STEM Abbreviation for scanning-transmission *electron microscope.

step-down, step-up transformer *See* transformer.

stepped-index device *See* fibre-optics system.

stepped leader stroke Of lightning. *See* leader stroke.

step wedge A block or sheet of material having a series of layers successively more opaque to a given radiation in steps of definite value. Step wedges are employed, for example, in *photometry and X-ray studies.

steradian Symbol: sr. A unit of *solid angle. One steradian is subtended at the centre of a sphere of radius *r* by a portion of its surface of area r^2. 1 sphere = 4π sr. The steradian is a dimensionless derived *SI unit.

stereographic projection The projection of any points or figures on the surface of a sphere, centre C, from a pole P on the surface of the sphere onto a plane through C normal to PC.

stereophonic reproduction Reproduction of sound so as to give an illusion of location and direction from which a sound has originated. A single-channel sound system gives no illusion of sound location, all sounds appearing to come from a single source – the loudspeaker system. Multiple loudspeakers or a diffusion of the sound by broken reflectors give only a small improvement on the single sound source. Two or more channels of communication are essential for an illusion of sounds located in space.

stereoscope An instrument by which an impression of depth can be obtained from photographs or similar two-dimensional images by presenting to the eyes two images of a scene taken from two slightly different viewpoints, thus imitating binocular vision.

stereoscopic microscope *See* microscope.

Stern–Gerlach experiment An experiment, first performed by Stern and Gerlach (1921), that demonstrates the existence of the magnetic moment of an electron, particularly that due to its spin. When a sharply bounded stream of atoms is shot through a nonuniform magnetic field, the stream is split up into distinct components, dependent on the magnetic properties of the atoms. The atoms take up definite orientations relative to the field, and, in consequence of its nonuniformity, are deflected by different amounts. The experiment provides proof that there exist only certain permitted orientations; otherwise instead of splitting up, there would merely be a broadening of the beam since the atoms would be randomly orientated. This in turn provides proof of the quantization of the angular momentum of the electron. *See* spin.

stiffness The restoring force per unit displacement in a vibrating system. It is the product of angular frequency and mechanical (or acoustic) stiffness *reactance.

stilb Symbol: sb. A unit of *luminance equal to 1 *candela per square centimetre.

stimulated emission *See* laser.

STM Abbreviation for *scanning tunnelling microscope.

stochastic process A process resulting from the random behaviour of its generators.

stokes Symbol: St. The *CGS unit of *kinematic viscosity. 1 St = 10^{-4} metre squared per second.

Stokes's law 1. Of fluid resistance. The drag *D* of a sphere of radius *r* moving with a velocity *V* through a fluid of infinite extent is $D = 6\pi\eta rV$, where η is the *viscosity. The law holds only for a restricted range of conditions.

2. Of fluorescent light. The wavelength of light emitted during fluorescence is longer than that of the absorbed light. (*See* fluorescence.) Einstein cited Stokes's law as his first example to illustrate the argument that electromagnetic radiation generally interacts with matter in quanta.

stop A perforated screen or diaphragm that limits the width of pencils of light traversing a system (*aperture stop*) or limits the field of view (*field stop*). Sometimes stops are used to reduce the blur of *spherical aberration, at other times, to reduce the illumination of the image, to cut off the indistinct portions of a field, to prevent reflections from the inside of the tube, etc. *See* apertures and stops in optical systems.

stop number *See f*-number.

stopping power A measure of the effect of a substance upon the kinetic energy E, of a charged particle passing through it.

1. The *linear stopping power*, S_i, is the energy loss per unit distance: $S_i = -dE/dx$, expressed in MeV/cm or keV/m.

2. The *mass stopping power*, S_m, for a substance of density, ρ, is the energy loss per unit surface density:

$$S_m = S_i/\rho = S_i L/nA,$$

where L is the Avogadro constant, A the relative atomic mass (atomic weight), and n is the number of atoms per unit volume.

3. The *atomic stopping power*, S_a, is the energy loss per atom per unit area of the substance normal to the motion of the charged particle:

$$S_a = S_i/n = S_m A/L.$$

Stopping power is often expressed relative to that of a standard substance, such as air or aluminium.

storage capacity *See* capacity.

storage device A device in a *computer system that can receive data and retain it for subsequent use. Storage devices are either *backing store or semiconductor devices used as *main memory, and vary widely in the amount of data that can be stored and the speed of access to a particular item.

storage oscilloscope An oscilloscope that can capture a signal, especially a fast nonrepetitive signal, and continue to display it until reset. A digital storage oscilloscope samples the incoming signal, stores these samples, and displays them. Other storage oscilloscopes use a special cathode-ray tube that retains the image by mapping it as a charge pattern on a target electrode behind the screen. The information is extracted by flooding the target with electrons, which pass through the target to the screen. The deposited charge modulates the numbers of electrons reaching the screen and the light image produced is proportional to the captured signal.

storage ring *See* accelerator; intersecting storage ring.

storage time 1. In general, the time for which information may be stored in any device, without significant loss of information.

2. Of a *p-n junction. The time interval observed between application of *reverse bias to the junction and cessation of a reverse current surge. The latter results from *carrier storage*: under conditions of *forward bias, excess minority carriers are injected across and remain near the junction. The reverse current surge continues until these stored charges are removed, either by recombination or by recrossing the junction under the influence of the reverse bias.

STP (or s.t.p.) Abbreviation for standard temperature and pressure.

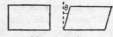

Shear strain

strain The change of volume and/or shape of a body, or part of a body, due to applied stresses. The three simplest strains are as follows. (1) *Linear* (or *longitudinal*) *strain*: the change in length per unit length (e.g. when stretching a wire). (2) *Volume* (or *bulk*) *strain*: the change in volume per unit volume (e.g. when a hydrostatic pressure is applied to a body). (3) *Shear strain* (or *shear*): angular deformation without change in volume (e.g. a rectangular block strained so that two opposite faces become parallelograms, the others not changing shape). The radian measure of the change in angle, θ, at one corner is a measure of the strain (*see* diagram). θ is small in practice and is equal to the tangential displacement of two planes unit distance apart. *See* homogeneous strain; stress.

strain gauge An instrument for measuring strain at the surface of a solid body in terms of, for example, the change in electrical resistance, capacitance, or inductance or the piezoelectric or magnetostriction effects produced by the strain.

strange matter Matter composed of a mixture of up, down, and strange *quarks (rather than the up and down quarks of protons and neutrons). It has been postulated that stable nuggets of strange matter might have been formed in the extreme conditions of the big bang. Some astrophysicists have also

suggested that it could be formed in supernova explosions. There is also a possibility that drops of strange matter (known as *S-drops*) could be produced in particle accelerators by bombarding a target with a beam of heavy nuclei. So far, no evidence for its existence has been found.

strangeness Symbol: S. A *quantum number associated with *elementary particles (specifically *hadrons) that is conserved in *strong and *electromagnetic interactions but not in *weak interactions. S changes by ± 1 in weak interactions. Its existence was postulated in order to explain the fact that some elementary particles (e.g. kaons, Σ, and Λ) that were expected to decay very rapidly by strong interaction (since they could do so without violating any of the known conservation laws), had much longer lifetimes than expected. Strangeness is associated with the presence of one or more strange *quarks in the particle. The strange quark(s) has strangeness -1 and its antiquark $\bar{s}$ has strangeness $+1$. All other quark flavours have strangeness 0. The strangeness of a particle is the sum of the number of $\bar{s}$ quarks minus the number of s quarks.

stratosphere *See* atmospheric layers.

stray capacitance Any capacitance in a circuit or device due to interconnections, electrodes, or the proximity of elements in the circuit in addition to the intentional capacitance provided.

stream function *Syn.* current-function. Symbol: ψ. In two-dimensional motion of a fluid the stream function at any point P is defined as the flow across a curve AP where A is a fixed point in the two-dimensional plane. If the plane is the Cartesian plane xy then the component velocities at any point (x, y) are:

$$V_x = -\partial\psi/\partial y \text{ and } V_y = \partial\psi/\partial x.$$

In axisymmetrical three-dimensional motion of a fluid – motion that is the same in every plane through a certain axis of symmetry – there is a similar function (ψ) whose value at a point P is defined as $1/2\pi$ of the flow out of a surface of revolution formed by rotating a curve AP about the axis, where A is any fixed point in the meridian plane of P. This function is called *Stokes's stream function*. In both cases the curves: ψ = constant, give the streamlines of the fluid motion.

streaming potential A difference of potential set up between the two sides of a porous material such as clay when water is forced through. It is the potential set up between the ends of a capillary tube when an electrolyte is forced through it. This may be regarded as the reverse of *electrosmosis.

streamline A line drawn in a fluid so that the tangent at any point is in the direction of the fluid velocity at that point. The aggregate of streamlines at any instant of time forms the flow pattern.

Shear stress

stress A system of forces in equilibrium producing *strain in a body or part of a body. The stresses may be regarded as the forces applied to deform the body or as the equal and opposite forces with which the body resists. In all cases a stress is measured as a force per unit area. The simplest stresses are: (1) *tensional* or *compressive stress* (i.e. *normal stress*), e.g. the force per unit area of cross section applied to each end of a rod to extend or compress it; (2) *hydrostatic pressure*, e.g. the force per unit area applied to a body by immersion in a fluid; (3) *shear stress*, e.g. the system of four tangential stresses applied to the surfaces of a rectangular block (each force being parallel to one edge) tending to produce shear (*see* strain). *See also* modulus of elasticity; stress components.

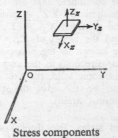

Stress components

stress components The internal forces per unit area arising between contiguous parts of a body due to applied surface and body forces. Consider an infinitesimal plane area within the body; the force exerted by the matter on one side to that on the other can be resolved into components normal and tangential to the area that are called the *normal stress component* and the *shear stress component* respectively, at the area. Except in the special case of hydrostatic pressure, the stress at a point depends on the orientation of the area used in defining it.

If the component stresses at three infinitesimal areas at the point, each being parallel to one of the planes defined by Cartesian axes outside the body, are known, it is possible to calculate the stress across an infinitesimal area at the point however orientated. The tangential component in the plane of the infinitesimal area perpendicular to OZ (see diagram) may be resolved into components X_z and Y_z (the subscript indicates the axis perpendicular to the plane across which the force acts). The system of 9 stress components (3 normal and 6 shear) finally reduces to 6, for by considering the equilibrium of a small rectangular solid at the point, it is found that $X_y = Y_x$, $Z_x = X_z$, $Y_z = Z_y$. These six, X_x, Y_y, Z_z and X_y, Y_z, Z_x are the stress components at the point (x, y, z).

If the coordinate axes are rotated to new positions $OX'Y'Z'$, a position can be found at which $X'_{y'} = Y'_{z'} = Z'_{x'} = 0$ and the values of $X'_{x'}$, $Y'_{y'}$, and $Z'_{z'}$ are known as the *principal stresses* and their directions as the *axes of stress*. In an isotropic solid, the axes of stress will also be the axes of strain (*see* homogeneous strain).

stretched string The theory of *transverse vibrations in a stretched string assumes it to be uniform, perfectly flexible, and of practically constant length while in vibration. These conditions are nearly fulfilled when a long thin wire is stretched between two rigid supports. *Standing waves set up in a stretched string can be considered to be due to the superposition of two transverse progressive waves travelling in opposite directions with velocity $v = \sqrt{(T/m)}$, where T is the tension and m the mass per unit length. Ideally, in the fundamental mode there is a single loop with an antinode at the centre and nodes at the ends. In this case the wavelength is twice the length of the string l, and the fundamental frequency is

$$f = \sqrt{(T/m)}(2l)^{-1}.$$

The various *partials are produced when the string vibrates in several loops. All the partials are *harmonic, their frequencies being obtained by multiplying that of the fundamental by the number of loops on the string. A string may vibrate with several partials at the same time, their number and magnitude depending on the method of excitation. In practice the ends are not perfect displacement nodes, thus the quality of a musical note is affected.

A transversely vibrating string is used as a source of sound in many musical instruments when it is attached to a soundboard that increases its ability to radiate sound energy.

string galvanometer *See* Einthoven galvanometer.

string theory A theory of *elementary particles based on the idea that the fundamental entities are not point-like particles, but finite lines (*strings*) or closed loops formed by strings. The original idea was that an elementary particle was the result of a standing wave in a string.

A considerable amount of theoretical effort has been put into developing string theories. In particular, combining the idea of strings with that of *supersymmetry leads to the idea of *superstrings*. This theory may be a more useful route to a unified theory of fundamental interactions than *quantum field theory because it probably avoids the infinities that arise when gravitational interactions are introduced into field theories. Thus, superstring theory inevitably leads to particles of spin 2, identified as *gravitons. String theory also shows why particles violate *parity conservation in weak interactions.

Superstring theories involve the idea of higher dimensional spaces: 10 dimensions for fermions and 26 dimensions for bosons. It has been suggested that there are the normal 4 space–time dimensions, with the extra dimensions being tightly 'curled' up.

There is no direct experimental evidence for superstrings. They are thought to have a length of about 10^{-35} m and energies of 10^{19} GeV, which is well above the energy of any accelerator. An extension of the theory postulates that the fundamental entities are not one-dimensional but two-dimensional, i.e. they are *supermembranes*.

stripping A *nuclear reaction in which a *nucleon of the bombarding nucleus is captured by the struck nucleus without the nuclei merging to form a compound nucleus. The *Oppenheimer–Phillips (O–P) process* is a reaction in which a low-energy deuteron gives its neutron to a nucleus without entering it.

stroboscope An instrument producing an intense flashing light whose frequency can be adjusted to synchronize with some multiple of the frequency of rotation or vibration of a moving object or part, or some other periodic phenomenon, making it appear stationary. It is used to study the motion and also to determine rotation or vibration speeds.

strong interactions Interactions between *elementary particles involving the strong interaction force. This force is about one hundred times greater than the *electromagnetic force between charged elementary particles. However, it is a short-range force – it is only important for particles separated by a distance of less than about 10^{-15} m – and is the force that holds protons and neutrons together in atomic nuclei. For 'soft' interactions between *hadrons, where relatively small transfers of momentum are involved, the strong interactions may be described in terms of the exchange of virtual hadrons (*see* virtual particle), just as electromagnetic interactions between charged particles may be described in terms of the exchange of virtual photons. At a more fundamental level, the strong interaction arises as the result of the exchange of gluons between quarks and/or antiquarks as described by *quantum chromodynamics (QCD).

In the hadron exchange picture, any hadron can act as the exchanged particle provided certain *quantum numbers are conserved. These quantum numbers are the total angular momentum, charge, *baryon number, *isospin (both I and I_3), *strangeness, *parity, *charge conjugation parity, and *G-parity. Strong interactions are investigated experimentally by observing how beams of high-energy hadrons are scattered when they collide with other hadrons. Two hadrons colliding at high energy will only remain near to each other for a very short time. However, during the collision they may come sufficiently close to each other for a strong interaction to occur by the exchange of a virtual particle. As a result of this interaction, the two colliding particles will be deflected (scattered) from their original paths. If the virtual hadron exchanged during the interaction carries some quantum numbers from one particle to the other, the particles found after the collision may differ from those before it. Sometimes the number of particles is increased in a collision. An example of a strong interaction process is $\pi^-p \to \rho^\circ n$. Here the colliding π^- and proton become a ρ° and neutron after the collision. This process is thought to take place by the proton emitting a virtual π^+, which combines with the π^- to give a ρ° as illustrated in Fig. *a* below. This diagram is often simplified to the form shown in Fig. *b*. Where all the necessary quantum numbers can be conserved, elementary particles can decay by a strong interaction. An example of this is the decay of the ρ-meson into two pions. Here the ρ-meson can be thought of as breaking up into two virtual pions that are bound by the strong interaction force. However, since this can occur without violating the law of *conservation of mass and energy (the mass of two pions is less than the mass of a ρ) these pions can

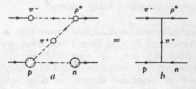

become physical particles and separate from each other.

In high energy hadron–hadron interactions, the number of hadrons produced increases approximately logarithmically with the total centre of mass energy, reaching about 50 particles for proton–antiproton collisions at 900 GeV, for example. In some of these collisions, two oppositely directed collimated *jets* of hadrons are produced, which are interpreted as being due to an underlying interaction involving the exchange of an energetic gluon between, for example, a quark from the proton and an antiquark from the antiproton. The scattered quark and antiquark cannot exist as free particles but instead *fragment* into a large number of hadrons (mostly pions and kaons) travelling approximately along the original quark or antiquark direction. This results in collimated jets of hadrons which can be detected experimentally. Studies of this and other similar processes are in good agreement with QCD predictions. *See* bootstrap theory; exchange force; Regge pole model; resonances.

SU₃ *See* unitary symmetry.

subcarrier A *carrier wave used to modulate another carrier wave.

subcritical *See* chain reaction.

subharmonic vibration A vibration of a frequency that is a whole-number submultiple of the fundamental frequency.

sublimation The direct transition from solid to vapour, or conversely, without any liquid phase being involved.

submillimetre waves Radio waves with a wavelength ranging from 1 mm down to about 0.1 mm, i.e. with a frequency between 300 and about 3000 GHz. They are of particular interest in radio astronomy because of the large number of molecular emission lines to be found in this wavelength range.

subshell *See* atomic orbital; electron shell.

subsonic Denoting an object, airflow, etc., moving at less than the speed of sound, i.e. at less than Mach 1.

substandard A standard measuring device not quite as accurate as the primary standard, used as the intermediate link between the primary standard and the device being calibrated or checked.

substation A complete assemblage of plant, equipment, and the necessary buildings at a place where electrical power is received (from one or more *power stations) for conversion (e.g. from alternating current to direct current), for stepping-up or down by means of transformers, used for control purposes.

substrate 1. A single body of material on or in which one or more electronic circuit elements or integrated circuits are fabricated. It may be a semiconductor crystal, which plays an active role in the device, or it may act as a support, as with the insulating layer in a printed-circuit board.

2. Any surface or layer used as a basis for some process.

subtractive process A process by which colours can be produced or reproduced by mixing absorbing media (or *filters) of three different dyes or pigments, called *subtractive primaries*. The colour of light reflected by (or passing through) the mixture is determined by the *absorption, or subtraction, of specific colours by each medium. The three dyes or pigments are usually yellow, magenta, and cyan (greenish-blue), an approxmately equal mixture of which will appear black. *Compare* additive process.

summation tone *See* combination tones.

sunspots Dark patches seen on the sun's visible surface (its photosphere), usually in groups and with a lifetime of several weeks. All but the smallest have a dark inner region (the umbra) surrounded by a less dark edge (the penumbra). Sunspots are regions of relatively cool gas and their presence is connected with local variations in the sun's magnetic field. The number of sunspots fluctuates over a period of about 11 years.

superconductivity A phenomenon occurring in many metals including tin, aluminium, zinc, mercury, and cadmium, many alloys, and intermetallic compounds. If these substances are cooled below a *transition temperature*, T_c, the electrical resistance vanishes. (There is also a marked difference in the variation with temperature of the *specific heat capacity below T_c.) For pure metals the transition temperature is usually a few kelvin but for some compounds much higher values are obtained. Current research is aimed at developing superconductors with T_c near to, or even above, room temperatures.

Since a compound of two nonsuperconducting metals can be superconducting, the phenomenon is not a property of the atom but of the free electrons in the metal. In the superconducting state the electrons do not move independently. There is a dynamic pairing of electrons (a *Cooper pair*) such that if the quantum state with *wavenumber σ and *spin $\frac{1}{2}$ is occupied by an electron, then so is the state with wavenumber $-\sigma$ and spin $-\frac{1}{2}$. These pairs are superimposed in phase. The two electrons interact through lattice vibrations, and the formation of these bound pairs is not prevented by the presence of other elec-

trons. Cooper pairs are the basis of the *BCS theory* (1957, named after Bardeen, Cooper, and Schrieffer). This accounts for many of the properties of conventional superconductors but is less successful for the recently discovered high-temperature superconductors; these rely on *heavy-fermion systems in which the transition temperature can be as high as 100 K. The practical advantage of these *high-temperature superconductors* is that they can operate at liquid-nitrogen temperatures rather than the liquid-helium temperatures required by BCS superconductors. An example of such a superconductor is $YBa_2Cu_3O_{1-7}$. The theory of these superconductors has not been established yet but various models have been proposed and tested.

The magnetic behaviour of superconductors is extremely complicated. When a superconductor, in a weak magnetic field, is cooled below its transition temperature, the magnetic flux inside the substance is expelled except for a thin surface layer. This is the *Meissner effect*. A bar magnet dropped onto such a conductor will be repelled and will hover above it, exhibiting *levitation*. The Meissner effect implies that a superconductor exhibits perfect *diamagnetism. This in turn implies the existence of a large energy gap between ground state and first excited state so that all the superconducting electrons are in a particular ground state; this is possible for Cooper pairs. Superconductivity can be destroyed by a magnetic field – either an external field or one produced by a current flowing in the metal. This is used in the *cryotron. A current induced in a closed ring of superconducting material by a magnetic field will continue to flow after the removal of the field for a considerable time, without diminution in strength, if the temperature is kept below T_c. This effect has been used in *superconducting magnets* in which very large magnetic field strengths can be produced without the expenditure of large amounts of electrical power or the production of heat. The superconducting electrons form an *energy band below that of the normal conduction band and do not take part in heat conduction. Hence the thermal conductivity of metals is usually less in the superconducting state. At very low temperatures, however, it may rise because of increased *phonon conductivity.

See also Josephson effect.

supercooling The process by which liquids, by slow and continuous cooling, are reduced to a temperature below the normal freezing point. A supercooled liquid is a *metastable state and the introduction of the smallest quantity of the solid at once starts solidification. Small mechanical disturbance may also initiate solidification, which, once started, will continue with the evolution of heat until the normal freezing point is reached; subsequently further solidification will take place only as heat is lost from the liquid. *Compare* superheating.

Clouds very commonly consist of supercooled water droplets. These may freeze on contact with solids. If ice particles form on suitable nuclei in a cloud of supercooled droplets, they grow at the expense of the latter to form *snow flakes because of the difference in saturation vapour pressure over solid and supercooled liquid surfaces (*see* diagram for *triple point).

supercritical *See* chain reaction.

superfluid A fluid that flows without any resistance. The electrons in *superconductivity constitute a superfluid. Another example is the form of liquid helium-4 below a certain transition temperature dependent upon pressure. This has its highest value, 2.186 K, at the equilib-

rium vapour pressure, and is called the *lambda point* because of the shape of the peak in the graph of specific heat capacity against temperature. The phenomenon is associated with exceptionally high thermal conductivity, which rises to 10^6 times the value above the lambda point.

Superfluidity is attributed to *Bose-Einstein condensation*, when a large number of helium atoms (which are *bosons) are in the translational state of lowest energy. These atoms constitute a superfluid that is homogeneously mixed with normal fluid consisting of atoms with higher translational kinetic energies.

supergiant *See* Hertzsprung–Russell diagram.

supergravity An unproved *unified field theory involving *supersymmetry that encompasses all four fundamental *interactions. By means of introducing supersymmetry the number of infinities in the calculations is fewer than in other quantum-theory based unified theories that include the gravitational interaction, but the theory still contains infinities that cannot be removed. Some theorists believe that a unified theory to include gravity is unlikely to be a *quantum field theory and seek instead a theory based on superstrings (*see* string theory).

superheating The heating of a liquid above its normal boiling point without boiling occurring. *See also* supercooling.

superheterodyne receiver The most widely used type of *radio receiver, in which the incoming signal is fed into a *mixer and mixed with a locally generated signal from a *local oscillator*. The output consists of a signal of *carrier frequency equal to the difference between the locally generated signal and the carrier frequencies, but containing all the original modulation. This signal (the *intermediate frequency* (or i.f.) signal) is amplified and detected in an i.f. amplifier, and passed to the audiofrequency amplifier. The high-gain amplification and great selectivity of the superhet receiver are directly attributable to the use of the intermediate frequency.

superhigh frequency (SHF) *See* frequency bands.

superlattice 1. *See* solid solution.

2. In electronics, a semiconducting crystal in which two semiconductors with different electronic properties are interleaved in ultrathin layers. Such crystals are produced by depositing the two materials in alternating layers or by introducing impurities into layers of a single semiconductor; viable devices have only recently been produced. The two semiconductor materials are selected so that the energy difference (or band gap) between their valence and conduction bands is different (*see* energy bands). An example is a superlattice consisting of gallium arsenide and gallium aluminium arsenide, the latter having the larger band gap. The great advantage of a superlattice device is that its electronic and optical properties can be tailored to a particular function by an appropriate choice of the semiconductors and the width and number of the layers.

supermembrane *See* string theory.

supernova A star that explodes as a result of instabilities following the exhaustion of its nuclear fuel. The explosion involves an enormous energy release: a supernova can become 10^9 times as bright as the sun (reaching an absolute *magnitude of -17). All or most of the star's matter is ejected at relativistic speeds, forming an expanding shell of debris called a *supernova remnant*. A

*pulsar can be formed at the core of the supernova.

superposition principle A principle that holds generally in physics wherever linear phenomena occur. In elasticity, the principle states that each stress is accompanied by the same strains whether it acts alone or in conjunction with others; it is true so long as the total stress does not exceed the limit of proportionality. In vibrations and wave motion the principle asserts that one set of vibrations or waves is unaffected by the presence of another set. For example, two sets of ripples on water will pass through one another without mutual interaction. so that, at a particular instant, the resultant disturbance at any point traversed by both sets of waves is the sum of the two component disturbances.

The superposition of two vibrations, y_1 and y_2, both of frequency f, produces a resultant vibration of the same frequency, its amplitude and phase being functions of the component amplitudes and phases. Thus if

$$y_1 = a_1 \sin(2\pi ft + \delta_1)$$
$$y_2 = a_2 \sin(2\pi ft + \delta_2),$$

then the resultant vibration, y, is given by:

$$y_1 + y_2 = A \sin(2\pi ft + \Delta),$$

where amplitude A and phase Δ are both functions of a_1, a_2, δ_1, and δ_2.

super-regenerative reception A method of reception of ultrahigh frequencies by means of a radio receiver employing an oscillating *detector, the oscillations of which are periodically stopped (or quenched) at a frequency dependent on the input frequency. There is very great amplification but the selectivity is rather poor compared to a *superheterodyne receiver.

supersaturated vapour A vapour, the pressure of which exceeds the saturation vapour pressure at that temperature. It is unstable and condensation occurs in the presence of suitable nuclei or surfaces.

supersonic flow The movement of a fluid at a speed exceeding the speed of sound in the fluid. In such a case, changes of density in the flow can no longer be neglected. As the speed of an object moving through a fluid is increased through the speed of sound and beyond, the resistance (drag) rises due to the formation of *shock waves. *See also* sonic boom.

superstring *See* string theory.

supersymmetry A *symmetry including both *bosons and *fermions. In theories based on supersymmetry every boson has a corresponding fermion and every fermion has a corresponding boson. The boson partners of existing fermions have names formed by prefacing the name of the fermion with an "s" (e.g. selectron, squark, slepton). The names of the fermion partners of existing bosons are obtained by changing the terminal -on of the boson to -ino (e.g. photino, gluino, wino, zino).

Although supersymmetries have not been observed experimentally, they may prove important in the search for a *unified theory of the fundamental *interactions.

surface acoustic wave devices (SAW devices) Miniature devices that are used for signal processing and employ an analogue rather than a digital representation of information. In such a device, an ultrasonic *acoustic wave propagates along the plane surface of a solid. The frequency ranges from a few megahertz to a few gigahertz so high transmission rates are possible. Electric signals can be converted to surface acoustic waves (and vice versa) by means of transducers

based on the *piezoelectric effect. Piezoelectric crystals are therefore used as the substrate along which the waves travel. SAW devices can be fabricated to perform a variety of functions; they are particularly important as filters (*SAW filters*).

surface-barrier transistor A *transistor in which the usual p-n junctions are replaced by metal-semiconductor contacts called Schottky barriers (*see* Schottky diode). Carrier storage under saturation conditions is zero with the Schottky barriers (*see* storage time) and the transistors are useful for high-frequency switching applications.

surface colour Coloured light reflected by a surface as distinct from the more common body colour that arises from reflection after some penetration into the medium. Transmitted light by bodies showing surface colour is complementary to the reflected colour.

surface density The quantity per unit area of anything distributed on a surface.

surface energy The energy per unit area of exposed surface. The (total) surface energy in general exceeds the *surface tension, which is the *free* surface energy, concerned in isothermal changes.

surface resistivity The resistance between two opposite sides of a unit square of the surface of a material. Its reciprocal is the *surface conductivity*.

surface tension Symbol: γ. Intermolecular forces are repulsive at small separations, decrease to zero at about 10^{-10} m, then become attractive at slightly greater distances, rise to a maximum and then fall off rapidly to zero. Liquids are usually under pressure so any molecule in the interior must be, on average, repelled by the molecules on each side. The separation between near neighbours is therefore in the region of repulsion. In the plane of the surface however the separation of near neighbours is greater, i.e. in the region of attractive forces increasing with distance. Consequently, the liquid surface is in tension and contracts as far as is possible; thus a small free drop is spherical. The increased separation in the surface results from the fact that surface molecules have fewer near neighbours, and so have less negative potential energy, than those in the interior. The work done in creating unit area of surface against this tension at constant temperature is called the *free surface energy*. If the liquid surface is assumed to be in tension in all directions, the force required to hold the straight edge of a plane liquid surface is the surface tension, γ (usually expressed in newtons per metre). If the surface is stretched isothermally so that the edge moves unit distance so creating unit area of new surface, the work done is γ (in joules). Thus the surface tension expressed in $N\,m^{-1}$ is numerically and dimensionally equal to the free surface energy in $J\,m^{-2}$ but is not the total surface energy. Small additions of foreign substances often profoundly affect the surface tension.

Due to the surface tension there is a pressure difference between the two sides of a liquid surface equal to $\gamma(1/R_1 + 1/R_2)$, where R_1 and R_2 are the radii of curvature of two perpendicular normal sections. Thus the pressure inside a soap bubble (which has *two* surfaces of radius R) is $4\gamma/R$.

surface wave 1. *See* ripples; water waves.
   2. *See* ground wave.

surge An abnormal transient electrical disturbance in a conductor. Surges result, for example, from lightning, sudden

faults in electrical equipment or transmission lines, or switching operations.

susceptance Symbol: B. The imaginary part of the *admittance, Y, i.e.
$$Y = G + iB,$$
where G is the *conductance. It is the reciprocal of the reactance and is measured in *siemens.

susceptibility 1. (magnetic) Symbol: χ_m. The quantity $\mu_r - 1$, where μ_r is the relative *permeability.
   2. (electric) Symbol: χ_e. The quantity $\varepsilon_r - 1$, where ε_r is the relative *permittivity.

Sutherland's formula One of several formulae proposed to show the variation of viscosity η of a gas with thermodynamic temperature T:
$$\eta = \eta_0(T/273)^{3/2}(273 + k)/(T + k),$$
where k is constant for a given gas, and η_0 is the viscosity at 0 °C.

SVP Abbreviation for *saturated vapour pressure.

sweep *See* time base.

switch A device for opening or closing a circuit, or for changing its operating conditions between specified levels. It is also used to select from two or more components, circuits, etc., the desired element for a particular mode of operation. Switches may consist of a mechanical device, such as a *circuit breaker, or a solid-state device such as a *transistor, *Schottky diode, or *field-effect transistor.

symmetry The set of invariances of a system. A symmetry operation on a system is an operation that does not change the system. It is studied mathematically using *group theory. Some symmetries are directly physical, for example reflections and rotations for molecules and translations in crystal lattices. More abstract symmetries involve changing properties, as in the *CPT theorem and the symmetries associated with *gauge theories.

synchrocyclotron *Syn.* frequency-modulated cyclotron. A modification of the *cyclotron in which the magnetic field remains constant but the frequency of the accelerating electric field is slowly decreased. In the cyclotron, as the velocity becomes relativistic an increase in mass occurs and as a result the particle gets out of phase with the alternating electric field. To counteract this, the alternating frequency in the synchrocyclotron is slowly decreased so that the particles remain in phase with the field. Energies up to 700 MeV for protons can be obtained. *Compare* synchrotron.

synchronous alternating-current generator *Syn.* alternator; synchronous generator. An electrical machine for generating alternating current. It has a number of *field magnets, which are usually excited by means of *field windings carrying direct current obtained from an independent source. The frequency of the generated e.m.f.s and currents is determined by the number of magnetic poles in the machine and the speed at which it is driven (*see* synchronous speed). This type of generator can operate and deliver its output independently of any other source of alternating current. Power stations usually employ generators of this type.

synchronous clock A *clock in which a *synchronous motor drives the mechanism that advances the hands. The timekeeping is determined entirely by the frequency of the a.c. electricity supply to which the motor is connected.

synchronous induction motor Basically, an *induction motor having a slip-ring

rotor and a direct-coupled d.c. exciter. It is started as a normal induction motor with consequent high starting torque. When the motor is running with a small *slip, direct current from the exciter is injected into the rotor circuit. The motor then runs at *synchronous speed as a normal *synchronous motor, with consequent constant speed and high or leading *power factor.

synchronous motor An a.c. electric motor, the mean running speed of which is independent of the load and is determined by the number of its magnetic poles and the frequency of the electric supply. (*See* synchronous speed.) A typical industrial motor of this type consists of a stator carrying a winding that, when connected to the a.c. supply, produces a magnetic field that rotates in space, and a rotor excited by direct current. The rotor is, fundamentally, an electromagnet that locks with the field produced by the stator and rotates at the same speed as the field. By overexciting the rotor, the motor can be made to operate at a leading *power factor. A plain synchronous motor is, fundamentally, not self-starting.

synchronous orbit *See* geostationary orbit.

synchronous speed The speed of rotation of the magnetic flux in an a.c. machine. It is given by the ratio f/p r.p.s., where f = frequency in hertz of the a.c. supply, and p = number of pairs of magnetic poles for which the a.c. winding has been designed.

synchrotron *Syn.* electron synchrotron. A cyclic *accelerator that is based on the *betatron but uses a constant-frequency electric field in addition to a changing magnetic field. As in the betatron, the increasing magnetic flux density B counteracts the relativistic increase in

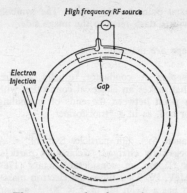

High frequency RF source

Electron Injection

Gap

The vacuum tube of a synchrotron

mass at high velocities. A high-frequency electric field from a radio-frequency oscillator can be applied across a gap in a metallic cavity inside the circular chamber (*see* diagram). The frequency is in synchronism with the constant angular frequency of the electrons, which are thus accelerated inside the cavity. There are usually several RF cavities interspersed between the magnets that bend the particle beam and that focus it.

Initially the machine acts as a betatron until the electrons have an energy of several MeV. The high-frequency field is then applied while the magnetic field is increasing. At the required energy the magnetic flux condition for a stable orbit is destroyed and the electrons are deflected from the path. Very high energies in the GeV range have been achieved. *See also* proton synchrotron.

synchrotron radiation *Syn.* magneto-bremsstrahlung. Electromagnetic radiation that is emitted by high-energy charged particles moving at relativistic speeds in a strong magnetic field. The radiation is emitted as the particles follow a circular path through the field. Such emission occurs, for example, in *synchrotrons and *storage rings. There is a smooth distribution of wavelengths,

ranging from microwaves to hard X-rays. The exact profile of the spectrum depends on the radius of the particle orbit and the particle energy. It is not *thermal radiation but is strongly polarized.

Synchrotron radiation can be a significant problem, in terms of energy loss from the charged particles. The radiation can, however, be used as a tool. Synchrotron sources are being built to produce intense beams of radiation, primarily of X-ray and UV frequencies.

Because many regions of the universe are associated with very high magnetic fields, the radiation emitted from electrons moving in these regions is called synchrotron radiation. It is the mechanism thought most likely to explain radio emission from extragalactic *radio sources and the emission from *supernova remnants.

synoptic chart A map showing wind, barometric pressure, etc., at a particular time. It is used in weather forecasting.

systematic error *See* errors of measurement.

Système International d'Unités *See* SI units.

systems software *See* software.

T

tachometer An instrument for measuring angular speeds. There are many types available.

tachyon A particle postulated to move with velocity greater than that of electromagnetic radiation. Of the two properties rest mass and energy, one must be real and the other imaginary. If it exists, it may be detected through the emission of *Cerenkov radiation.

Talbot's law The apparent intensity, I, of a light source, flashing at a frequency greater than 10 hertz, is given by:

$$I = I_0(t/t_0),$$

where I_0 is the actual intensity, t is the duration of the flash, and t_0 is the total time. The light appears steady due to the persistence of vision.

tandem generator A modification of the *Van de Graaff generator in which a doubling of the energy of the particles is achieved for the same accelerating potential. Negative ions are accelerated from earth potential, the electrons are stripped off, and the resulting positive ions are accelerated back to earth potential. This is produced by connecting two generators in series.

tangent galvanometer A *galvanometer in which a short magnetic needle similar to that of a *magnetometer is suspended at the centre of a circular coil of wire of radius large compared with the length of the needle. The plane of the coil is placed along the magnetic meridian so that a current in the coil deflects the needle against the controlling couple of the earth's magnetic field. The tangent law of the magnetometer applies, i.e. the current through the coil is proportional to tan θ, where θ is the angle of deflection of the needle. The galvanometer can be used to compare currents or to determine the earth's field if the current is known.

tangential coma *See* coma.

tangent law The law

$$ny \tan \alpha = n'y' \tan \alpha',$$

where n is the refractive index of object space, y the size of object, α the angle of inclination of a paraxial ray from an axial point on the object. The symbols with a dash refer to the image side.

tape *See* magnetic tape.

tapping A conductor, usually a wire, that makes an electrical connection with a point between the ends of a winding or coil, as in a *transformer.

tauon *Syn.* tau particle. Symbol: τ. A negatively charged *elementary particle, a *lepton, of considerable mass (1784 MeV, i.e. 3560 times the electron mass). It has a mean life of 3×10^{-13} second and three principal decay modes. Its antiparticle is called the *positive tauon*. The tauon is assumed to have an associated neutrino, ν_τ.

telecommunications The study and practice of the transfer of information by any kind of electromagnetic system, e.g. wire, radio waves, etc. There are many types of telecommunication system including telephony, *television, *radio, and communication *satellites.

telemetry A means of making measurements, in which the measured quantity is distant from the recording apparatus and the data are transmitted over a particular *telecommunication system from the measuring position to the recording position. Particular examples of telemetry systems are space exploration and physiological monitoring in hospitals.

telephony A communications system designed to transmit speech and other information such as fax and electronic mail. A complete system consists of all the circuits, switching apparatus, and other equipment necessary to establish a *communications channel between any two users. The communication may take place along suitable electric cables or optical fibres, and/or by means of radio links (as in cellular radio) or satellite links.

The use of *digital signals rather than *analogue signals to transmit the information provides a faster more reliable telephone connection; *analogue digital converters are used to convert speech into digital form. Conversion to digital systems is under way worldwide. In the case of an analogue system, digital input and output (e.g. from a computer) is achieved using *modems.

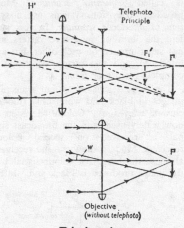

Telephoto Principle

Objective
(*without telephoto*)

Telephoto lens

telephoto lens A photographic lens that can produce a large image of a distant object with normal camera extensions. It has a large effective focal length, and hence a narrow angle of view and small depth of field. It consists of a converging lens followed by a diverging lens after the manner of the Galilean telescope (*see* refracting telescope), but with a separation such that convergence to a real image occurs. Appreciable magnifications can thereby be obtained with constant back focal length, since the second principal plane of the combination

can be placed at relatively great distances in front of the back lens.

telescope 1. An optical device for collecting light in order to form images of distant objects. Its light-gathering power allows fainter objects to be discerned than with the unaided eye. The images of distant extended objects, such as the moon, are magnified (*see* magnifying power). Point sources, such as stars, are easier to distinguish (*see* resolving power). The light is collected and focused by a primary mirror in a *reflecting telescope or an objective lens in a *refracting telescope, or by a combination of a lens and mirror as in the *Schmidt or *Maksutov telescopes. An optical telescope is usually described in terms of the *aperture of its main mirror or lens. As the aperture increases, the light-gathering power and resolving power both increase.

Telescopes are used principally in astronomy. The structure that rigidly supports a telescope is called the *mounting*. It is designed so that the telescope tube can be turned about two axes at right angles, allowing almost any part of the sky to be observed. In an *equatorial mounting*, one axis (the polar axis) is parallel to the earth's axis. The telescope is clamped at the required angle on its second axis and is rotated about its polar axis once in 24 hours in the opposite direction to the earth's rotation. Any astronomical object then remains stationary in the field of view. In the *altazimuth mounting*, one axis is vertical and the other horizontal. To follow an object's apparent daily motion across the sky, the telescope must be turned about both axes simultaneously and at different rates. This requires computer-controlled drive mechanisms.

A *coudé system* usually consists of a reflector or refractor having an equatorial mounting. The image is formed, after an additional reflection, at a point on the polar axis (the *coudé focus*). As this point remains fixed with respect to the earth, and thus to the observer, light can be analysed with a permanently installed spectrograph, etc.

In the *meridian circle*, the telescope is mounted on an east-west axis so that it turns in the meridian plane. It is used to give the altitude of a star at the moment it crosses the meridian and for deducing right ascension and declination (*see* celestial sphere).

2. Any astronomical device by which radiation from a particular region of the spectrum is collected and brought to a suitable recording system, usually electronic, for analysis. Ground-based *radio telescopes and infrared telescopes collect radiation penetrating the radio and infrared *atmospheric windows. Satellite-borne instruments must be used to detect ultraviolet radiation, X-rays, and gamma rays from space.

television A *telecommunication system in which visual and aural information is transmitted for reproduction at a receiver. The basic elements of the system are as follows: *television cameras plus microphones to convert the information into electrical signals, i.e. into *video* and *audio signals*; amplifying, control, and transmission circuits to transmit the information; broadcast information using a modulated radio-frequency *carrier wave; a *television receiver* that detects the signals and produces an image on the screen of a specially designed cathode-ray tube.

The information on the target in the television *camera tube is extracted by *scanning, and the spot on the screen of the receiver tube is scanned in synchronism with it to produce the final image. A process of rectilinear scanning is used in which the electron beam traverses the target area in both the horizontal and vertical directions; this is known as *raster scanning*. *Sawtooth

waveforms are used to produce the deflections of the beam. The number of horizontal scans is made larger than the number of vertical scans so that as much of the target area of the receiver is covered as possible. Each horizontal traverse is a *line* and the repetition rate is the *line frequency*.

The vertical direction is the *field*, the number of vertical scans per second being the *field frequency*. Each vertical scan is a *raster*. If the entire picture is produced in a single raster, the scanning is called *sequential scanning*. Most broadcast television systems use a system of *interlaced scanning*. In this system the lines of successive rasters are not superimposed on each other, but are interlaced, and two rasters constitute a complete picture or *frame*.

The basic television system transmits images in black and white only (monochrome television). *Colour television is now widely used, and the broadcast signal is received on special colour receivers. Modern monochrome receivers also use the broadcast colour signal but the image produced is black and white.

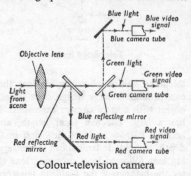

Colour-television camera

television camera The device used in a *television system to convert the optical images from a lens into electrical *video signals*. The optical image formed by the lens system of the camera enters the *camera tube (which forms part of the camera) and falls onto photosensitive material. This is scanned, usually by a low-velocity electron beam, and the resulting output is modulated with video information obtained from the target area. The resulting output is amplified before transmission.

The camera used in *colour television consists of three camera tubes, each of which receives information that has been selectively filtered to provide it with light from a different portion of the spectrum. Light from the optical lens system is directed at an arrangement of dichroic mirrors, each of which reflects one colour and allows other frequencies to pass through. The original multicoloured signal is split into red, green, and blue components, and the video output from the three camera tubes represents the red, green, and blue components of the image (*see* diagram). The scanning systems in the three tubes are driven simultaneously by a master oscillator to ensure that the output of each tube corresponds to the same image point.

television receiver *See* television.

TE modes *See* waveguide.

temperament The adjustment of tuning of the notes of a keyboard instrument to give a near diatonic scale (*see* musical scale) for all keys. In the diatonic scale the frequency of each note is a fixed multiple of that of the key note, i.e. $1\frac{1}{8}$, $1\frac{1}{4}$, $1\frac{1}{3}$, etc. If diatonic scales were built up on all keys, the number of finger keys to the octave would be very large. In order to keep to the traditional number of 13 finger keys with a minimum of mistuning, modern keyboard instruments are tuned to the *equitempered scale* (or scale of *equal temperament*) in which the mistuning is evened out over the whole octave. The octave is divided into twelve intervals with a semitone ratio of $2^{1/12}$. Only the octaves are true

but the errors are not such that an ear accustomed to the system feels distress. Such a tuning permits the use of the same keys for different scales. The table gives a comparison between the scales of *just intonation* (diatonic) and equitemperament. A (440 hertz) has been taken as the standard in both scales quoted.

Just intonation		Equal temperament
264	C	261.6
297	D	293.7
330	E	329.6
352	F	349.2
396	G	392.0
440	A	440
495	B	493.9
528	C	523.3

temperature Symbol: T. The property of an object that determines the direction of heat flow when the object is brought into thermal contact with other objects: heat flows from regions of higher to those of lower temperatures. (*See* thermodynamics (second law).)

This definition merely allows one to place in order of temperature a number of objects. To assign numerical values, it is necessary to establish a scale of temperature. The only temperature scale now in use for scientific purposes is the *International Practical Temperature Scale, temperature being expressed in *degrees Celsius or *kelvins. Temperature is a measure of the kinetic energy of the molecules, atoms, or ions of which a body or substance is composed. The *thermodynamic temperature of a body is now treated as a physical quantity and is measured in kelvins.

The measurement of low and moderate temperatures (roughly up to 500 °C) is usually classed as *thermometry* while *pyrometry* covers the high-temperature ranges.

temperature coefficient of resistance For any material, the small change in the resistance of the material for changes in the thermodynamic temperature of the material.

At a given Celsius temperature t the resistance of a body, R_t, can be expressed as a power series in t. For a moderate range of temperature,

$$R_t = R_0 (1 + \alpha t + \beta t^2),$$

where R_0 is the resistance at 0 °C and α and β are constants characteristic of the material. For limited range of t or moderate precision, β is usually negligible and α is called the temperature coefficient of resistance. It is measured in reciprocal degrees.

In general, conductors have a positive coefficient of resistance, semiconductors and insulators have a negative coefficient of resistance. This results from the energy distribution of electrons in the material (*see* energy bands). Conductors always have quantum states available for conduction, and increasing the temperature above absolute zero causes ions to vibrate less uniformly about their lattice positions and scatter the conduction electrons more as they drift through the material. This has the effect of increasing the resistance.

In semiconductors and insulators, where a forbidden band exists between the valence bands and the conduction bands, increasing the temperature increases the number of electrons that can cross the forbidden band and decreases the resistance. Scattering by ions will normally make only a small contribution to the resistance.

temperature inversion 1. A phenomenon that occurs in the troposphere when the rate of decrease in temperature with height is less than the adiabatic *lapse rate.

2. A level in the atmosphere in which the temperature gradient changes sign.

temporal coherence *See* coherence.

tensile strength The *ultimate strength of a material as measured under tension.

tensor An abstract mathematical entity, an operator, by which one *vector can be converted into another vector by a linear modification of components. Tensor analysis is a generalization of vector analysis. Simple tensors are fundamental in all problems of *anisotropy – electric, magnetic, elastic, and thermal. For example, in an isotropic medium the vectors electric displacement D and field strength E have the same direction and are related by the scalar quantity *permittivity. In an anisotropic medium the vectors are not in the same direction and have to be related by a tensor to effect the required change in direction and magnitude. More complex tensors occur in, for example, the transformation of space-time coordinates from one observer to another in the theory of relativity.

Tensors are involved with transformations from one set of coordinates to another. A point in a space of n dimensions can be specified by a set of n coordinates, $x_1, x_2, \ldots x_n$. The set can be written in the form x_i where i takes the values 1, 2, 3, $\ldots n$. If a transformation of coordinates is made, the x_i becomes x_i' in the new coordinate system. A set of n magnitudes can be formed,

$$A_1, A_2, \ldots A_n,$$

each one being a function of the original coordinates x_i. The set is denoted A_i. When the transformation occurs these change to A_i'. Such a set is a tensor if certain transformation laws hold. A tensor of rank r has n to the power r components. A tensor of rank zero is a *scalar and a tensor of rank 1 is a *vector.

tera- Symbol: T. A prefix denoting 10^{12}; for example, one terametre (1 TM) is equal to 10^{12} metres.

terminal 1. A device remote from and linked to a *computer, providing *input /output facilities. The most common terminal is a *visual-display unit paired with a keyboard. This has an additional built-in capacity to store and manipulate data, and is thus described as an *intelligent terminal*. *See also* time sharing.
2. Any of the points on an electronic circuit or device at which interconnecting leads may be attached and signals fed in and out.

terminal velocity The velocity with which a body moves relative to a fluid if the resultant force on the body is zero. From *Stokes' law the terminal velocity of a sphere falling in a fluid under gravity is:
$$2(\sigma - \rho)r^2 g / 9\eta,$$
where σ is the body density, ρ is the fluid density, r is the radius of the sphere, and η is the coefficient of the viscosity. *See also* Reynolds number.

termination Of a *transmission line. A load impedance placed at the end of the transmission line to ensure *impedance matching and prevent unwanted reflection.

tesla Symbol: T. The *SI unit of *magnetic flux density, defined as one weber of magnetic flux per metre squared.

Tesla coil An apparatus for generating very high frequency currents at high potential. An induction coil, discharging across a spark gap, feeds the primary of a transformer through two large capacitors. The transformer is wound on a large open frame, the primary having only a few turns.

tetragonal system *See* crystal systems.

tetrode Any electronic device with four electrodes, in particular a *thermionic valve.

TFT Abbreviation for *thin-film transistor.

theodolite A telescope fitted with spirit levels and angular scales for measuring altitude and azimuth; it is used in surveying, etc.

theorem 1. A universal or general proposition or statement, not self-evident (*compare* axiom) but demonstrable by argument.
2. A proposition embodying merely something to be proved as distinct from a problem that embodies something to be done.

theorem of parallel axes If the moment of inertia of a body of mass M about an axis through the centre of mass is I, the moment of inertia about a parallel axis distance h from the first axis is $I + Mh^2$. If the radius of gyration is k about the first axis, it is $\sqrt{(k^2 + h^2)}$ about the second.

therm A unit of heat used in the gas industry, equal to 100 000 *British thermal units.

thermal agitation The random movement of the molecules of a substance, the total energy of which (kinetic and potential) is the internal *energy.

thermal capacity A former name for *heat capacity.

thermal conductance *See* heat-transfer coefficient.

thermal conductivity Symbol: λ, K, or k. The rate at which heat passes through a small area A inside the body is given by:

$$dQ/dT = -\lambda A\, dT/dx,$$

where dT/dx is the temperature gradient normal to the area A in the direction of heat flow, and λ is the thermal conductivity of the body at temperature T. The units are $J\,s^{-1}\,m^{-1}\,K^{-1}$.

On the *kinetic theory, the thermal conductivity of a gas is independent of the pressure. This is true for moderate pressures but at very low pressures the conductivity becomes proportional to the pressure. The thermal conductivity of a solid metal is related to the electrical conductivity by the *Wiedemann–Franz–Lorenz law.

thermal diffusivity Symbol: a. The quantity defined by the expression $\lambda/\rho c_p$, where λ is the *thermal conductivity, ρ is the density, and c_p is the specific heat capacity at constant pressure. It has the units metre squared per second.

thermal effusion *See* thermal transpiration.

thermal equilibrium The condition of a system in which the net rate of exchange of heat between the components is zero.

thermal imaging The production of images using the infrared radiation emitted by objects. The radiation is usually collected by a two-dimensional array of infrared detectors, whose electrical signals are extracted by a scanning method and converted into a visible image.

thermalize To bring neutrons into thermal equilibrium with their surroundings. *Thermal neutrons can be produced by passing *fast neutrons through a *moderator.

thermal neutrons Neutrons that are approximately in thermal equilibrium with matter within which they are diffusing.

They roughly obey the Maxwell *distribution of speeds, giving average kinetic energy $(3/2)kT$, most probable kinetic energy $(1/2)kT$, and most probable speed $v = \sqrt{(2kT/m)}$, where T is the thermodynamic temperature of the matter, k is the Boltzmann constant, and m is the mass of the neutron. At 20 °C v is $2200\ m\,s^{-1}$, and the value of kT is 4.05×10^{-21} J (0.0253 eV). Values of the *cross sections for nuclear reactions are commonly tabulated for the standard speed $2200\ m\,s^{-1}$.

thermal noise *See* noise.

thermal power station 1. A *power station in which electrical power is produced by combustion of a fuel such as coal, coke, or oil.
2. A nuclear power station in which electrical power is produced by a thermal reactor (*see* nuclear reactors).

thermal radiation At all temperatures bodies emit radiant energy whose quantity and quality depend on the thermodynamic temperature T of the body. For any given body the radiation that depends only on temperature is known as thermal radiation. It is excited by the *thermal agitation of molecules or atoms, and its spectrum is continuous from the far infrared to the extreme ultraviolet. *See* black-body radiation.

thermal reactor *See* nuclear reactors.

thermal resistance *See* resistance (thermal).

thermal transpiration *Syn.* thermal effusion. A phenomenon occurring when a temperature gradient exists in a tube containing gas at such a pressure that the *mean free path of the molecules is not negligible compared with the tube diameter. The pressure then is no longer uniform but is greatest at the high-temperature end. The term also applies to the case of a gas contained in two vessels at different temperatures connected by a porous medium whose holes are small compared with the mean free path of the molecules. In the steady state, the ratio of the pressures in the vessels equals the ratio of the square roots of the thermodynamic temperatures of the vessels.

thermionic cathode A *cathode that provides a source of electrons due to *thermionic emission. *Compare* photocathode.

thermionic emission The spontaneous emission of electrons from solids and liquids observed at high temperatures. According to the *Fermi–Dirac distribution function, a small number of the valence electrons have kinetic energies larger in magnitude than the negative potential energy within the substance, thus their total energies are positive. Such high-energy electrons can leave the surface spontaneously. The current emitted rises sharply with temperature (*see* Richardson–Dushman equation).

In addition to electrons both positive and negative atomic or molecular ions may be emitted as a result of the presence of impurities within the substance or on its surface. Such particles and electrons are collectively called *thermions*.

See also Schottky effect.

thermionic valve Any of a group of electronic devices once extensively used for a variety of purposes. Each one is a multielectrode evacuated *electron tube, containing a *thermionic cathode as the source of electrons. Thermionic valves containing three or more electrodes are capable of voltage amplification, the current flowing through the valve between two electrodes (usually the anode and the cathode) being modulated by a voltage applied to one or more of

the other electrodes. Thermionic valves have rectifying characteristics, i.e. current will flow in one direction only (the forward direction) when positive potential is applied to the anode.

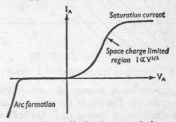

a Vacuum-diode characteristic

The simplest type of thermionic valve is the *diode*, which has been most often used in rectifying circuits. Electrons are released from the heated cathode by *thermionic emission. Under zero-bias conditions electrons released by the cathode form a *space charge region in the vacuum surrounding the cathode, and exist in dynamic equilibrium with the electrons being emitted. If a positive potential is applied to the anode, electrons are attracted across the valve to the anode and current flows. The maximum available current (the saturation current) is a function of the cathode temperature. The current does not rise rapidly to the saturation value as the anode voltage is increased, but is limited by the mutual repulsion of electrons in the interelectrode region. This is the *space-charge limited* portion of the characteristic and the current is approximately proportional to $V_A^{3/2}$, where V_A is the anode voltage. Under conditions of reverse bias, no current flows until the field across the valve is sufficient to cause *field emission from the anode or *arc formation, when *breakdown of the device occurs. The characteristics of a simple diode are shown (Fig. *a*).

The diode characteristic can be modified by interposing extra electrodes, called *grids* as they are usually in the form

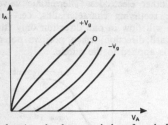

b Anode characteristics of a triode

of a wire mesh, between the anode and the cathode. The *triode* has only one extra electrode, a *control grid*. Application of a voltage to the grid affects the electric field at the cathode, and hence the current flowing in the valve. A family of characteristics is generated for different values of grid voltage, similar in shape to the diode characteristic. The anode current, at a given value of anode voltage, is a function of grid voltage. Amplification may thus be achieved by feeding a varying voltage to the grid; comparatively small changes of grid voltage cause large changes in the anode current (Figs. *b* and *c*). In normal operation the grid is held at a negative potential, and therefore no current flows in the grid, as no electrons are collected by it. Triodes have been extensively used in amplifying and oscillatory circuits.

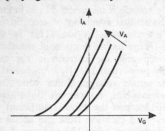

c Transfer characteristics of a triode

A disadvantage of the triode is the large grid-anode capacitance, which allows a.c. transmission, and extra electrodes have been added to reduce this

effect. The *tetrode* has one extra grid electrode, the *screen grid*, placed between the control grid and the anode and held at a fixed positive potential. Some electrons will be collected by the screen grid, the number of electrons being a function of anode voltage. At high anode voltages the majority of electrons pass through the screen grid to the anode. An undesirable kink in the characteristics is observed in a tetrode due to *secondary emission of electrons from the anode, these secondary electrons being collected by the screen grid. Secondary electrons are prevented from reaching the screen grid in the *pentode* by introducing another grid (the *suppressor grid*) between the screen grid and the anode and maintaining it at a fixed negative potential (usually cathode potential). The pentode characteristics (Fig. *d*) are similar to those observed in *field-effect transistors, which are the solid-state analogues. Thermionic valves with even more electrodes have been designed to produce particular characteristics.

Thermionic valves have been almost completely replaced by their solid-state equivalents. In applications requiring high voltages and currents, valves are still used, but these are special-purpose valves such as *cathode-ray tubes, *magnetrons, and *klystrons. For most applications, solid-state devices such as the p-n junction *diode, bipolar junction *transistor, and field-effect transistor (frequently in the form of *integrated circuits) have the advantages of small physical size, cheapness, robustness, and safety as the power required is very much less than for valves.

thermions *See* thermionic emission.

thermistor A *semiconductor device that has a large negative *temperature coefficient of resistance, and can be used for temperature measurement, or as a controlling element in electronic control circuits. It can also be used to compensate for temperature variations in other components.

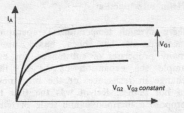

d Characteristics of a pentode

thermoammeter An *ammeter that measures a current (a.c. or d.c.) in terms of its *heating effect. For instance, the heating effect can be used to raise the temperature of a thermocouple connected to a sensitive galvanometer.

thermocouple An electrical circuit consisting of two dissimilar metals (or a metal and a semiconductor) joined at each end, in which an e.m.f. is produced when the two junctions are at different temperatures. This is due to the Seebeck effect (*see* thermoelectric effects). The thermocouple is often used as a temperature-measuring device: it can be used over a very wide temperature range, and is a conveniently usable device that can give temperatures at a very small point, and present readings at a considerable distance away if necessary. For the measurement of temperatures up to about 500 °C copper/constantan or iron/constantan thermocouples are used; at temperatures up to 1500 °C chromel/alumel or platinum/platinum plus 10% rhodium alloy is used, and at still higher temperatures, iridium/iridium-rhodium alloy.

The sensitivity of a thermocouple instrument is increased by connecting a number of junctions, in series, forming a *thermopile.

thermodynamic potential A measure of the energy level of a system that represents the amount of work obtainable when the system undergoes a change. The main types of potential are the internal *energy (U), the *Helmholtz function defined as ($U - TS$), the *enthalpy given by ($U + pV$), and the *Gibbs function defined as ($U + pV - TS$); the latter is sometimes referred to as the thermodynamic potential.

thermodynamics The study of the interrelation between *heat, *work, and internal *energy. The thermodynamic state of a body is defined in terms of certain thermodynamic variables, for example, in the case of a simple homogeneous body such as a gas or a solid, in terms of pressure p, volume V, and thermodynamic temperature, T. There is generally an *equation of state.

Thermodynamics deals with systems consisting of very large numbers of particles and not with the behaviour of individual molecules, hence p, V, T, etc., are statistical quantities. The *kinetic theory of heat is an attempt to relate the thermodynamic variables with the dynamic variables of the individual molecules. Thermodynamics is only concerned with changes of energy and not with the mechanism by which that change is brought about and its methods and results are therefore very general. It is based on two fundamental laws, the 1st and 2nd laws, to which is sometimes added a 3rd law, more generally known as the *Nernst heat theorem (*see also* zeroth law of thermodynamics).

1st law of thermodynamics. This states simply that heat is a process of energy transfer and that in a closed system the total amount of energy of all kinds is constant. It is therefore the application of the principle of conservation of energy to include heat transfer. An alternative statement of the law is that it is impossible to construct a continuously operating machine that does work without obtaining energy from an external source. To obtain a mathematical interpretation of the law it is necessary to introduce the internal energy U of the system as:

$$\delta Q = dU + \delta W,$$

where δQ is the heat absorbed by the system, dU is the increase in internal energy, δW is the work done by the system.

2nd law of thermodynamics. This deals with the question of the direction in which any chemical or physical process involving energy takes place. The formulation due to Lord Kelvin is that "it is impossible to construct a continuously operating machine which does mechanical work and which cools a source of heat without producing any other effects". In nature, heat is never found to proceed up a temperature gradient of its own accord, this being one special case of the general truth expressed by the law, which may in fact be stated in the form: no self-acting machine can transfer heat continuously from a colder to a hotter body and produce no other external effect.

*Carnot's theorem, based on the 2nd law, and consideration of the *Carnot cycle leads to the concept of *thermodynamic temperature. When a working substance is taken through a complete Carnot cycle the total change in *entropy of the universe is zero, and the entropy, S, of a substance is a definite function of its condition, just as its pressure, volume, temperature, or its internal energy. This result is a direct deduction from the second law, and may be regarded as a statement of that law. Thus for a perfectly reversible process:

$$\delta Q = TdS = dU + \delta W.$$

If, as is usually the case, the only external work results from a uniform pressure p, then:

$$TdS = dU + pdV.$$

This is a convenient mathematical statement of the first and second laws taken together.

See also enthalpy; Gibbs function; Helmholtz function.

thermodynamic temperature Temperature measurement has been based upon many properties of substances, for example the expansion of a gas, the change in resistance, or the brightness of hot bodies. Kelvin was the first to propose a thermodynamic scale of temperature (1848) in which changes of temperature are independent of the working substance used.

Temperatures on this scale are defined so that if a reversible engine working on a *Carnot cycle takes up a quantity of heat q_1 at a temperature T_1 and rejects a quantity q_2 at T_2, then:

$$T_1/T_2 = q_1/q_2.$$

Kelvin showed that the scale so defined was identical to one based upon an *ideal gas. Using this concept, thermodynamic temperature is regarded as a physical quantity that can be expressed in the unit called the *kelvin, the triple point of water being defined as 273.16 kelvin. In practice thermodynamic temperature is measured on the *International Practical Temperature Scale.

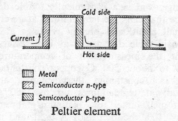

Cold side

Current

Hot side

▦ Metal
▨ Semiconductor n-type
▧ Semiconductor p-type

Peltier element

thermoelectric effects A series of phenomena occurring when temperature differences exist in an electrical circuit.

(1) *Seebeck effect*: If two different metals (or a metal and a semiconductor) are joined and the two junctions are

kept at different temperatures, an electromotive force is developed in the circuit. The circuit constitutes a *thermocouple. The e.m.f. is not affected by the presence of other junctions in the circuit if they are all maintained at the same temperature. The e.m.f. is given by the equation:

$$E = \alpha + \beta\theta + \gamma\theta^2,$$

where θ is the temperature difference between the hot and cold junctions, and α, β, and γ are constants that depend on the substances comprising the circuit. γ is normally quite small so that for a small temperature difference the e.m.f. change is directly proportional to this difference.

(2) *Peltier effect*: This is the converse of the Seebeck effect. If a current is passed through a metallic junction or a metal-semiconductor junction, the junction is either warmed or cooled according to the direction of flow. The effect is reversible, i.e. reversing the current causes the cool junction to become hot and the hot junction to cool. Larger temperature differences are produced with metal-semiconductor junctions than with metal-metal junctions. An n-type semiconductor produces a temperature difference in the opposite sense to a p-type semiconductor. A *Peltier element* consists of a number of such junctions in series, with n-type and p-type semiconductor alternating (*see* diagram). It acts as a heating or cooling element.

(3) *Kelvin* (or *Thomson*) *effects*: A potential difference is developed between different parts of a single conductor if there is a temperature difference between them; for a temperature difference dT between two points, the e.m.f. in this element is $\mu\text{d}T$ where μ is the *Thomson coefficient* (positive when directed from points of lower to points of higher temperature). Also if a current passes through a wire in which a temperature gradient exists, this current causes a flow of heat from one part to the other – the direction of flow depending on the substance concerned.

thermoelectric generator A device that converts heat directly into electric power by means of a *thermoelectric effect. An e.m.f. is developed between the junctions of a thermocouple, or between two regions of a metal, when a temperature difference exists between them. The e.m.f. is used to power an external circuit. The hot junction or region can be heated, for example, by the decay of a radionuclide.

thermoelectric series A series of metals arranged so that if a thermocouple is made from two of them, current flows at the hot junction from the metal occurring earlier in the series to the other metal.

thermogalvanometer *See* galvanometer.

thermograph A recording thermometer. *See* Bourdon tube.

thermoluminescence *See* luminescence.

thermomagnetic effect *See* magnetocaloric effect.

thermometer A device for measuring the temperature of a body, from a measurement of some property of the thermometric substance that depends on temperature. A fluid suitable for use in a thermometer should possess a marked degree of expansion for small temperature rise, a uniform expansion rate, good thermal conductivity, chemical stability, high boiling point (if liquid), and low freezing point (if liquid) or liquefaction point (if gaseous). The various types of thermometers include the *mercury in glass, *maximum and minimum, and platinum *resistance thermometers, and devices such as the *thermocouple and *bimetallic strip. *See also* pyrometer.

thermometry *See* temperature.

thermonuclear reaction A reaction that involves the *nuclear fusion of particles and nuclei possessing enough kinetic energy to initiate and sustain the process. *Thermonuclear energy* is released during the reaction. The reaction rate increases rapidly with temperature, the required temperature being in the million degrees range. The energy in stars is produced by such processes. *See also* fusion reactor.

thermophone A source of sound consisting of a very thin strip of platinum or gold mounted between two terminal blocks. Alternating current passing through this strip causes periodic variations in its temperature and expansions and contractions of the air surrounding it. The corresponding pressure variations are radiated in the form of sound waves. The heating strip must have a very small heat capacity so that its temperature will accurately follow rapid current variations. The sound output is low but may be amplified with a resonator.

thermopile A device consisting essentially of a large number of *thermocouples connected in series to give an easily measurable e.m.f. when heat radiation is allowed to fall on one set of junctions, the other set being shielded from the radiation. Such an instrument gives a current that is easily detectable.

thermosphere *See* atmospheric layers.

thermostat A device that responds to changes of temperature and automatically actuates a mechanical valve, electric switch, etc. Common types depend for their action upon the variation with temperature of the expansion of a metal rod, the shape of a spring, or the pressure (and/or volume) of a gas. Thermostats are usually employed to regulate the supply of heat in situations where a substantially constant temperature is to be maintained.

theta pinch *See* fusion reactor.

thick-film circuit A circuit usually consisting only of interconnections and *passive components (e.g. resistors and inductors) that is fabricated on a film, up to about 20 μm thick, deposited on a glass or ceramic substrate. The film itself is a suitable glaze or cement, such as a ceramic/metal alloy. *Active components or devices on silicon chips may be wire-bonded to the thick-film circuit to produce a form of hybrid *integrated circuit.

thick lens A real lens, as distinct from a hypothetical infinitely thin lens. In a thick lens the separation between the two surfaces cannot be ignored; this may require the calculation of focal lengths, positions of principal planes, etc.

thick mirror *See* mirror.

thin-film circuit A circuit consisting of interconnections and components deposited on a glass or ceramic substrate usually by *vacuum evaporation or *sputtering. The deposited layers are up to a few micrometers thick. Components are usually passive, but active components such as the *thin-film transistor have been made.

thin-film transistor (TFT) An insulated-gate *field-effect transistor constructed by thin-film circuit techniques (i.e. *vacuum evaporation or *sputtering) on an insulating substrate rather than a semiconductor chip. The insulating substrate leads to fast switching speeds. The technique was originally used to produce discrete cadmium sulphide transistors but is now used mainly to construct silicon-on-sapphire MOS logic circuits.

Thomson effects, Thomson coefficient *See* thermoelectric effects.

Thomson scattering *See* scattering.

thorium-lead dating *See* dating.

thorium series *See* radioactive series.

three-body problem The most important example of the *n*-body problem. The determination of the positions and motions of three bodies in a mutual gravitational field. No analytical solution can be obtained (except in special cases, e.g. where the bodies occupy the vertices of an equilateral triangle), but positions are determined from their previous values.

three-colour process *See* additive process; subtractive process.

three-phase Of an electrical system or device. Having three equal alternating voltages between which there are relative *phase differences of 120°. *See* polyphase system.

threshold energy The least energy required to bring about a certain process, in particular a reaction in nuclear or particle physics. It is often important to distinguish between the energies required in the laboratory and in centre-of-mass coordinates.

threshold frequency The minimum frequency giving rise to a particular phenomenon, such as *photoconductivity or the *photoelectric effect.

threshold of hearing That minimum intensity level of a sound wave that is

audible. It occurs at a loudness of about 4 phons (equal to 4 decibels at 1 kHz).

threshold voltage The voltage at which a particular characteristic of an electronic device first occurs.

thyratron *Syn.* gas-filled relay. A gas-filled *electron tube with three electrodes in which the voltage on one electrode (the grid) controls the starting of the discharge in the tube. A positive potential is applied to the anode, the potential being greater than the *ionization potential of the gas. A negative potential is applied to the grid, and, if sufficiently large, this neutralizes the effect of the anode potential at the thermionic cathode and prevents any current flowing. If the grid voltage is made less negative, the field at the cathode increases until the discharge starts. This is called *striking* the tube. Once the discharge has started, the grid has no further effect on the anode current even if the voltage is made very negative. At the instant of striking, the anode potential falls to approximately the ionization potential of the gas, and the discharge may only be stopped by reducing the anode potential below this value.

Thyratrons were formerly often used as relays and for counting radioactive particles. They have now been largely superceded by their solid-state analogue, the *silicon controlled rectifier.

thyristor *See* silicon controlled rectifier.

tidal energy *See* renewable energy sources.

timbre The distinguishing quality, other than pitch or intensity, of a note produced by a musical instrument, voice, etc. The *quality is generally stated to be dependent upon the relative amplitude and number of the *partials, the resulting waveform being determined by the *superposition of the component partials.

time Symbol: *t*. A fundamental quantity usually indicating duration – a period or interval of time – or a precise moment. Time is measured in *seconds in *SI units, but the minute, hour, day, and year may still be used. One day is defined as 24 hours, i.e. as 86 400 seconds. The length of the year depends on how it is measured. The practical measurement of time formerly depended on determining the period of rotation of the earth relative to the astronomical bodies. An atomic *clock and other types of clocks are now used for more precise measurements of time

Apparent solar time is measured by successive intervals between transits of the sun across the meridian, and is shown on a sundial. *Mean solar time* averages out this interval over the course of one year, and is measured with reference to the motion of the *mean sun*. This is a point that moves uniformly around the celestial equator (*see* celestial sphere) in the same total time as the real sun takes in its apparent motion (which is not uniform) round the ecliptic. The difference on any day between apparent and mean solar time, up to 16 minutes, is known as the *equation of time*. The provision of a uniform timescale using mean solar time was based on the assumption that the earth's rotation rate is constant. The rotation rate is now known to vary very slightly and irregularly.

Sidereal time, used in astronomy, is measured in terms of the *sidereal day*; this is the interval between successive transits of a point (related to the vernal *equinox) across a given meridian. It is thus also based on the earth's rotation rate. The sidereal day is about 3 minutes 56 seconds shorter than the 24-hour day. A star will be on the meridian of an observatory when the local sidereal time becomes equal to the star's right ascension (*see* celestial sphere).

Greenwich Mean Time (GMT) is determined from transits of certain stars across the Prime Meridian (of zero longitude) at Greenwich. The coordinates of the stars are known, and this allows a correction to be made to the time given by a sidereal clock and hence obtain GMT.

Universal coordinated time is the mean of that provided by sets of atomic clocks distributed around the world. It thus provides a timescale that changes uniformly. It is kept within one second of GMT by the insertion or deletion of *leap seconds* when the earth's irregular rotation necessitates this (usually made at the end of December).

The *year* is the time taken by the earth to make one complete orbit of the sun, measured with respect to a given point. The *sidereal year* is measured with respect to a particular star regarded as fixed in position; it is equal to about 365.256 36 days. The *tropical year* is the interval between successive passages of the sun through the vernal equinox; it is equal to about 365.242 19 days. Since the equinox is not a fixed point, the tropical and sidereal years are different.

The *calendar year* (or *civil year*) is adjusted so that its average length is very close to that of the tropical year; for practical reasons the calendar year must contain a whole number of days. In the *Gregorian calendar* used almost worldwide, the calendar year has 365 days, plus an extra day in leap years, but the century years are not leap years unless the number of the century is exactly divisible by 400. On average, the calendar year is equal to 365.2425 days. *See also* arrow of time; time reversal.

time base A voltage that is a predetermined function of time and is used to deflect the electron beam of a *cathode-ray tube so that the luminous spot traverses the screen in the desired manner. One complete traverse of the screen is called a *sweep* (or sometimes a time base). The most common type of time base is one that produces a linear sweep; a *sawtooth waveform is usually employed for this so that at the end of each useful trace the luminous spot is returned rapidly to its starting point. The return of the spot to its starting point is called the *flyback*.

time constant Physical quantities such as voltage, current, and temperature sometimes decrease with time in such a manner that, at any instant, the rate of decrease of the quantity is given by:

$$-(dv/dt) = v/T \qquad \text{(i)}$$

where v = the instantaneous value of the quantity, T = the time constant. In this case, the time constant is the time taken for the quantity to decrease to $1/e$ (approximately 0.368) of its initial value. Alternatively, a quantity may increase with time in such a manner that, at any instant:

$$dv/dt = (V - v)/T \qquad \text{(ii)}$$

where V = the ultimate value of the quantity (a constant). In this case, the time constant is the time taken for the quantity to increase from zero to $1 - 1/e$ (approximately 0.632) of its ultimate value.

The time constant is particularly important in connection with electrical circuits. For example, in a circuit containing either (*a*) a resistance in series with a capacitance, or (*b*) a resistance in series with an inductance, which is connected suddenly to a constant-voltage d.c. supply, the component voltages, current, and charge (as appropriate) vary as in (i) or (ii) above. The time constant in seconds for circuit (*a*) may be calculated by multiplying the resistance by the capacitance, and for circuit (*b*) by dividing the inductance by the resistance.

time delay *See* time lag.

time dilation *See* relativity.

time-division multiplexing A method of *multiplex operation in which each of the input signals is sampled and transmitted sequentially; the transmission channel is thus shared by allocating specific time intervals to each signal. The time of switching from one to the next must be such that each signal is sampled many times in the course of a cycle. *Pulse modulation is frequently used for time-division multiplexing.

time lag *Syn.* time delay. The time that elapses between the closing of one circuit in a circuit breaker, relay, or similar apparatus, and the response of the current in the main circuit.

time-lapse photography A means of obtaining a speeded-up version of a slow process, such as a flower opening, by recording single exposures, taken at regular intervals, on ciné film without moving the camera and then projecting the film at normal speed.

time reversal Symbol: T. The substitution of time t by time $-t$, the symmetry of which is known as T invariance. As with CP violation (*see* CP invariance), T violation occurs in *weak interactions involving kaon decay. *See also* CPT theorem.

time sharing A technique whereby the processing time of a *computer is shared among several jobs by means of rapid switching between them. For example, in a *multiaccess system* a number of computer users communicate with the machine seemingly simultaneously via individual *terminals. The speed of the machine gives each user the impression that he has sole use of it. *See also* interactive.

time switch A switch that incorporates a type of clock mechanism for making and/or breaking an electric circuit at times that are predetermined by the setting of the mechanism.

tint *See* colour.

TM modes *See* waveguide.

T-number A modified (increased) *f-number that a real lens would have to have in order to transmit an amount of light corresponding to its calculated f-number. The T-number takes into account the reflection and absorption losses that occur in practice with a lens.

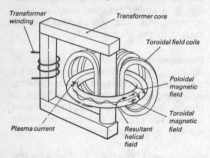

Principal magnetic fields of a tokamak

tokamak An arrangement for confining plasma in a *fusion reactor, which shows promise for achieving controlled energy release from nuclear fusion. Devised in the USSR in the 1960s, the device combines two principles: the torus, which has no ends through which the plasma can escape, and a magnetic field, which spirals around this toroidal plasma.

A spiral (or helical) magnetic field is achieved by adding together magnetic fields produced by toroidal coils (along the plasma) with the magnetic field (around the plasma) resulting from the strong plasma current driven by a trans-

former core. The principle of the tokamak is illustrated in the diagram.

Whether or not a demonstration power-producing tokamak will be built will depend upon the results of major experiments being carried out in the UK, USA, Japan, and the Soviet Union.

tone 1. An audible note containing no *partials.
2. The quality of a musical sound, e.g. soft tone, thin tone.
3. The interval of a major second, e.g. C to D, as opposed to a *semitone.

tone control A means of adjusting the relative frequency response of an audiofrequency amplifier used in the reception or production of sound in order to achieve a more pleasing result.

tonne *Syn.* metric ton. Symbol: t. A metric unit of mass: $1 t = 1000$ kg. It differs in size from an imperial ton (1016 kg) by about $1\frac{1}{2}\%$.

topology The branch of geometry concerned with continuity in geometrical figures, i.e. with those properties of a figure that remain unchanged (*topologically equivalent*) when a figure is bent, stretched, or shrunk, but not when it is torn or deformed by the fusion of points on it. For example, a torus and a cup are topologically equivalent (because the cup's handle and the torus both have a single hole through them), whereas a torus and a sphere are not. Topology has many applications, including the theory of liquid crystals and the classification of *magnetic monopoles in *gauge theory.

toric lens A lens with a toroidal surface, i.e. a surface generated by a circular arc rotating about an axis that does not pass through the centre of the circle. The curvature in one plane is then different from that in an orthogonal plane.

These lenses can be used to correct *astigmatism in the eye.

toroidal winding See ring winding.

torque Syn. moment of a force about an axis or of a couple. See moment.

torquemeter Syn. dynamometer. An apparatus for measuring the torque exerted by the rotating part of a prime mover (such as a petrol engine), electric motor, etc., under dynamic conditions.

torr A unit of pressure formerly used in vacuum technology. It is equal to 133.322 pascals.

Torricellian vacuum See barometer.

Torricelli's law The velocity of efflux of a fluid from an orifice in a reservoir is $\sqrt{(2gH)}$, where H is the depth of the orifice – more precisely of the *vena contracta below the free surface of the reservoir.

torsional vibrations Vibrations in a body, usually a cylindrical bar or tube, in which the displacement is in the form of a twist due to the application of an alternating torque to one end of the body, the other end being clamped.

torsional waves Waves formed in a medium, usually a cylindrical bar or tube, as the result of the application of torsional vibrations to one or more parts of the medium. The velocity, c, of torsional waves is given by $c = \sqrt{(G/\rho)}$, where G is the modulus of rigidity (see modulus of elasticity), and ρ is the density.

torsion balance A very sensitive balance consisting of an arm attached to a fibre; when a force is applied to the arm, the fibre twists until the torque on the arm is balanced by that in the fibre. Since

the latter is proportional to the angle of twist, this quantity is determined by the use of a light pointer. Torsion balances are employed in the measurement of small forces such as those associated with surface tension and static charges.

total angular momentum quantum number A number, integral or half-integral, that characterizes the total angular momentum (*orbital angular momentum plus *spin) of an atom, nucleus, particle, etc. The symbol j is used for a single entity, J or j_i for a whole system.

total heat Former name for enthalpy.

total internal reflection See total reflection.

total radiation pyrometer See pyrometer.

total reflection Syn. total internal reflection. A phenomenon occurring when light strikes the surface of an optically less dense medium at an angle of incidence greater than the *critical angle: instead of emerging into the less dense medium, it is reflected back into the optically denser (incident) medium. Total-reflection prisms have a wide application in instruments for changing direction of rays, for producing lateral inversion, or for completing inversion of the image. Although the reflection is total, the light actually penetrates a small distance into the rarer medium.

Total reflection also occurs with other waves: e.g. with sound, at moderately oblique incidence from air to water, and with X-rays, at almost grazing incidence from air to a solid or liquid.

tourmaline A mineral crystal that exhibits dichroism.

trace A figure traced out by the luminous spot on the screen of a *cathode-ray tube.

tracer *See* radioactive tracer.

tracking 1. Of any two electronic devices or circuits. An arrangement by which an electrical parameter of one device or circuit varies in sympathy with the same or another parameter of the second one, when both are subjected to a common stimulus. In particular, it is the maintenance of a constant difference in the resonant frequencies of two *ganged *tuned circuits.

2. The formation of unwanted electrically conducting paths (often by carbonization) on the surface of solid dielectrics and insulators, when subjected to high electrical fields.

transconductance *See* mutual conductance.

transducer *Syn.* sensor. Any device for converting a nonelectrical signal into electrical signals (or vice versa), the variations in the electrical signal being a function of the input. Transducers are used as measuring instruments and in the electroacoustic field, the term being applied to gramophone pick-ups, microphones, and loudspeakers.

The physical quantity measured by the transducer is the *measurand*, and the portion of the transducer in which the output originates is the *transduction element*. The device in the transducer that responds directly to the measurand is the *sensing element* and the upper and lower limits of the measurand value for which the transducer provides a useful output is the *dynamic range*.

Several basic transduction elements can be used in transducers for different measurands. They include capacitive, electromagnetic, inductive, photoconductive, photovoltaic, and piezoelectric elements. Most transducers require external electrical excitation for their operation; exceptions are self-excited transducers such as piezoelectric crystals, photovoltaic, and electromagnetic types.

Most transducers provide linear (analogue) output, i.e. the output is a continuous function of the measurand, but some provide digital output in the form of discrete values. Most transducers are designed to provide output that is a linear function of the measurand as this allows easier data handling. If the measurand varies over a stated frequency range, the output of the transducer varies with frequency.

transductor *Syn.* saturable reactor. A device that is used in control circuits and consists of a number of windings on a magnetic core. Usually a standing current in one winding is adjusted to bring the core to a magnetic state in which small changes in the current of one of the other windings (the signal winding) can control large powers in coupled circuits. Variations in the control circuit supplying the signal winding must be slow relative to the frequency of supply current, but it is possible to use up to 2000 hertz and thus to control signals in the lower audiofrequency range.

transfer characteristic The relation between the current (or the voltage) at one electrode of an amplifier, transducer, or other electronic device or network and the voltage (or current) at a different electrode; it is usually shown in graphical form. The tangent at any point on a given transfer-characteristic curve gives the associated *transfer parameter* at that point. The transfer parameter most commonly used is the *mutual conductance, determined from the characteristic relating output current and input voltage.

transference number *Syn.* transport number. *See* migration of ions.

transformation *See* transition.

transformer An apparatus without moving parts for transforming electrical power at one alternating voltage into electrical power at another (usually different) alternating voltage, without change of frequency. It depends for its action upon mutual induction (*see* electromagnetic induction) and consists essentially of two electric circuits coupled together magnetically; the usual construction is of two coils (or windings) with a laminated core suitably arranged between them. One of these circuits, called the *primary*, receives power from an a.c. supply at one voltage, and the other circuit, called the *secondary*, delivers power to the load at (usually) a different voltage.

If core losses are ignored, the ratio of primary to secondary voltage is equal to the ratio, n, of the number of turns in the primary winding to the number in the secondary winding. The transformer is described as *step-up* or *step-down* according to whether the secondary voltage is respectively greater or less than the primary voltage. Apart from the property of voltage transformation, it has the property of current transformation: if core losses are ignored, the ratio of primary to secondary current is equal to $1/n$. *See* autotransformer; voltage transformer; current transformer.

transients Temporary disturbances in a system resulting from the sudden incidence of an impulse voltage (or current) or the application or removal of a driving force. The form of any such transient is characteristic of the system, but its magnitude is a function of the magnitude of the impulse or driving force. The persistence of the transient is controlled by the dissipative components of the system. An example of the production of transients is that of *forced vibrations.

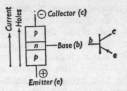

a A p–n–p bipolar junction transistor

transistor A multielectrode *semiconductor device in which the current flowing between two specified electrodes is modulated by the voltage or current applied to one or more specified electrodes. The semiconductor material is usually silicon. Transistors are small robust cheap devices requiring small supply voltages. They have now replaced *thermionic valves as the general-purpose *active electronic device except for some specialized uses.

The first transistors, invented in 1948, were *point-contact transistors*, which are now obsolete. The *junction transistor* was developed in 1949 and its action was fully described by Shockley in 1950. *Field-effect transistors were developed more recently and have a different principle of operation.

Modern transistors fall into two main classes: bipolar devices, which depend on the flow of both *minority and *majority carriers through the device, and unipolar transistors (*see* field-effect transistors) in which current is carried by majority carriers only.

Bipolar junction transistors (usually simply called *transistors*). The basic device consists of two *p-n junctions in close proximity, with either the n or p regions common to both junctions (Figs. *a* and *b*) and so forming either *p-n-p* or *n-p-n* transistors respectively; the latter is most commonly used. The central region is called the *base* and the electrode attached to it is the *base electrode*.

Consider the n-p-n transistor. If a voltage is applied across the transistor,

493

one junction becomes *forward biased and the other junction *reverse biased. Current flows across the forward-biased junction; electrons (the majority carriers) from the n-type region cross into the base, and *holes (the minority carriers) from the base cross into the n-region. This junction is called the emitter-base junction, the n-region being the *emitter*. The electrons entering the base diffuse across it towards the reverse-biased junction. Once they enter the *depletion layers associated with this junction, they are swept across into the other n-region, which is called the *collector*.

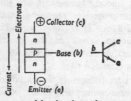

b An n–p–n bipolar junction transistor

The hole concentration in the base falls due to holes entering the emitter across the forward-biased junction and to holes recombining with injected electrons from the emitter. This will reduce the forward voltage across the junction until the current ceases. If the base region is connected to a suitable point in the circuit so that the emitter-base junction remains forward biased, electrons flow out of the base region to maintain the hole concentration and current continues to flow across the device, from collector to emitter (i.e. in the opposite direction to the flow of electrons). The total current flowing will be related by:

$$I_e = I_b + I_c,$$

where I_e is emitter current, I_b base current, and I_c collector current.

Voltage amplification is possible using the emitter as the input terminal. This operation is called *common-base connection; the characteristics are shown in Fig. *c*. For efficient amplification, the collector current must be as nearly equal to the emitter current as possible. This may be achieved by reducing the base current to as low a value as is practicable, for example by using a high *doping level in the emitter and a narrow base region. The *common-base current gain* or *collector efficiency*, α, is given by I_c/I_e, and is a function of a particular device. The base current I_b is equal to $(1 - \alpha)I_e$. For any given transistor the ratio

$$I_e:I_b:I_c = 1:(1 - \alpha):\alpha$$

is constant.

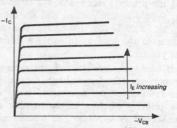

c Common-base output characteristics of a junction transistor

Current amplification is possible by applying the input signal to the base rather than to the emitter. Provided the emitter-base junction is always forward biased, carriers travel through the device as described above, but a variation in the base current causes a corresponding change in the collector current to maintain the ratio $\alpha:(1 - \alpha)$. This is known as *common-emitter connection, and is the most usual method of operating the device. The characteristics are shown in Fig. *d*. The *beta current gain factor is defined as:

$$\beta = I_c/I_b = \alpha/(1 - \alpha).$$

Since α can approach unity, β can be very large. Large collector currents are therefore generated by small base currents. This leads to saturation of the transistor as the current that can flow out of the collector is limited by the components in the external circuit.

When the collector current is saturated, the collector-emitter voltage drops to a small value; this is the *saturation voltage of the device. It is dependent on the value of base current and the external circuit components. The transistor may be used as a switch by driving it into saturation using the base current as the driver. It has many switching applications in *digital circuits.

One of the most important differences between holes and electrons is the difference between their mobilities, electrons being about three times as mobile as holes. This makes devices depending mainly on electron flow much faster in operation than those depending on the flow of holes, and capable of being used at higher frequencies. This accounts for some of the small differences between n-p-n and p-n-p transistors. However, in principle they are the same, except that holes replace electrons in the above description.

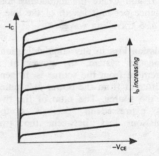

d Common-emitter output characteristics
junction transistor

transistor parameters A transistor is a nonlinear device whose behaviour is difficult to represent exactly by a set of mathematical equations. When designing transistor circuits the behaviour of the transistor is represented approximately by *equivalent circuits that act as models of the device. The particular equivalent circuit used will be the one that is most appropriate for the type of circuit being designed (i.e. for use with large signals, small signals, as switches, etc.).

Matrix parameters. The transistor is represented by an equivalent circuit with two input terminals and two output terminals. This is a quadripole network. Over a small portion of its operating characteristic the device is assumed to behave linearly. This is particularly true in small-signal operation, but is only an approximation for large-signal operation. The input and output voltages and currents are related by two simultaneous equations of the general matrix form:

$$[A] = [p][B],$$

where A and B represent current or voltage and p the particular transistor parameter used.

Other equivalent circuits are made up of components relating to the actual physical nature of the device rather than the more abstract linear networks involved in the matrix parameter treatment.

transistor-transistor logic (TTL) A family of high-speed *logic circuits that is produced in integrated-circuit form and whose principal switching components are bipolar transistors. TTL is widely used for high-speed applications, and is also characterized by medium power dissipation and *fan-out and good immunity to noise. A low-power higher-speed version (Schottky TTL) is also available.

transit circle *Syn.* meridian circle. *See* telescope.

transition 1. Any change accompanied by a marked alteration of physical properties, especially a change of phase (*see* transition temperature). (*See also* bridge transition; shunt transition).

2. A sudden change in the energy state of an atom or nucleus between two of its *energy levels. A nuclear

495

transition in which an alpha or beta particle is emitted is called a *transformation*. *See also* selection rules.

transition temperature The temperature at which a change of phase occurs, namely the freezing point, boiling point, or sublimation point. The term is also used for the temperature at which a substance becomes superconducting.

transit time In an electronic device, the time taken for a charge *carrier to pass directly from one specified point to another under given operating conditions.

translation The movement of a body or system in such a way that all points are moved in parallel directions through equal distances.

transmission coefficient 1. Symbol: τ. The ratio P_{tr}/P_0, where P_0 and P_{tr} are the *sound power (or more generally the acoustic power) incident on and transmitted by a body respectively. The *dissipation factor* is $(\alpha_a - \tau)$ where α_a is the *acoustic absorption coefficient.
  2. Former name for *transmittance.

transmission density *Syn.* optical density. Symbol: D. The logarithm to base ten of the reciprocal of the *transmittance, i.e. $D = \log_{10} 1/\tau$.

transmission line 1. *Syn.* power line. An electric line, often an overhead wire not surrounded by insulation, that conveys electric power from a power station or substation to other stations or substations.
  2. An electric *cable or *waveguide that conveys electric signals from one point in a communications system to another, and forms a continuous path between the points.
  3. *Syn.* feeder. The one or more conductors – wires or waveguides – that connect an aerial to a transmitter or re-

ceiver, the conductor(s) being substantially nonradiative.

Any of the above transmission lines is described as *uniform* if its electrical parameters, e.g. series resistance, are distributed uniformly along its length. A *balanced* line has conductors of the same type, equal values of resistance per unit length, and equal impedances from each conductor to earth and to other electric circuits; an example is the overhead line. A coaxial cable is an example of an *unbalanced* line. (*See also* balun.)

The most efficient transfer of signal power from a transmission line to the *load occurs when the line is terminated by a load impedance equal to the *characteristic impedance*, Z_0, of the transmission line. All the power travelling down the line is then absorbed by the load and none is reflected; the line is said to be *matched*. In a uniform line carrying high (e.g. radio) frequencies, Z_0 tends to the value $\sqrt{(L/C)}$, i.e. a pure resistance, where L and C are the inductance and capacitance per unit length of the line.

See also travelling wave.

transmission loss In any communications or acoustics system, the power at a point remote from the source is, in general, different from the power at a point nearer the source. The ratio of the two powers, expressed in *decibels, is the transmission loss between the two points.

transmission modes *See* waveguide.

transmissivity A measure of the ability of a material to transmit radiation as measured by the *internal transmittance of a layer of substance when the path of radiation is of unit length and the boundaries of the material have no influence.

transmittance Symbol: τ. A measure of the ability of a body or substance to transmit electromagnetic radiation, as expressed by the ratio Φ_{tr}/Φ_0 of the flux (*radiant or *luminous) transmitted to the incident flux.

Translucent bodies transmit light by diffuse transmission – the path of the light is independent, on the macroscopic scale, of the laws of *refraction. Transparent bodies on the other hand have regular transmission with no diffusion of light. In general, a body exhibits mixed transmission and the total transmittance can then be divided into a regular transmittance (τ_r) and a diffuse transmittance (τ_d), where $\tau = \tau_r + \tau_d$. *See also* internal transmittance; transmission density.

transmitter In any communications system, the device, apparatus, or circuits by means of which the signal is transmitted to the receiving parts of the system. *See also* aerial.

transmutation The formation of one element from another, generally naturally as a result of radioactive decay or artificially as a result of radioactive bombardment with particles or electromagnetic radiation.

transponder A combined transmitter and receiver system that automatically transmits a signal on reception of a predetermined *trigger. The trigger is usually in the form of a pulse, which must have a minimum amplitude.

transport number *Syn.* transference number. *See* migration of ions.

transport phenomena The class of phenomena due to the transfer of mass, momentum, or energy in a system as a result of molecular agitation, including such properties as thermal conduction, viscosity, diffusion, etc.

transuranic elements *See* radioactivity.

transverse mass The relativistic mass in a direction perpendicular to the motion of the particle relative to the observer (*see* relativity). It is given by:

$$m_0 / \sqrt{(1 - \beta^2)}$$

for a particle of rest mass m_0 moving with relative speed $\beta = v/c$, expressed as a fraction of the speed of light. *Compare* longitudinal mass.

transverse vibrations Vibrations in which the displacement is perpendicular to the main axis or direction of the vibrating body or system and so to the direction in which waves are travelling. Typical examples are the vibrations of a *stretched string or a *tuning fork. A vibrating bar can be regarded as an extension in theory of a transverse vibration in a string whose *stiffness has been increased so that the restoring forces are due to the *bending moment rather than the tension.

transverse waves Waves in which the displacement of the transmitting medium is perpendicular to the direction of propagation. Examples of transverse wave motion are *electromagnetic radiation and waves travelling along a *stretched string. Waves on the surface of liquids are half transverse and half longitudinal.

travelling microscope A *microscope with a low magnifying power (normally about × 10) and a *graticule placed in the plane of the eyepiece. It is mounted on rails to enable it to travel in the horizontal or vertical, or both, and is used to make very accurate determinations of length, e.g. on a photographic plate. Measurements are correct to 0.01 mm or better, over a distance of perhaps 0.2 m or more.

travelling wave Consider (hypothetically) a loss-free uniform *transmission line of infinite length situated in a medium of relative permittivity ε_r and relative permeability μ_r. If this line is connected to a sinusoidal a.c. supply at its sending end, electric power is transmitted along the line and the instantaneous values of the current and voltage at all points along the line are distributed in space as sine waves. As time increases, these sine-wave space distributions travel in the direction from the sending end to the receiving end with a velocity given by $c/\sqrt{(\varepsilon_r\mu_r)}$, where c is the speed of light. Either of these sine waves may be regarded as a travelling wave (of current or voltage respectively). More generally, it is an electromagnetic wave that travels along, and is guided by, a transmission line: sinusoidal conditions are not implicit (e.g. a surge propagated along a conductor or transmission line is also a travelling wave). Losses in the line cause a reduction in the velocity given above and also give rise to *attenuation. The above also applies to a line of finite length if it is terminated at its receiving end with an impedance equal to its *characteristic impedance. If an impedance discontinuity occurs at any point in the line (i.e. if there is an abrupt change in the characteristic impedance), reflection takes place at that point and the initial travelling wave (incident wave) is divided into the transmitted wave that travels towards the receiving end and the reflected wave that travels from the point towards the sending end.

travelling-wave tube An *electron tube that depends on the interaction of a *velocity-modulated electron beam with a radio-frequency electromagnetic field to produce amplification at microwaves frequencies. One form of tube is a modification of the multicavity *klystron tube. Several resonant cavities are placed along the length of the tube and coupled together with a *transmission line. The electron beam is velocity

modulated by the radio-frequency input signal at the first resonant cavity, and induces radio-frequency voltages in each subsequent cavity. If the spacing of the cavities is correctly adjusted, the voltages at each cavity induced by the modulated electron beam are in phase and travel along the transmission line to the output, with an additive effect, so that the power output is much greater than the power input. Some power travels in the other direction along the transmission line (the backward waves) but these contributions are designed to be out of phase with each other and cancel out. The effect of the transmission-line coupling leads to a greater bandwidth for the amplifier, compared to a klystron.

In the *backward-wave oscillator* the backward waves are in phase with each other and provide positive *feedback to the input cavity, allowing the tube to oscillate.

triac *See* silicon controlled rectifier.

triboelectricity *See* frictional electricity.

triboluminescence *Luminescence resulting from *friction.

trichromatic theory of vision *See* colour vision.

triclinic system *See* crystal systems.

trigger Any stimulus that initiates operation of an electronic circuit or device.

trimmer *Syn.* trimming capacitor. A variable capacitor of relatively small maximum capacitance used in parallel with a fixed capacitor to enable the capacitance of the combination to be finely adjusted.

Trinitron *See* colour picture tube.

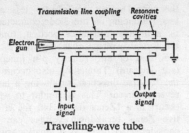

Travelling-wave tube

triode Any electronic device with three electrodes, usually a three-electrode evacuated *thermionic valve.

triple point The point on a pressure-temperature diagram representing the condi-

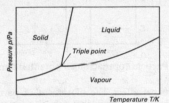

Typical pure substance

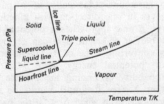

Water

Triple point

tion that the solid, liquid, and vapour phases can be together in equilibrium. The diagram for water is unusual in that the line showing equilibrium between solid and liquid has a negative slope. This is because water has the rare property of contracting on melting, so $(v_2 - v_1)$ is negative in the *Clausius-Clapeyron equation. The lower equilibrium pressure for hoar frost compared

with supercooled water is significant in the formation of snow.

The temperature of the triple point for pure water is defined to be 273.16 kelvin. This is the fixed point for the *thermodynamic temperature scale and it, together with the triple points of hydrogen and oxygen, are fixed points for the *International Practical Temperature Scale.

For some substances, notably carbon dioxide, the pressure at the triple point is above normal atmospheric pressure. These substances only have the liquid phase when in closed vessels at high pressures.

tristimulus values *See* chromaticity.

triton The nucleus of a tritium atom, containing one proton and two neutrons.

tropical year *See* time.

troposphere *See* atmospheric layers.

tsunami *See* water wave.

Tube of flux

tube of flux *Syn.* tube of force; field tube. The vector field of an electric field can be divided into tubes that are bundles of lines of electric field strength. The lateral surface of the tube is made up of lines of force so that the field strength at any point of that surface is tangential to it and the *flux is the same through all cross sections. The field strength at any part of the tube is inversely proportional to the section at that place, taken normal to the lines, the tube narrowing as the field strength increases. A unit tube is one through which unit flux flows. An analogous concept can be applied to magnetic flux density.

tuned circuit A *resonant circuit whose resonant frequency can be varied. The process of adjusting the circuit to a condition of *resonance is known as *tuning*, and may be carried out by adjusting the capacitance of the circuit (capacitive tuning) or by adjusting the inductance of the circuit (inductive tuning).

tuning fork A suitably proportioned metal bar bent into the shape of a U and mounted upon a stem at the base of the U. If excited by bowing, striking, or pressing the prongs together, a note consisting almost entirely of a fundamental is faintly heard. The tuning fork maintains a constant frequency for long periods, the frequency varying only very slightly with temperature. It can therefore be used as a standard of pitch in the tuning of musical instruments. Forks, which have been made for frequencies as high as 90 kHz, are generally made of steel, invar, or elinvar. The vibrating system is a dynamically balanced one – the prongs moving in together and out together.

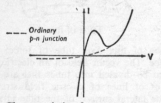

Characteristic of a tunnel diode

tunnel diode *Syn*. Esaki diode. A *p-n junction *diode that has extremely high *doping levels on each side of the junction. As a result electrons can tunnel across the junction (*see* tunnel effect) in the forward direction, i.e. with positive voltage applied to the p-region. With increasing forward bias on the junction,

the tunnel effect contributes less and less, producing on the diode characteristic a negative-resistance portion; finally the forward-voltage characteristic resembles that of an ordinary p-n junction (*see* diagram). Tunnelling also occurs in the *reverse direction, producing a large reverse current. This is similar to what occurs in a *Zener diode, but the effective Zener breakdown voltage can be considered to occur at a small positive voltage.

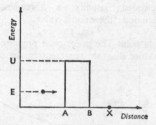

Particle approaching a potential barrier

tunnel effect The movement of particles through barriers that, on classical theory, they would have insufficient energy to surmount. Classically, if a particle moves in one direction with kinetic energy E and approaches a potential-energy barrier of height U, it can get over it by converting some of its kinetic energy into potential energy. If $E < U$ the probability of finding the particle at a point X (*see* diagram) would be zero. However, even when $E < U$ the particle can tunnel through the barrier. This is explained by wave mechanics. If the *wave function, ψ, of the particle is considered inside the region of the barrier, i.e. taking point A as $x = 0$, the *Schrödinger equation has the form:

$$\frac{\mathrm{d}^2\psi}{\mathrm{d}x^2} + \frac{8\pi^2 m}{h^2}(E - U)\psi = 0,$$

and a general solution:

$$\psi = A\exp[(2\pi ix/h)\sqrt{(2m(E - U))}],$$

where A is a constant, m the particle mass, h the Planck constant, x the distance into the barrier, and i $= \sqrt{-1}$. If $E < U$ the solution is:

$$\psi = A \exp[-(2\pi x/h)\sqrt{(2m(U - E))}].$$

Provided the barrier is not infinitely thick or wide there is thus a probability of the particle crossing this region and reaching X. The effect is too unlikely to occur in macroscpic systems but is the basis of *alpha decay and *field emission.

turbine An engine in which a shaft is rotated by fluid impinging upon a system of blades or buckets mounted upon it. *Water turbines* are used in *hydroelectric power stations; *steam turbines* are used for ship propulsion and for electric power stations using coal, oil, nuclear, or other fuel; *gas turbines* have been used particularly for aircraft propulsion but have other applications. According to the system of nozzles or blades fixed upon the framework of a steam turbine, it is conventionally classified as an *impulse turbine, reaction turbine* or *impulse-reaction turbine.*

turbo-alternator An alternator (e.g. a *synchronous alternating-current generator) intended for steam-turbine drive. Turbo-alternators are employed in thermal power stations for generating alternating current.

turbulence The state of a fluid possessing a nonregular motion such that the velocity at any point may vary in both direction and magnitude with time. Turbulent motion is accompanied by the formation of eddies and the rapid interchange of momentum in the fluid. The change from laminar to turbulent motion occurs at a critical value of the *Reynolds number. The drag resistance to a body for turbulent flow is proportional to the square of the velocity; re-

sistance for laminar flow is proportional to the velocity.

turns ratio *See* ratio.

tweeter A loudspeaker of small dimensions designed to reproduce sounds of relatively high frequency in a hi-fi system. *Compare* woofer.

twin cable *See* paired cable.

twin paradox A concept that has been discussed in terms of both the special and general theories of *relativity. It is supposed that a pair of identical twins are initially together in an *inertial system. One twin is then accelerated to a high speed and departs for a long space journey, finally returning to rest beside the other. It is predicted that the twin who has made the journey will have aged less than the one who has remained at home.

twisted pair A pair of insulated wires, twisted together to form a *transmission line; the twisting improves signal transmission. Twisted pairs are often used in high-frequency circuits as an alternative to *coaxial cable. *See* paired cable.

two-phase *Syn.* quarter-phase. Of an electrical system or device. Having two equal alternating voltages between which there is a relative *phase difference of 90°. *See* polyphase system.

Tyndall effect *See* Rayleigh scattering.

U

UHF Abbreviation for ultrahigh frequency. *See* frequency bands.

ultimate strength The limiting stress (in terms of force per original unit area of cross section) at which a material completely breaks down (i e. fractures or crushes).

ultracentrifuge A centrifuge operating at a very high angular velocity, suitable for use with colloidal solutions. It is used to separate colloidal particles and also can be used to estimate their size and measure the relative molecular masses of very large molecules, such as proteins. Quantitative instruments have a transparent cell and the formation of sediment is photographically recorded. *See* sedimentation.

ultrahigh frequency (UHF) *See* frequency bands.

ultramicrobalance A very sensitive balance for accurate weighing to 10^{-8} grams.

ultrasonic imaging *Syn.* ultrasonography. The use in medicine of ultrasonic waves (*see* ultrasonics) to produce images of soft tissues in the body. (These tissues cannot be imaged by X-rays.) At each tissue interface, some of an incident ultrasonic beam is reflected and may be detected for use in forming the image. The source and detector is often the same *piezoelectric crystal, the frequency being in the range 1–15 MHz. The technique is used routinely to check foetal growth, measure the motion and condition of heart valves, detect brain tumours, etc.

ultrasonics The study and application of mechanical vibrations with frequencies beyond the limits of hearing of the human ear, i.e. with frequencies about 20 kilohertz and upwards. There is no theoretical upper limit to the ultrasonic frequency, although ultrasonic applications are usually restricted to about 20 MHz.

Ultrasonic waves have the same basic properties as *sound waves, both being examples of *acoustic waves.

Ultrasonic waves can be generated by *piezoelectric oscillators and by *magnetostriction oscillators, both forms transforming high-frequency electric oscillations into powerful mechanical oscillations. They can also be generated mechanically by the *Galton whistle or *Hartmann generator. The piezoelectric oscillator may also be used to detect ultrasound and measure the amplitude and frequency of the waves.

An alternative method of detection is based on the *diffraction of light that occurs in a liquid traversed by ultrasonic waves. The series of compressions and rarefactions that constitute ultrasonic waves in a liquid act as a grating of light and cause optical diffraction. A narrow parallel beam of laser light, wavelength λ, emerging from a slit passes through a trough of liquid at right angles to the ultrasonic beam. Looking along the laser beam a central image of the slit is seen together with its diffraction images. If λ_s is the wavelength of the sound in the liquid and α_k is the angle of diffraction for the kth order, then:

$$\sin \alpha_k = k\lambda/\lambda_s.$$

It is a very accurate method for measuring the wavelength of sound and hence the speed of sound at ultrasonic frequency in liquids.

The applications of ultrasonics are numerous in many branches of applied and pure science. They include *sonar, *ultrasonic imaging, flaw detection, cleaning small objects by vibrating them in a solvent, welding dissimilar materials, and soldering aluminium.

ultrasonography *See* ultrasonic imaging.

ultrasound Acoustic waves of ultrasonic frequency. *See* ultrasonics.

ultraviolet astronomy *See* ultraviolet radiation.

ultraviolet catastrophe *See* quantum theory.

ultraviolet microscopy *See* resolving power; ultraviolet radiation.

ultraviolet radiation (UV radiation) *Electromagnetic radiation that is produced by a variety of sources, including the sun, and lies in the wavelength range extending from just beyond the violet end of the visible spectrum up to and overlapping longer-wavelength X-rays, i.e. about 380–13 nm. UV radiation can be grouped according to wavelength into, for example, *near ultraviolet* (approx. 380–300 nm), *extreme ultraviolet* (approx. 120–13 nm), and *vacuum ultraviolet* (< 200 nm).

Substances that may be transparent to light absorb strongly as the wavelength is decreased in the ultraviolet, e.g. ordinary crown glass at 300 nm, vita glass 252 nm, quartz 180 nm, fluorspar 120 nm. In the extreme ultraviolet, most substances become opaque or show selective absorption, and even small paths in air or gas at low pressure may show considerable absorption. Atmospheric absorption by ozone in the stratosphere restricts the solar ultraviolet at about 300 nm. Incandescent bodies even at high temperature yield a relatively small proportion of ultraviolet to total radiation. Arc and spark discharges and vacuum-tube discharge with enclosures of transparent media (quartz) are the main sources.

UV radiation can be detected and investigated by its imaging action on photographic film and plates, and also by the *photoelectric effects and *fluorescence that it produces. In *ultraviolet spectroscopy,* according to the region investigated, spectrographs with quartz or fluorite lenses and prisms, reflecting gratings enclosed in vacuum, plastic windows, and special photographic plates are used. The UV band includes many of the resonant absorption and emission lines (resulting from transitions of electrons to and from atomic ground states) of most of the more abundant chemical elements. *Ultraviolet microscopy* uses special quartz lenses and commonly works with a monochromatic source (wavelength 275 nm). This enables higher resolution and magnifications (3600 times) than can be obtained with visual observation. *Ultraviolet astronomy* can only be carried out using satellites, or balloons or rockets, due to strong absorption in the earth's atmosphere. Celestial sources include high-temperature stars, the outer atmospheres of cooler stars, and interstellar gas.

ultraviolet spectroscopy *See* ultraviolet radiation.

umbra *See* shadows.

Umklapp process A collision process between *phonons, or between phonons and electrons. The crystal momentum is not conserved (no conservation principle applies). The process is responsible for thermal resistance in nonconducting materials.

uncertainty principle *Syn.* Heisenberg uncertainty relation; indeterminacy principle. The principle, enunciated by Heisenberg, that the product of the uncertainty in the measured value of a component of momentum (p_x) and the uncertainty in the corresponding coordinate of position (x) is of the same order of magnitude as the *Planck constant. In its most precise form:

$$\Delta p_x \times \Delta x \geqslant h/4\pi,$$

where Δx represents the root-mean-square value of the uncertainty. For most purposes one can assume:

$$\Delta p_x \times \Delta x \simeq h/2\pi.$$

503

The principle can be derived exactly from *quantum mechanics but is most easily understood as a consequence of the fact that any measurement of a system must disturb the system under investigation, with a resulting lack of precision in measurement. For example, if it were possible to see an electron and thus measure its position, photons would have to be reflected from the electron. If a single photon could be used and detected with a microscope, the collision between the electron and photon would change the electron's momentum (*see* Compton effect). A similar relationship applies to the determination of energy and time, thus:

$$\Delta E \times \Delta t \geqslant h/4\pi.$$

The effects of the uncertainty principle are not apparent with large systems because of the small size of h. However the principle is of fundamental importance in the behaviour of systems on the atomic scale. For example, the principle explains the inherent width of spectral lines; if the lifetime of an atom in an *excited state is very short there is a large uncertainty in its energy and a line resulting from a transition is broad.

One consequence of the uncertainty principle is that it is impossible fully to predict the behaviour of a system and the macroscopic principle of *causality cannot apply at the atomic level. *Quantum mechanics gives a statistical description of the behaviour of physical systems.

undercurrent (or undervoltage) release A switch, circuit-breaker, or other *tripping device that operates when the current (or voltage) in a circuit falls below a predetermined value.

underdamped *See* damped.

uniaxial crystals Crystals belonging to the tetragonal, rhombohedral, and hexagonal systems. They are double refracting except for light travelling through them in the direction of the principal crystallographic axis. In this direction only they are singly refracting. Iceland spar (calcite) and quartz are uniaxial; the former is an *optically negative crystal, the latter is a positive crystal. *See* double refraction.

unified atomic mass unit *See* atomic mass unit.

unified field theory A theory that seeks to unite the properties of gravitational, electromagnetic, weak, and strong interactions so that a single set of equations can be used to predict all their characteristics. At present it is not known whether such a theory can be developed, or whether the physical universe is amenable to a single analysis in terms of the current concepts of physics. There are unsolved problems in using the framework of a relativistic *quantum field theory to encompass the four fundamental interactions and all the elementary particles. It may be that using extended objects (*see* superstrings; supersymmetry) will enable this synthesis to be achieved. *See also* grand unified theory.

unifilar suspension A type of suspension used in electrical instruments in which the moving part is suspended on a single thread, wire, or strip. The controlling (or restoring) torque is produced by the twisting of the thread, etc., which takes place when the moving part is moved from its initial position of rest.

uniform temperature enclosure An enclosure whose walls are maintained at the same constant temperature. The amount and kind of radiation within such an enclosure is independent of the nature of the walls and the contents of the enclosure, depending only on the temperature of the walls. Such radiation is identical with that emitted by a *black body at the same temperature.

unijunction transistor *Syn.* double-base diode. A *transistor consisting of a bar

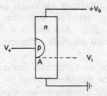

a Unijunction transistor

of lightly doped (high-resistivity) *semiconductor, usually n-type, with a region of highly doped semiconductor of opposite polarity located near the centre (Fig. *a*). Ohmic contacts are formed to each end of the bar (base 1 and base 2) and the central region (the emitter).

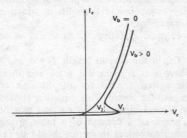

b Unijunction transistor characteristics

If a voltage V_b is applied across the bar a potential drop will be present through the bar. Let the potential on the less positive side of the junction at point A be V_1. If a voltage V_e is applied to the emitter, for $V_e < V_1$ the junction will be *reverse biased and very little current will flow. If V_e is increased to V_1, the junction will become *forward biased at point A. *Holes will be injected into the bar and attracted to

the less positive terminal (base 1), lowering the resistivity between A and base 1. Point A becomes less positive and more of the junction becomes forward biased. This causes the emitter current, I_e, to increase rapidly. As I_e increases, the decreased resistivity causes V_e to drop and the device exhibits negative resistance. The typical *characteristics are shown in Fig. *b*. The emitter voltage V_2 is the point at which the device ceases to show negative resistance. The switching time between V_1 (at which the device starts to conduct) and V_2 depends on the device geometry and the biasing voltage V_b. As V_b is increased, V_1 increases and the emitter current at V_2 also increases.

The most common use of the unijunction transistor is in *relaxation oscillator circuits.

unipolar transistor A *transistor in which current flow is due to the movement of *majority carriers only. *See* field-effect transistor.

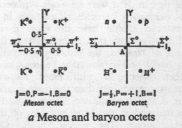

a Meson and baryon octets

unitary symmetry A generalization of *isospin theory. In theory, *group isospin is concerned with a group called SU₂ (the special unitary group of 2×2 matrices). Unitary symmetry is concerned with a group called SU₃. It predicts that as far as *strong interactions are concerned *elementary particles can be grouped into multiplets containing 1, 8, 10, or 27 particles and that the particles in each multiplet may be regarded as different states of the same particle. Unitary symmetry multiplets contain one or more isospin multiplets. All particles in a multiplet have the same spin (J), parity (P), and baryon number (B).

The multiplets are most clearly illustrated by plotting the constituent particles on a graph of *hypercharge (Y) against isospin quantum number (I_3). Examples of a meson and baryon octet (multiplet of 8 particles) and a baryon decuplet (10 particle multiplet) are illustrated in Figs. *a* and *b*.

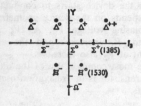

$$J=3/2, P=+1, B=1$$

b Baryon decuplet

The predictions of SU_3 theory do not agree with experiment as well as those of isospin. If it were an exact symmetry of elementary particles, all particles in a multiplet should have the same mass; this is far from true. In addition, in the case of *quarks, the resulting symmetry is badly broken now that hadrons have been discovered with more massive quark components than the three quarks (u, d, s) first postulated. The u, d, and s correspond to the basic multiplet of the SU_3 group. The singlet, octet, and decuplet are obtained by either combining three quark multiplets (for baryon multiplets) or a quark and antiquark multiplet (for meson multiplets). However, SU_3 theory has had some outstanding successes, including the prediction of the existence of the Ω^- (needed to complete the baryon decuplet).

unit cell The smallest crystal unit possessing the entire symmetry of the whole periodic structure. It is defined in terms of six elements or parameters: *a*, *b*, *c*, the lengths of the cell edges (taken as axes), and α, β, γ, the angles between the axial directions.

unit pole A magnetic pole that when placed 1 cm from an equal pole, in a vacuum, is repelled with the force of one dyne. This CGS unit is no longer in use. *See also* magnetic monopole.

units Since its earliest beginnings physical science has been beset by entirely unnecessary complications arising from the variety and diversity of the units that have been used to express the magnitudes of physical quantities. These complications have finally been resolved by the adoption of the internationally agreed system of units known as *SI units (Système Internationale d'Unités). This system is based on the *MKS system and replaces the *CGS and *f.p.s. systems. SI units make quite clear the relationship between a physical quantity (say length l) and the units in which it is expressed, i.e.

Physical quantity = numerical value × unit

Or, in the case of length, $l = n$ m, where n is a numerical value and m is the agreed symbol for the SI unit of length, the metre. This equation should be treated algebraically, e.g. the axis of a graph or a column of a table giving numerical values in metres should be labelled l/m. If the axis or column gives the values of $1/l$, it should be labelled m/l. *See also* rationalization of electric and magnetic quantities.

unit vector *Vectors, usually written i, j, and k, that have unit length and lie along the x-, y-, and z-axes respectively. A vector function, F, can therefore be written:

$$F = xi + yj + zk.$$

As the angles between these vectors are 90°, the scalar products and vector products (*see* vector) are either equal to zero or one.

universal coordinated time *See* time.

universal gas constant Former name for *molar gas constant.

universal motor An electric motor that is suitable for use with direct current or alternating current. It incorporates a *commutator and is usually *series-wound. Small motors of this type are commonly used in electrical appliances such as vacuum cleaners, portable drills, etc.

universal shunt A galvanometer shunt that is tapped so that it can pass 1/10, 1/100, 1/1000, etc., of the main current through the galvanometer, whatever resistance this has.

unsaturated vapour A vapour at a certain temperature that contains less than the equilibrium amount of the substance in the gaseous phase. Such a vapour may undergo slight isothermal compression without condensation occurring, and obeys the *ideal gas laws approximately.

unstable equilibrium *See* equilibrium.

unstable oscillation Any oscillation – mechanical, electrical, etc. – that tends to increase in amplitude with time.

Unwin coefficients *See* Reynolds's law.

upper atmosphere The outer layers of the gaseous envelope surrounding the earth above about 30 km. It includes part of the stratosphere. *See* atmospheric layers.

uranium–lead dating *See* dating.

uranium series *See* radioactive series.

UV Abbreviation for ultraviolet.

V

vacancy A position in a crystal lattice that is not occupied by an atomic nucleus. A vacancy should not be confused with a *hole. *See* defect.

vacuum 1. Strictly, a physical state totally devoid of particles – either particles of matter or photons of radiation (*compare* free space). Such a state does not exist in practice.
2. In vacuum technology, any space in which the pressure is below normal atmospheric pressure. The degree or quality of the vacuum attained – coarse, medium, high, very high, ultrahigh – is indicated by the total pressure of the residual gases. The highest vacuums achieved on earth have a pressure of less than 10^{-8} N m^{-2}.

vacuum evaporation A technique used for producing a coating of one solid on another, usually a coating of metal or semiconductor. Evaporation occurs from a solid or liquid at high temperature in a vacuum. At low pressures atoms leaving the hot surface suffer few collisions in the gas and can travel directly to a nearby cool surface where they condense, thus forming a thin film.

vacuum flask *See* Dewar vessel.

vacuum gauge *See* pressure gauges.

vacuum pump *See* pumps, vacuum.

vacuum tube An *electron tube evacuated to a sufficiently low pressure that its electrical characteristics are independent of any residual gas. *Compare* gas-filled tube.

vacuum ultraviolet Part of the ultraviolet region of the spectrum, in which the radiation is absorbed by air and experiments have thus to be performed in a vacuum. It is the part of the ultraviolet region having a wavelength less than about 200 nanometres and the highest quantum energies.

valence band *See* energy band.

valence electrons Electrons in the outermost shell of an atom that are involved in chemical changes. *See also* energy bands.

valency The number of hydrogen (or equivalent) atoms that an atom will combine with or displace.

valve *Syn.* electron tube. A device in which two or more electrodes are enclosed in an envelope usually of glass, one of the electrodes being a primary source of electrons. The electrons are most often provided by thermionic emission (*thermionic valve), and the device may be either evacuated (*vacuum tube) or gas-filled (*gas-filled tube). The word valve is becoming obsolete and is being replaced by the term electron tube.

Van Allen belts *Syn.* radiation belts. Two belts of energetic charged particles, mainly electrons and protons, lying around the earth, across the equatorial plane. The particles are confined there by the earth's magnetic field; they oscillate back and forth between the magnetic poles, spiralling around the field lines and emitting radiation as they do so. The inner belt is approximately 1000 to 5000 km above the equator and contains more energetic particles. These are electrons and protons captured from the *solar wind or produced as secondary products of *cosmic-ray bombardment in the upper atmosphere. The outer belt is approximately 15 000 to 25 000 km

above the equator, curving down towards the magnetic poles. It contains mainly electrons from the solar wind.

Van de Graaff accelerator A type of *accelerator in which the high-voltage terminal of a *Van de Graaff generator is used as a source of charged particles.

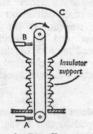

Van de Graaff generator

Van de Graaff generator A high-voltage electrostatic generator that can produce potentials of millions of volts. It consists essentially of an endless insulated fabric belt moving vertically (*see* diagram). A charge is applied to the belt by the needle points A from an external source of up to 100 kV. This charge is carried continuously up into a large hollow sphere C where collector points B remove it. The potential of the sphere continues to increase (no charge residing on the interior) and is limited only by the leakage rate of the supporting insulators and the surrounding gas. The fabric belt may be replaced by a series of metal beads connected by insulating string. *See also* tandem generator.

van der Waals equation of state An *equation of state describing both the gaseous and liquid phases of a substance and widely used for the approximate analysis of fluid properties. It can be written for one mole of fluid:
$$(p + a/V^2)(V - b) = RT,$$
where p is pressure, V volume, T the thermodynamic temperature, R the mo-

lar gas constant, and *a* and *b* are constants characteristic of a given substance. The quantity *b* allows for the volume effectively occupied by molecules because of the very short-range repulsive forces and a/V^2 allows for the short-range attractive forces. The equation gives a more accurate account of the properties of gases than does the *ideal gas equation. It is less accurate for the liquid phase.

van der Waals forces Intermolecular and interatomic forces that are electrostatic in origin. If two identical molecules have permanent *dipole moments and are in random thermal motion then some of their relative orientations cause repulsion and some attraction. On average there will be a net attraction. A molecule with a permanent dipole can also induce a dipole in a similar neighbouring molecule and cause mutual attraction. These dipole-dipole and dipole-induced dipole interactions cannot occur between atoms.

The van der Waals forces between single atoms arise because of small instantaneous dipole moments in the atoms themselves. For example hydrogen, with one proton and one electron, has a symmetrical electron cloud about the nucleus. However, at any instant there is an asymmetric charge distribution that depends on the position of the electron. Thus the atom can be considered to have a fluctuating rotating dipole moment. There is no net attraction between dipoles of neighbouring atoms because there is insufficient time for orientation to occur. However, an instantaneous dipole in one atom can polarize a similar neighbouring atom and hence cause attraction.

The three types of interaction all have the form: $E_p = -A/r^6$, where E_p is the potential energy of the two molecules or atoms, *r* is the molecular or atomic separation, and *A* is a constant for a particular atom or molecule. The full expression for potential energy includes a term representing repulsion between the atoms. E_p is called the *Lennard-Jones potential*. Van der Waals forces are responsible for departures from *ideal gas behaviour in real gases. They are also the forces between molecules or atoms in liquids and in nonionic solids.

van't Hoff factor *See* osmosis.

vaporization The process of change into a *vapour or into the gaseous state. The rate of vaporization per unit surface area of solid or liquid at temperature *T* is given by:
$$\mathrm{d}m/\mathrm{d}t = \alpha p\sqrt{(M/2\pi RT)},$$
where *p* is the vapour pressure of the vapour of relative molecular mass *M*, *R* is the molar gas constant, and α is the *vaporization coefficient*, which is necessarily less than unity.

vapour A substance, in gaseous form but below its *critical temperature so that it could be liquefied by pressure alone, without cooling to a lower temperature.

vapour-absorption cycle *See* refrigerator.

vapour-compression cycle *See* refrigerator.

vapour concentration *See* humidity.

vapour density The mass per unit volume of a vapour under specified conditions of temperature and pressure.

vapour pressure The pressure exerted by a *vapour. For a vapour in equilibrium with its liquid, the vapour pressure depends only on the temperature of the liquid for a plane surface and is known as the *saturated vapour pressure (SVP) at that temperature. The variation of vapour pressure with temperature over a small range is given by the *Kirchhoff formula.

vapour-pressure thermometer A thermometer that uses the fact that the (saturation) vapour pressure of a liquid is a function only of temperature. Thermometers of this type are the most reliable for the measurement of temperatures below the boiling point of helium ($-268\,°C$). For this application, He, NH_3, SO_2, CO_2, CH_4, C_2H_4, O_2, and H_2 may be used as the working liquid in the thermometer; their range is very limited since very high or very low pressures cannot be conveniently measured.

var A unit of *power identical to the *watt but used for the reactive power of an alternating current. One var is one volt times one ampere.

varactor A semiconductor *diode, operated with reverse bias so that it acts as a voltage-dependent capacitor. The *depletion layer at the junction acts as the dielectric, the n- and p-regions act as the plates. A diode used in this way is usually designed to have an unusually large capacitance. The depletion-layer width, and therefore the capacitance, depend on the voltage across the junction. If the semiconductor type changes abruptly from n-type to p-type then $C \propto V^{-1/2}$.

If it changes gradually (linearly graded junctions) then $C \propto V^{-1/3}$. A *Schottky diode can be used as a varactor in a similar way.

varactor tuning A means of tuning employed in receivers (e.g. television receivers) in which *varactors are used as the variable-capacitance elements.

variable star A star whose physical properties, most noticeably brightness, vary either regularly or irregularly with time. The variation in an *intrinsic variable* is caused by changes in internal conditions. A *pulsating star* is of this kind, the light variation being due to expansion and contraction of the surface layers of the star. The variation can also be due to external causes. Two stars rotating around a common centre of gravity (binary stars) can sometimes eclipse one another if the plane of the orbit lies in the line of sight. These are known as *eclipsing binaries*. *Novas are an example of *catacysmic variables*, the brightness increasing by an enormous amount in a very short time.

variance The square of the standard *deviation.

variometer A variable inductor that usually consists of a fixed coil connected in series with a movable coil so that by moving (rotating) the latter the coupling between the two coils, and hence also the self-inductance of the series combination, may be varied.

varistor A *resistor with characteristics that do not follow Ohm's law. It may be formed from a semiconductor *diode. A symmetrical varistor consists of two diodes connected in parallel with opposite polarity. This arrangement exhibits the forward current-voltage characteristic of a diode in either direction of applied voltage, and may be used as a *voltage limiter*.

V-beam radar *See* radar.

VDU Abbreviation for *visual-display unit.

vector A quantity with magnitude and direction. It can be represented by a line whose length is proportional to the magnitude and whose direction is that of the vector, or by three components in a rectangular coordinate system. (*See also* unit vector.)

A true vector, or *polar vector*, involves a displacement or virtual displacement. Polar vectors include velocity, acceleration, force, electric and magnetic field strength. The signs of their components are reversed on reversing the coordinate axes. Their dimensions include length to an odd power.

A *pseudovector*, or *axial vector*, involves the orientation of an axis in space. The direction is conventionally obtained in a right-handed system by sighting along the axis so that the rotation appears clockwise. Pseudovectors include angular velocity, vector area, and magnetic flux density. The signs of their components are unchanged on reversing the coordinate axes. Their dimensions include length to an even power.

Polar vectors and axial vectors obey the same laws of vector analysis.

(*a*) *Vector addition.* If two vectors *A* and *B* are represented in magnitude and direction by the adjacent sides of a parallelogram, the diagonal represents the vector sum (*A* + *B*) in magnitude and direction. Forces, velocities, etc., combine in this way.

(*b*) *Vector multiplication.* There are two ways of multiplying vectors.

(i) The *scalar product* of two vectors equals the product of their magnitudes and the cosine of the angle between them, and is a scalar quantity. It is usually written

$A \cdot B$ (read as A dot B).

(ii) The *vector product* of two vectors *A* and *B* is defined as a pseudovector of magnitude $AB \sin \theta$, having a direction perpendicular to the plane containing them. The sense of the product along this perpendicular is defined by the rule: if *A* is turned towards *B* through the smaller angle, this rotation appears clockwise when sighting along the direction of the vector product. A vector product is usually written

$A \times B$ (read as A cross B).

Vectors should be distinguished from scalars by printing the symbols in bold italic letters.

vector field A *field, such as a gravitational field or magnetic field, in which the magnitude and direction of the vector quantity are one-valued functions of position. They can be mapped by curved lines whose direction at any point is that of the vector and whose density (i.e. number per unit area crossing an infinitesimally small area perpendicular to the lines) is proportional to the magnitude of the vector at the point. These lines are called lines of flux (or force).

vector product *See* vector.

velocity 1. Linear velocity, symbol: *v*. The average velocity is the displacement divided by the time taken. Instantaneous velocity is the rate of change of displacement. Linear velocity is a polar *vector. The magnitude of the instantaneous velocity is called *speed*, which is a scalar. In precise technical usage velocity and speed are clearly distinguished, the former only being used when the direction of motion is specified. The distinction is particularly significant in the case of a body moving with constant speed in a curved path. As the direction is changing the velocity is not constant although its magnitude is, so the body has an *acceleration and must be subject to a resultant force (*see* Newton's laws of motion). The term velocity is however often used loosely to mean speed.

2. *See* angular velocity.

velocity modulation If a beam of electrons passes through a sharply defined region, such as a *cavity resonator, and is subjected to a radio-frequency field, the individual electrons will be retarded or accelerated according to the half-cycle prevailing when they enter the region. If retarded, the electrons following will catch up (and the converse) so that the total effect will be a *bunching* of the electron beam into a series of pulses similar to the rarefactions and compressions of a sound wave. Such a beam is said to be velocity modulated. Velocity modulation is employed for the amplification and generation of microwave frequencies in *electron tubes such as the *klystron and *travelling-wave tube.

velocity potential If the velocity of a point (x, y, z) of a fluid has components (Cartesian) u, v, w, and there exists a scalar function ϕ such that:

$$u = -\partial\phi/\partial x, \qquad v = -\partial\phi/\partial y,$$
$$w = -\partial\phi/\partial z,$$

then the motion is irrotational and ϕ is called the velocity potential. The negative sign is conventional and is sometimes omitted.

velocity ratio *See* machine.

vena contracta When a jet issues from an orifice in a reservoir tank the change in direction of the stream lines is not completed at the orifice but continues past the orifice causing the subsequent jet section to be smaller than that of the orifice. The point of the jet where the contraction is complete is called the vena contracta, and at this point the stream lines are all parallel and the pressure of the fluid is approximately that of the surrounding medium.

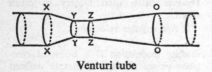

Venturi tube

venturi tube A device used for measuring the quantity of fluid flowing through a pipe. The principal features are: the inlet section, XY (*see* diagram), which converges to the throat, YZ, which consists of a short straight portion of the pipe. From the throat the pipe diverges again, ZO, usually to the original size. The quantity of fluid flowing per second through the pipe is given by:

$$Q = \frac{A_1 A_2}{\sqrt{A_1^2 - A_2^2}} \sqrt{\frac{2(p_1 - p_2)}{\rho}}$$

where A is the area of cross section, p the pressure (static) of the fluid, ρ the fluid density; the suffixes 1 and 2 refer to the inlet and throat respectively.

vergence The convergence and divergence of rays. Reciprocal distances measure vergence. *Reduced vergence* is the reciprocal of a *reduced distance. The change of vergence of rays is equal to the focal power of an optical element.

vernal equinox *See* equinoxes.

vernier scale A short scale sliding on the main scale of a length- or angle-measuring instrument. It is used for determining the fraction of the smallest interval into which the main scale is divided by the instrument pointer, which is the zero division of the vernier scale. For example, in the diagram the vernier scale has 10 intervals equal to 9 intervals on the main scale. The zero division of the vernier (the pointer) reads 101.4 divisions on the main scale. The fractional part of the reading is found from the coincidence of divisions on the two scales.

vertex focal length The distance measured from the last surface of a thick lens or combination of lenses to the principal focus. Its reciprocal is the *vertex power*.

vertical polarization *See* horizontal polarization.

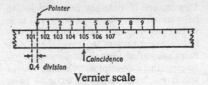

Vernier scale

very high frequency (VHF) *See* frequency bands.

very low frequency (VLF) *See* frequency bands.

VHF Abbreviation for very high frequency. *See* frequency bands.

vibration The rapid to and fro motion characteristic of an elastic solid (e.g. a tuning fork) or a fluid medium influenced by such a solid. The time occupied in each to and fro motion is constant and is called the *period. The frequency is the number of vibrations per unit time and is, therefore, the inverse of the period.

vibrational quantum number *See* energy level.

vibration galvanometer A moving-coil *galvanometer having the coil suspended on a wire, under tension. It can be tuned for frequencies between 40 and 1000 Hz, by varying the tension of the wire. Small a.c. currents at the resonant frequency can produce a large response, so that the device can be used as a current detector in a.c. bridges, etc.

vibration-rotation spectrum *See* spectrum.

vibrator A device for producing an alternating current by periodically interrupting or reversing the current obtained from a direct-current source. It is operated electromagnetically and has a vibrating armature that alternately makes and breaks one or more pairs of contacts. It is most commonly used in a power-supply unit that is required to produce direct current at high voltage from a low-voltage d.c. source, such as a battery. The vibrator produces a low-voltage a.c. supply that a transformer converts into a high-voltage a.c. supply, and the latter is then rectified to produce a high-voltage d.c. supply.

video amplifier *See* video frequency.

videocassette, videocassette recorder *See* videotape.

video frequency The frequency of any component of the signal produced by a *television camera. Video frequencies lie between 10 Hz and 2 MHz. An amplifier designed to amplify video-frequency signals is called a *video amplifier*.

video signal *See* television camera.

videotape A form of magnetic tape that is suitable for use with a *television camera. Simultaneous recording of the video signal from the TV camera and the audio signal from the microphone system is carried out on separate tracks on the videotape. Many TV programmes are recorded on videotape before they are transmitted.

A form of videotape recorder is available for use with domestic TV receivers. The tape is enclosed in a container for protection and easy handling; the package is called a *videocassette*. The *videocassette recorder* (VCR) can be used to record a TV program on tape, which can then be subsequently replayed directly into the TV receiver, or it can replay a prerecorded videocassette.

vidicon *See* camera tube.

vignetting The progressive reduction in the cross-sectional area of a beam of light passing through an optical system as the obliquity of the beam is increased. It is due to obstruction of the beam by mechanical apertures, lens mounts, etc., of the system.

virgin neutrons Neutrons from any source that have not been involved in any collisions.

virial In a system in which an atom having coordinates (x, y, z) is acted upon by a force having components X, Y, Z parallel to these axes, the virial is defined as the average value with respect to time of the sum of all expressions of the form:

$$-\tfrac{1}{2}(xX + yY + zZ).$$

virial expansion A relation expressing the behaviour of a real gas:

$$pV = RT + Bp + Cp^2 + Dp^3 + \\ \cdots$$

The empirical constants B, C, D $\cdots$ are known as the 2nd, 3rd, 4th, $\cdots$ *virial coefficients*.

virial law (due to Clausius) The mean kinetic energy of a system is equal to its *virial, which depends only on the forces acting on the atoms and not on their motions.

virtual cathode In a *thermionic valve. The surface, situated in the region of a *space charge, at which the potential is a mathematical minimum and the potential gradient (or electric force) is zero. It acts as if it were a source of electrons.

virtual image *See* image.

virtual particle Because of the *uncertainty principle it is possible for the law of *conservation of mass and energy to be broken by an amount ΔE providing this only occurs for a time Δt such that:

$$\Delta E \Delta t \leqslant h/4\pi.$$

This makes it possible for particles to be created for short periods of time where their creation would normally violate conservation of energy. These particles are called virtual particles. For example, in a complete vacuum – in which no "real" particles exist – pairs of virtual electrons and positrons are continuously forming and rapidly disappearing (in less than 10^{-23} second). Other conservation laws such as those applying to angular momentum, *isospin, etc., cannot be violated even for short periods of time.

virtual work principle A system with workless constraints is in equilibrium under applied forces if, and only if, zero (virtual) work is done by the applied forces in an arbitrary infinitesimal displacement satisfying the constraints. *See* constrain.

viscometer A device for measuring the viscosity of a fluid (liquid or gas). The main types include those based on the flow of fluids through capillary tubes (*see* Poiseuille flow), the torque needed to keep two coaxial cylinders in rotation when the space between is filled with the liquid under test, and the rate of damping of a vibrating body by the fluid. *See also* Ostwald viscometer.

viscosity The property of fluids by virtue of which they offer a resistance to flow for low values of the *Reynolds number. Newton's law of viscous flow for streamline, as opposed to turbulent, liquid motion is:

$$F = \eta A \, dv/dx,$$

where F is the tangential force between two parallel layers of liquid of area A, dx apart, moving with a relative velocity dv. The quantity η is called the *coefficient of viscosity* (or just the viscosity) of the liquid and is measured in $\mathrm{kg\,m^{-1}\,s^{-1}}$ or $\mathrm{N\,s\,m^{-2}}$. Viscosity of a liquid usually decreases with temperature but that of a gas increases. A very large number of liquids obey Newton's law in that the viscosity is independent of the velocity gradient (dv/dx); these are called *Newtonian fluids*.

On the *kinetic theory, the viscosity of a gas is given by the equation:

$$\eta = \frac{1}{3}\rho\bar{C}L,$$

where ρ is the density, $\bar{C}$ the mean velocity, and L the mean free path. Since the product (ρL) is independent of the pressure, on this theory the viscosity of a gas should be independent of the pressure (Maxwell's law). This is confirmed except at very low pressures when the law breaks down, the effective viscosity becoming proportional to the pressure.

See also anomalous viscosity; kinematic viscosity.

viscosity gauge *See* molecular gauge.

viscosity manometer *See* molecular gauge.

viscous damping Damping in which the opposing force is proportional to velocity, as with the damping resulting from *viscosity of a fluid or from *eddy currents.

viscous flow *See* Newton's laws of fluid friction; viscosity.

visible spectrum The continuous *spectrum of visible radiation, i.e. radiation lying in the wavelength range between approximately 380 and 780 nm. It is seen in the *rainbow and in the display of colours produced when a beam of white light is dispersed by a prism or *diffraction grating. There is a continuous variation of wavelength but six colours are usually distinguished: violet, blue, green, yellow, orange, and red; red is the component of longest wavelength. The value of the longest wavelength to which the eye is sensitive depends on the brightness.

visual acuity Keenness of vision, measured as the minimum angular separation of two points of light at which they can just be resolved by the eye.

visual angle The angle subtended by an object at the nodal point of the eye.

visual-display unit (VDU) A device that displays the output from a *computer temporarily on a screen, generally that of a *cathode-ray tube. The information may be in the form of letters, numbers, and other characters or may be graphical, e.g. diagrams, graphs, etc. A VDU is usually paired with a keyboard, by which information can be fed into the computer, and there are often other input devices such as a mouse or light pen.

VLF Abbreviation for very low frequency. *See* frequency bands.

VLSI *See* integrated circuit.

V-number *See* Abbe number.

voice frequency *See* audiofrequency.

Voigt effect The *double refraction produced when light traverses a vapour acted on by a transverse magnetic field, the vapour acting as a *uniaxial crystal with axis parallel to the field direction.

volt Symbol: V. The *SI unit of electric *potential, *potential difference, and *electromotive force, defined as the potential difference between two points on a conductor carrying a current of one ampere when the power dissipated is one watt. In practice voltages are measured by comparison with the electromotive force of a *Weston standard cell using a *potentiometer.

voltage Symbol: *V*. A term loosely used for the potential difference between two specified points in a circuit or device or for electromotive force. It is expressed in volts.

voltage amplifier *See* amplifier.

voltage between lines *Syn.* line voltage; voltage between phases. Of an electrical power system. The voltage between the two lines of a *single-phase system, or between any two lines of a symmetrical *three-phase system.

voltage divider *See* potential divider.

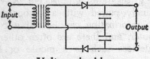

Voltage doubler

voltage doubler An arrangement of two rectifiers to give double the voltage output of a single rectifier. The diagram shows a typical circuit for *diode rectifiers.

voltage drop The voltage between any two specified points of an electrical conductor (such as the terminals of a circuit element or component) due to the flow of current between them. The voltage drop is equal to the product of the current and the resistance between the two points (for direct current) or the product of the current and the impedance between the two points (for alternating current). In the case of a.c. the product of the current and the resistance gives the *resistance drop*, which is in phase with the current, whereas the product of the current and the reactance gives the *reactance drop*, which is in *quadrature with the current.

voltage limiter *See* varistor.

voltage ratio *See* ratio.

voltage stabilizer A device or circuit designed to maintain a voltage at its output terminals that is substantially constant and independent of either variations in the input voltage or in the load current. A typical circuit uses a *Zener diode connected across the output load to regulate the voltage.

voltage transformer *Syn.* potential transformer. An instrument transformer utilizing the voltage-transformation property of a *transformer. The primary winding is connected to the main circuit and the secondary winding is connected to a measuring instrument (e.g. voltmeter). Voltage transformers are extensively used to extend the range of a.c. instruments and to isolate instruments from high-voltage circuits.

voltaic cell *See* cell.

voltameter Former name for coulombmeter.

volt-amperes Symbol: VA. The unit of apparent electric *power, defined as the product of *root-mean-square values of voltage and current in an alternating-current circuit. *See* active volt-amperes; reactive volt-amperes.

voltmeter A device for measuring *potential differences. Voltmeters in common use include *digital voltmeters, *cathode-ray oscilloscopes, and d.c. instruments such as permanent-magnet moving-coil devices. Voltmeters should cause no appreciable disturbance in the circuit to which they are connected. They therefore require very high input impedances so that very little current is taken from the circuit. Digital voltmeters and oscilloscopes comply with this requirement, but a large series impedance is used with moving-coil instruments to increase their input impedances.

volume 1. Symbol: V. The amount of space occupied by a body. It is measured in cubic metres or litres.
2. The general loudness of sounds, or the magnitude of transmitted audiofrequency signals giving rise to sounds. *See* automatic gain control; volume compressors (and expanders).

volume charge density *Syn.* volume density of charge. *See* charge density.

volume compressors (and expanders) A compressor is an electrical device that automatically reduces the range of amplitude variations of an audiofrequency (a.f.) signal in a transmission system: it decreases the amplification when the signal amplitude exceeds a predetermined value and increases the amplification when the signal amplitude is less than a second predetermined value. A volume expander is a device that produces the opposite effect to a compressor, i.e. it automatically extends the range of the amplitude variations of the transmitted a.f. signal. With suitable design, an expander included in one part of the system can be made to compensate for the effect of a compressor in another part of the system. A compressor and an expander used in this manner are together described as a *compandor*.

volume density of charge *Syn.* volume charge density. *See* charge.

volume elasticity *See* modulus of elasticity.

volume expander *See* volume compressors (and expanders).

volume strain *Syn.* bulk strain. *See* strain.

vortical field *See* curl.

vorticity In the three-dimensional motion of a fluid the velocity of an element of fluid at the point (x, y, z) has Cartesian components u, v, w. The general motion of the element is three-part: (a) a general translation; (b) a pure strain motion; (c) a rotational motion of the whole element about an instantaneous axis. The component angular velocities in (c) are given by $\frac{1}{2}\xi$, $\frac{1}{2}\eta$, $\frac{1}{2}\zeta$, where

$$\xi = \frac{\partial w}{\partial y} - \frac{\partial v}{\partial z}, \eta = \frac{\partial u}{\partial z} - \frac{\partial w}{\partial x}, \zeta = \frac{\partial v}{\partial x} - \frac{\partial u}{\partial y}$$

The vector having the components ξ, η, ζ is called the vorticity at the point (x, y, z).

W

Wadsworth prism An equilateral glass prism with a plane mirror at 45° with the base. A ray passing through the prism at minimum deviation is reflected by the mirror to 90° deviation from the incident ray.

wafer *See* chip.

wall effect Any significant effect of the inside wall of a container or reaction vessel on the behaviour of the enclosed system. Some examples are: (i) The contribution to the current in an *ionization chamber made by electrons that are liberated from the inside walls rather than from the enclosed gas. (ii) The loss of ionization in a *counter when some of the energy of the primary ionizing radiation is absorbed in the chamber wall rather than in the enclosed gas.

wall energy The energy per unit area of the boundary between the *domains in

517

a ferromagnetic material (*see* ferromagnetism).

warble tone A tone in which the frequency varies cyclically between two limits. The frequency variation is usually small compared with the actual frequency of the note and the warble occurs several times per second. A warble tone can be produced from an oscillator by using a small variable capacitor in the tuned circuit.

water equivalent The mass of water that would have the same *heat capacity as a given body. It is numerically equal to the product of the body's mass and its specific heat capacity.

water waves Waves may be set up in the free surface of a liquid or in the interface between two liquids by a disturbance of the plane surface, as when the wind blows over a sea or lake. The restoring force is gravity.

In the simplest form of surface wave that can be set up in deep water, the individual particles in the surface trace out circles with a frequency f. From the aspect of an observer travelling with the waves at their velocity c, the flow is steady and *Bernoulli's theorem can be applied. Hence it can be shown that:
$$c = g/2\pi f = \sqrt{(g\lambda/2\pi)},$$
where g is the acceleration of free fall. This treatment ignores surface tension, which is important for short waves and modifies their velocity. (*See* ripples.)

The amplitude A_x of the wave at depth x is given by:
$$A_x = A_0 \exp(-2\pi x/\lambda),$$
where A_0 is the amplitude at the surface.

For a liquid that is not deep compared with the wavelength, the speed of the wave is affected by the depth. For very shallow liquid of depth h the speed is given by:
$$c = \sqrt{hg}.$$

An earthquake under the ocean can generate a large wave of extremely long wavelength. As this reaches shallow water the speed decreases and the amplitude increases correspondingly. The wave, which can be very destructive, is known by the Japanese name *tsunami*.

watt Symbol: W. The *SI unit of *power (mechanical, thermal, and electrical), defined as the power when work of one joule is done in one second, or an equal heat transfer occurs in one second. In electrical circuits one watt is the product of one ampere and one volt.

watt-hour A unit of work or energy, equal to one watt operating for one hour (equal to 3.6×10^3 joules).

wattmeter An instrument used for measuring electric *power – active power – and having a scale graduated in watts, multiples of a watt or submultiples of a watt. It is commonly an *electrodynamometer connected so that the couple produced is proportional to the active power in the main a.c. or d.c. circuit. A *quadrant electrometer can also be connected up to measure power directly.

wave A time-varying quantity that is also a function of position. It is a disturbance, either continuous (e.g. sinusoidal) or transient, travelling through a medium by virtue of the elastic and inertia factors of the medium, or of magnetic and electric properties of space, the resulting displacements (mechanical, electric, etc.) of the medium being relatively small and returning to zero when the disturbance has passed. The general impression therefore in the transmission of a wave is that the particles of the medium vibrate relatively to each other in such a way that the wave appears to travel bodily forward with a velocity given by the velocity of the wave motion. *See* progressive wave; travelling wave; standing wave.

wave analyser An instrument, such as a *spectrum analyser, that resolves a given waveform into its fundamental and harmonic components. The analysis may be made manually or it may be made automatically according to the design of the instrument. The result is expressed in the form of frequencies and amplitudes of the various components.

waveband A range of wavelengths in the electromagnetic spectrum, defined according to some property of the radiation, or some requirement or functional aspect of a detecting or transmitting system.

wave energy *See* renewable energy sources.

wave equation The partial differential equation:

$$\frac{\partial^2 U}{\partial x^2} + \frac{\partial^2 U}{\partial y^2} + \frac{\partial^2 U}{\partial z^2} = \frac{1}{c^2}\frac{\partial^2 U}{\partial t^2}$$

(or its counterpart in one or two dimensions or in other coordinates), the solution of which represents the propagation of displacements U as waves with velocity c. *See also* Schrödinger wave equation.

waveform *Syn.* waveshape. Of a periodic quantity. The shape of the graph obtained by plotting the instantaneous values of the quantity against time. The waveform is usually described as being distorted if it is not *sinusoidal. In acoustics the waveform determines the *quality of the sound.

wavefront 1. A surface over which the oscillations in a wave have the same phase. The surface is normal to rays in isotropic media; in doubly refracting media a pair of wavefronts progress (forming the *wave surface*) and it is only for the ordinary wave that the wavefront is normal to the ordinary ray. The optical path between two successive positions of a wavefront, measured along rays, is constant.

2. *See* impulse voltage (or current).

wave function Symbol: ψ. A mathematical quantity analogous to the amplitude of a wave which appears in the equations of *wave mechanics, particularly the *Schrödinger wave equation. The most generally accepted interpretation is that $|\psi|^2 dV$ represents the probability that a particle is located within the volume-element dV (*see* de Broglie waves). ψ is often a complex quantity.

The analogy between ψ and the amplitude of a wave is purely formal. There is no macroscopic physical quantity with which ψ can be identified (in contrast with, for example, the amplitude of an electromagnetic wave, which is expressed in terms of electric and magnetic field intensities).

In general there is an infinite number of functions satisfying a wave equation but only some of these will satisfy the boundary conditions. ψ must be finite and single-valued at every point, and the spatial derivatives must be continuous at an interface. For a particle subject to a law of conservation of numbers (*see* fermion), the integral of $|\psi|^2 dV$ over all space must remain equal to 1, since this is the probability that it exists somewhere. To satisfy this condition the wave equation must be of the first order in $(d\psi/dt)$. Wave functions obtained when these conditions are applied form a set of *characteristic functions of the Schrödinger wave equation. These are often called *eigenfunctions* and correspond to a set of fixed energy values in which the system may exist, called *eigenvalues*. Energy eigenfunctions describe *stationary states of the system.

For certain bound states of a system the eigenfunctions do not change sign on reversing the coordinate axes. These states are said to have *even *parity*. For other states the sign changes on space reversal and the parity is said to be *odd*.

waveguide A hollow metal conductor containing a dielectric (usually air) down which *travelling waves are propagated. Waveguides are thus used as *transmission lines, especially for UHF radio waves. They have much lower attenuation than coaxial cables at these high frequencies, and also have a higher power-carrying capacity and a simpler construction. The conductor is usually of rectangular or circular cross section, but irregular shapes are also used for special applications.

Electromagnetic waves can be excited in a waveguide by the electric and magnetic fields associated with waves present in another device, such as a *cavity resonator or *microwave tube. Source and waveguide are connected so as to achieve the optimum transfer of energy. Energy can be extracted in a similar manner. Energy may also be transferred by using a probe to which a voltage is applied or a coil that carries a current. Again, energy can be similarly extracted.

The electromagnetic wave in a waveguide can have an infinite number of *transmission modes*, characterized by the electric and magnetic field patterns of the wave. In general these modes are of two kinds: in *transverse electric (TE) modes* the electric vector is always perpendicular to the direction of propagation; in *transverse magnetic (TM) modes* the magnetic vector is perpendicular to the propagation direction. Physical constraints and the frequency of the wave usually limit the number of modes. For each mode there is a cut-off frequency, determined by the size and shape of the guide; waves below this frequency cannot be propagated by that mode. For any transmission frequency it is usually possible to choose the dimensions of the waveguide so that only one mode is above the cut-off frequency and all other modes are rapidly attenuated. In a rectangular guide this *dominant mode* is a TE mode for which the cut-off wavelength is twice the wide dimension.

A waveguide is a completely shielded transmission line, and may be bent and twisted with no radiation loss as long as the cross section remains uniform. A change in dimensions amounts to a change in characteristic impedance.

wavelength Symbol: λ. The least distance in a progressive wave between two surfaces with the same phase. If v is the *phase speed and ν the frequency, the wavelength is given by $v = \nu\lambda$. For *electromagnetic waves the phase speed and wavelength in a material medium are equal to their values in free space divided by the *refractive index. The wavelengths of spectral lines are normally specified for free space.

Optical wavelengths are measured absolutely using interferometers or diffraction gratings, or comparatively using a prism *spectrometer.

The wavelength can only have an exact value for an infinite wave train. If an atomic body emits a quantum in the form of a train of waves of duration τ the fractional uncertainty of the wavelength, $\Delta\lambda/\lambda$, is approximately $\lambda/2\pi c\tau$, where c is the speed in free space. This is associated with the indeterminacy of the energy given by the *uncertainty principle.

See also Doppler effect; redshift.

wave mechanics One of the forms of *quantum mechanics, due to de Broglie and extended by Schrödinger, Dirac, and many others. It originated in the suggestion that light consists of corpuscles as well as of waves and the consequent suggestion that all elementary particles are associated with waves. Wave mechanics is based on the *Schrödinger wave equation describing the wave properties of matter. It relates the energy of a system to a *wave function, and in general it is found that a system (such as an atom or molecule) can only have certain allowed wave functions (eigenfunctions) and certain allowed energies (eigenvalues). In wave mechanics the quantum conditions arise in a natural way from the basic postulates as solutions of the wave equation.

wavemeter An apparatus for measuring the frequency or wavelength of a radio wave. It consists essentially of a capacitively *tuned circuit and a current-detecting instrument. The variable capacitor is calibrated in terms of frequency or wavelength. A current maximum is obtained when the resonant frequency of the circuit corresponds to the frequency of the radio wave.

wave motion The process of transmitting *waves. Wave motion appears naturally in many different forms, the principal ones being surface waves (e.g. water waves), longitudinal waves (e.g. sound), transverse waves (e.g. electromagnetic radiation and waves in a vibrating string), and torsional waves (e.g. waves in a bar due to torsional vibrations at one end). All types are governed by a single equation, the equation of wave motion (*see* progressive wave), and have the property of transmitting energy over considerable distances.

wavenumber Symbol: σ. The reciprocal of the *wavelength, i.e. the number of waves per unit path length. It is expressed in m^{-1}. The *angular wavenumber*, symbol: k, is given by $2\pi\sigma$, i.e. $2\pi/\lambda$. This is the magnitude of the *angular wave vector* or *propagation vector*, symbol: $\mathbf{k}$.

wave packet *See* wavetrain.

wave-particle duality The phenomenon whereby electromagnetic radiation and particles can exhibit either wave-like or particle-like behaviour, but not both. *See* corpuscular theory; complementarity.

wave plate *See* half-plate; quarter-wave plate.

waveshape *See* waveform.

wave surface *See* wavefront.

wavetail *See* impulse voltage (or current).

wave theory of light *See* corpuscular theory.

wavetrain A succession of waves, especially a group of waves of limited duration (also called a *wave packet*).

wavetrap A *tuned circuit, usually a *rejector, incorporated in a radio receiver to reduce interference at a particular radio frequency.

weak interaction A kind of interaction between *elementary particles that is weaker than the *strong interaction force by a factor of about 10^{12}. When strong interactions can occur in reactions involving elementary particles the weak interactions are usually unobservable. However, sometimes strong and *electromagnetic interactions are prevented because they would violate the conservation of some *quantum number (e.g. *strangeness) that has to be conserved in such reactions. When this happens weak interactions may still occur.

The weak interaction operates over an extremely short range (about 2×10^{-18} m). It is mediated by the exchange of a very heavy particle (a *gauge boson) that may be the charged W^+ or W^- particle (mass about 80 GeV/c^2) or the neutral Z^0 particle (mass about 91 GeV/c^2). The gauge bosons that mediate the weak interactions are analogous to the photon that mediates the *electromagnetic interaction. Weak interactions mediated by W particles involve a change in the charge and hence the identity of the reacting particle. The neutral Z^0 does not lead to such a change in identity. Both sorts of weak interaction can violate *parity.

Most of the long-lived elementary particles decay as a result of weak interactions. For example, the kaon decay $K^+ \rightarrow \mu^+\nu_\mu$ may be thought of as being due to the annihilation of the u quark and $\bar{s}$ antiquark in the K^+ to produce a virtual W^+ boson (*see* virtual particle), which then converts into a positive muon and a neutrino. This decay process cannot take place by a strong or electromagnetic interaction because strangeness is not conserved. *Beta decay is the most common example of a weak interaction decay. Because it is so weak, particles that can only decay by weak interactions do so relatively slowly, i.e. they have relatively long lifetimes. Other examples of weak interactions include the scattering of the neutrino by other particles and certain very small effects on electrons within the atom.

Understanding of weak interactions is based on the *electroweak theory, in which it is proposed that the weak and electromagnetic interactions are different manifestations of a single underlying force, known as the electroweak force. Many of the predictions of the theory have been confirmed experimentally.

weber Symbol: Wb. The *SI unit of *magnetic flux, defined as the flux that, linking a circuit of one turn, produces an electromotive force of one volt when the flux is reduced to zero at a uniform rate in one second. 1 Wb = 10^8 maxwell.

Weber–Fechner law To make a sensation (such as *loudness or brightness) increase in arithmetical progression, the stimulus must increase in geometrical progression.

wedge A strip of material, such as a photographic plate or piece of gelatin, that shows a gradation of transmission from clear to opaque along its length. The gradation may be continuous or in steps, and may be in neutral tones or a single colour.

weight 1. In mechanics, a term used loosely with different meanings that are often confused. (*a*) The actual force of gravitation acting on a body near to the surface of the earth or another astronomical body. (*b*) The apparent force of gravitation equal to the mass of the body times the acceleration of *free fall *g*; this differs slightly from the former since *g* is measured with respect to a nearby point on the surface of the rotating planet, not with respect to the centre. (*c*) For a body remaining supported on the surface, the force exerted by the body on the support; this is equal to (*b*) but acts on a different body at a different point by a different mechanism. (*d*) The force exerted by the support on the body in the last case; this force is sometimes called the "reaction" to the weight and is often wrongly supposed to be related to the force of gravity by Newton's third law of motion. (*e*) The mass of a body, in particular a standard mass used in weighing. In all cases except (*e*) weight means a force so it is measured in newtons. *See also* weightlessness.

2. *See* weighted mean.

weighted mean Of a number of values $x_1, x_2, x_3 \ldots x_n$. The quantity given by:

$$\frac{w_1 x_1 + w_2 x_2 + \cdots w_n x_n}{w_1 + w_2 + \cdots w_n},$$

where w is the *weight* of an observation; w is a measure of the reliability of the corresponding x, and can either be assigned intuitively or calculated from w = (probable error)$^{-2}$.

weightlessness The condition of a body in *free fall. Since there is no support the body is not acted upon by a force, as described in *weight **1** (*d*). The condition of weightlessness does not imply that the body is not subject to gravitation, but that no other force acts on it. Since gravity acts uniformly throughout a body it does not by itself cause any stress. Any force exerted by a support on the surface of a body necessarily causes deformation, so the normal condition of living organisms on earth is one of stress, which is removed when in free fall in an orbit.

Weiss constant *See* Curie–Weiss law.

Weissenberg photography A crystal-diffraction method in which a single crystal is allowed to rotate about an axis normal to the incident beam of monochromatic radiation, while the cylindrical photographic film moves in a synchronized way, to and fro parallel to the rotation axis, screens being arranged so that only one layer line is recorded at a time. It is used for the measurement of intensities of diffracted spectra.

well *See* potential well.

well counter A radiation counter used in connection with radioactive fluids. The fluid is held in a cylindrical container within the detecting device.

Wertheim effects *See* Wiedemann effects.

Weston standard cell *Syn.* cadmium cell. A cell that is a portable standard of

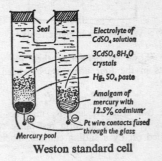

Weston standard cell

electromotive force and is used for calibrating potentiometers and hence all other voltage-measuring instruments. It is constructed in an H-shaped glass vessel, the constituents being shown in the diagram. The cell has a very low temperature coefficient of e.m.f. Its e.m.f. at t °C is given as:

$$E_t = 1.018\,58 - 4.06 \times 10^{-5}(t - 20)$$
$$- 9.5 \times 10^{-7}(t - 20)^2 +$$
$$1 \times 10^{-8}(t - 20)^3.$$

See Clark cell; Josephson effect.

wet and dry bulb hygrometer A simple hygrometer consisting of an ordinary thermometer side by side with another thermometer whose bulb is surrounded by fibres dipping into water. Evaporation of water from the fibres cools the wet bulb. The rate of evaporation depends on the relative *humidity of the air. Tables are used to calculate the relative humidity at the dry-bulb temperature from readings of the two thermometers.

Wheatstone bridge A *bridge used to measure resistance. A network of resistors is arranged as in the diagram, R_1 and R_2 being the unknown and the reference resistance. When the galvanom-

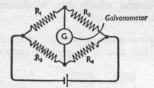

Wheatstone bridge circuit

eter shows no deflection then the currents in the four arms are balanced so that:

$$R_1/R_2 = R_3/R_4.$$

The two arms R_3 and R_4 may be sections l_1, l_2 of a uniform resistance wire tapped off by a sliding contact. Then:

$$R_1/R_2 = l_1/l_2.$$

Several forms of Wheatstone bridge exist.

white dwarf Any of a large class of very faint stars that are thought to be low-mass stars in the last stage of stellar evolution. Their mass lies below the *Chandrasekhar limit* (about 1.4 solar masses). Their nuclear fuel (hydrogen) has been completely exhausted and they have undergone *gravitational collapse to form small but very dense bodies, thought to consist of helium nuclei and a degenerate gas of electrons. *See also* Hertzsprung–Russell diagram.

white light Light, such as daylight, containing all wavelengths of the visible spectrum at normal intensities so that no coloration is apparent.

white noise *See* noise.

wholetone scale *See* musical scale.

wide-angle lens A camera lens with a relatively short focal length compared with a standard lens and a large field of view – typically 80° to 100° for still cameras.

Wiedemann effects *Syn.* Wertheim effects. Circular magnetic effects: (1) the twist produced in a rod due to interaction of longitudinal and circular magnetic fields; (2) the longitudinal magnetization produced by twisting a circular magnetized rod; and (3) the circular magnetic field produced by twisting a longitudinally magnetized rod.

Wiedemann–Franz–Lorenz law An approximate relationship between thermal and electrical conductivities of metals and alloys that holds roughly over a wide range of temperatures, but becomes less accurate at very low temperatures:

$$\lambda/\sigma T = (2.0 \pm 0.5) \times 10^{-8} \text{ V}^2 \text{ K}^{-2},$$

where λ is the thermal conductivity, σ is the electrical conductivity, and T is the thermodynamic temperature.

The relationship $\lambda/\sigma \simeq$ constant, at constant T, was published in 1853 by Wiedemann and Franz. Later work has established the more general form of the law over a wide range of temperatures.

Wein bridge A four-arm a.c. *bridge that can be used to measure capacitance, inductance, or *power factor.

Wien displacement law *See* black-body radiation.

Wien effect The increase in conductivity of an electrolyte under a high voltage gradient (of the order of 2 MV m⁻¹). Under this condition, the rate of movement of an ion in solution is such that it passes completely out of its ionic atmosphere during the *relaxation time, and is thus free from the retarding effect normally encountered.

Wien radiation law *See* black-body radiation.

Wigner effect *Syn.* discomposition effect. A change in the physical or chemical properties of a solid as a result of radiation damage. The effect is caused by the displacement of atoms from their normal lattice positions as a result of the impact of nuclear particles.

Wigner nuclides *See* mirror nuclides.

Wilson cloud chamber *See* cloud chamber.

Wilson effect When an insulating material is moved through a region of magnetic flux an induced potential difference is set up across the material. Because the creation of an electric current is inhibited by the nonconducting properties of the material, it becomes electrically polarized, a phenomenon known as the Wilson effect.

Wimshurst machine An early electrostatic generator consisting of two parallel plates rotating in opposite directions. Charges, induced on sections round the two perimeters, are collected by systems of pointed combs and are used to produce a spark.

windage loss The power loss, usually expressed in watts, that occurs in an electrical machine as a result of the motion imparted to the gas or vapour (commonly air) surrounding the moving parts, by the latter. This loss is inherent in all electrical machines since ventilation is required for cooling purposes.

wind energy *See* renewable energy sources.

winding Of an electrical machine, transformer, or other piece of apparatus. A complete group of insulated conductors designed to produce a magnetic field or to be acted upon by a magnetic field. A winding may consist of a number of separate conductors connected together electrically at their ends or may consist of a single conductor (wire or strip) that has been shaped or bent to form a number of loops or turns.

window 1. *See* atmospheric windows.

2. The thin sheet of material (often mica) covering the end of a radiation detector or counter, through which the radiation is received.

wind tunnel Essentially a hollow tube through which a uniform flow of air is passed. Observations made on a scale model of all or part of an object, such as an aircraft, allow the aerodynamic behaviour of the object to be estimated.

wire-wound resistor *See* resistor.

wobbulator A *signal generator whose output frequency can be automatically varied over a definite range of values. It is used in testing the frequency response of electronic circuits and devices.

Wollaston prism

Wollaston prism A polarizing beam splitter made from two prisms of calcite or quartz, cemented or in optical contact along their diagonals. Their optic axes are arranged so as to separate the ordinary and extraordinary components of a ray of unpolarized light at the diagonal interface (*see* double refraction). The diagram shows the refractive action of a calcite device. For prism P the optic axis is parallel to AB; for prism Q it is normal to the plane of the paper. The angle of deviation of the emerging beams is determined by the wedge angle between AB and the diagonal. The two beams are orthogonally polarized.

Wollaston wire Exceedingly fine wire produced by encasing platinum wire in a silver sheath, drawing them together, and then dissolving away the silver by acid. It is used in electrical instruments.

Wood's glass A glass having a high transmission factor in the ultraviolet range of the spectrum but relatively opaque to visible radiation.

woofer A loudspeaker of large dimensions, used to reproduce sounds of relatively low frequency in a hi-fi system. *Compare* tweeter.

word A string of *bits used to store an item of information in a *computer. Word length depends on the machine but typically consists of 32 or 16 bits.

work Symbol: W; unit: joule. **1.** If a constant force F acts at a point on a body while it undergoes a displacement D, the work done by the body that exerts the force is the scalar product $F \cdot D$, i.e. $W = FD \cos \theta$, where θ is the angle between F and D. This can be expressed as the product of the force times the component of displacement in the direction of the force, or as the displacement times the component of the force in the direction of the displacement.

2. If a constant torque G acts on a body while it undergoes a rotation through an angle H, the work done is the scalar product $G \cdot H$ (H measured in radian). If the torque and angular displacement have the same direction this gives the simple product GH. If the axes of torque and rotation are perpendicular (as in *precession) no work is done.

3. If a surface is displaced sweeping out a volume ΔV against a pressure p, the work done is $p\Delta V$.

4. If a charge Q is displaced between two points with potential difference U, the work done electrically is QU. The unit of potential difference, the volt, is

so defined that if work is in joules and charge in coulombs, then U is in volts.

5. Work may also be done in changes of the state of magnetization or electrification of a body. For example, a magnetized paramagnetic substance does work at the expense of its internal energy on demagnetizing itself on removal of an external field (*see* adiabatic demagnetization).

See also energy; machine; power; virtual work principle.

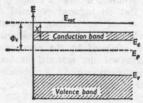

Energy bands

Work function and electron affinity of a semiconductor

work function Symbol: Φ. The difference in energy between the *Fermi level of a solid and the energy of the free space outside the solid (the vacuum level). At the absolute zero of temperature the work function is the minimum energy required to remove an electron from a solid. In a metal there is a contribution to the work function from the *image potential that an electron would experience outside the metal.

In a *semiconductor (*see* diagram) the *electron affinity* (symbol: χ) is defined by the energy difference between the vacuum level and the bottom of the conduction band.

Work function and electron affinity are usually defined as energies and measured in electronvolts, although volts are sometimes used.

work hardening The hardening of a metal when it is strained considerably by a stress above the *elastic limit. It is attributed principally to the locking together of dislocations (*see* defect). The *ductility and *malleability of a metal can be restored after work hardening by *annealing.

wow An undesirable form of *frequency modulation heard in the reproduction of high-fidelity sound and characterized by variations in pitch up to about 10 Hz. In the case of a gramophone record it is often due to nonuniform rotation of the turntable. *See also* flutter.

W particle *Syn.* W boson. The extremely massive charged particle, symbol: W^+ or W^-, that mediates certain types of *weak interaction. The neutral *Z particle* (or *Z boson*), symbol: Z^0, mediates the other types. Both are *gauge bosons. The W and Z particles were first detected at CERN (1983) by studying collisions between protons and antiprotons with total energy 540 GeV in centre-of-mass coordinates. The rest masses were determined as about 80 GeV/c^2 and 91 GeV/c^2 for the W and Z particles, respectively, as had been predicted by the *electroweak theory.

wrench A force together with a couple whose axis is the line of action of the force. The quotient of the couple by the force is the pitch of the wrench and has the dimensions of a length. The magnitude of the force is the intensity and the line of action of the force is the axis of the wrench. In general, any system of forces can be reduced to a wrench.

X

xerography A photographic process in which the image is formed by electrical effects rather than chemical effects. Ultraviolet radiation is passed through,

for example, a document to be copied and falls on an electrostatically charged plate, usually coated with selenium. This is discharged to an extent that depends on the intensity of the incident radiation. A powder with an opposite electric charge is then sprayed on the plate and sticks to the "dark" areas where the plate has not been discharged by the radiation. The powder, which is a mixture of graphite and a thermoplastic resin, is then transferred from the plate to a charged paper where it is fixed by heat treatment.

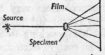

a Transmission method

X-ray analysis Analysis of the structure of crystalline substances, or of those that have crystalline phases, based on the diffraction of *X-rays. X-rays are diffracted by crystals in a manner dependent on the wavelength of the rays and the *Bravais lattice of the crystal. The analytical method adopted depends on the form in which the substance is available. With large crystals *Laue diagrams can provide useful characterization, but more frequently the crystal is rotated when mounted at the centre of a cylindrical film, thus bringing successive sets of crystalline planes into position. The *Debye–Scherrer ring or powder method is used when the specimen consists of a number of small crystals. Because of the number of crystals, randomly distributed, some are usually available in each plane to diffract the X-ray beam. The plate may be set up as in Fig. *a* or *b*, or may be rotated in a cylindrical camera, which is the most usual method. *See also* Bragg's law; X-ray spectrum.

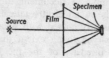

b Back reflection method

X-ray astronomy The study of X-ray emission from astronomical sources both in and beyond our Galaxy. Since X-rays are absorbed by the earth's atmosphere, observations must be made at altitudes above about 150 km using instruments mounted in satellites, rockets, and balloons. The X-rays may be *thermal radiation produced from very high temperature gas (about 10^6 to 10^8 K), or nonthermal X-rays arising from interaction of high-energy electrons with a magnetic field (*synchrotron radiation) or with low-energy photons (inverse *Compton effect). The X-rays may be detected, recorded, and analysed by various instruments, including *proportional counters, *CCDs (charge-coupled devices), and grazing-incidence X-ray telescopes.

The most common and luminous sources of X-rays in our Galaxy are *X-ray binaries*, in which gas is flowing from a normal star to a close companion – either a *white dwarf, *neutron star, or even a *black hole. *Supernova remnants, such as the Crab nebula, are another source. Fainter but intrinsically much more powerful X-ray emission is detected from many extragalactic objects, especially active galaxies – such as *Seyfert galaxies, *quasars, and powerful radio galaxies.

X-ray binary *See* X-ray astronomy; pulsar.

X-ray crystallography The study of crystal structure, texture, and behaviour, and the identification of crystals, by methods involving the use of X-rays. *See* X-ray analysis.

X-ray diffraction *See* X-rays; X-ray analysis.

X-ray microscopy *See* microscope.

X-rays A form of energetic *electromagnetic radiation. There is no universally agreed classification for electromagnetic radiation of high quantum energy. The name X-radiation was given by Röntgen in 1895 to the unidentified radiation that he observed when electrons of high energy (10^3 eV upwards) strike matter. For many years research was limited to quantum energies less than about 10^5 eV, so the term X-ray came to be used by some workers for radiations with quantum energies up to about this value (corresponding to a wavelength about 10^{-11} m). Thus radiations in this range produced by other mechanisms are sometimes called X-rays. (*See* synchrotron radiation.) In 1896 Becquerel discovered *gamma rays, which were later shown to originate in the nucleus. Early research detected only those radiations with quantum energies of the order 10^5 to 10^6 eV, hence it was said that gamma rays were of shorter wavelength than X-rays; the term gamma ray is thus sometimes applied to a radiation of this very high quantum energy irrespective of origin.

Such a distinction between X- and γ-rays is now invalid: nuclei often emit radiations of much lower quantum energy than 10^5 eV while machines have been developed giving very much shorter wavelength radiation than that emitted from any nucleus. In this article the term X-ray is used to mean radiation produced in a machine in which electrons are accelerated to any high energy and strike a target.

An *X-ray tube* is an evacuated vessel in which electrons from a hot filament are focused onto a cooled target; this is usually a metal of high melting point, normally tungsten. X-rays are emitted from the area struck by the electron beam, and emerge from the tube through a window of material with a low atomic number that transmits the X-rays.

The spectrum of X-rays emitted from an X-ray tube consists of lines characteristic of the target material and a continuum that has a short-wave limit λ_m, given by hc/eV, where h is the Planck constant, c the speed of light, e the electron charge, and V the potential difference across the tube. The *characteristic X-rays* are caused by the transitions of electrons between the various shells of the atom; the continuous band of wavelengths is caused by the acceleration of electrons in the vicinity of nuclei.

X-rays can be reflected, refracted, and polarized by suitable materials. They also show interference and diffraction effects. The wavelengths of X-rays were first determined (Bragg 1911) using a crystal as a three-dimensional diffraction grating. The atoms in regular array in a crystal act as point sources when placed in an X-ray beam and constructive interference takes place in preferential directions dependent on wavelength, the crystal, and its setting. The wavelength is typically 10^{-10} to 10^{-11} metre. In *quantum theory the energy, E, of X-ray photons is related to the frequency ν, by the expression: $E = h\nu$.

X-rays ionize gases but the process is a secondary one caused by the electrons set free when X-rays interact with matter. They penetrate matter to a degree dependent on the wavelength of the rays. In general, *hard X-rays* (small wavelengths) are able to penetrate a given substance more easily than *soft X-rays* (longer wavelengths). The intensity of homogeneous radiation transmitted through a given material of thickness t, is related to the initial intensity I_0, by the relation:

$$I = I_0 e^{-\mu t},$$

where μ is the *absorption coefficient* of the material. The incident beam is weakened by (a) scattering caused by atoms, (b) the *photoelectric effect, and (c) the *Compton effect. In the latter two processes, free electrons are ejected from the atoms of the material. Photographic plates are blackened by X-rays to a degree dependent on the intensity of the radiation and the wavelength. *See also* ionization chamber; counter.

X-ray spectrum When X-rays are scattered by atomic centres arranged at regular intervals, interference phenomena occur, crystals providing gratings of suitable small interval. The interference effects may be used to provide a spectrum of the beam of X-rays, since, according to *Bragg's law, the angle of reflection of X-rays from a crystal depends on the wavelength of the rays. For lower-energy X-rays mechanically ruled gratings can be used. Each chemical element emits *characteristic X-rays* in sharply defined groups in more widely separated regions. They are known as the K, L, M, N, etc., series. There is a regular displacement of the lines of any series towards shorter wavelengths as the atomic number of the element concerned increases. *See* Moseley's law; spectroscopy; X-rays.

X-ray tube *See* X-rays.

Y

Yagi aerial A sharply directional *aerial array used especially for television that consists of one or two dipoles, connected to the transmitting or receiving circuits, a parallel reflector, and a series of directors. The directors are parallel and spaced from 0.15 to 0.25 of a wavelength apart such that, in transmission, energy is absorbed from the field of the dipole and re-radiated so as to reinforce the field in a forward direction and oppose it in the reverse direction; in receiving, the signals are focused onto the dipole.

Y connection *See* star connection.

year *See* time.

yield point A point on a graph of *stress versus *strain for a material at which the strain becomes dependent on time and the material begins to flow. *See* yield value.

yield stress The minimum stress for *creep to take place. Below this value any deformation produced by an external force will be purely elastic.

yield value The minimum value of stress that must be applied to a material in order that it shall flow.

YIG Abbreviation for yttrium iron garnet. A synthetic *ferrite widely used for microwave applications. The magnetic properties are altered by the amount of trace elements present.

yoke A piece of ferromagnetic material that is used to connect permanently two or more magnetic *cores and thus complete a magnetic circuit without surrounding it by a winding of any type.

Young modulus *See* modulus of elasticity.

Young's fringes *See* interference.

Yukawa potential A potential used to explain the forces between nucleons. Two particles of equal and opposite charge

are attracted by an *electromagnetic field and their mutual potential energy is $-e^2/r$, where e is their charge and r their distance apart. In the nucleus, stronger short-range forces are acting (*see* strong interactions) and Yukawa assumed that their potential energy varied according to $e^{-\mu r}/r$ rather than $1/r$, where μ is a constant. The interaction is assumed to be caused by the virtual production of a boson (a pion) that is exchanged between the nucleons.

Z

Zeeman effect An effect occurring when atoms emit or absorb radiation in the presence of a moderately strong magnetic field. Each spectral line is split into closely spaced polarized components: when the source is viewed at right angles to the field there are three components, the middle one having the same frequency as the unmodified line; when the source is viewed parallel to the field there are two components, the undisplaced line being absent. This is the "normal" Zeeman effect. With most spectral lines, however, the *anomalous Zeeman effect* occurs, where there are a greater number of symmetrically arranged polarized components. In both effects the displacement of the components is a measure of the magnetic field strength. In some cases the components cannot be resolved and the spectral line appears broadened.

The Zeeman effect occurs because the energies of individual electron states depend on their inclination to the direction of the magnetic field, and because quantum energy requirements impose conditions such that the plane of an electron orbit can only set itself at certain definite angles to the applied field. These angles are such that the projection of the total angular momentum on the field direction is an integral multiple of $h/2\pi$ (h is the *Planck constant). The Zeeman effect is observed with moderately strong fields where the precession of the orbital angular momentum and the spin angular momentum (*see* spin) of the electrons about each other is much faster than the total precession around the field direction. For stronger fields the *Paschen–Back effect predominates. The normal Zeeman effect is observed when the conditions are such that the *Landé factor is unity, otherwise the anomalous effect is found. This anomaly was one of the factors contributing to the discovery of electron spin.

Zener breakdown A type of *breakdown, observed in reverse-biased p-n junctions in which very high *doping levels exist: the built-in potential across the junction is therefore high and the *depletion layer narrow. The application of a small reverse voltage is sufficient to cause the electrons to tunnel directly from the valence band to the conduction band (*see* energy bands; tunnel effect). No multiplication of charge *carriers occurs (*compare* avalanche breakdown) and the breakdown is reversible. A very sharp increase in the reverse current is observed at the breakdown potential. *See also* Zener diode.

Zener diode A p-n junction *diode that has sufficiently high *doping levels on each side of the junction for *Zener breakdown to occur. The diode thus has a well-defined reverse *breakdown voltage (about a few volts) and can be used as a voltage regulator. It behaves like a normal diode in the forward direction.

The term is also applied to less highly doped diodes that have higher values of breakdown voltage (up to 200 V) and whose characteristics depend on *avalanche breakdown.

zenith The point, infinitely distant, above an observer on the earth's surface, in the direction in which gravity acts. *See* celestial sphere.

zero error *See* index error.

zero-point energy In classical physics it was assumed that at absolute zero all particles would be at rest, hence the translational and rotational kinetic energies and the vibrational energy of molecules would be zero. *Quantum theory shows that the lowest energy state of a system is often nonzero, hence although at absolute zero all particles would be in the lowest energy states available their energies may be significant.

From the *uncertainty principle a particle cannot have zero momentum unless the uncertainty in its position is infinite. Thus the translational kinetic energy of molecules at absolute zero can only be zero for an *ideal gas, which is in principle realized by extrapolation to infinite volume.

In a condensed substance each atom can be considered as being equivalent to three linear oscillators that are very nearly simple harmonic except at high temperatures. By *wave mechanics the lowest energy of a linear simple-harmonic oscillator of frequency v is $\frac{1}{2}hv$, where h is the Planck constant. The *Debye theory of specific heat capacities shows from this that the zero-point energy of a mole of solid resulting from vibrations is $9R\Theta_D/8$, where R is the *molar gas constant and Θ_D is the Debye characteristic temperature.

The zero-point energy of the valence electrons in a solid or liquid is relatively very large (*see* energy bands).

The zero-point energy affects the *saturated vapour pressure and the *latent heats of vaporization and sublimation.

531

zeroth law of thermodynamics The law of *thermodynamics that is fundamental to the three basic laws of thermodynamics (hence its name). It states that if two bodies are each in thermal equilibrium with a third body, then all three bodies are in thermal equilibrium with each other.

zeta pinch *See* fusion reactor.

z-modulation *See* intensity modulation.

zone plate *See* diffraction of light.

zoom lens A lens system consisting of converging and diverging elements, one or more of which can be moved so that the focal length, which depends on the separation of the elements, can be continuously adjusted. By connecting two or more of the elements together so that they both move through the same distance, the image remains sharp at the different focal lengths. It is also desirable that the *f-number should not need resetting as the focal length changes. To avoid this the lens system is usually in two parts: the basic imaging system, for which the f-number remains constant, and a variable-focus attachment.

Z particle *Syn.* Z boson. *See* W particle.

APPENDIX

Table 1. Conversion Factors

SI, CGS, and FPS units

Length	m	cm	in	ft	yd
1 metre	1	100	39·3701	3·280 84	1·093 61
1 centimetre	0·01	1	0·393 701	0·032 808 4	0·010 936 1
1 inch	0·0254	2·54	1	0·083 333 3	0·027 777 8
1 foot	0·3048	30·48	12	1	0·333 333
1 yard	0·9144	91·44	36	3	1

	km	mile	n mile
1 kilometre	1	0·621 371	0·539 957
1 mile	1·609 34	1	0·868 976
1 nautical mile	1·852 00	1·150 78	1

1 light year $= 9·460\ 70 \times 10^{15}$ metres $= 5·878\ 48 \times 10^{12}$ miles.
1 astronomical unit $= 1·496 \times 10^{11}$ metres.
1 parsec $= 3·0857 \times 10^{16}$ metres $= 3·2616$ light years.

Velocity	m s^{-1}	km h^{-1}	mile h^{-1}	ft s^{-1}
1 metre per second	1	3·6	2·236 94	3·280 84
1 kilometre per hour	0·277 778	1	0·621 371	0·911 346
1 mile per hour	0·447 04	1·609 344	1	1·466 67
1 foot per second	0·3048	1·097 28	0·681 817	1

1 knot $= 1$ nautical mile per hour $= 0·514\ 444$ metre per second.

Mass	kg	g	lb	long ton
1 kilogram	1	1000	2·204 62	$9·842\ 07 \times 10^{-4}$
1 gram	10^{-3}	1	$2·204\ 62 \times 10^{-3}$	$9·842\ 07 \times 10^{-7}$
1 pound	0·453 592	453·592	1	$4·464\ 29 \times 10^{-4}$
1 long ton	1016·047	$1·016\ 047 \times 10^{6}$	2240	1

Force	N	kg	dyne	poundal	lb
1 newton	1	0·101 972	10^{5}	7·233 00	0·224 809
1 kilogram force	9·806 65	1	$9·806\ 65 \times 10^{5}$	70·9316	2·204 62
1 dyne	10^{-5}	$1·019\ 72 \times 10^{-6}$	1	$7·233\ 00 \times 10^{-5}$	$2·248\ 09 \times 10^{-6}$
1 poundal	0·138 255	$1·409\ 81 \times 10^{-2}$	$1·382\ 55 \times 10^{4}$	1	0·031 081
1 pound force	4·448 22	0·453 592	$4·448\ 23 \times 10^{5}$	32·174	1

Pressure	Pa	kg/cm^2	lb/in^2	atm
1 pascal	1	$1·019\ 72 \times 10^{-5}$	$1·450\ 38 \times 10^{-4}$	$9·869\ 23 \times 10^{-6}$
1 kilogram per square centimetre	$980·665 \times 10^{2}$	1	14·2234	0·967 841
1 pound per square inch	$6·894\ 76 \times 10^{3}$	0·070 306 8	1	0·068 046
1 atmosphere	$1·013\ 25 \times 10^{5}$	1·033 23	14·6959	1

1 pascal per square metre $= 10$ dynes per square centimetre.
1 bar $= 10^{5}$ pascals per square metre $= 0·986\ 923$ atmosphere.
1 torr $= 133·322$ pascals per square metre $= 1/760$ atmosphere.
1 atmosphere $= 760$ mmHg $= 29·92$ in Hg $= 33·90$ ft water (all at 0 °C).

Table 1. Conversion Factors (*continued*)

Work and Energy	J	cal$_{IT}$	kWhr	btu$_{IT}$
1 joule	1	0 238 846	$2{\cdot}777\,78 \times 10^{-7}$	$9{\cdot}478\,13 \times 10^{-4}$
1 calorie (IT)	4·1868	1	$1{\cdot}163\,00 \times 10^{-6}$	$3{\cdot}968\,31 \times 10^{-3}$
1 kilowatt hour	$3{\cdot}6 \times 10^{6}$	$8{\cdot}598\,45 \times 10^{5}$	1	3412·14
1 British Thermal Unit (IT)	1055·06	251·997	$2{\cdot}930\,71 \times 10^{-4}$	1

1 joule = 1 newton metre = 1 watt second = 10^{7} erg = 0·737 561 ft lb.
1 electronvolt = $1{\cdot}602\,10 \times 10^{-19}$ joule.

Table 2. Base SI Units

Physical quantity	Name	Symbol
length	metre	m
mass	kilogram	kg
time	second	s
electric current	ampere	A
thermodynamic temperature	kelvin	K
amount of substance	mole	mol
luminous intensity	candela	cd

Table 3. Prefixes used with SI Units

Factor	Name of Prefix	Symbol	Factor	Name of Prefix	Symbol
10	deca-	da	10^{-1}	deci-	d
10^{2}	hecto-	h	10^{-2}	centi-	c
10^{3}	kilo-	k	10^{-3}	milli-	m
10^{6}	mega-	M	10^{-6}	micro-	μ
10^{9}	giga-	G	10^{-9}	nano-	n
10^{12}	tera-	T	10^{-12}	pico-	p
10^{15}	peta-	P	10^{-15}	femto-	f
10^{18}	exa-	E	10^{-18}	atto-	a

Table 4. Derived SI Units with Special Names

Physical quantity	Name	Symbol
frequency	hertz	Hz
force	newton	N
pressure, stress	pascal	Pa
energy, work, quantity of heat	joule	J
power	watt	W
electric charge, quantity of electricity	coulomb	C
electric potential, potential difference, tension, electro-motive force	volt	V
electric capacitance	farad	F
electric resistance	ohm	Ω
electric conductance	siemens	S
flux of magnetic induction, magnetic flux	weber	Wb
magnetic flux density, magnetic induction	tesla	T
inductance	henry	H
Celsius temperature	degree Celsius	°C
luminous flux	lumen	lm
illuminance	lux	lx
activity (of a radionuclide)	becquerel	Bq
absorbed dose, specific energy imparted, kerma, absorbed dose index	gray	Gy
dose equivalent	sievert	Sv
plane angle	radian	rad
solid angle	steradian	sr

Table 5. **Fundamental Constants**

Constant	Symbol	Value (with estimated error)
speed of light	c	2.99792458×10^8 m s^{-1} exact by definition
magnetic constant (permeability of free space)	μ_0	$4\pi \times 10^{-7} = 1.25663706144 \times 10^{-6}$ H m^{-1}
electric constant (permittivity of free space)	$\varepsilon_0 = \mu_0{}^{-1}c^{-2}$	$8.854187817 \times 10^{-12}$ F m^{-1}
charge of electron or proton	e	$\pm 1.60217733 \times 10^{-19}$ C
rest mass of electron	m_e	$9.1093897 \times 10^{-31}$ kg
rest mass of proton	m_p	$1.6726231 \times 10^{-27}$ kg
rest mass of neutron	m_n	1.674929×10^{-27} kg
electronic radius	$r_e = \dfrac{e^2}{4\pi\varepsilon_0 m_e c^2}$	$2.81794092 \times 10^{-15}$ m
Planck constant	h	6.626076×10^{-34} J s
Boltzmann constant	$k = \dfrac{R}{L}$	1.380658×10^{-23} J K^{-1}
Avogadro constant	L, N_A	6.0221367×10^{23} mol^{-1}
Loschmidt constant	N_L, n_o	2.686763×10^{25} m^{-3}
molar gas constant	$R = Lk$	8.314510 J K^{-1} mol^{-1}
Faraday constant	$F = Le$	9.6484531×10^4 C mol^{-1}
Stefan-Boltzmann constant	$\sigma = \dfrac{2\pi^5 k^4}{15h^3 c^2}$	5.67051×10^{-8} W m^{-2} K^{-4}
fine structure constant	$\alpha = \dfrac{e^2}{2\varepsilon_0 hc}$	7.2973531×10^{-3}
Rydberg constant	$R = \dfrac{m_e e^4}{8\varepsilon_0{}^2 h^3 c}$	1.0973731534×10^7 m^{-1}
gravitational constant	G	6.67259×10^{-11} N m^2 kg^{-2}
acceleration of free fall (standard value)	g_n	9.80665 m s^{-2}

Table 6. Spectrum of Electromagnetic Radiation

Wavelength/m		Frequency/kHz
10^{-13}		10^{19}
10^{-12}	gamma rays	10^{18}
10^{-11}		10^{17}
10^{-10}	X-rays	10^{16}
10^{-9}		10^{15}
10^{-8}	ultraviolet radiation	10^{14}
10^{-7}		10^{13}
10^{-6}	visible light	10^{12}
10^{-5}	infrared radiation	10^{11}
10^{-4}		10^{10}
10^{-3}		10^{9}
10^{-2}	EHF	10^{8}
10^{-1}	SHF — radio frequencies	10^{7}
1	UHF	10^{6}
10	VHF	10^{5}
10^{2}	HF	10^{4}
10^{3}	MF	10^{3}
10^{4}	LF	10^{2}
10^{5}	VLF	10
		1

Table 7. Elementary Particles

	Particle	Quark content	Mass MeV/c²	Isospin I	J^{PC}	Lifetime/s
gauge bosons	γ		0		1^-	stable
	$W^\pm$		80000		1	
	Z^0		91000		1	
leptons	ν		0		$\frac{1}{2}$	stable
	e		0.511		$\frac{1}{2}$	stable
	μ		105.7		$\frac{1}{2}$	2.2×10^{-6}
	τ		1784.1		$\frac{1}{2}$	3.0×10^{-13}
mesons	$\pi^\pm$	$u\bar{d},\bar{u}d$	139.6	1	0^-	2.6×10^{-8}
	π°	$u\bar{u},d\bar{d}$	105.7	1	0^{-+}	8.4×10^{-17}
	$K^\pm$	$u\bar{s},s\bar{u}$	493.6	$\frac{1}{2}$	0^-	1.2×10^{-8}
	K°	$d\bar{s}$	497.7	$\frac{1}{2}$	0^-	
	K_S°		497.7	$\frac{1}{2}$	0^-	8.9×10^{-11}
	K_L°		497.7	$\frac{1}{2}$	0^-	5.2×10^{-8}
	η°	$u\bar{u},d\bar{d},s\bar{s}$	548.8	0	0^{---}	
	$D^\pm$	$c\bar{d},d\bar{c}$	1869	$\frac{1}{2}$	0^-	1.1×10^{-12}
	D°	$c\bar{u}$	1865	$\frac{1}{2}$	0^-	4×10^{-13}
	$D_S^\pm$	$c\bar{s},s\bar{c}$	1969	0	0^-	4×10^{-13}
	$B^\perp$	$u\bar{b},b\bar{u}$	5278	$\frac{1}{2}$	0^-	1×10^{-12}
	B°	$d\bar{b}$	5279	$\frac{1}{2}$	0^-	1×10^{-12}
baryons	p	uud	938.3	$\frac{1}{2}$	$\frac{1}{2}^+$	stable
	n	udd	939.6	$\frac{1}{2}$	$\frac{1}{2}^+$	896
	Λ°	uds	1115.6	0	$\frac{1}{2}^+$	2.6×10^{-10}
	Σ^+	uus	1189.4	1	$\frac{1}{2}^+$	8.0×10^{-10}
	Σ°	uds	1192.5	1	$\frac{1}{2}^+$	7.4×10^{-20}
	Σ^-	dds	1197.4	1	$\frac{1}{2}^+$	1.5×10^{-10}
	Ξ°	uss	1314.9	$\frac{1}{2}$	$\frac{1}{2}^+$	2.9×10^{-10}
	Ξ^-	dss	1321.3	$\frac{1}{2}$	$\frac{1}{2}^+$	1.7×10^{-10}
	Ω^-	sss	1672.5	0	$\frac{3}{2}^+$	1.3×10^{-10}
	Λ_c^+	udc	2285	0	$\frac{1}{2}^+$	2×10^{-13}

Table 8. Periodic Table of the Elements

1A	2A	3B	4B	5B	6B	7B		8		1B	2B	3A	4A	5A	6A	7A	0
1 H																	2 He
3 Li	4 Be											5 B	6 C	7 N	8 O	9 F	10 Ne
11 Na	12 Mg	◄———————transition elements———————►										13 Al	14 Si	15 P	16 S	17 Cl	18 Ar
19 K	20 Ca	21 Sc	22 Ti	23 V	24 Cr	25 Mn	26 Fe	27 Co	28 Ni	29 Cu	30 Zn	31 Ga	32 Ge	33 As	34 Se	35 Br	36 Kr
37 Rb	38 Sr	39 Y	40 Zr	41 Nb	42 Mo	43 Tc	44 Ru	45 Rh	46 Pd	47 Ag	48 Cd	49 In	50 Sn	51 Sb	52 Te	53 I	54 Xe
55 Cs	56 Ba	57* La	72 Hf	73 Ta	74 W	75 Re	76 Os	77 Ir	78 Pt	79 Au	80 Hg	81 Tl	82 Pb	83 Bi	84 Po	85 At	86 Rn
87 Fr	88 Ra	89† Ac															

*lanthanides		57 La	58 Ce	59 Pr	60 Nd	61 Pm	62 Sm	63 Eu	64 Gd	65 Tb	66 Dy	67 Ho	68 Er	69 Tm	70 Yb	71 Lu
†actinides		89 Ac	90 Th	91 Pa	92 U	93 Np	94 Pu	95 Am	96 Cm	97 Bk	98 Cf	99 Es	100 Fm	101 Md	102 No	103 Lr

Table 9. **Symbols for Physical Quantities**

Name of quantity	Symbol	Name of quantity	Symbol
absorptance	α	Fermi energy	E_F, ε_F
acceleration	a	force	F
activity, radioactivity	A	frequency	ν, f
admittance	Y		
amount of substance	n	Gibbs function	G
angular acceleration	α		
angular frequency	ω		
angular momentum	L	half life	$T_{\frac{1}{2}}, t_{\frac{1}{2}}$
angular velocity	ω	Hamiltonian function	H
area	A, S	heat capacity: at constant	
atomic mass constant	m_u	pressure	C_p
atomic number, proton		heat capacity: at constant	
number	Z	volume	C_v
		heat flow rate	Θ
Bragg angle	θ	height	h
bulk modulus	K	Helmholtz function	A, F
capacitance	C	illuminance, illumination	E_V, E
characteristic temperature	Θ	impedance	Z
charge density	ρ	internal energy	U
coefficient of friction	μ	irradiance	E_e, E
compressibility	κ		
concentration	c	Joule–Thomson coefficient	μ, μ_{JT}
conductance	G		
conductivity	γ, σ	kinematic viscosity	ν
cross section	σ	kinetic energy	T, E_k, K
cubic expansion coefficient	α_v		
Curie temperature	T_C	Lagrangian function	L
		linear absorption coefficient	a
decay constant	λ	linear attenuation (extinction)	
density	ρ	coefficient	μ
		linear expansion coefficient	α
efficiency	η	linear strain (relative	
electric charge	Q	elongation)	ε
electric current	I	loss angle	δ
electric current density	j, J	luminance	L_V, L
electric dipole moment	p	luminous exitance	M_V, M
electric displacement	D	luminous flux	Φ_V, Φ
electric field strength	E	luminous intensity	I_V, I
electric flux	Ψ		
electric polarization	P	magnetic field strength	H
electric potential	V	magnetic flux	Θ
electric susceptibility	χ_e	magnetic flux density, magnetic	
electromotive force	E	induction	B
electron mass	m, m_e	magnetic moment	m
elementary charge, charge of		magnetic moment of particle	μ
proton	e	magnetic quantum number	M, m_l
emissivity	ε	magnetic susceptibility	χ, χ_m
energy	E	magnetization	M
enthalpy	H	magnetomotive force	F_m
entropy	S	mass	m
equilibrium constant	K	mass excess	Δ

Table 9. **Symbols for Physical Quantities** (*continued*)

Name of quantity	Symbol	Name of quantity	Symbol
mass number, nucleon number	A	radiance	L_e, L
mean free path	λ, l	radiant exitance	M_e, M
mean life	τ	radiant flux, radiant power	Φ_e, Φ
molality	m_A	radiant intensity	I_e, I
molecular momentum	$p(p_x, p_y, p_z)$	radius	r
molecular position	$r(r_x, r_y, r_z)$	ratio of heat capacities, C_p/C_V	γ, κ
molecular velocity	$u(u_x, u_y, u_z)$	reactance	X
moment of force	M	reduced mass	μ
moment of inertia	I	reflectance	ρ
momentum	p	refractive index	n
most probable speed	$\hat{u}$	relative atomic mass	A_r
mutual inductance	M, L_{12}	relative density	d
		relative permeability	μ_r
Néel temperature	T_N	relative permittivity (dielectric constant)	ε_r
neutron mass	m_n	relaxation time	τ
neutron number	N	resistance	R
nuclear magneton	μ_N	resistivity	ρ
number of molecules	N	Reynolds number	Re
number of turns	N	rotational quantum number	J, K
orbital angular momentum quantum number	L, l_1	self-inductance	L
osmotic pressure	Π	shear modulus	G
		solid angle	Ω, ω
packing fraction	f	specific heat capacity: at constant pressure	c_p
period	T	specific heat capacity: at constant volume	c_V
permeability	μ	specific volume	v
permittivity	ε	speed	u, v
Planck function	Y	spin quantum number	S, s
plane angle	$\alpha, \beta, \gamma, \theta, \varphi$	strain, linear	ε, e
polarizability	α, γ	strain, shear	γ
position vector, radius vector	r	stress, shear	τ
potential difference	U, V	surface charge density	σ
potential energy	E_p, V, Φ	surface tension	γ, σ
power	P	susceptance	B
pressure	p		
principal quantum number	n	temperature	T, t
propagation coefficient	P, γ	thermal conductivity	λ
proton mass	m_p	thermal diffusion factor	α_T
proton number, atomic number	Z	thermal diffusion ratio	k_T
quantity of heat	Q	thermal diffusivity	a
quantum number of electron spin	S	thermodynamic temperature	T
		time	t
quantum number of nuclear spin	I	torque	T
quantum number of total angular momentum	N	transmission coefficient (acoustics)	τ
quantum number of vibrational mode	v	velocity	v
		vibrational quantum number	v

Table 9. **Symbols for Physical Quantities** (*continued*)

Name of quantity	Symbol	Name of quantity	Symbol
viscosity	η	weight	G, W
volume	V, v	work	W
volume (bulk) strain	θ	work function.	Φ
wavelength	λ	Young modulus (modulus of	
wavenumber	σ	elasticity)	E

Table 10. **Symbols used in Electronics**

Device or concept	Symbol	Device or concept	Symbol
Qualifying graphical symbols		plug & socket (male & female)	
alternating current		earth	
variability (noninherent)		primary cell or accumulator (longer line represents +ve pole)	
variability in steps			
thermal effect		battery of accumulators or primary cells	
electromagnetic effect			
radiation, electromagnetic nonionizing		switch, general symbol; make contact	
coherent radiation		resistor, general symbol (first form preferred)	
ionizing radiation			
positive-going pulse		variable resistor	
negative-going pulse		resistor with sliding contact	
pulse of a.c.		capacitor, general symbol (first form preferred)	
positive-going step function			
negative-going step function		inductor, coil, winding, choke, general symbol	
fault		inductor with magnetic core	
Graphical symbols		transformer, 2 windings	
connection of conductors		piezoelectric crystal, 2 electrodes	
terminal (circle may be filled in)			
junction of conductors		semiconductor diode, general symbol	

543

Table 10. **Symbols used in Electronics** (*continued*)

Device or concept	Symbol	Device or concept	Symbol
light-emitting diode, general symbol		OR gate, general symbol	
photodiode		inverter (NOT gate)	
pnp transistor		NAND gate (negated AND)	
npn transistor		NOR gate (negated OR)	
JFET, n-type channel		exclusive-OR gate	
JFET, p-type channel		indicating instrument (first form) & recording instrument; asterisk is replaced by symbol of unit of quantity being measured (e.g. V for voltmeter, A for ammeter, or by some other appropriate symbol)	
IGFET, enhancement type, single gate, p-type channel without substrate connection			
amplifier, general symbol			
AND gate, general symbol		antenna, general symbol	

Table 11. **The Greek Alphabet**

Letters		Name	Letters		Name
A	α	alpha	N	ν	nu
B	β	beta	Ξ	ξ	xi
Γ	γ	gamma	O	ο	omicron
Δ	δ	delta	Π	π	pi
E	ε	epsilon	P	ρ	rho
Z	ζ	zeta	Σ	σ	sigma
H	η	eta	T	τ	tau
Θ	θ	theta	Y	υ	upsilon
I	ι	iota	Φ	φ	phi
K	κ	kappa	X	χ	chi
Λ	λ	lambda	Ψ	ψ	psi
M	μ	mu	Ω	ω	omega

FOR THE BEST IN PAPERBACKS, LOOK FOR THE

FOR THE BEST IN PAPERBACKS, LOOK FOR THE 🐧

PENGUIN PHILOSOPHY

I: The Philosophy and Psychology of Personal Identity Jonathan Glover

From cases of split brains and multiple personalities to the importance of memory and recognition by others, the author of *Causing Death and Saving Lives* tackles the vexed questions of personal identity. 'Fascinating ... the ideas which Glover pours forth in profusion deserve more detailed consideration' – Anthony Storr

Minds, Brains and Science John Searle

Based on Professor Searle's acclaimed series of Reith Lectures, *Minds, Brains and Science* is 'punchy and engaging ... a timely exposé of those woolly-minded computer-lovers who believe that computers can think, and indeed that the human mind is just a biological computer' – *The Times Literary Supplement*

Ethics Inventing Right and Wrong J. L. Mackie

Widely used as a text, Mackie's complete and clear treatise on moral theory deals with the status and content of ethics, sketches a practical moral system and examines the frontiers at which ethics touches psychology, theology, law and politics.

The Penguin History of Western Philosophy D. W. Hamlyn

'Well-crafted and readable ... neither laden with footnotes nor weighed down with technical language ... a general guide to three millennia of philosophizing in the West' – *The Times Literary Supplement*

Science and Philosophy: Past and Present Derek Gjertsen

Philosophy and science, once intimately connected, are today often seen as widely different disciplines. Ranging from Aristotle to Einstein, from quantum theory to renaissance magic, Confucius and parapsychology, this penetrating and original study shows such a view to be both naive and ill-informed.

The Problem of Knowledge A. J. Ayer

How do you *know* that this is a book? How do you *know* that you know? In *The Problem of Knowledge* A. J. Ayer presented the sceptic's arguments as forcefully as possible, investigating the extent to which they can be met. 'Thorough ... penetrating, vigorous ... readable and manageable' – *Spectator*

PENGUIN SCIENCE AND MATHEMATICS

Facts from Figures   M. J. Moroney

Starting from the very first principles of the laws of chance, this authoritative 'conducted tour of the statistician's workshop' provides an essential introduction to the major techniques and concepts used in statistics today.

God and the New Physics   Paul Davies

Can science, now come of age, offer a surer path to God than religion? This 'very interesting' (*New Scientist*) book suggests it can.

Descartes' Dream   Philip J. Davis and Reuben Hersh

All of us are 'drowning in digits' and depend constantly on mathematics for our high-tech lifestyle. But is so much mathematics really good for us? This major book takes a sharp look at the ethical issues raised by our computerized society.

The Blind Watchmaker   Richard Dawkins

'An enchantingly witty and persuasive neo-Darwinist attack on the anti-evolutionists, pleasurably intelligible to the scientifically illiterate' – Hermione Lee in the *Observer* Books of the Year

Microbes and Man   John Postgate

From mining to wine-making, microbes play a crucial role in human life. This clear, non-specialist book introduces us to microbes in all their astounding versatility – and to the latest and most exciting developments in microbiology and immunology.

Asimov's New Guide to Science   Isaac Asimov

A classic work brought up to date – far and away the best one-volume survey of all the physical and biological sciences.

PENGUIN SCIENCE AND MATHEMATICS

The Panda's Thumb Stephen Jay Gould

More reflections on natural history from the author of *Ever Since Darwin*. 'A quirky and provocative exploration of the nature of evolution ... wonderfully entertaining' – *Sunday Telegraph*

Genetic Engineering for Almost Everybody William Bains

Now that the genetic engineering revolution has most certainly arrived, we all need to understand its ethical and practical implications. This book sets them out in accessible language.

The Double Helix James D. Watson

Watson's vivid and outspoken account of how he and Crick discovered the structure of DNA (and won themselves a Nobel Prize) – one of the greatest scientific achievements of the century.

The Quantum World J. C. Polkinghorne

Quantum mechanics has revolutionized our views about the structure of the physical world – yet after more than fifty years it remains controversial. This 'delightful book' (*The Times Educational Supplement*) succeeds superbly in rendering an important and complex debate both clear and fascinating.

Einstein's Universe Nigel Calder

'A valuable contribution to the demystification of relativity' – *Nature*

Mathematical Circus Martin Gardner

A mind-bending collection of puzzles and paradoxes, games and diversions from the undisputed master of recreational mathematics.

PENGUIN REFERENCE BOOKS

The Penguin Guide to the Law

This acclaimed reference book is designed for everyday use and forms the most comprehensive handbook ever published on the law as it affects the individual.

The Penguin Medical Encyclopedia

Covers the body and mind in sickness and in health, including drugs, surgery, medical history, medical vocabulary and many other aspects. 'Highly commendable' – *Journal of the Institute of Health Education*

The Slang Thesaurus

Do you make the public bar sound like a gentleman's club? Do you need help in understanding *Minder*? The miraculous *Slang Thesaurus* will liven up your language in no time. You won't Adam and Eve it! A mine of funny, witty, acid and vulgar synonyms for the words you use every day.

The Penguin Dictionary of Troublesome Words   Bill Bryson

Why should you avoid discussing the *weather conditions*? Can a married woman be *celibate*? Why is it eccentric to talk about the *aroma* of a cowshed? A straightforward guide to the pitfalls and hotly disputed issues in standard written English.

A Dictionary of Literary Terms

Defines over 2,000 literary terms (including lesser known, foreign language and technical terms), explained with illustrations from literature past and present.

The Concise Cambridge Italian Dictionary

Compiled by Barbara Reynolds, this work is notable for the range of examples provided to illustrate the exact meaning of Italian words and phrases. It also contains a pronunciation guide and a reference grammar.

FOR THE BEST IN PAPERBACKS, LOOK FOR THE 🐧

PENGUIN DICTIONARIES